D1243758

Wind Energy Explained

Wind Energy Explained
Theory, Design and Application

J.F. Manwell, J.G. McGowan and A.L. Rogers
University of Massachusetts, Amherst, USA

JOHN WILEY & SONS, LTD

Copyright © 2002 John Wiley & Sons Ltd,
The Atrium, Southern Gate, Chichester,
West Sussex PO19 8SQ, England

Telephone (+44) 1243 779777

Email (for orders and customer service enquiries): cs-books@wiley.co.uk
Visit our Home Page on www.wileyeurope.com or www.wiley.com

Reprinted September 2002, September 2005, November 2006, October 2007, November 2007, July 2006, February 2008
Reprinted with corrections August 2003

All Rights Reserved. No part of this publication may be reproduced, stored in a retrieval system or transmitted in any form or by any means, electronic, mechanical, photocopying, recording, scanning or otherwise, except under the terms of the Copyright, Designs and Patents Act 1988 or under the terms of a licence issued by the Copyright Licensing Agency Ltd, 90 Tottenham Court Road, London W1T 4LP, UK, without the permission in writing of the Publisher. Requests to the Publisher should be addressed to the Permissions Department, John Wiley & Sons Ltd, The Atrium, Southern Gate, Chichester, West Sussex PO19 8SQ, England, or emailed to permreq@wiley.co.uk, or faxed to (+44) 1243 770571.

This publication is designed to provide accurate and authoritative information in regard to the subject matter covered. It is sold on the understanding that the Publisher is not engaged in rendering professional services. If professional advice or other expert assistance is required, the services of a competent professional should be sought.

Other Wiley Editorial Offices

John Wiley & Sons Inc., 111 River Street, Hoboken, NJ 07030, USA

Jossey-Bass, 989 Market Street, San Francisco, CA 94103-1741, USA

Wiley-VCH Verlag GmbH, Boschstr. 12, D-69469 Weinheim, Germany

John Wiley & Sons Australia Ltd, 33 Park Road, Milton, Queensland 4064, Australia

John Wiley & Sons (Asia) Pte Ltd, 2 Clementi Loop #02-01, Jin Xing Distripark, Singapore 129809

John Wiley & Sons Canada Ltd, 22 Worcester Road, Etobicoke, Ontario, Canada M9W 1L1

British Library Cataloguing in Publication Data

A catalogue record for this book is available from the British Library

ISBN 13: 978-0-471-49972-5 (H/B)

Printed from files supplied by the author
Printed and bound in Great Britain by Antony Rowe Ltd, Chippenham, Wiltshire

Contents

Preface

The technology of extracting energy from the wind has evolved dramatically over the last few decades, and there have, up until now, been relatively few attempts to describe that technology in a single textbook. The lack of such a text, together with a perceived need, provided the impetus for writing this book.

The material in this text has evolved from course notes from Wind Energy Engineering, a course which has been taught at the University of Massachusetts since the mid 1970s. These notes were later substantially revised and expanded with the support of the US Department of Energy's National Renewable Energy Laboratory.

This book provides a description of the topics which are fundamental to understanding the conversion of wind energy to electricity and its eventual use by society. These topics span a wide range, from meteorology through many fields of engineering to economics and environmental concerns. The book begins with an introduction which provides an overview of the technology, and explains how it came to take the form it has today. The next chapter describes the wind resource and how it relates to energy production. Chapter 3 discusses aerodynamic principles and explains how the wind's energy will cause a wind turbine's rotor to turn. Chapter 4 delves into the dynamic and mechanical aspects of the turbine in more detail, and considers the relation of the rotor to the rest of the machine. Chapter 5 provides a summary of the electrical aspects of wind energy conversion, particularly regarding the actual generation and conversion of the electricity. Chapter 6 discusses the design of wind turbines and the key issues involved. Chapter 7 examines wind turbine and wind system control. Chapter 8 discusses siting of wind turbines and their integration into electrical systems both large and small. Chapter 9 concerns the economics of wind energy. It describes economic analysis methods and shows how wind energy can be compared with conventional forms of generation. Finally, Chapter 10 describes the environmental aspects of wind energy generation.

This book is intended primarily as a textbook for engineering students and for professionals in related fields who are just getting into wind energy. It is also intended to be used by anyone with a good background in math and physics who wants to gain familiarity with the subject. It should be useful for those interested in wind turbine design *per se*. For others, it should provide enough understanding of the underlying principles of wind turbine operation and design to appreciate more fully those aspects in which they have a particular interest. These areas include turbine siting, grid integration, environmental issues, economics, and public policy.

The study of wind energy spans such a wide range of fields. Since it is likely that many readers would not have a background in all of them, most of the chapters include some introductory material. Where appropriate, the reader is referred to other sources for more details. Solutions to the problems set at the end of the book can be found at: www.wiley.co.uk/ windenergy.

Preface

Acknowledgements

We would like to acknowledge Professor William Heronemus (emeritus), founder of the renewable energy program at the University of Massachusetts. Without his vision and tenacity, this program would never have existed, and this book would never have been written. We are also indebted to the numerous staff and students, past and present, at the University of Massachusetts who have contributed to this program. We would like to express special thanks to Sally Wright for her help with the graphics and to Clint Johnson and George Grills for carefully reviewing the final draft of the text.

We would also like to acknowledge the contribution of the National Wind Technology Center at the National Renewable Energy Laboratory, particularly Bob Thresher and Darrell Dodge, in their supporting the revision and expansion of the notes on which this text is based.

Finally, we would like to acknowledge the support of our families: our wives (Joanne, Suzanne, and Anne) and our sons (Nate, Gerry and Ned and Josh and Brian) who have all inspired our work.

1

Introduction: Modern Wind Energy and its Origins

The re-emergence of the wind as a significant source of the world's energy must rank as one of the significant developments of the late 20th century. The advent of the steam engine, followed by the appearance of other technologies for converting fossil fuels to useful energy, would seem to have forever relegated to insignificance the role of the wind in energy generation. In fact, by the mid 1950s that appeared to be what had already happened. By the late 1960s, however, the first signs of a reversal could be discerned, and by the early 1990s it was becoming apparent that a fundamental reversal was underway.

To understand what was happening, it is necessary to consider five main factors. First of all there was a need. An emerging awareness of the finiteness of the earth's fossil fuel reserves as well as of the adverse effects of burning those fuels for energy had caused many people to look for alternatives. Second, there was the potential. Wind exists everywhere on the earth, and in some places with considerable energy density. Wind had been widely used in the past, for mechanical power as well as transportation. Certainly, it was conceivable to use it again. Third, there was the technological capacity. In particular, there had been developments in other fields, which, when applied to wind turbines, could revolutionize they way they could be used. These first three factors were necessary to foster the re-emergence of wind energy, but not sufficient. There needed to be two more factors, first of all a vision of a new way to use the wind, and second the political will to make it happen. The vision began well before the 1960s with such individuals as Poul la Cour, Albert Betz, Palmer Putnam, and Percy Thomas. It was continued by Johannes Juul, E. W. Golding, Ulrich Hütter, and William Heronemus, but soon spread to others too numerous to mention. At the beginning of wind's re-emergence, cost of energy from wind turbines was far higher than that from fossil fuels. Government support was required to carry out research, development, and testing; to provide regulatory reform to allow wind turbines to interconnect with electrical networks; and to offer incentives to help hasten the deployment of the new technology. The necessary political will for this support appeared at different times and to varying degrees, in a number of countries: first in the United States, Denmark, and Germany, and now in much of the rest of the world.

The purpose of this chapter is to provide an overview of wind energy technology today, so as to set a context for the rest of the book. It addresses such questions as: What does modern wind technology look like? What is it used for? How did it get this way? Where is it going?

1.1 Modern Wind Turbines

A wind turbine, as described in this book, is a machine which converts the power in the wind into electricity. This is in contrast to a 'windmill', which is a machine which converts the wind's power into mechanical power. As electricity generators, wind turbines are connected to some electrical network. These networks include battery charging circuits, residential scale power systems, isolated or island networks, and large utility grids. In terms of total numbers, the most frequently found wind turbines are actually quite small - on the order of 10 kW or less. In terms of total generating capacity, the turbines that make up the majority of the capacity are in general rather large - in the range of 500 kW to 2 MW. These larger turbines are used primarily in large utility grids, mostly in Europe and the United States. A typical modern wind turbine, connected to a utility network, is illustrated in Figure 1.1.

Figure 1.1 Modern wind turbine. Reproduced by permission of NEG Micon

To understand how wind turbines are used, it is useful to briefly consider some of the fundamental facts underlying their operation. In modern wind turbines, the actual conversion process uses the basic aerodynamic force of lift to produce a net positive torque on a rotating shaft, resulting first in the production of mechanical power and then in its transformation to electricity in a generator. Wind turbines, unlike almost every other generator, can produce energy only in response to the wind that is immediately available. It is not possible to store the wind and use it a later time. The output of a wind turbine is thus inherently fluctuating and non-dispatchable. (The most one can do is to limit production below what the wind could produce.) Any system to which a wind turbine is connected must in some way take this variability into account. In larger networks, the wind turbine serves to reduce the total electrical load and thus results in a decrease in either the number of conventional generators being used or in the fuel use of those that are running. In smaller networks, there may be energy storage, backup generators, and some specialized control systems. A further fact is that the wind is not transportable: it can only be converted where it is blowing. Historically, a product such as ground wheat was made at the windmill and then transported to its point of use. Today, the possibility of conveying electrical energy via power lines compensates to some extent for wind's inability to be transported. In the future, hydrogen-based energy systems may add to this possibility.

1.1.1 Modern wind turbine design

Today, the most common design of wind turbine, and the only kind discussed in any detail in this book, is the horizontal axis wind turbine (HAWT). That is, the axis of rotation is parallel to the ground. HAWT rotors are usually classified according to the rotor orientation (upwind or downwind of the tower), hub design (rigid or teetering), rotor control (pitch vs. stall), number of blades (usually two or three blades), and how they are aligned with the wind (free yaw or active yaw). Figure 1.2 shows the upwind and downwind configurations.

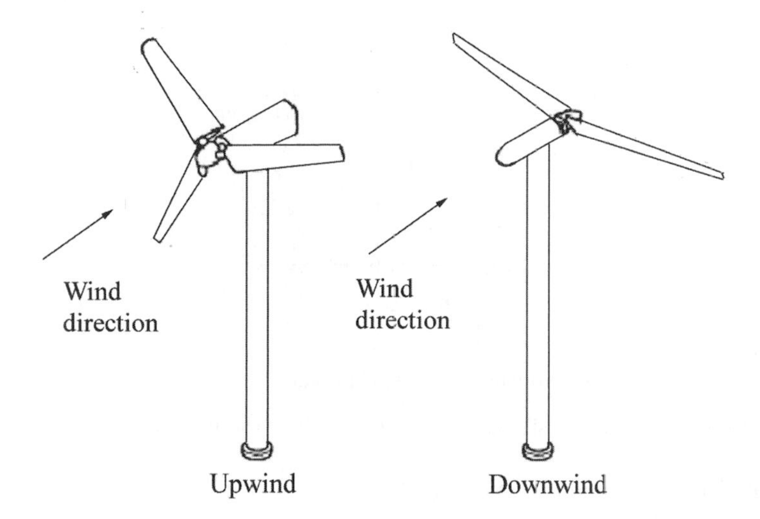

Figure 1.2 HAWT rotor configurations

The principal subsystems of a typical horizontal axis wind turbine are shown in Figure 1.3. These include:

- The rotor, consisting of the blades and the supporting hub
- The drive train, which includes the rotating parts of the wind turbine (exclusive of the rotor); it usually consists of shafts, gearbox, coupling, a mechanical brake, and the generator
- The nacelle and main frame, including wind turbine housing, bedplate, and the yaw system
- The tower and the foundation
- The machine controls
- The balance of the electrical system, including cables, switchgear, transformers, and possibly electronic power converters

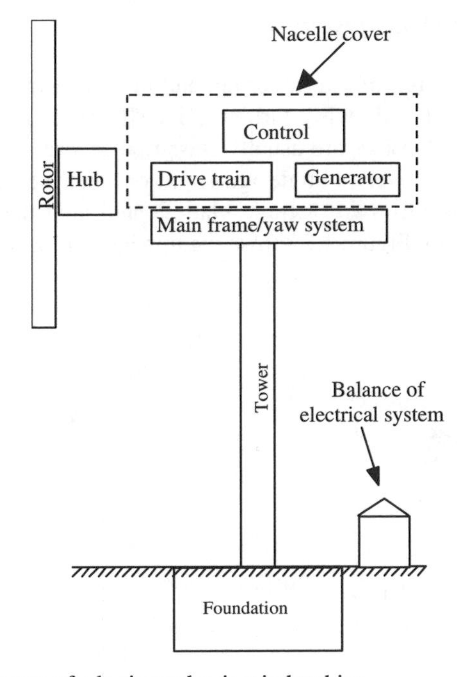

Figure 1.3 Major components of a horizontal axis wind turbine

The main options in wind machine design and construction include:

- Number of blades (commonly two or three)
- Rotor orientation: downwind or upwind of tower
- Blade material, construction method, and profile
- Hub design: rigid, teetering or hinged
- Power control via aerodynamic control (stall control) or variable pitch blades (pitch control)
- Fixed or variable rotor speed
- Orientation by self aligning action (free yaw), or direct control (active yaw)
- Synchronous or induction generator
- Gearbox or direct drive generator

A short introduction to and overview of some of the most important components follows. A more detailed discussion of the overall design aspects of these components, and other important parts of a wind turbine system, is contained in Chapters 3, 4, 5, 6, and 7 of this book.

1.1.1.1 Rotor

The rotor consists of the hub and blades of the wind turbine. These are often considered to be its most important components from both a performance and overall cost standpoint.

Most turbines today have upwind rotors with three blades. There are some downwind rotors and a few designs with two blades. Single-blade turbines have been built in the past, but are no longer in production. Most of the intermediate sized turbines, especially from Denmark, have used fixed blade pitch and stall control (described in Chapters 3, 6, and 7). A number of US manufacturers have used pitch control, and the general trend now seems to be an increased use of pitch control, especially in larger machines. The blades on the majority of turbines are made from composites, primarily fiberglass reinforced plastics (GRP), but sometimes wood/epoxy laminates are used. These subjects are addressed in more detail in the aerodynamics chapter (Chapter 3) and the design chapter (Chapter 6).

1.1.1.2 Drive train

The drive train consists of the rotating parts of the wind turbine. These typically include a low-speed shaft (on the rotor side), a gearbox, and a high-speed shaft (on the generator side). Other drive train components include the support bearings, one or more couplings, a brake, and the rotating parts of the generator (discussed separately in the next section). The purpose of the gearbox is to speed up the rate of rotation of the rotor from a low value (tens of rpm) to a rate suitable for driving a standard generator (hundreds or thousands of rpm). Two types of gearboxes are used in wind turbines: parallel shaft and planetary. For larger machines (over approximately 500 kW), the weight and size advantages of planetary gearboxes become more pronounced. Some wind turbine designs use specially designed, low-speed generators requiring no gearbox.

While the design of wind turbine drive train components usually follows conventional mechanical engineering machine design practice, the unique loading of wind turbine drive trains requires special consideration. Fluctuating winds and the dynamics of large rotating rotors impose significant varying loads on drive train components.

1.1.1.3 Generator

Nearly all wind turbines use either induction or synchronous generators. Both of these designs entail a constant or near-constant rotational speed of the generator when the generator is directly connected to a utility network.

The majority of wind turbines installed in grid connected applications use induction generators. An induction generator operates within a narrow range of speeds slightly higher than its synchronous speed (a four-pole generator operating in a 60 Hz grid has a synchronous speed of 1800 rpm). The main advantage of induction generators is that they are rugged, inexpensive, and easy to connect to an electrical network.

An option for electrical power generation involves the use of a variable speed wind turbine. There are a number of benefits that such a system offers, including the reduction of wear and tear on the wind turbine and potential operation of the wind turbine at maximum efficiency over a wind range of wind speeds, yielding increased energy capture. Although there are a large number of potential hardware options for variable speed operation of wind turbines, power electronic components are used in most variable speed machines currently being designed. When used with suitable power electronic converters, either synchronous or induction generators can run at variable speed.

1.1.1.4 Nacelle and yaw system

This category includes the wind turbine housing, the machine bedplate or main frame, and the yaw orientation system. The main frame provides for the mounting and proper alignment of the drive train components. The nacelle cover protects the contents from the weather.

A yaw orientation system is required to keep the rotor shaft properly aligned with the wind. The primary component is a large bearing that connects the main frame to the tower. An active yaw drive, generally used with an upwind wind turbine, contains one or more yaw motors, each of which drives a pinion gear against a bull gear attached to the yaw bearing. This mechanism is controlled by an automatic yaw control system with its wind direction sensor usually mounted on the nacelle of the wind turbine. Sometimes yaw brakes are used with this type of design to hold the nacelle in position. Free yaw systems (meaning that they can self-align with the wind) are commonly used on downwind wind machines.

1.1.1.5 Tower and foundation

This category includes the tower structure and the supporting foundation. The principal types of tower design currently in use are the free standing type using steel tubes, lattice (or truss) towers, and concrete towers. For smaller turbines, guyed towers are also used. Tower height is typically 1 to 1.5 times the rotor diameter, but in any case is normally at least 20 m. Tower selection is greatly influenced by the characteristics of the site. The stiffness of the tower is a major factor in wind turbine system dynamics because of the possibility of coupled vibrations between the rotor and tower. For turbines with downwind rotors, the effect of tower shadow (the wake created by airflow around a tower) on turbine dynamics, power fluctuations, and noise generation must be considered. For example, because of the tower shadow, downwind turbines are typically noisier than their upwind counterparts.

1.1.1.6 Controls

The control system for a wind turbine is important with respect to both machine operation and power production. A wind turbine control system includes the following components:

- Sensors - speed, position, flow, temperature, current, voltage, etc.
- Controllers - mechanical mechanisms, electrical circuits, and computers
- Power amplifiers - switches, electrical amplifiers, hydraulic pumps and valves
- Actuators - motors, pistons, magnets, and solenoids

The design of control systems for wind turbine application follows traditional control engineering practices. Many aspects, however, are quite specific to wind turbines, and are discussed in Chapter 7. Wind turbine control involves the following three major aspects and the judicious balancing of their requirements:

- Setting upper bounds on and limiting the torque and power experienced by the drive train.

- Maximizing the fatigue life of the rotor drive train and other structural components in the presence of changes in the wind direction, speed (including gusts), and turbulence, as well as start–stop cycles of the wind turbine.
- Maximizing the energy production.

1.1.1.7 Balance of electrical system

In addition to the generator, the wind turbine system utilizes a number of other electrical components. Some examples are cables, switchgear, transformers, power electronic converters, power factor correction capacitors, yaw and pitch motors. Details of the electrical aspects of wind turbines themselves are contained in Chapter 5. Interconnection with electrical networks is discussed in Chapter 8.

1.1.2 Power output prediction

The power output of a wind turbine varies with wind speed and every wind turbine has a characteristic power performance curve. With such a curve it is possible to predict the energy production of a wind turbine without considering the technical details of its various components. The power curve gives the electrical power output as a function of the hub height wind speed. Figure 1.4 presents an example of a power curve for a hypothetical wind turbine.

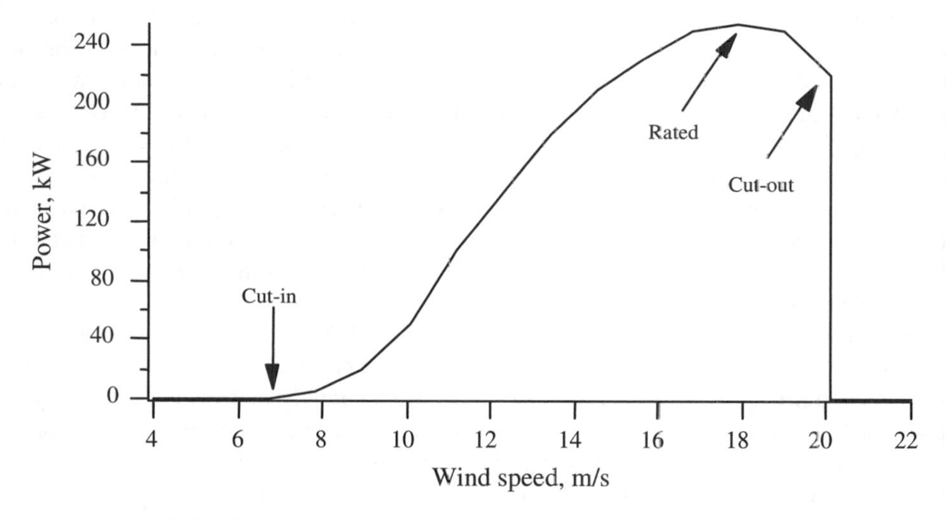

Figure 1.4 Typical wind turbine power curve

The performance of a given wind turbine generator can be related to three key points on the velocity scale:

- Cut-in speed: the minimum wind speed at which the machine will deliver useful power
- Rated wind speed: the wind speed at which the rated power (generally the maximum power output of the electrical generator) is reached
- Cut-out speed: the maximum wind speed at which the turbine is allowed to deliver power (usually limited by engineering design and safety constraints)

Power curves for existing machines can normally be obtained from the manufacturer. The curves are derived from field tests, using standardized testing methods. As is discussed in Chapter 6, it is also possible to estimate the approximate shape of the power curve for a given machine. Such a process, however, is by no means a simple task because it involves determination of the power characteristics of the wind turbine rotor and electrical generator, gearbox gear ratios, and component efficiencies.

1.1.3 Other wind turbine concepts

The wind turbine overview provided above assumed a topology of a basic type, namely one that employs a horizontal axis rotor, driven by lift forces. It is worth noting that a vast number of other topologies have been proposed, and in some cases built. None of these has met with the same degree of success as those with a horizontal axis, lift driven rotor. A few words are in order, however, to summarize briefly some of these other concepts. The closest runner up to the HAWT is the Darrieus vertical axis wind turbine. This concept was studied extensively in the both the United States and Canada in the 1970s and 1980s. Despite some appealing features, it was never able to match corresponding HAWTs in cost of energy. However, it is possible that the concept could emerge again for some applications.

Another concept that appears periodically is the concentrator. The idea is to channel the wind to increase the productivity of the rotor. The problem is that the cost of building an effective concentrator which can also withstand occasional extreme winds has always been more than the device was worth.

Finally, a number of rotors using drag instead of lift have been proposed. One concept, the Savonius rotor, has been used for some small water pumping applications. There are two fundamental problems with such rotors: (1) they are inherently inefficient (see comments on drag machines in Chapter 3), and (2) it is difficult to protect them from extreme winds. It is doubtful whether such rotors will ever achieve widespread use in wind turbines.

The reader interested in some of the variety of wind turbine concepts may wish to consult Nelson (1996). This book provides a description of a number of innovative wind systems. Reviews of various types of wind machines are given in Eldridge (1980) and Le Gourieres (1982). Some of the more innovative designs are documented in work supported by the US Department of Energy (1979, 1980). A few of the many interesting wind turbine concepts are illustrated in Figures 1.5 and 1.6.

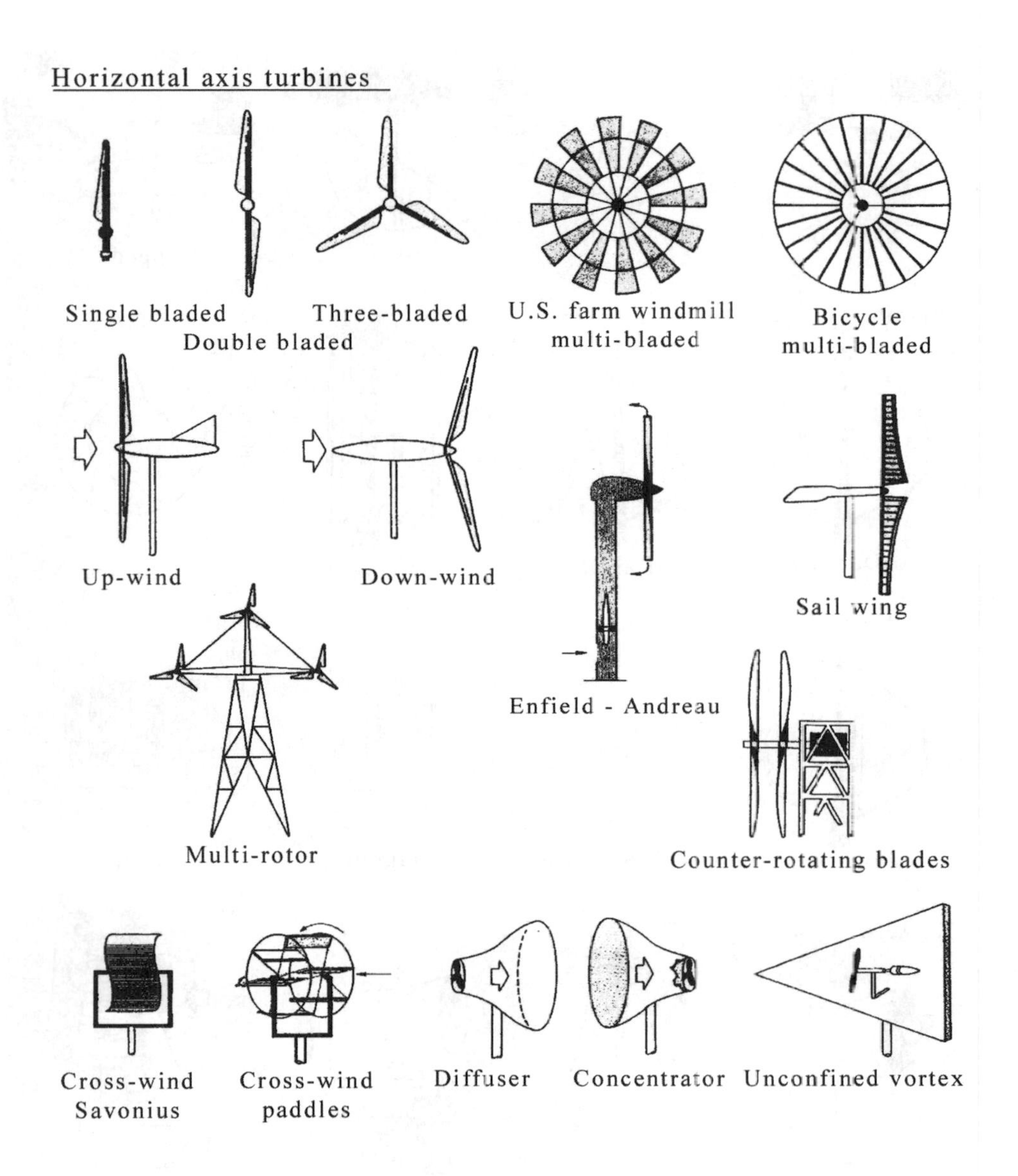

Figure 1.5 Various concepts for horizontal axis turbines (Eldridge, 1980)

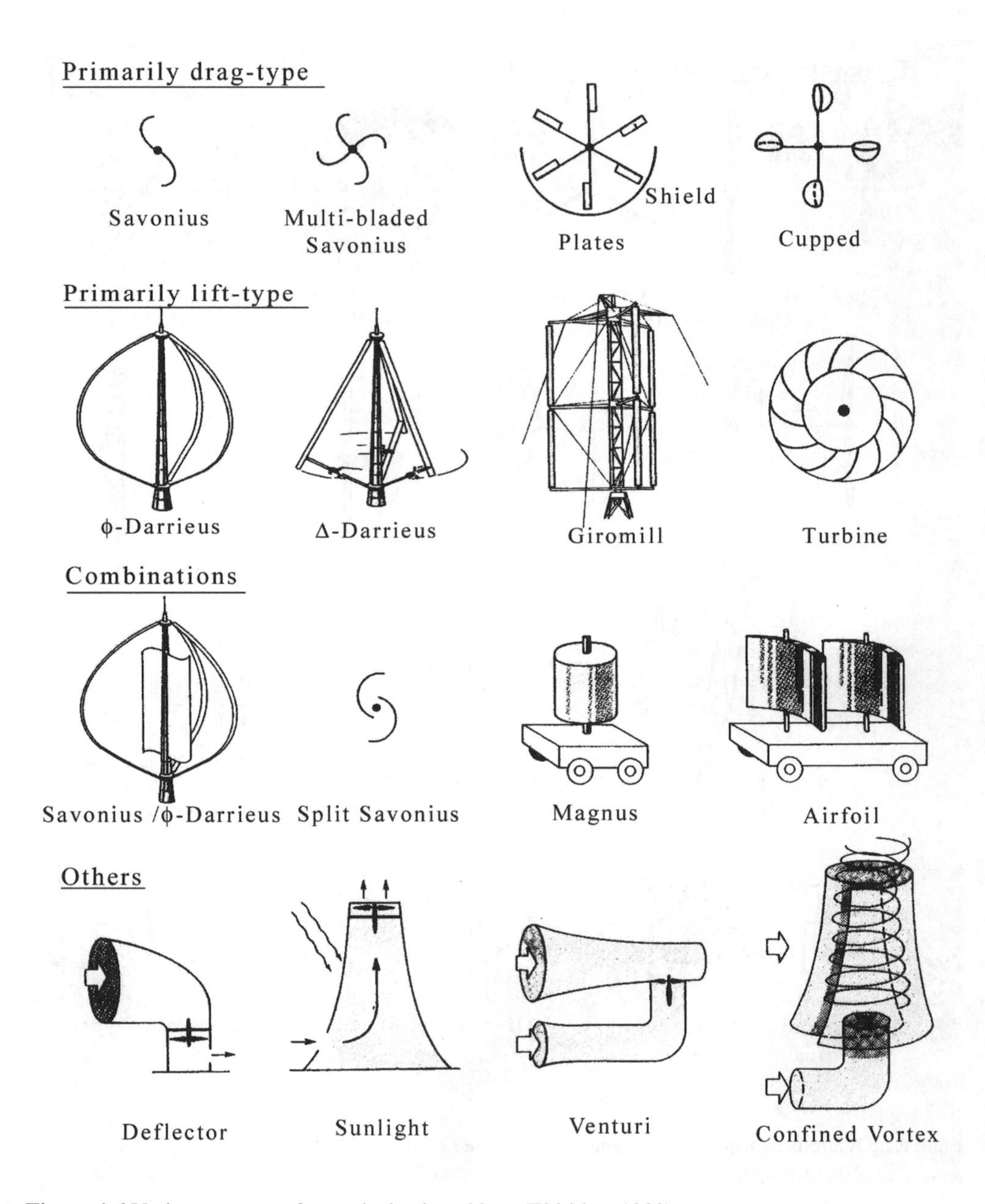

Figure 1.6 Various concepts for vertical axis turbines (Eldridge, 1980)

1.2 History of Wind Energy

It is worthwhile to consider some of the history of wind energy. The history serves to illustrate the issues that wind energy systems still face today, and provides insight into why turbines look the way they do. In the following summary, emphasis is given to those concepts which have particular relevance today.

The reader interested in a fuller description of the history of wind energy is referred to Park (1981) , Eldridge (1980), Inglis (1978), Freris (1990), Dodge (2000), and Ackermann and Soder (2000). Golding (1977) presents a history of wind turbine design from the ancient Persians to the mid-1950s. In addition to a summary of the historic uses of wind power, Johnson (1985) presents a history of wind electric generation, and the US research work of the 1970–85 period on horizontal axis, vertical axis and innovative types of wind turbines. The most recent comprehensive historical reviews of wind energy systems and wind turbines are contained in the books of Spera (1994), Gipe (1995), and Harrison et al. (2000). Eggleston and Stoddard (1987) give a historical perspective of some of the key components of modern wind turbines. Berger (1997) provides a fascinating picture of the early days of wind's re-emergence, particularly of the California wind farms.

1.2.1 A brief history of windmills

The first windmills on record were built by the Persians in approximately 900 AD. It is of note that these windmills (illustrated in Figure 1.7) had vertical axes and were drag type devices. As such they were inherently inefficient, and particularly susceptible to damage in high winds.

Figure 1.7 Early Persian windmill (Gipe, 1995)

Wind energy made its appearance in Europe during the Middle Ages. These windmills all had horizontal axes. They were used for nearly any mechanical task, including water pumping, grinding grain, sawing wood, and powering tools. The early mills were built on posts, so that the entire mill could be turned to face the wind (or yaw) when its direction changed. These mills normally had four blades. The number and size of blades presumably was based on ease of construction as well as an empirically determined efficient solidity (ratio of blade area to swept area). A typical European windmill is illustrated in Figure 1.8 (Hills, 1994).

Figure 1.8 European smock mill (Hills, 1994). Reproduced by permission of Cambridge University Press

The wind continued to be a major source of energy in Europe through the period just prior to the Industrial Revolution, but began to recede in importance after that time. The reason that wind energy began to disappear is primarily attributable to its non-dispatchability and its non-transportability. Coal had many advantages which the wind did not possess. Coal could be transported to wherever it was needed and used whenever it was desired. When coal was used to fuel a steam engine, the output of the engine could be adjusted to suit the load. Waterpower, which has some similarities to wind energy, was not eclipsed so dramatically. This is no doubt because waterpower is to some extent transportable (via canals) and dispatchable (by using ponds as storage).

Prior to its demise, the European windmill had reached a high level of design sophistication. In the later mills (or 'smock mills'), the majority of the mill was stationary. Only the top would be moved to face the wind. Yaw mechanisms included both manually operated arms, and separate yaw rotors. Blades had acquired somewhat of an airfoil shape and included some twist. The power output of some machines could be adjusted by an automatic control system. This was the forerunner of the system used by James Watt on steam engines. In the windmill's case a fly ball governor would sense when the rotor was speeding up. The 'skin' over the blade framework consisted of multiple small flaps, resembling Venetian blinds. The governor was attached via linkages to the flaps. It would open the flaps, thereby reducing the power, and hence speed, when the governor was going faster than nominal. It would close the flaps when more power was needed.

One significant development in the 18th century was the introduction of scientific testing and evaluation of windmills. The Englishman John Smeaton, using such apparatus as illustrated in Figure 1.9, discovered three basic rules that are still applicable:

- The speed of the blade tips is ideally proportional to the speed of wind
- The maximum torque is proportional to the speed of wind squared
- The maximum power is proportional to the speed of wind cubed

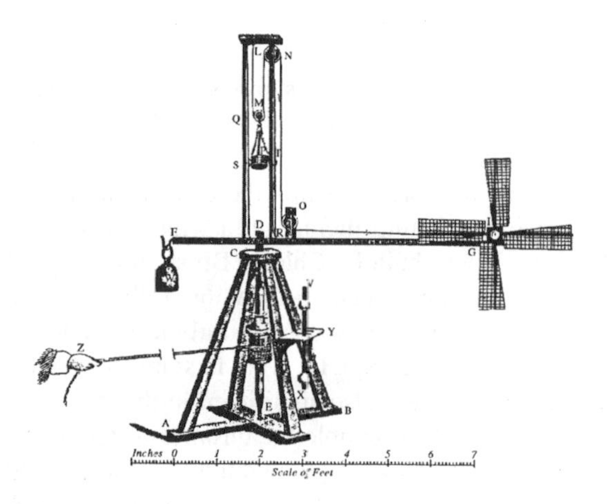

Figure 1.9 Smeaton's laboratory windmill testing apparatus

The 18th century European windmills represented the culmination of one approach to using wind for mechanical power and included a number of features which were later incorporated into some early electricity generating wind turbines.

As the European windmills were entering their final years, another variant of windmill came into widespread use in the United States. This type of windmill, illustrated in Figure 1.10, was most notably used for pumping water, particularly in the West. They were used on ranches for cattle and to supply water for the steam railroads. These mills were distinctive for their multiple blades and are often referred to as 'fan mills'. One of their most significant features was a simple but effective regulating system. This allowed the turbines to run unattended for long periods. Such regulating systems foreshadowed the automatic control systems which are now an integral part of modern wind turbines.

Figure 1.10 American water pumping windmill design (US Department of Agriculture)

1.2.2 Early wind generation of electricity

The initial use of wind for electric generation, as opposed to for mechanical power, included the successful commercial development of small wind generators and research and experiments using large wind turbines.

When electrical generators appeared towards the end of the nineteenth century, it was reasonable that people would try to turn them with a windmill rotor. In the United States, the most notable early example was built by Charles Brush in Cleveland, Ohio in 1888. The Brush turbine did not result in any trend, but in the following years, small electrical generators did become widespread. These small turbines, pioneered most notably by Marcellus Jacobs and illustrated in Figure 1.11, were in some ways the logical successors to the water pumping fan mill. They were also significant in that their rotors had three blades with true airfoil shapes and began to resemble the turbines of today. Another feature of the Jacobs turbine was that it was typically incorporated into a complete, residential scale power system, including battery storage. The Jacobs turbine is considered to be a direct forerunner of such modern small turbines as the Bergey and Southwest Windpower machines. The expansion of the central electrical grid under the auspices of the Rural Electrification Administration during the 1930s marked the beginning of the end of the widespread use of small wind electric generators, at least for the time being.

Figure 1.11 Jacobs turbine (Jacobs, 1961)

The first half of the 20th century also saw the construction or conceptualization of a number of larger wind turbines which substantially influenced the development of today's technology. Probably the most important sequence of turbines was in Denmark. Between 1891 and 1918 Poul La Cour built more than 100 electricity generating turbines in the 20–35 kW size range. His design was based on the latest generation of Danish smock mills. One of its more remarkable features was that the electricity was used to produce hydrogen, and the hydrogen gas was then used for lighting. La Cour's turbines were followed by a number of turbines made by Lykkegaard Ltd. and F. L. Smidth & Co prior to World War II. These ranged in size from 30 to 60 kW. Just after the war, Johannes Juul erected the 200 kW Gedser turbine, illustrated in Figure 1.12, in southeastern Denmark. This three-bladed machine was particularly innovative in that it employed aerodynamic stall for power control and used an induction generator, rather than the more conventional (at the time) synchronous generator. An induction generator is much simpler to connect to the grid than is a synchronous generator. Stall is also a simple way to control power. These two concepts formed the core of the strong Danish presence in wind energy in the 1980s (see http://www.risoe.dk/ and http://www.windpower.dk/. for more details on wind energy in Denmark). One of the pioneers in wind energy in the 1950s was Ulrich Hütter in Germany. His work focused on applying modern aerodynamic principles to wind turbine design. Many of the concepts he worked with are still in use in some form today.

Figure 1.12 Danish Gedser wind turbine. Reproduced by permission of Danish Wind Turbine Manufacturers

In the United States, the most significant early large turbine was the Smith–Putnam machine, built at Grandpa's Knob in Vermont in the late 1930s (Putnam, 1948). With a diameter of 53.3 m and a power rating of 1.25 MW, this was the largest wind turbine ever built up until that time and for many years thereafter. This turbine, illustrated in Figure 1.13, was also significant in that it was the first large turbine with two blades. In this sense it was a predecessor for the two-bladed turbines built by the US Department of Energy in the late 1970s and early 1980s. The turbine was also notable in that the company that built it, S. Morgan Smith, had long experience in hydroelectric generation and intended to produce a commercial line of wind machines. Unfortunately, the Smith–Putnam turbine was too large, too early, given the level of understanding of wind energy engineering. It suffered a blade failure in 1945, and the project was abandoned.

1.2.3 The re-emergence of wind energy

The re-emergence of wind energy can be considered to have begun in the late 1960s. The book *Silent Spring* (Carson, 1962) made many people aware of the environmental consequences of industrial development. *Limits to Growth* (Meadows et al., 1972) followed in the same vein, arguing that unfettered growth would inevitably lead to either disaster or change. Among the culprits identified were fossil fuels. The potential dangers of nuclear energy also became more public at this time. Discussion of these topics formed the backdrop for an environmental movement which began to advocate cleaner sources of energy.

Figure 1.13 Smith–Putnam wind turbine (Eldridge, 1980)

In the United States, in spite of growing concern for environmental issues, not much new happened in wind energy development until the Oil Crises of the mid-1970s. Under the Carter administration, a new effort was begun to develop 'alternative' sources of energy, one of which was wind energy. The US Department of Energy (DOE) sponsored a number of projects to foster the development of the technology. Most of the resources were allocated to large machines, with mixed results. These machines ranged from the 100 kW (38 m diameter) NASA MOD-0 to the 3.2 MW Boeing MOD-5B with its 98 m diameter. Much interesting data was generated but none of the large turbines led to commercial projects. DOE also supported development of some small wind turbines and built a test facility for small machines at Rocky Flats, Colorado. A number of small manufacturers of wind turbines also began to spring up, but there was not a lot activity until the late 1970s.

The big opportunities occurred as the result of changes in the utility regulatory structure and the provision of incentives. The US federal government, through the Public Utility Regulatory Policy Act of 1978 required utilities (1) to allow wind turbines to connect with the grid and (2) to pay the 'avoided cost' for each kWh the turbines generated and fed into the grid. The actual avoided cost was debatable, but in many states utilities would pay enough that wind generation began to make economic sense. In addition, the federal government and some states provided investment tax credits to those who installed wind turbines. The state which provided the best incentives, and which also had regions with good winds, was California. It was now possible to install a number of small turbines together in a group ('wind farm'), connect them to the grid, and make some money.

The California wind rush was on. Over a period of a few years, thousands of wind turbines were installed in California, particularly in the Altamont Pass, San Gorgonio Pass, and Tehachipi. A typical installation is shown in Figure 1.14. The installed capacity reached approximately 1500 MW. The early years of the California wind rush were fraught with difficulties, however. Many of the machines were essentially still prototypes, and not yet up to the task. An investment tax credit (as opposed to a production tax credit) is arguably not the best way to encourage the development and deployment of productive machines, especially when there is no means for certifying that machines will actually perform as the manufacturer claims. When the federal tax credits were withdrawn by the Reagan administration in the early 1980s, the wind rush collapsed.

Figure 1.14 California wind farm (National Renewable Energy Laboratory)

Wind turbines installed in California were not limited to those made in the United States. In fact, it was not long before Danish turbines began to have a major presence in the California wind farms. The Danish machines also had some teething problems in California, but in general they were closer to production quality than were their US counterparts. When all the dust had settled after the wind rush had ended, the majority of US manufacturers had gone out of business. The Danish manufacturers had restructured or merged, but had in some way survived.

During the 1990s, a decade which saw the demise (in 1996) of the largest US manufacturer, Kennetech Windpower, the focal point of wind turbine manufacturing definitively moved to Europe, particularly Denmark and Germany. Concerns about global warming and continued apprehensions about nuclear power have resulted in a strong demand for more wind generation there and in other countries as well. Some of the major European suppliers are setting up manufacturing plants in other countries, such as Spain, India and the United States.

Over the last 25 years, the size of the largest commercial wind turbines, as illustrated in Figure 1.15, has increased from approximately 50 kW to 2 MW, with machines up to 5 MW under design. The total installed capacity in the world as of the year 2001 was approximately 20,000 MW, with the majority of installations in Europe. Offshore wind energy systems are also under active development in Europe. Design standards and machine certification procedures have been established, so that the reliability and performance are far superior to those of the 1970s and 1980s. The cost of energy from wind has dropped to the point that in some sites it is nearly competitive with conventional sources, even without incentives. In those countries where incentives are in place, the rate of development is quite strong.

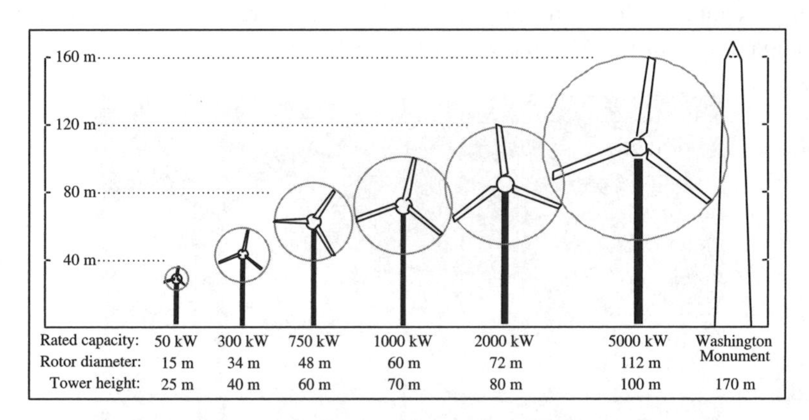

Figure 1.15 Representative size, height, and diameter of wind turbines

1.2.4 Technological underpinnings of modern wind turbines

Wind turbine technology, dormant for many years, awoke at the end of the 20th century to a world of new opportunities. Developments in many other areas of technology were adapted to wind turbines and have helped to hasten their re-emergence. A few of the many areas which have contributed to the new generation of wind turbines include materials science, computer science, aerodynamics, analytical methods, testing, and power electronics. Materials science has brought new composites for the blades, and alloys for the metal components. Developments in computer science facilitate design, analysis, monitoring, and control. Aerodynamic design methods, originally developed for the aerospace industry, have now been adapted to wind turbines. Analytical methods have now developed to the point where it is possible to have a much clearer understanding of how a new design should perform than was previously possible. Testing using a vast array of commercially available sensors and data collection and analysis equipment allows designers to better understand how the new turbines actually perform. Power electronics is a relatively new area which is just beginning to be used with wind turbines. Power electronic devices can help connect the turbine's generator smoothly to the electrical network; allow the turbine to run at variable speed, producing more energy, reducing fatigue damage, and benefiting the utility in the process; facilitate operation in a small, isolated network; and transfer energy to and from storage.

1.2.5 Trends

Wind turbines have evolved a great deal over the last 25 years. They are more reliable, more cost effective, and quieter. It cannot be concluded that the evolutionary period is over, however. It should still be possible to reduce the cost of energy at sites with lower wind speeds. Turbines for use in remote communities still remain to be made commercially viable. The world of offshore wind energy is just in its infancy. There are tremendous opportunities in offshore locations but many difficulties to be overcome. As wind energy comes to supply an ever larger fraction of the world's electricity, the issues of intermittency, transmission, and storage must be revisited.

There will be continuing pressure for designers to improve the cost effectiveness of turbines for all applications. Improved engineering methods for the analysis, design, and for mass-produced manufacturing will be required. Opportunities also exist for the development of new materials to increase wind turbine life. Increased consideration will need to be given to the requirements of specialized applications. In all cases, the advancement of the wind industry represents an opportunity and challenge for a wide range of disciplines, especially including mechanical, electrical, materials, aeronautical, controls and civil engineering as well as computer science.

References

Ackermann, T., Soder, L. (2000) Wind Energy Technology and Current Status: A Review *Renewable and Sustainable Energy Reviews*, 4, 315–374.

Berger, J. J. (1997) *Charging Ahead: The Business of Rewnewable Energy and What it Means for America*, Univ. of California Press, Berkeley, CA.

Carson, R. (1962) *Silent Spring*, Houghton Mifflin, New York.

Danish Wind Turbine Manufacturers (2001) http://www.windpower.dk.

Dodge, D. M. (2000) Illustrated History of Wind Power Development, http://telosnet.com/wind.

Eggleston, D. M., Stoddard, F. S. (1987) *Wind Turbine Engineering Design.* Van Nostrand Reinhold, New York.

Eldridge, F. R. (1980) *Wind Machines*, 2nd Edition, Van Nostrand Reinhold, New York.

Freris, L. L. (1990) *Wind Energy Conversion Systems.* Prentice Hall, New York.

Gipe, P. (1995) *Wind Energy Comes of Age.* Wiley, New York.

Golding, E. W. (1977) *The Generation of Electricity by Wind Power.* E. & F. N. Spon, London.

Harrison, R., Hau, E., Snel, H. (2000) *Large Wind Turbines: Design and Economics.* Wiley, Chichester.

Hills, R. L. (1994) *Power from Wind.* Cambridge University Press, Cambridge, UK.

Inglis, D. R. (1978) *Windpower and Other Energy Options.* University of Michigan Press, Ann Arbor, MI.

Jacobs, M. L. (1961) Experience with Jacobs Wind-Driven Electric Generating Plant, 1931–1957, *Proc. of the United Nations Conference on New Sources of Energy*, Vol. 7, Rome, 337–339.

Johnson, G. L. (1985) *Wind Energy Systems.* Prentice Hall, Englewood Cliffs, NJ.

Le Gourieres, D. (1982) *Wind Power Plants.* Pergamon Press, Oxford.

Meadows, D.H., Meadows, D.L., Randers, J., Behrens III, W.W. (1972) *The Limits to Growth.* Universe Books, New York.

Nelson, V. (1996) *Wind Energy and Wind Turbines.* Alternative Energy Institute, Canyon, TX.

Park, J. (1981) *The Wind Power Book.* Chesire Books, Palo Alto, CA.

Putnam, P. C. (1948) *Power From the Wind.* Van Nostrand Reinhold, New York.

Spera, D. A. (ed.) (1994) *Wind Turbine Technology: Fundamental Concepts of Wind Turbine Engineering.* ASME Press, New York.

US Department of Energy (1979) *Wind Energy Innovative Systems Conference Proceedings.* Solar Energy Research Institute (SERI).

US Department of Energy (1980), *SERI Second Wind Energy Innovative Systems Conference Proceedings*, Solar Energy Research Institute (SERI).

2

Wind Characteristics and Resources

2.1 Introduction

This chapter will review an important topic in wind energy: wind resources and characteristics. The material covered in this chapter can be of direct use to other aspects of wind energy which are discussed in the other sections of this book. For example, knowledge of the wind characteristics at a particular site is relevant to the following topics:

- Systems design - System design requires knowledge of representative average wind conditions, as well as information on the turbulent nature of the wind. This information is used in the design of a wind turbine intended for a particular site.
- Performance Evaluation - Performance evaluation requires determining the expected energy productivity and cost effectiveness of a particular wind energy system based on the wind resource.
- Siting - Siting requirements can include the assessment or prediction of the relative desirability of candidate sites for one or more wind turbines.
- Operations - Operation requirements include the need for wind resource information that can be used for load management, operational procedures (such as start-up and shut-down), and the prediction of maintenance or system life.

The chapter starts with a general discussion of wind resource characteristics, followed by a section on the characteristics of the atmospheric boundary layer that are directly applicable to wind energy applications. The next two sections present a number of topics that enable one to analyze wind data, make resource estimates, and determine wind turbine power production from wind resource data, or from a limited amount of wind data (such as average wind speed). Next, a summary of available worldwide wind resource assessment data is given. The following section reviews wind resource measurement techniques and instrumentation. The chapter concludes with a summary of a number of advanced topics in the area of wind resource characterization.

The importance of these subjects is emphasized by the size of sections on wind characteristics and resources in wind energy reference textbooks. These include the classic references of Golding (1977) and Putnam (1948) as well as books by Eldridge (1980),

Johnson (1985), Freris (1990) and Spera (1994). In addition, use will be made of the wind resource material included in several books devoted to this subject. These include the work of Justus (1978), Hiester and Pennell (1981), and the text of Rohatgi and Nelson (1994).

2.2 General Characteristics of the Wind Resource

In discussing the general characteristics of the wind resource it is important to consider such topics as the global origins of the wind resource, the general characteristics of the wind, and estimates of the wind resource potential.

2.2.1 Wind resource: global origins

2.2.1.1 Overall global patterns

Global winds are caused by pressure differences across the earth's surface due to the uneven heating of the earth by solar radiation. For example, the amount of solar radiation absorbed at the earth's surface is greater at the equator than at the poles. The variation in incoming energy sets up convective cells in the lower layers of the atmosphere (the troposphere). Thus, in a simple flow model, air rises at the equator and sinks at the poles. The circulation of the atmosphere that results from uneven heating is greatly influenced by the effects of the rotation of the earth (at a speed of about 600 kilometers per hour at the equator, decreasing to zero at the poles). In addition, seasonal variations in the distribution of solar energy give rise to variations in the circulation.

The spatial variations in heat transfer to the earth's atmosphere create variations in the atmospheric pressure field that cause air to move from high to low pressure. The pressure gradient force in the vertical direction is usually cancelled by the downward gravitational force. Thus, the winds blow predominately in the horizontal plane, responding to horizontal pressure gradients. At the same time, there are forces that strive to mix the different temperature and pressure air masses distributed across the earth's surface. In addition to the pressure gradient and gravitational forces, inertia of the air, the earth's rotation, and friction with the earth's surface (resulting in turbulence), affect the atmospheric winds. The influence of each of these forces on atmospheric wind systems differs depending on the scale of motion considered.

As shown in Figure 2.1, worldwide wind circulation involves large-scale wind patterns, affecting prevailing near surface winds, that cover the entire planet. It should be noted that this model is an oversimplification because it does not reflect the effect that land masses have on the wind distribution.

2.2.1.2 Mechanics of wind motion

In one of the simplest models for the mechanics of the atmosphere's wind motion, four atmospheric forces can be considered. These include pressure forces, the Coriolis force caused by the rotation of the earth, inertial forces due to large-scale circular motion, and frictional forces at the earth's surface.

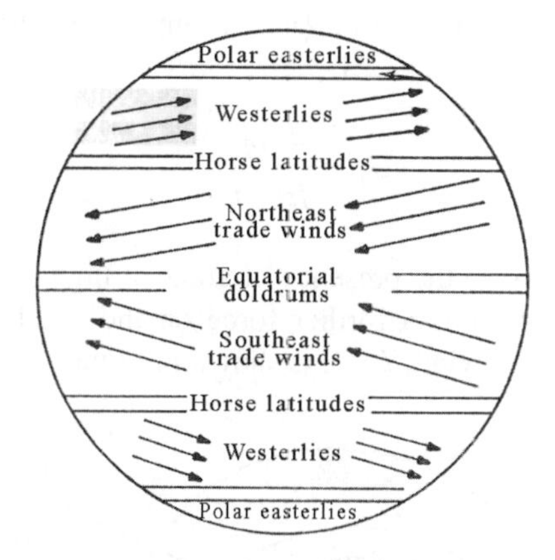

Figure 2.1 Surface winds of worldwide circulation pattern (Hiester and Pennell, 1981)

The pressure force on the air (per unit mass), F_p is given by:

$$F_p = \frac{-1}{\rho}\frac{\partial p}{\partial n} \tag{2.2.1}$$

where ρ is the density of the air and n is the direction normal to lines of constant pressure. Also, $\partial p/\partial n$ is defined as the pressure gradient normal to the lines of constant pressure, or isobars. The Coriolis force (per unit mass), F_c, a fictitious force caused by measurements with respect to a rotating reference frame (the earth), is expressed as:

$$F_c = fU \tag{2.2.2}$$

where U is the wind speed, and f is the Coriolis parameter [$f = 2\omega \sin(\phi)$]. ϕ represents the latitude and ω the angular rotation of the earth. Thus, the magnitude of the Coriolis force depends on wind speed and latitude. The direction of the Coriolis force is perpendicular to the direction of motion of the air. The result of these two forces, called the geostrophic wind, tends to be parallel to isobars (see Figure 2.2).

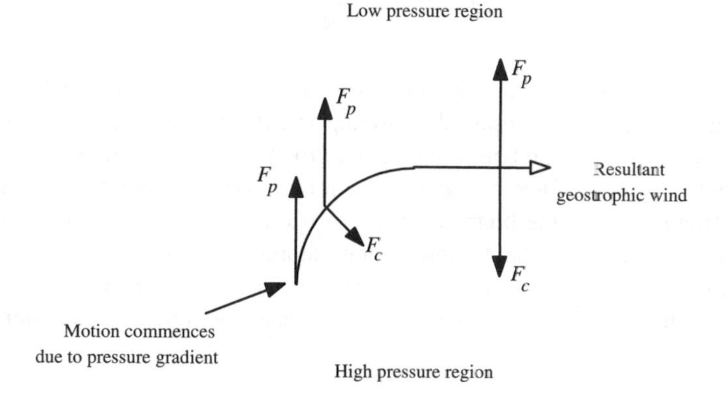

Figure 2.2 Illustration of the geostrophic wind; F_p, pressure force on the air; F_p, Coriolis force

The magnitude of the geostrophic wind, U_g, is a function of the balance of forces and is given by:

$$U_g = \frac{-1}{f\rho}\frac{\partial p}{\partial n} \tag{2.2.3}$$

This is an idealized case since the presence of areas of high and low pressures cause the isobars to be curved. This imposes a further force on the wind, a centrifugal force. The resulting wind, called a gradient wind, U_{gr}, is shown in Figure 2.3.

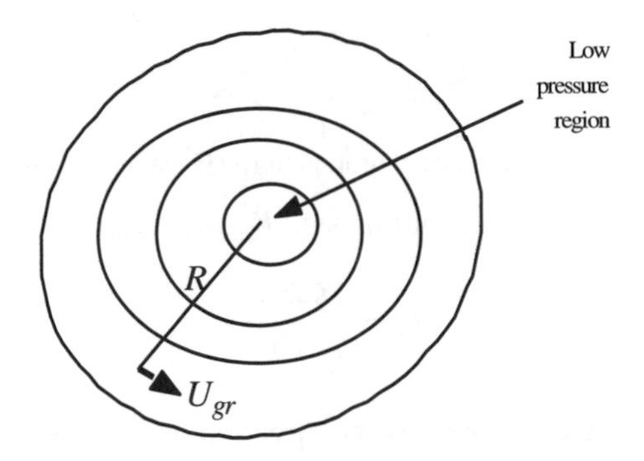

Figure 2.3 Illustration of the gradient wind; U_{gr}; R, radius of curvature

The gradient wind is also parallel to the isobars and is the result of the balance of the forces:

$$\frac{U_{gr}^2}{R} = -fU_{gr} - \frac{1}{\rho}\frac{\partial p}{\partial n} \tag{2.2.4}$$

where R is the radius of curvature of the path of the air particles, and

$$U_{gr} = U_g - \frac{U_{gr}^2}{fR} \tag{2.2.5}$$

A final force on the wind is due to friction at the earth's surface. That is, the earth's surface exerts a horizontal force upon the moving air, the effect of which is to retard the flow. This force decreases as the height above the ground increases and becomes negligible above the boundary layer (defined as the near earth region of the atmosphere where viscous forces are important). Above the boundary layer, a frictionless wind balance is established, and the wind flows with the gradient wind velocity along the isobars. Friction at the surface causes the wind to be diverted more toward the low-pressure region. More details concerning the earth's boundary layer and its characteristics will be given in later sections.

2.2.1.3 Other atmospheric circulation patterns

The general circulation flow pattern described previously best represents a model for a smooth spherical surface. In reality, the earth's surface varies considerably, with large ocean and land masses. These different surfaces can affect the flow of air due to variations in pressure fields, the absorption of solar radiation, and the amount of moisture available.

The oceans act as a large sink for energy. Therefore, the movement of air is often coupled with the ocean circulation. All these effects lead to differential pressures which affect the global winds and many of the persistent regional winds, such as occur during monsoons. In addition, local heating or cooling may cause persistent local winds to occur on a seasonal or daily basis. These include sea breezes and mountain winds.

Smaller scale atmospheric circulation can be divided into secondary and tertiary circulation (see Rohatgi and Nelson, 1994). Secondary circulation occurs if the centers of high or low pressure are caused by heating or cooling of the lower atmosphere. Secondary circulations include the following:

- Hurricanes
- Monsoon circulation
- Extratropical cyclones

Tertiary circulations are small-scale, local circulations characterized by local winds. These include the following:

- Land and sea breezes
- Valley and mountain winds
- Monsoon-like flow (example: flow in California passes)
- Foehn winds (dry high-temperature winds on the downwind side of mountain ranges)
- Thunderstorms
- Tornadoes

Examples of tertiary circulation, valley and mountain winds, are shown in Figure 2.4. During the day, the warmer air of the mountain slope rises and replaces the heavier cool air above it. The direction reverses at night, as cold air drains down the slopes and stagnates in the valley floor.

An understanding of these wind patterns, and other local effects, is important for the evaluation of potential wind energy sites.

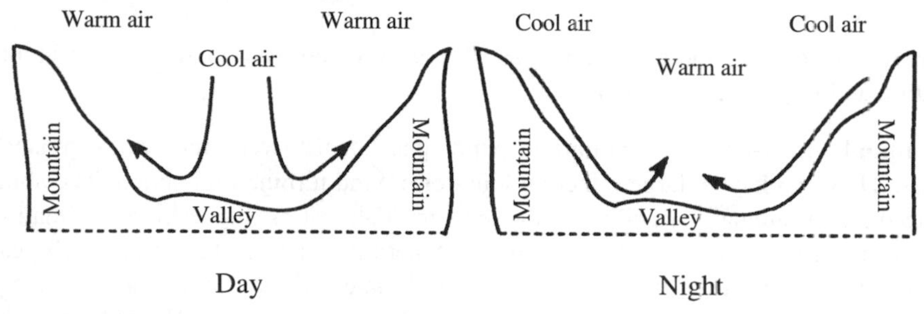

Figure 2.4 Diurnal valley and mountain wind (Rohatgi and Nelson, 1994). Reproduced by permission of Alternative Energy Institute

2.2.2 *General Characteristics of Wind*

Atmospheric motions vary in both time (seconds to months) and space (centimeters to thousands of kilometers). Figure 2.5 summarizes the time and space variations of atmospheric motion as applied to wind energy. As will be discussed in later sections, space variations are generally dependent on height above the ground and global and local geographical conditions.

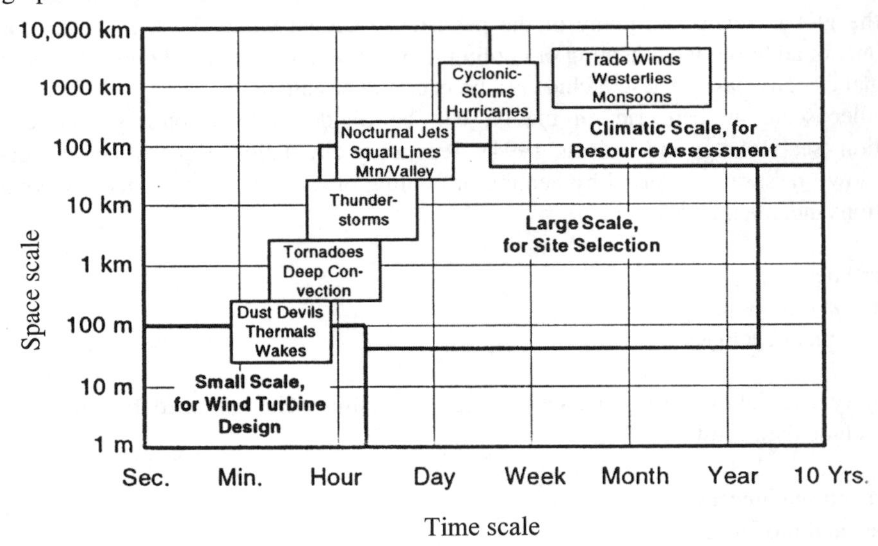

Figure 2.5 Time and space scales of atmospheric motion (Spera, 1994). Reproduced by permission of ASME.

2.2.2.1 Variations in time

Following conventional practice, variations of wind speed in time can be divided into the following categories:

- Inter-annual
- Annual
- Diurnal
- Short-term (gusts and turbulence)

A review of each of these categories as well as comments on wind speed variation due to location and wind direction follows.

Inter-annual Inter-annual variations in wind speed occur over time scales greater than one year. They can have a large effect on long-term wind turbine production. The ability to estimate the inter-annual variability at a given site is almost as important as estimating the long-term mean wind at a site. Meteorologists generally conclude that it takes 30 years of data to determine long-term values of weather or climate and that it takes at least five years to arrive at a reliable average annual wind speed at a given location. Nevertheless, shorter data records can be useful. Aspliden et al. (1986) note that one statistically developed rule of thumb is that one year of record data is generally sufficient to predict long-term seasonal mean wind speeds within an accuracy of 10% with a confidence level of 90%.

Researchers are still looking for reliable prediction models for long-term mean wind speed. The complexities of the interactions of the meteorological and topographical factors that cause its variation make the task difficult.

Annual Significant variations in seasonal or monthly averaged wind speeds are common over most of the world. For example, for the eastern one-third of the United States, maximum wind speeds occur during the winter and early spring. Spring maximums occur over the Great Plains, the North Central States, the Texas Coast, in the basins and valleys of the West, and the coastal areas of Central and Southern California. Winter maximums occur over all US mountainous regions, except for some areas in the lower Southwest, where spring maximums occur. Spring and summer maximums occur in the wind corridors of Oregon, Washington, and California.

Figure 2.6 illustrates seasonal changes of monthly wind speed for Billings, Montana. It is interesting to note that this figure clearly shows that the typical behavior of monthly variation is not defined by a single year of data.

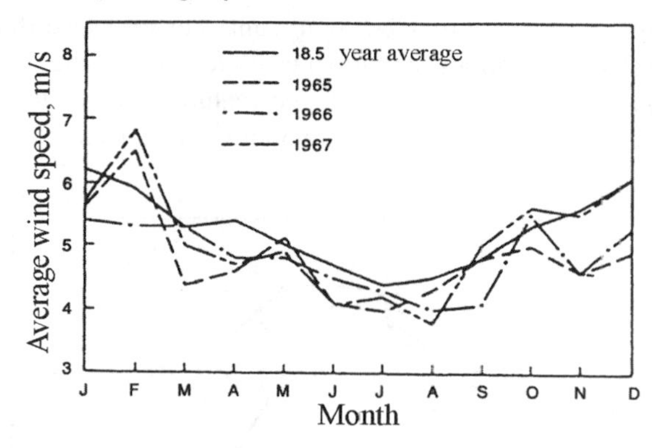

Figure 2.6 Seasonal changes of monthly average wind speeds (Hiester and Pennell, 1981)

Similarly, Figure 2.7 provides an illustration of the importance of annual wind speed variation, and its effect on available wind power.

Diurnal (time of day) In both tropical and temperate latitudes, large wind variations also can occur on a diurnal or daily time scale. This type of wind speed variation is due to differential heating of the earth's surface during the daily radiation cycle. A typical diurnal variation is an increase in wind speed during the day with the wind speeds lowest during the hours from midnight to sunrise. Daily variations in solar radiation are responsible for diurnal wind variations in temperate latitudes over relatively flat land areas. The largest diurnal changes generally occur in spring and summer, and the smallest in winter. Furthermore, the diurnal variation in wind speed may vary with location and altitude above sea level. For example, at altitudes high above surrounding terrain, e.g., mountains or ridges, the diurnal pattern may be very different. This variation can be explained by mixing or transfer of momentum from the upper air to the lower air.

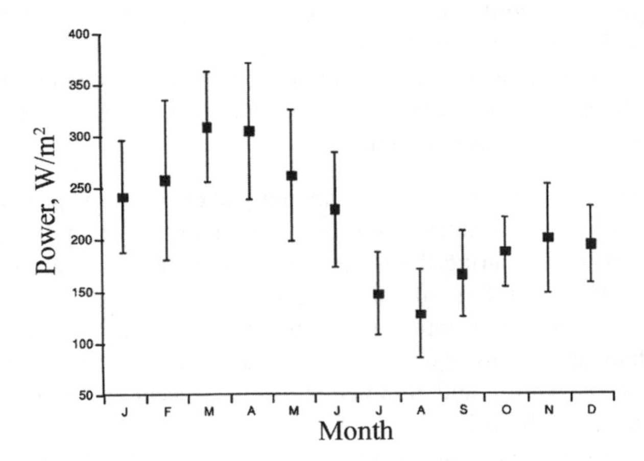

Figure 2.7 Seasonal variation of available wind power per unit area for Amarillo, Texas (Rohatgi and Nelson, 1994). Reproduced by permission of Alternative Energy Institute

As illustrated in Figure 2.8, there may be significant year-to-year differences in diurnal behavior, even at fairly windy locations. Although gross features of the diurnal cycle can be established with a single year of data, more detailed features such as the amplitude of the diurnal oscillation and the time of day that the maximum winds occur cannot be determined precisely.

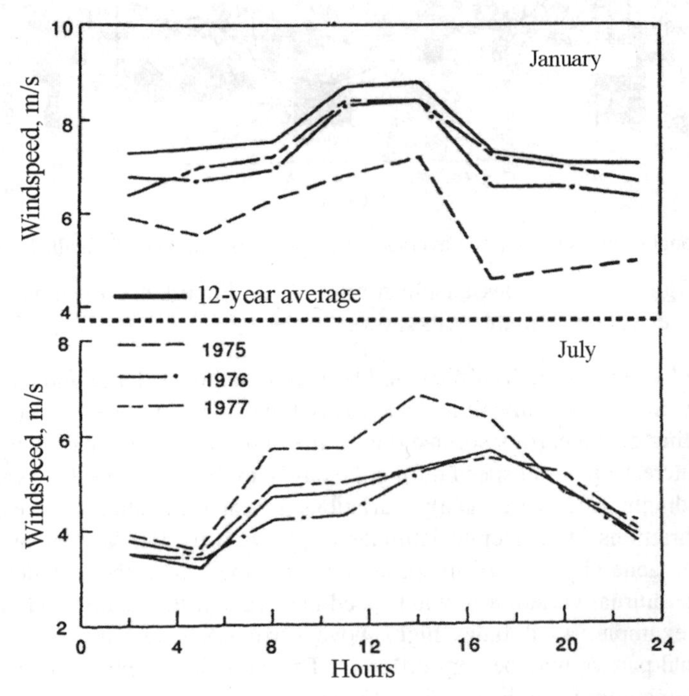

Figure 2.8 Monthly mean diurnal wind speeds for January and July for Casper, Wyoming (Hiester and Pennell, 1981)

Short-term Short-term wind speed variations of interest include turbulence and gusts. Figure 2.9, output from an anemometer (described later), shows the type of short-term wind speed variations that normally exist.

Short-term variations usually mean variations over time intervals of 10 minutes or less. Ten-minute averages are typically determined using a sampling rate of about 1 second. It is generally accepted that variations in wind speed with periods from less than a second to 10 minutes and that have a stochastic character are considered to represent turbulence. For wind energy applications, turbulent fluctuations in the flow need to be quantified for the turbine design considerations based on maximum load and fatigue prediction, structural excitations, control, system operation, and power quality. More details on these factors as related to turbine design are discussed in Chapter 6 of this text.

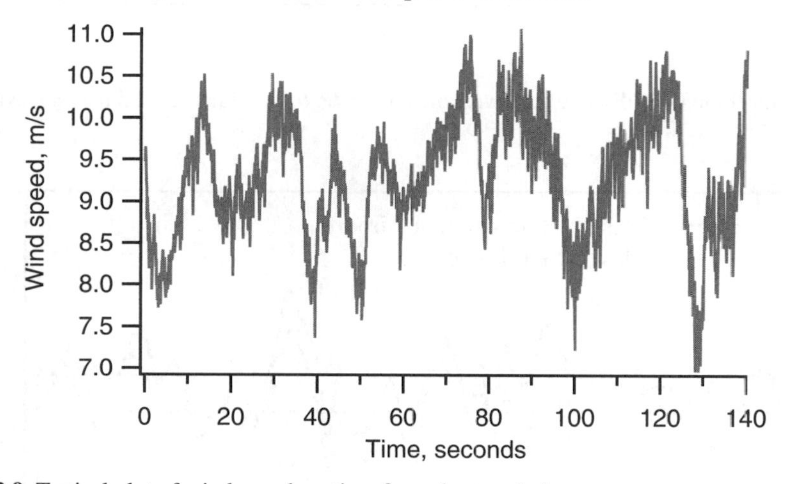

Figure 2.9 Typical plot of wind speed vs. time for a short period

Turbulence can be thought of as random wind speed fluctuations imposed on the mean wind speed. These fluctuations occur in all three directions: longitudinal (in the direction of the wind), lateral (perpendicular to the average wind), and vertical. Turbulence and its effects will be discussed in more detail in other sections of this text.

A gust is a discrete event within a turbulent wind field. As illustrated in Figure 2.10, one way to characterize the attributes of a gust is to measure: (a) amplitude, (b) rise time, (c) maximum gust variation, and (d) lapse time. Wind turbine loads caused by gusts can be determined as functions of these four attributes. For example, extreme loads can be analyzed by determining the response to the largest gust during a wind machine's expected lifetime.

2.2.2.2 Variations due to location and wind direction

Variations due to location Wind speed is also very dependent on local topographical and ground cover variations. For example as shown in Figure 2.11 (Hiester and Pennell. 1981), differences between two sites close to each other can be significant. The graph shows monthly and five-year mean wind speeds for two sites 21 km (13 miles) apart. The five-year average mean wind speeds differ by about 12% (4.75 and 4.25 m/s annual averages).

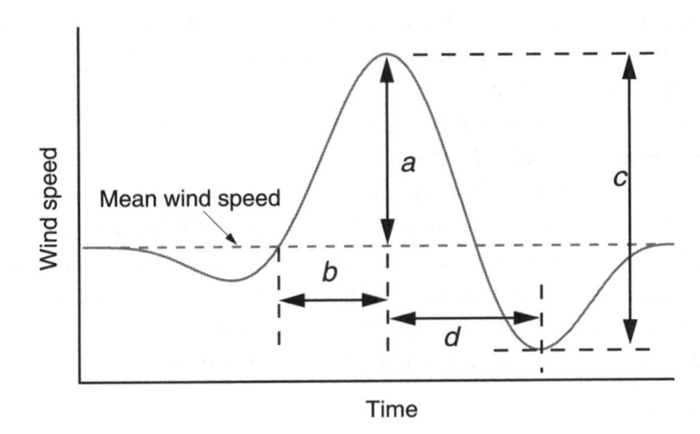

Figure 2.10 Illustration of a discrete gust event; *a*, amplitude; *b*, rise time; *c*, maximum gust variation; *d*, lapse time

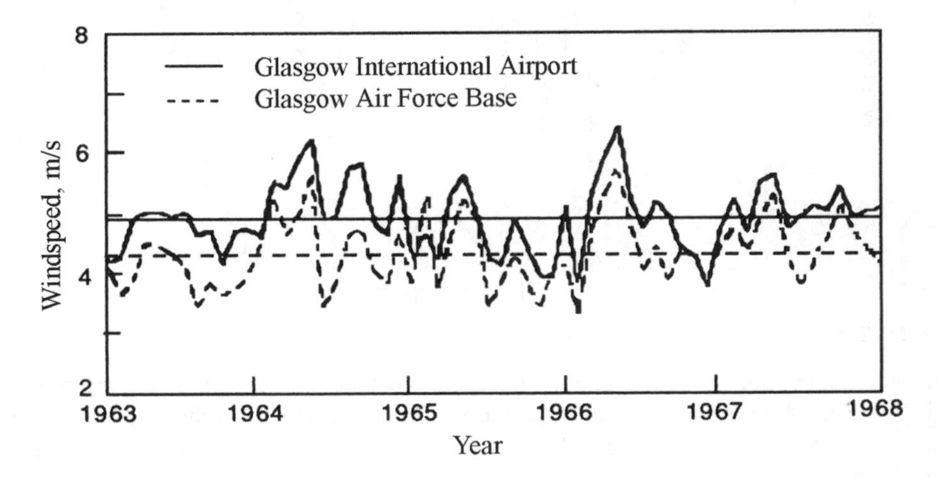

Figure 2.11 Time series of monthly wind speeds for Glasgow, Montana International Airport and Air Force Base (AFB) (Hiester and Pennell, 1981)

Variations in wind direction Wind direction also varies over the same time scales over which wind speeds vary. Seasonal variations may be small, on the order of 30 degrees, or the average monthly winds may change direction by 180 degrees over a year. Short-term direction variations are the result of the turbulent nature of the wind. These short-term variations in wind direction need to be considered in wind turbine design and siting. Horizontal axis wind turbines must rotate (yaw) with changes in wind direction. Yawing causes gyroscopic loads throughout the turbine structure and exercises any mechanism involved in the yawing motion. Crosswind due to changes in wind direction affect blade loads. Thus, as will be discussed in Chapter 4, short term variations in wind direction and the associated motion affect the fatigue life of components such as blades and yaw drives.

2.2.3 *Estimation of Potential Wind Resource*

In this section the energy potential of the wind resource and its power production capabilities will be presented.

2.2.3.1 Available wind power

As illustrated in Figure 2.12, one can determine the mass flow of air, d*m*/d*t*, through a rotor disk of area *A*. From the continuity equation of fluid mechanics, the mass flow rate is a function of air density, ρ , and air velocity (assumed uniform), *U*, and is given by:

$$\frac{\mathrm{d}m}{\mathrm{d}t} = \rho A U \qquad (2.2.6)$$

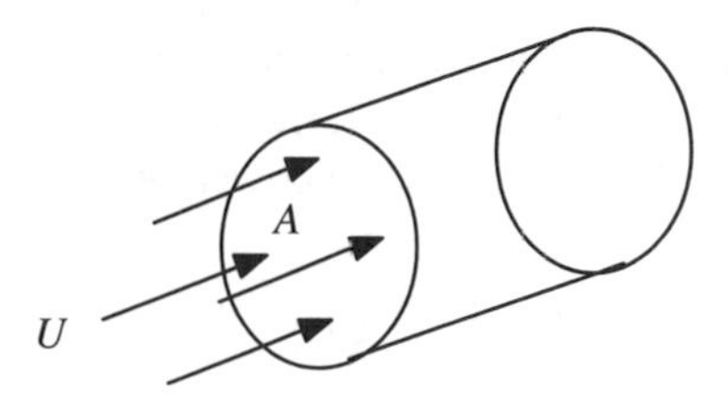

Figure 2.12 Flow of air through a rotor disk; *A*, area; *U*, wind velocity

The kinetic energy per unit time, or power, of the flow is given by:

$$P = \frac{1}{2}\frac{\mathrm{d}m}{\mathrm{d}t}U^2 = \frac{1}{2}\rho A U^3 \qquad (2.2.7)$$

The wind power per unit area, P/A or wind power density is:

$$\frac{P}{A} = \frac{1}{2}\rho U^3 \qquad (2.2.8)$$

One should note that:

- The wind power density is proportional to the density of the air. For standard conditions (sea-level, 15°C) the density of air is 1.225 kg/m^3
- Power from the wind is proportional to the area swept by the rotor (or the rotor diameter squared for a conventional horizontal axis wind machine)
- The wind power density is proportional to the cube of the wind velocity

The actual power production potential of a wind turbine must take into account the fluid mechanics of the flow passing through a power producing rotor, and the aerodynamics and efficiency of the rotor/generator combination. In practice, a maximum of about 45% of the available wind energy is harvested by the best modern horizontal axis wind turbines (this will be discussed in Chapter 3).

Table 2.1 shows that the wind velocity is an important parameter and significantly influences the power per unit area available from the wind.

Table 2.1 Power per unit area available from steady wind (air density = 1.225 kg/m^3)

Wind Speed (m/s)	Power/Area (W/m^2)
0	0
5	80
10	610
15	2,070
20	4,900
25	9,560
30	16,550

If annual average wind speeds are known for certain regions, one can develop maps that show average wind power density over these regions. More accurate estimates can be made if hourly averages, U_i, are available for a year. Then the average of power estimates for each hour can be determined. The average wind power density, based on hourly averages is:

$$\overline{P}/A = \frac{1}{2}\rho\overline{U}^3 K_e \tag{2.2.9}$$

where $\overline{U}$ is the annual average wind speed and K_e is called the energy pattern factor. The energy pattern factor is calculated from:

$$K_e = \frac{1}{N\overline{U}^3}\sum_{i=1}^{N} U_i^3 \tag{2.2.10}$$

where N is the number of hours in a year, 8760.

Some sample qualitative magnitude evaluations of the wind resource are:

$\overline{P}/A < 100$ W/m^2 – poor
$\overline{P}/A \approx 400$ W/m^2 – good
$\overline{P}/A > 700$ W/m^2 – great

2.2.3.2 Estimates of world wide resource

Based on wind resource data and an estimate of the real efficiency of actual wind turbines, numerous investigators have made estimates of the wind power or energy potential of regions of the earth and of the entire earth itself. It will be shown in Chapter 3 that the maximum power producing potential that can be theoretically realized from the kinetic energy contained in the wind is about 60% of the available power. Also, this factor is reduced to about 40% for the best horizontal axis wind machines that can be practically built.

Using estimates for regional wind resources, one can estimate the (electrical) power producing potential of wind energy. It is important to distinguish between the different types of wind energy potential that can be estimated. One such estimate (World Energy Council, 1993) identified the following five categories:

1. Meteorological potential. This is equivalent to the available wind resource.
2. Site potential. This is based on the meteorological potential, but is restricted to those sites that are geographically available for power production.
3. Technical potential. The technical potential is calculated from the site potential, accounting for the available technology.
4. Economic potential. The economic potential is the technical potential that can be realized economically.
5. Implementation potential. Implementation potential takes into account constraints and incentives to assess the wind turbine capacity that can be implemented within a certain time frame.

For worldwide wind resource assessments, at least the first three categories have been considered. For example, one of the earliest global wind energy resource assessments was carried out by Gustavson (1979). In this study Gustavson based his resource estimate on the input of the solar energy reaching the earth and how much of this energy was transformed into useful wind energy. On a global basis, his estimate was that the global resource was about 1000 x 10^{12} kWh/yr. In comparison, the global consumption of electricity at that time was about 55 x 10^{12} kWh/yr.

In a more recent study, the World Energy Council (1993) using a world average estimate for the meteorological potential of wind energy, and including machine efficiency and availability (percent of time on line), made a global estimate of the wind resource. They estimated that the inland wind power resource was about 20 x 10^{12} kWh/yr, still a considerable resource.

Numerous wind resource estimates have been made for the potential of wind energy in the United States. The estimates published in the 1990's are much more realistic than earlier ones, since they feature practical assumptions about machine characteristics and land exclusions (technical and site potential considerations). They also featured expanded data collection methods and improved analysis techniques. Elliot et al. (1991) used this type of improved analysis in conjunction with US wind resource data (Elliot et al., 1987). Their work concluded that wind energy could supply at least 20% of the US electrical needs, and that wind energy could be exploited in locations where the average wind speed was at least 7.3 m/s (16 mph) at a height of 30 m. In order to provide this fraction of the US electrical demand (about 600 billion kWh per year), 0.6% of the land (about 18,000 square miles) in the lower 48 states would have to be developed. The majority of this land is in the West, however, and far from major population centers. Therefore, the actual use of this land involves other siting considerations such as transmission line access.

2.3 Characteristics of the Atmospheric Boundary Layer

An important parameter in the characterization of the wind resource is the variation of horizontal wind speed with height above the ground. One would expect the horizontal wind speed to be zero at the earth's surface and to increase with height in the atmospheric boundary layer. This variation of wind speed with elevation is called the vertical profile of the wind speed or vertical wind shear. In wind energy engineering the determination of vertical wind shear is an important design parameter, since: (1) It directly determines the productivity of a wind turbine on a tower of certain height, and (2) It can strongly influence the lifetime of a turbine rotor blade. Rotor blade fatigue life is influenced by the cyclic loads resulting from rotation through a wind field that varies in the vertical direction.

There are at least two basic problems of interest with the determination of vertical wind profiles for wind energy applications:

- Instantaneous variation of wind speeds as a function of height (e.g., time scale on the order of seconds).
- Seasonal variation of average wind speeds as a function of height (e.g., monthly or annual averages).

It should be noted that these are separate and distinct problems, and it is often erroneously assumed that a single methodology can be applied to both of them. That is, the variation of 'instantaneous' profiles is related via the similarity theory of boundary layers. On the other hand, the changes of long-term averages as a function of height is related to the statistics of occurrence of various influencing factors such as atmospheric stability (discussed next), and must rely on a more empirical approach (Justus, 1978).

In addition to variations due to the atmospheric stability, the variation of wind speed with height depends on surface roughness, and terrain. These factors will be discussed in the next sections.

2.3.1 Atmospheric boundary layer characteristics

A particularly important characteristic of the atmosphere is its stability - the tendency to resist vertical motion or to suppress existing turbulence. Atmospheric stability is usually classified as stable, neutrally stable, or unstable. The stability of the earth's atmosphere is governed by the vertical temperature distribution resulting from the radiative heating or cooling of its surface and the subsequent convective mixing of the air adjacent to the surface. A summary of how the atmospheric temperature changes with elevation (assuming an adiabatic expansion) follows.

2.3.1.1 Lapse rate

If the atmosphere is approximated as a dry (no water vapor in the mixture) ideal gas, the relationship between a change in pressure and a change in elevation for a fluid element in a gravitational field is given by:

$$dp = -\rho g \, dz \tag{2.3.1}$$

where p = atmospheric pressure, ρ = atmospheric density (assumed constant here), z = elevation, and g = local gravitational acceleration.

The negative sign results from the convention that height, z, is measured positive upward, and that the pressure, p, decreases in the positive z direction.

The first law of thermodynamics for an ideal gas closed system of unit mass undergoing a quasi-static change of state is given by:

$$\mathrm{d}q = \mathrm{d}u + p\mathrm{d}v = \mathrm{d}h - v\mathrm{d}p = c_p\,\mathrm{d}T - \frac{1}{\rho}\mathrm{d}p \tag{2.3.2}$$

where T = temperature, q = heat transferred, u = internal energy, h = enthalpy, v = specific volume, c_p = constant pressure specific heat.

For an adiabatic process (no heat transfer) $\mathrm{d}q = 0$, and Equation 2.3.2 becomes:

$$c_p\,\mathrm{d}T = \frac{1}{\rho}\mathrm{d}p \tag{2.3.3}$$

Substitution for $\mathrm{d}p$ in Equation 2.3.3 and rearrangement gives:

$$\left(\frac{\mathrm{d}T}{\mathrm{d}z}\right)_{Adiabatic} = g\frac{1}{c_p} \tag{2.3.4}$$

If the change in g and c_p with elevation are assumed negligible, then the change in temperature, under adiabatic conditions, is a constant. Using g = 9.81 m/s^2 and c_p = 1.005 kJ/kgK yields:

$$\left(\frac{\mathrm{d}T}{\mathrm{d}z}\right)_{Adiabatic} = -\frac{0.0098°\mathrm{K}}{\mathrm{m}} \tag{2.3.5}$$

Thus, the rate that temperature decreases with increase in height for a system with no heat transfer is about 1°K per 100 m (1°C per 100 m or about 5.4°F per 1000 ft). This is known as the dry adiabatic lapse rate. Using conventional nomenclature, the lapse rate, Γ, is defined as the negative of the temperature gradient in the atmosphere. Therefore, the dry adiabatic lapse rate is given by:

$$\Gamma = -\left(\frac{\mathrm{d}T}{\mathrm{d}z}\right)_{Adiabatic} \approx \frac{1°\mathrm{C}}{100\,\mathrm{m}} \tag{2.3.6}$$

The dry adiabatic lapse rate is extremely important in meteorological studies since a comparison of its value to the actual lapse rate in the lower atmosphere is a measure of the stability of the atmosphere. The international standard atmospheric lapse rate, based on meteorological data, has been defined and adopted for comparative purposes. Specifically, on the average, in the middle latitudes, the temperature decreases linearly with elevation up to about 10,000 m (for definition purposes 10.8 km). The temperature averages 288 °K at sea level and decreases to 216.7 °K at 10.8 km, giving the standard temperature gradient as:

$$\left(\frac{\mathrm{d}T}{\mathrm{d}z}\right)_{Standard} = \frac{(216.7 - 288)°\mathrm{C}}{10{,}800\ \mathrm{m}} = -\frac{0.0066°\mathrm{C}}{\mathrm{m}} = -\frac{0.00357°\mathrm{F}}{\mathrm{ft}} \tag{2.3.7}$$

Thus, the standard lapse rate, based on international convention is 0.66°C/100 m or 3.6°F/1000 ft.

As stated previously, different temperature gradients create different stability states in the atmosphere. Figure 2.13 illustrates that the temperature profiles change from day to night due to heating of the earth's surface. The temperature profile before sunrise (the solid line) decreases with height near the ground and reverses after sunrise (dashed line). The air is heated near the ground, and the temperature gradient close to the earth's surface increases with height, up to height z_i called the inversion height). The surface layer of air extending to z_i is called the convective or mixing layer. Above z_i the temperature profile reverses.

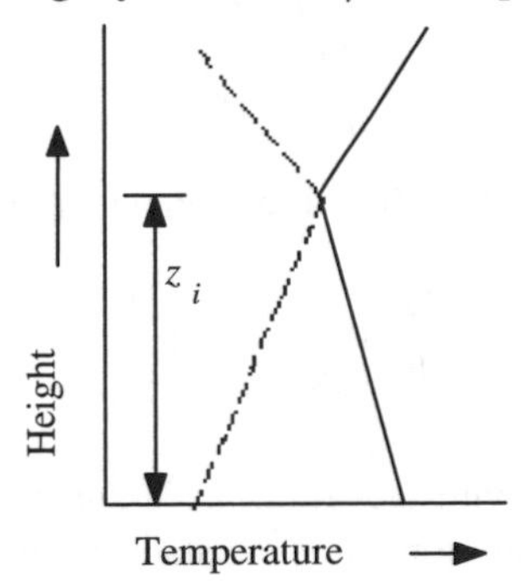

Figure 2.13 Temperature profile above the earth's surface, before (solid) and after (dashed) sunrise

2.3.1.2 Stability of the atmosphere

The concept of atmospheric stability is illustrated by considering the upward displacement of a small element of air to an altitude with a lower ambient pressure. Although there is great variability in the rate of fall of temperature of the surrounding air with altitude, one can assume the standard rate of 0.66°C/100 m. On the other hand, the small element of air being lifted in this example will cool at the dry adiabatic lapse rate (1°C per 100 m). If the test element of air had the same temperature as the surrounding air at the start, then after it had been raised 100 m, it will have cooled faster than the surrounding air and will be 0.34°C cooler than its surroundings. The sample will be more dense and will tend to return to its original level. This atmospheric state is called stable.

To generalize, any atmosphere whose dT/dz is greater than $(dT/dz)_{adiabatic}$ is a stable one. One should note that the standard international lapse rate seldom occurs in nature. This explains the need for the daily balloon soundings taken at major airports worldwide to determine the actual lapse rate. Also, in order to have stability, it is not necessary for an inversion (increase of temperature with height) to exist. When one does exist, however, the atmosphere is even more stable.

2.3.1.3 Atmospheric density and pressure

As demonstrated in Equation 2.2.8, the power in the wind is a function of air density. Air density is a function of temperature, T, and pressure, p, both of which vary with height. The density of dry air can be determined by applying the ideal gas law, which gives:

$$\rho = 3.4837 \frac{p}{T} \tag{2.3.8}$$

where the density is in kg/m^3, the pressure is in kPa (kN/m^2) and the temperature is in Kelvin. Moist air is slightly less dense than dry air, but corrections for air moisture are rarely used. Air density as a function of moisture content can be found in numerous books on thermodynamics such as Balmer (1990).

The international standard atmosphere assumes that the sea-level temperature and pressure are 288.15 K (15 C, 59 F) and 101.325 kPa (14.696 psi), resulting in a standard sea-level density of 1.225 kg/m^3 (see Avallone and Baumeister, 1978). Air pressure decreases with elevation above sea level. The pressure in the international standard atmosphere up to an elevation of 5000 m is very closely approximated by:

$$p = 101.29 - (0.011837)z + (4.793 \times 10^{-7})z^2 \tag{2.3.9}$$

where z is the elevation in meters and the density is in kg/m^3. Of course, the actual pressure may vary about the standard pressure as weather patterns change. In practice, at any location, the daily and seasonal temperature fluctuations have a much greater effect on air density than do daily and seasonal changes in pressure and air moisture.

2.3.2 Turbulence

Turbulence in the wind is caused by dissipation of the wind's kinetic energy into thermal energy via the creation and destruction of progressively smaller eddies (or gusts). Turbulent wind may have a relatively constant mean over time periods of an hour or more, but over shorter times (minutes or less) it may be quite variable. The wind's variability superficially appears to be quite random, but actually it has distinct features. These features are characterized by a number of statistical properties:

- Turbulence intensity
- Wind speed probability density functions
- Autocorrelation
- Integral time scale/length scale
- Power spectral density function

A summary and examples of these properties follows. More details concerning them are given in the texts of Rohatgi and Nelson (1994) and Bendat and Piersol (1993).

Turbulent wind consists of longitudinal, lateral and vertical components. The longitudinal component, in the prevailing wind direction, is designated $u(z,t)$. The lateral component (perpendicular to U) is $v(z,t)$ and the vertical component is $w(z,t)$. Each component is frequently conceived of as consisting of a short-term mean wind, for example, U, with a superimposed fluctuating wind of zero mean, $\tilde{u}$, added to it, thus:

$$u = U + \tilde{u} \tag{2.3.10}$$

where u = instantaneous longitudinal wind speed. The lateral and vertical components can be decomposed into a mean and fluctuating component in a similar manner.

Note that the short-term mean wind speed, in this case U, refers to mean wind speed averaged over some (short) time period, Δt, longer than the characteristic time of the fluctuations in the turbulence. This time period is usually taken to be 10 minutes, but can be as long as an hour. In equation form:

$$U = \frac{1}{\Delta t}\int_0^{\Delta t} u \ \mathrm{d}t \tag{2.3.11}$$

Instantaneous turbulent wind is not actually observed continuously; it is actually sampled at some relatively high rate. Assuming that the sample interval is δt, such that $\Delta t = N_s \delta t$ where N_s = number of samples during each short-term interval, then turbulent wind can be expressed as a sequence, u_i. The short-term mean wind speed can then be expressed in sampled form as:

$$U = \frac{1}{N}\sum_{i=1}^{N_s} u_i \tag{2.3.12}$$

The short-term average longitudinal wind speed, U, is the one most often used in time series observations and will be used henceforth in this text in that way.

2.3.2.1 Turbulence intensity

The most basic measure of turbulence is the turbulence intensity. It is defined by the ratio of the standard deviation of the wind speed to the mean. In this calculation both the mean and standard deviation are calculated over a time period longer than that of the turbulent fluctuations, but shorter than periods associated with other types of wind speed variations (such as diurnal effects). The length of this time period is normally no more than an hour, and by convention in wind energy engineering it is usually equal to 10 minutes. The sample rate is normally at least once per second (1 Hz). The turbulence intensity, *TI*, is defined by:

$$TI = \frac{\sigma_U}{U} \tag{2.3.13}$$

where σ_U is the standard deviation, given in sampled form by:

$$\sigma_U = \sqrt{\frac{1}{N_s - 1}\sum_{i=1}^{N_s}(u_i - U)^2} \tag{2.3.14}$$

Turbulence intensity is frequently in the range of 0.1 to 0.4. In general the highest value of turbulence intensities occur at the lowest wind speeds, but the lower limiting value at a given location will depend on the specific terrain features and surface conditions at the site. Figure 2.14 illustrates a graph of a typical segment of wind data sampled at 8 Hz. The data has a mean of 10.4 m/s and a standard deviation of 1.63 m/s. Thus, the turbulence intensity, over the 10 minute period, is 0.16.

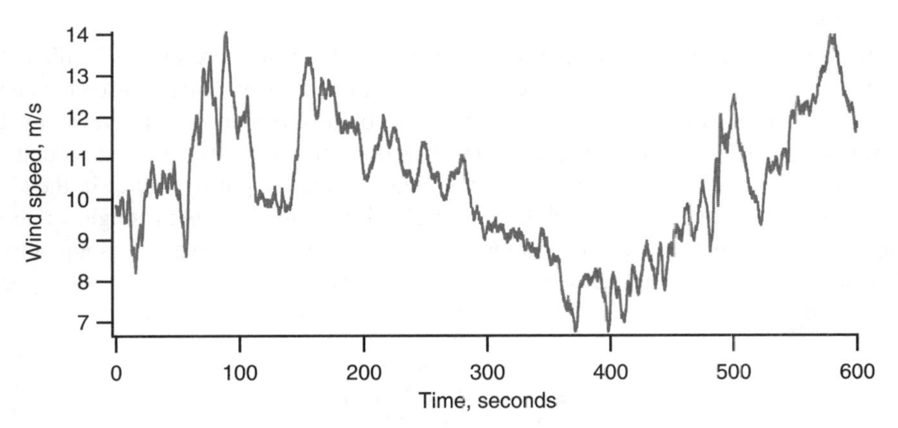

Figure 2.14 Sample wind data

2.3.2.2 Wind speed probability density functions

The likelihood that the wind speed has a particular value can be described in terms of a probability density function (pdf). Experience has shown that the wind speed is more likely to be close to the mean value than far from it, and that it is nearly as likely to be below the mean as above it. The probability density function that best describes this type of behavior for turbulence is the Gaussian or normal distribution. The normal probability density function for continuous data in terms of the variables used here is given by:

$$p(u) = \frac{1}{\sigma_u \sqrt{2\pi}} \exp\left[-\frac{(u-U)^2}{2\sigma_u^2}\right] \tag{2.3.15}$$

Figure 2.15 illustrates a histogram of the wind speeds about the mean wind speed in the sample data above (Figure 2.14). The Gaussian probability density function that represents the data is superimposed on the histogram.

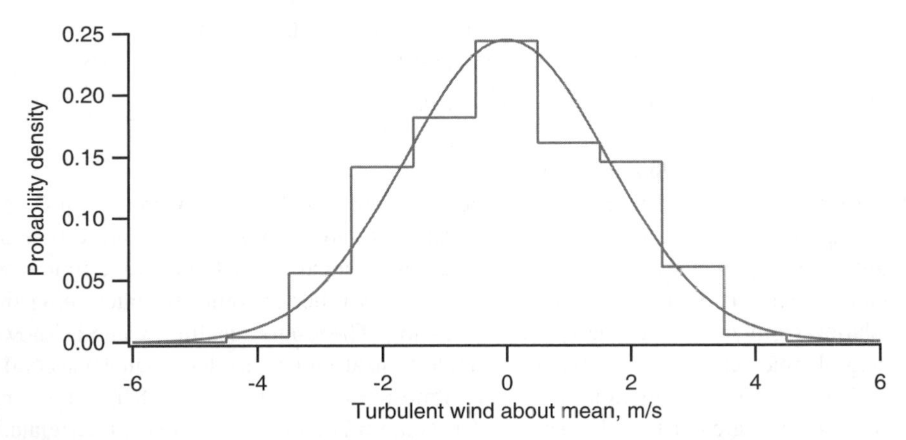

Figure 2.15 Gaussian probability density function and histogram of wind data

2.3.2.3 Autocorrelation

The probability density function of the wind speed provides a measure of the likelihood of particular values of wind speed. It provides no information, however, about what the speed is likely to be, given what it has been. A measure of that tendency is provided by the autocorrelation function. The autocorrelation function, for sampled data, is found by multiplying each value in a time series with the mean subtracted out by values in the same time series, offset by a time 'lag', and then summing the products to find a single value for each lag. The resulting sums are then normalized by the variance to give values equal to or less than one. The normalized autocorrelation function for sampled turbulent wind speed data is given by:

$$R(r\delta t)=\frac{1}{\sigma_u^2(N_s-r)}\sum_{i=1}^{N_s-r}u_i u_{i+r} \tag{2.3.16}$$

where r = lag number. Figure 2.16 shows a graph of the autocorrelation function of the data presented above in Figure 2.14.

The autocorrelation function can be used to determine the integral time scale of turbulence as described below.

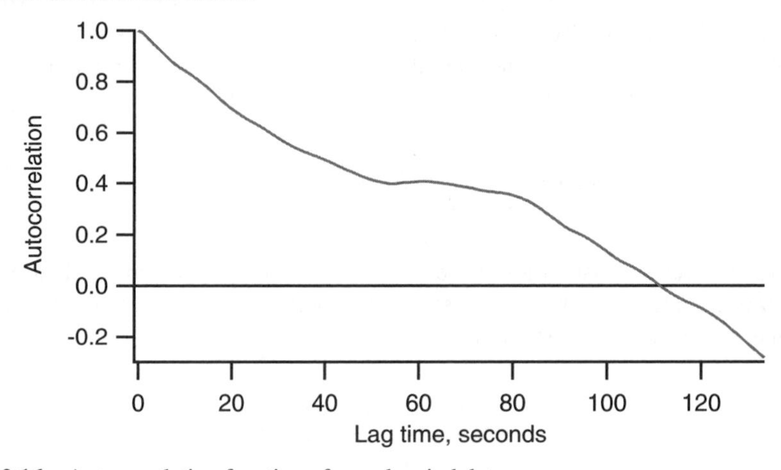

Figure 2.16 Autocorrelation function of sample wind data

2.3.2.4 Integral time scale/length scale

The autocorrelation function will, if any trends are removed before starting the process, decay from a value of 1.0 at a lag of zero to a value of zero, and will then tend to take on small positive or negative values as the lag increases. A measure of the average time over which wind speed fluctuations are correlated with each other is found by integrating the autocorrelation from zero lag to the first zero crossing. The single resulting value is known as the integral time scale of the turbulence. While typical values are less than 10 seconds, the integral time scale is a function of the site, atmospheric stability, and other factors and may be significantly greater than 10 seconds. Gusts are relatively coherent (well correlated) rises and falls in the wind, and have characteristic times on the same order as the integral time scale. Multiplying the integral time scale by the mean wind velocity gives the integral length scale. The integral length scale tends to be more constant over a range of wind speeds than is the integral time scale, and thus is somewhat more representative of a site.

Based on the autocorrelation function illustrated above, the integral time scale is 50.6 seconds. The mean wind velocity is 10.4 m/s. Thus, the size of the turbulent eddies in the mean flow, or the integral length scale, is on the order of 526 m.

2.3.2.5 Power spectral density function

The fluctuations in the wind can be thought of as resulting from a composite of sinusoidally varying winds imposed on the mean steady wind. These sinusoidal variations will have a variety of frequencies and amplitudes. The term 'spectrum' is used to describe functions of frequency. Thus the function that characterizes turbulence as a function of frequency is known as a 'spectral density' function. Since the average value of any sinusoid is zero, the amplitudes are characterized in terms of their mean square values. This type of analysis originated in electric power applications, where the square of the voltage or current is proportional to the power. The complete name for the function describing the relation between frequency and amplitudes of sinusoidally varying waves making up the fluctuating wind speed is therefore 'power spectral density'.

There are two points of particular importance to note regarding power spectral densities (psd's). The first is that the average power in the turbulence over a range of frequencies may be found by integrating the psd between the two frequencies. Secondly, the integral over all frequencies is equal to the total variance.

Power spectral densities are often used in dynamic analyses. A number of power spectral density functions are used as models in wind energy engineering when representative turbulence power spectral densities are unavailable for a given site. A suitable model that is similar to the one developed by von Karman for turbulence in wind tunnels (Freris, 1990) is given by:

$$S(f)=\frac{\sigma_u^2\, 4(L/U)}{\left[1+70.8(fL/U)^2\right]^{5/6}} \tag{2.3.17}$$

where f is the frequency (Hz), L is the integral length scale and U is the mean wind speed at the height of interest.

The power spectral density of the sample wind data above is illustrated in Figure 2.17. The graph also includes the von Karman power spectral density function described above for comparison.

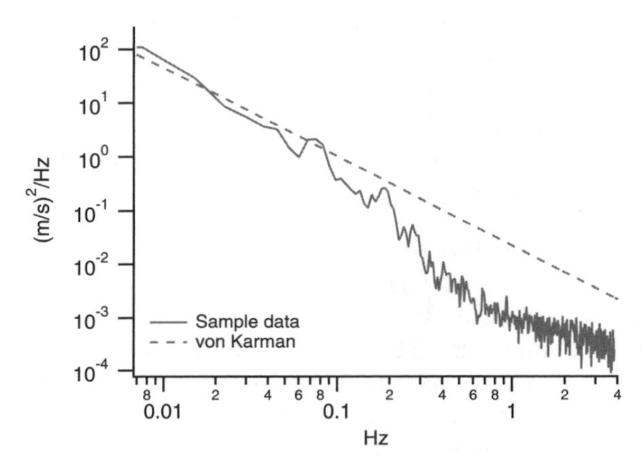

Figure 2.17 Wind data power spectral density functions

2.3.3 The steady wind: wind speed variation with height

The variation in wind speed with elevation influences both the assessment of wind resources and the design of wind turbines. First, the assessment of wind resources over a wide geographical area might require that the anemometer data from a number of sources be corrected to a common elevation. Second, from a design aspect, rotor blade fatigue life will be influenced by the cyclic loads resulting from rotation through a wind field that varies in the vertical direction. Thus, a model of the wind speed variation with height is required in wind energy applications. The summary that follows will present some of the current models that are used to predict the variation of wind speed with vertical elevation.

In wind energy studies, two mathematical models or 'laws' have generally been used to model the vertical profile of wind speed over regions of homogenous, flat terrain (e.g., fields, deserts, and prairies). The first approach, the log law, has its origins in boundary layer flow in fluid mechanics and in atmospheric research. It is based on a combination of theoretical and empirical research. The second approach, used by many wind energy researchers, is the power law. Both approaches are subject to uncertainty caused by the variable, complex nature of turbulent flows (Hiester and Pennell, 1981). A summary of each of these laws and their general application follows.

2.3.3.1 Logarithmic profile (log law)

Although there are a number of ways to arrive at a prediction of a logarithmic wind profile (e.g., mixing length theory, eddy viscosity theory, and similarity theory), a mixing length type analysis given by Wortman (1982) is summarized here.

Near the surface of the earth the momentum equation reduces to:

$$\frac{\partial p}{\partial x} = \frac{\partial}{\partial z} \tau_{xz} \tag{2.3.18}$$

where x and z are the horizontal and vertical coordinates, p is the pressure, and τ_{xz} is the shear stress in the direction of x whose normal coincides with z.

Near the surface the pressure is independent of z and integration yields:

$$\tau_{xz} = \tau_0 + z\frac{\partial p}{\partial x} \tag{2.3.19}$$

where τ_0 is the surface value of the shear stress. Near the surface the pressure gradient is small, so the second term on the right-hand side may be neglected. Also, using the Prandtl mixing length theory, the shear stress may be expressed as:

$$\tau_{xz} = \rho \ell^2 \left(\frac{\partial U}{\partial z}\right)^2 \tag{2.3.20}$$

where ρ is the density of the air, U the horizontal component of velocity, and ℓ the mixing length. Note that U is used here, signifying that the effects of turbulence have been averaged out.

Combining Equations 2.3.19 and 2.3.20 gives:

$$\frac{\partial U}{\partial z} = \frac{1}{\ell}\sqrt{\frac{\tau_0}{\rho}} = \frac{U^*}{\ell} \tag{2.3.21}$$

where U^* is defined as the friction velocity.

If one assumes a smooth surface, $\ell = kz$, with k = 0.4 (von Karman's constant), then Equation 2.3.21 can be integrated directly from z_0 to z where z_0 is the surface roughness length, which characterizes the roughness of the ground terrain. This yields:

$$U(z) = \frac{U^*}{k}\ln\left(\frac{z}{z_0}\right) \tag{2.3.22}$$

This equation is known as the logarithmic wind profile.

The integration is from the lower limit of z_0 instead of 0 because natural surfaces are never uniform and smooth. Table 2.2 gives example surface roughness lengths for various terrain types.

Equation 2.3.22 can also be written as:

$$\ln(z) = \left(\frac{k}{U^*}\right)U(z) + \ln(z_0) \tag{2.3.23}$$

This equation can be plotted as a straight line on semilog paper. The slope of this graph is k/U^*, and from a graph of experimental data, U^* and z_0 can be calculated. The log law is often used to estimate wind speed from a reference height, z_r, to another level using the following relationship:

Table 2.2 Values of surface roughness length for various types of terrain

Terrain Description	z_0 (mm)
Very smooth, ice or mud	0.01
Calm open sea	0.20
Blown sea	0.50
Snow surface	3.00
Lawn grass	8.00
Rough pasture	10.00
Fallow field	30.00
Crops	50.00
Few trees	100.00
Many trees, hedges, few buildings	250.00
Forest and woodlands	500.00
Suburbs	1500.00
Centers of cities with tall buildings	3000.00

$$U(z)/U(z_r)=\ln\left(\frac{z}{z_0}\right)\Big/\ln\left(\frac{z_r}{z_0}\right) \tag{2.3.24}$$

Sometimes, the log law is modified to consider mixing at the earth's surface, by expressing the mixing length as $\ell=k(z+z_0)$. When this is used, the log profile becomes:

$$U(z)=\frac{U^*}{k}\ln\left(\frac{z+z_0}{z_0}\right) \tag{2.3.25}$$

2.3.3.2 Power law profile

The power law represents a simple model for the vertical wind speed profile. Its basic form is:

$$\frac{U(z)}{U(z_r)}=\left(\frac{z}{z_r}\right)^{\alpha} \tag{2.3.26}$$

where $U(z)$ is the wind speed at height z, $U(z_r)$ is the reference wind speed at height z_r and α is the power law exponent.

The early work of von Karman showed that under certain conditions α is equal to 1/7, indicating a correspondence between wind profiles and flow over flat plates. In practice, the exponent α is a highly variable quantity.

The following example emphasizes the importance of a variation in α:

If U_0 = 5 m/s at 10 m, what is U at 30 m? Note that at 10 m, P/A = 75.6 W/m^2. The wind velocity at 30 m is tabulated below (Table 2.3) for three different values of α, and P/A is calculated assuming ρ = 1.225 kg/m^3.

Table 2.3 Effect of α on estimates of wind power density at higher elevations

	$\alpha = 0.1$	1/7	0.3
U_{30m} (m/s)	5.58	5.85	6.95
P/A (W/m^2)	106.4	122.6	205.6
% increase over 10 m	39.0	62.2	168.5

It has been found that α varies with such parameters as elevation, time of day, season, nature of the terrain, wind speed, temperature, and various thermal and mechanical mixing parameters. Some researchers have developed methods for calculating α from the parameters in the log law. Many researchers, however, feel that these complicated approximations reduce the simplicity and applicability of the general power law and that wind energy specialists should accept the empirical nature of the power law and choose values of α that best fit available wind data. A review of a few of the more popular empirical methods for determining representative power law exponents follows.

1. Correlation for the power law exponent as a function of velocity and height.
One way of handling this type of variation was proposed by Justus (1978). His expression had the form:

$$\alpha = \frac{0.37 - 0.088 \ln\left(U_{ref}\right)}{1 - 0.088 \ln\left(\frac{z_{ref}}{10}\right)} \tag{2.3.27}$$

where U is given in m/s and z_{ref} in m.

2. *Correlation dependent on surface roughness.*
The following form for this type of correlation was based on work of Counihan (1975):

$$\alpha = 0.096 \log_{10} z_0 + 0.016 \left(\log_{10} z_0\right)^2 + 0.24 \tag{2.3.28}$$

for 0.001 m < z_0 < 10 m, where z_0 represents the surface roughness in m (see Table 2.2 for example values).

3. *Correlations based on both surface roughness (z_0) and velocity.*
Wind researchers at NASA proposed equations for α based on both surface roughness and the wind speed at the reference elevation, U_{ref} (see Spera, 1994).

2.3.4 *Effect of terrain on wind characteristics*

The importance of terrain features on wind characteristics is discussed in various siting handbooks for wind systems (see Troen and Petersen, 1989; Hiester and Pennell, 1981; and Wegley, et al., 1980). Numerous researchers emphasize that the influence of terrain features on the energy output from a turbine may be so great that the economics of the whole project may depend on the proper selection of the site.

In the previous section, two methods were described (log profile and power law profile laws) for modeling the vertical wind speed profile. These were developed for flat and homogenous terrain. One can expect that any irregularities on the earth's surface will modify the wind flow, thus compromising the applicability of these prediction tools. This section presents a qualitative discussion of a few of the more important areas of interest on the subject of terrain effects.

2.3.4.1 Classification of terrain

The most basic classification of terrain divides it into flat and non-flat terrain. Many authors define non-flat terrain as complex terrain (this is defined as an area where terrain effects are significant on the flow over the land area being considered). Flat terrain is terrain with small irregularities such as forest, shelter belts, etc. (see Wegley et al., 1980). Non-flat terrain has large-scale elevations or depressions such as hills, ridges, valleys, and canyons. To quantify as flat terrain, the following conditions must hold. Note that some of these rules include wind turbine geometry:

- Elevation differences between the wind turbine site and the surrounding terrain are not greater than about 60 m anywhere in an 11.5 km diameter circle around the turbine site.
- No hill has an aspect ratio (height to width) greater than 1/50 within 4 km upstream and downstream of the site.
- The elevation difference between the lower end of the rotor disk and the lowest elevation on the terrain is greater than three times the maximum elevation difference (h) within 4 km. upstream (see Figure 2.18).

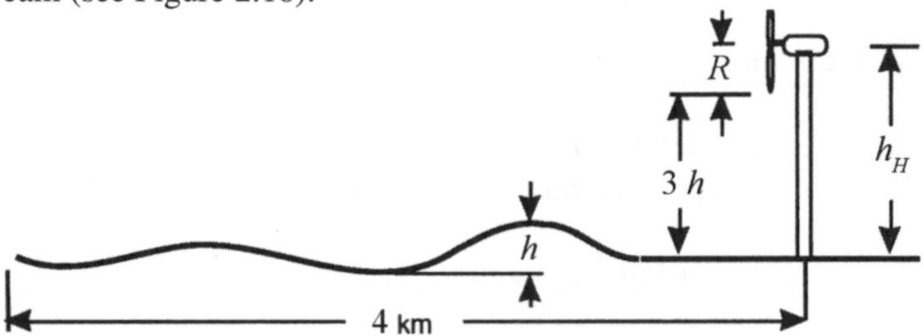

Figure 2.18 Determination of flat terrain (Wegley et al., 1980)

Non-flat or complex terrain, according Hiester and Pennell (1981), consists of a great variety of features, and one generally uses the following sub-classifications: (1) isolated elevation or depression, and (2) mountainous terrain. Flow conditions in mountainous terrain is complex because the elevations and depressions occur in a random fashion. Thus, flow in such terrain is divided into two classifications: small and large scales. The distinction between the two is made with comparison to the planetary boundary layer, which is assumed to be about 1 km. That is, a hill of a height which is a small fraction of the planetary boundary layer (approximately 10%) is considered to have small-scale terrain features.

An important point to be made here is that information on wind direction should be considered when defining the terrain classification. For example, if an isolated hill (200 m high and 1000 m wide) were situated 1 km south of a proposed site, the site could be classified as non-flat. If, however, the wind blows only 5% of the time from this direction with a low average speed, say 2 m/s, then this terrain should be classified as flat.

2.3.4.2 Flow over flat terrain with obstacles

Flow over flat terrain, especially with man-made and natural obstacles, has been studied extensively. Man-made obstacles are defined as buildings, silos, etc. Natural obstacles include rows of trees, shelter belts, etc. For man-made obstacles, a common approach is to consider the obstacle to be a rectangular block and to consider the flow to be two-dimensional. This type of flow, shown in Figure 2.19, produces a momentum wake, and, as illustrated, a free shear separates from the leading edge and reattaches downwind, forming a boundary between an inner recirculating flow region (eddy) and the outer flow region.

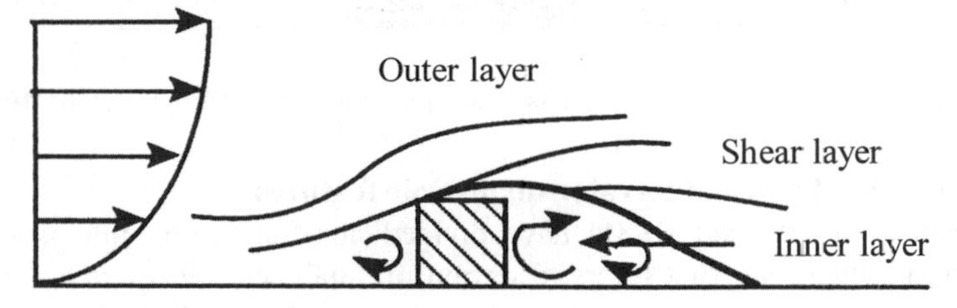

Figure 2.19 Schematic of a momentum wake (Rohatgi and Nelson, 1994). Reproduced by permission of Alternative Energy Institute

The results of an attempt to quantify data from man-made obstacles are shown in Figure 2.20, where the change in available power and turbulence is shown in the wake of a sloped-roof building. Note that the estimates in the figure apply at a level equal to one building height, h_s, above the ground, and that power losses become small downwind of the building after a distance equal to 15 h_s.

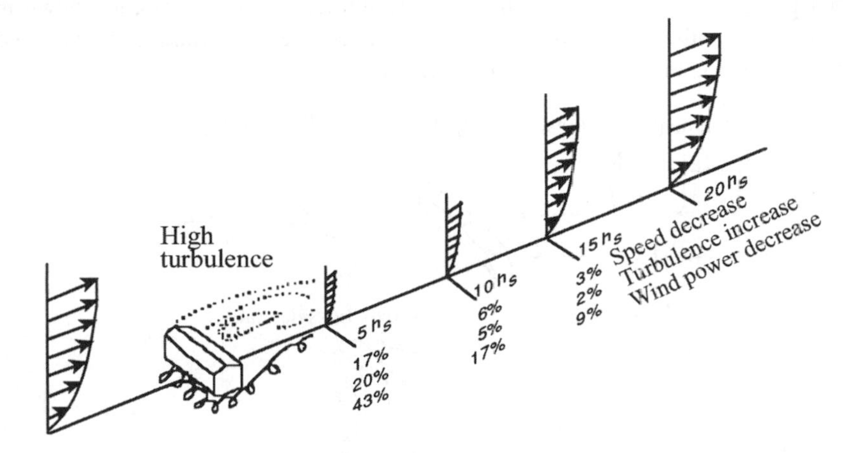

Figure 2.20 Speed, power, and turbulence effects downstream of a building (Wegley et al., 1980)

2.3.4.3 Flow in flat terrain with a change in surface roughness

In most natural terrain, the surface of the earth is not uniform and changes significantly from location to location. This affects the local wind profile. For example, Figure 2.21 shows that the downwind profile changes significantly in going from a smooth to a rough surface.

Figure 2.21 Effect of change in surface roughness from smooth to rough (Wegley et al., 1980)

2.3.4.4 Characteristics of non-flat terrain: small-scale features

Researchers (Hiester and Pennell, 1981) have divided non-flat terrain into isolated and mountainous terrain, where the first refers to terrain of small-scale features and the latter refers to large-scale features. For small-scale flows this classification is further divided into elevations and depressions. A summary of each follows.

Elevations Flow over elevated terrain features resembles flow around obstacles. Characterization studies of this type of flow in water and wind tunnels, especially for ridges and small cliffs, have been carried out. Examples of the results for ridges follow.

Ridges are elongated hills that are less than or equal to 600 m (2000 ft) above the surrounding terrain and have little or no flat area on the summit. The ratio of length to height should be at least 10. Figure 2.22 illustrates that, for wind turbine siting, the ideal prevailing wind direction should be perpendicular to the ridge axis. When the prevailing wind is not perpendicular, the ridge will not be as attractive a site. Also, as shown in this figure, concavity in the windward direction enhances speed-up, and convexity reduces speed-up by deflecting the wind flow around the ridge.

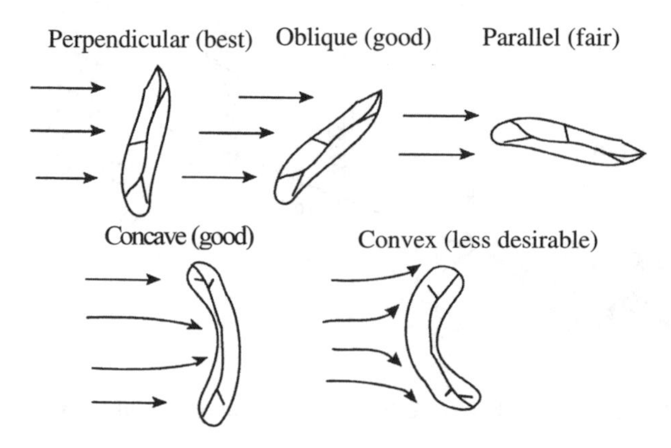

Figure 2.22 Effect of ridge orientation and shape on site suitability (Wegley et al., 1980)

The slope of a ridge is also an important parameter. Steeper slopes give rise to stronger wind flow, but on the lee of ridges steeper slopes give rise to high turbulence. Furthermore, as shown in Figure 2.23, a flat topped ridge creates a region of high wind shear due to the separation of the flow.

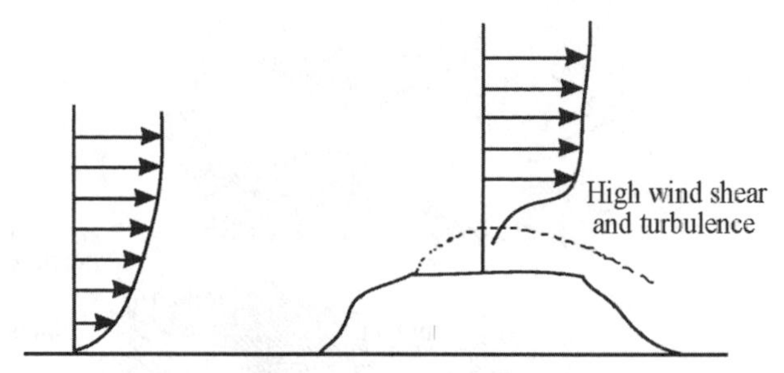

Figure 2.23 Region of high wind shear over a flat topped ridge (Wegley et al., 1980)

Depressions Depressions are characterized by a terrain feature lower than the surroundings. The speed-up of the wind is greatly increased if depressions can effectively channel the wind. This classification includes features such as valleys, canyons, basins, and passes. In addition to diurnal flow variations in certain depressions, there are many factors that influence the flow in depressions. These include orientation of the wind in relation to the depression, atmospheric stability, the width, length, slope, and roughness of the depression, and the regularity of the section of valley or canyon.

Shallow valleys and canyons (< 50 m) are considered small-scale depressions, and other features such as basins, gaps, etc. are considered as large-scale depressions. The large number of parameters affecting the wind characteristics in a valley, along with the variability of these parameters from valley to valley, make it almost impossible to draw specific conclusions valid for flow characterization.

2.3.4.5 Characteristics of non-flat terrain: large-scale features

Large-scale features are ones where the vertical dimension is significant in relation to the planetary boundary layer. They include mountains, ridges, high passes, large escarpments, mesas, deep valleys, and gorges. The flow over these features is the most complex, and flow predictions for this category of terrain classification are the least quantified. The following types of large depressions have been studied under this terrain classification:

- Valley and canyons
- Slope winds
- Prevailing winds in alignment
- Prevailing winds in non-alignment
- Gaps and gorges
- Passes and saddles
- Large basins

An example of a large depression with the prevailing winds in alignment is shown in Figure 2.24. This occurs when moderate to strong prevailing winds are parallel to or in alignment (within about 35 degrees) with the valley or canyon. Here the mountains can effectively channel and accelerate the flow.

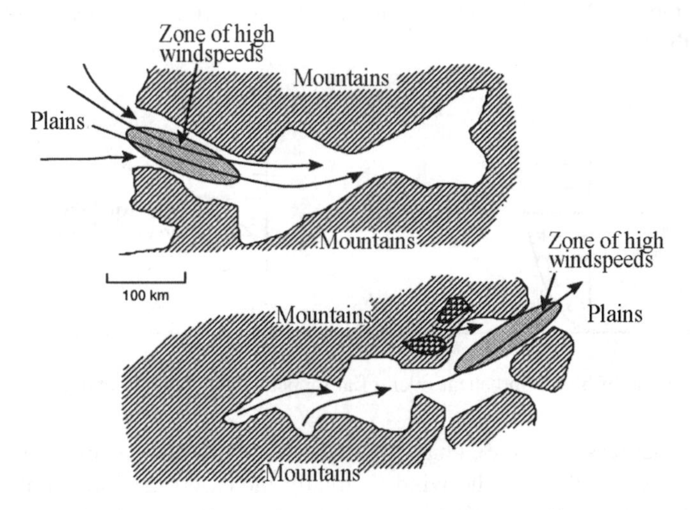

Figure 2.24 Increased wind speeds due to channelling of prevailing winds by mountains (Rohatgi and Nelson, 1994). Reproduced by permission of Alternative Energy Institute

2.4 Wind Data Analysis and Resource Estimation

Here it is assumed that a large quantity of wind data has been collected. (Wind measurements and instrumentation are discussed in a later section of this chapter.) This data could include direction data as well as wind speed data. There are a number of ways to summarize the data in a compact form so that one could evaluate the wind resource or wind power production potential of a particular site. These include both direct and statistical techniques. Furthermore, some of these techniques can be used with a limited amount of wind data (e.g., average wind speed only) from a given site. This section will review the following topics:

- Wind turbine energy production in general
- Direct (non-statistical) methods of data analysis and resource characterization
- Statistical analysis of wind data and resource characterization
- Statistically based wind turbine productivity estimates

2.4.1 General aspects of wind turbine energy production

Wind resource estimation consists of the determination of the productivity (both maximum energy potential and machine power output) of a given wind turbine at a given site where wind speed information is available in either time series format or in a summary format (average wind speed, standard deviation, etc.)

The power available from wind is $P = (1/2)\rho AU^3$ as shown in Section 2.2 (Equation 2.2.7). In practice, the power available from a wind turbine, P_w, can be shown by a machine power curve. Two typical curves, $P_w(U)$, simplified for this type of analysis, are shown in Figure 2.25. Later sections of this text will describe how such curves can be estimated from analytical models of the wind turbine system. Normally these curves are based on test data, as described in International Electrotechnical Commission (1998) or AWEA (1988).

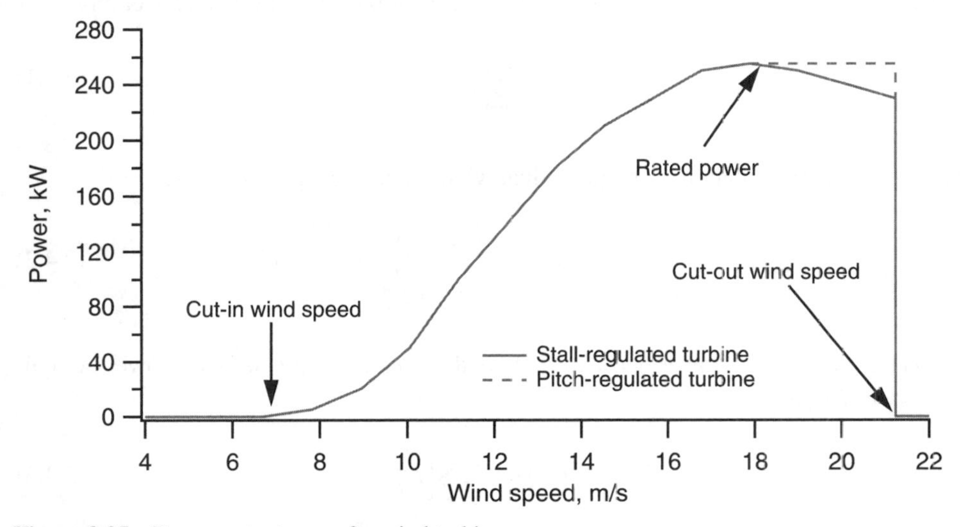

Figure 2.25 Power output curve for wind turbine

The power curve illustrates three important characteristic velocities:

- The cut-in velocity - the wind speed at which the turbine starts to generate power
- The rated velocity - the wind speed at which the wind turbine reaches rated turbine power. This is often, but not always, the maximum power
- The cut-out velocity - the wind speed at which the wind turbine is shut down to keep loads and generator power from reaching damaging levels

In the following sections, methods for the determination of machine production will be analyzed, as well as methods to summarize wind speed information from a given site, using the following four approaches:

- Direct use of data averaged over a short time interval
- The method of bins
- Development of power and distribution curves from data
- Statistical analysis using summary measures

The next section summarizes the use of the three non-statistical methods.

2.4.2 *Direct methods of data analysis, resource characterization, and turbine productivity*

2.4.2.1 Direct use of data

Suppose one is given a series of N wind speed observations, U_i, each averaged over the time interval Δt. These data can be used to calculate the following useful parameters:

(1) The long-term average wind speed, $\bar{U}$, over the total period of data collection is:

$$\bar{U} = \frac{1}{N}\sum_{i=1}^{N} U_i \tag{2.4.1}$$

(2) The standard deviation of the individual wind speed averages, σ_U, is:

$$\sigma_U = \sqrt{\frac{1}{N\text{-}1}\sum_{i=1}^{N}(U_i - \bar{U})^2} = \sqrt{\frac{1}{N\text{-}1}\left\{\sum_{i=1}^{N} U_i^2 - N\bar{U}^2\right\}} \tag{2.4.2}$$

(3) The average wind power density, $\bar{P}/A$, is the average available wind power per unit area and is given by:

$$\bar{P}/A = (1/2)\rho\,\frac{1}{N}\sum_{i=1}^{N} U_i^3 \tag{2.4.3}$$

Similarly, the wind energy density per unit area for a given extended time period $N\Delta t$ long is given by:

$$\bar{E}/A = (1/2)\rho\sum_{i=1}^{N} U_i^3 = (\bar{P}/A)(N\Delta t) \tag{2.4.4}$$

4) The average wind machine power, $\bar{P}_w$, is:

$$\bar{P}_w = \frac{1}{N}\sum_{i=1}^{N} P_w(U_i) \tag{2.4.5}$$

where $P_w(U_i)$ is the power output defined by a wind machine power curve.

5) The energy from a wind machine, E_w, is:

$$E_w = \sum_{i=1}^{N} P_w(U_i)(\Delta t) \tag{2.4.6}$$

2.4.2.2 Method of bins

The method of bins also provides a way to summarize wind data and to determine expected turbine productivity. The data must be separated into the wind speed intervals or bins in which it occurs. It is most convenient to use the same size bins. Suppose that the data are separated into N_B bins of width w_j, with midpoints m_j, and with f_j, the number of occurrences in each bin or frequency, such that:

$$N = \sum_{j=1}^{N_B} f_j \tag{2.4.7}$$

The values found from Equations 2.4.1–2.4.3, 2.4.5, and 2.4.6 can be determined from the following:

$$\bar{U} = \frac{1}{N}\sum_{j=1}^{N_B} m_j f_j \tag{2.4.8}$$

$$\sigma_U = \sqrt{\frac{1}{N\text{-}1}\left\{\sum_{j=1}^{N_B} m_j^2 f_j - N\left(\bar{U}\right)^2\right\}} = \sqrt{\frac{1}{N\text{-}1}\left\{\sum_{j=1}^{N_B} m_j^2 f_j - N\left(\frac{1}{N}\sum_{j=1}^{N_B} m_j f_j\right)^2\right\}} \tag{2.4.9}$$

$$\bar{P}/A = (1/2)\rho\,\frac{1}{N}\sum_{j=1}^{N_B} m_j^3 f_j \tag{2.4.10}$$

$$\bar{P}_w = \frac{1}{N}\sum_{j=1}^{N_B} P_w\left(m_j\right) f_j \tag{2.4.11}$$

$$E_w = \sum_{j=1}^{N_B} P_w\left(m_j\right) f_j\,\Delta t \tag{2.4.12}$$

A histogram (bar graph) showing the number of occurrences and bin widths is usually plotted when using this method.

2.4.2.3 Velocity and power duration curves from data

Velocity and power duration curves can be useful when comparing the energy potential of candidate wind sites. As defined in this text, the velocity duration curve is a graph with wind speed on the y axis and the number of hours in the year for which the speed equals or exceeds each particular value on the x axis. An example of velocity duration curves (Rohatgi and Nelson, 1994) for various parts of the world (with average wind speeds varying from about 4 to 11 m/s) is shown in Figure 2.26. This type of figure gives an approximate idea about the nature of the wind regime at each site. The flatter the curve, the more constant are the wind speeds (e.g., characteristic of the trade-wind regions of the earth). The steeper the curve, the more irregular the wind regime.

A velocity duration curve can be converted to a power duration curve by cubing the ordinates, which are then proportional to the available wind power for a given rotor swept area. The difference between the energy potential of different sites is visually apparent, because the areas under the curves are proportional to the annual energy available from the wind. The following steps must be carried out to construct velocity and power duration curves from data:

- Arrange the data in bins
- Find the number of hours that a given velocity (or power per unit area) is exceeded
- Plot the resulting curves

A machine productivity curve for a particular wind turbine at a given site may be constructed using the power duration curve in conjunction with a machine curve for a given wind turbine. An example of a curve of this type is shown in Figure 2.27. Note that the losses in energy production with the use of an actual wind turbine at this site can be identified.

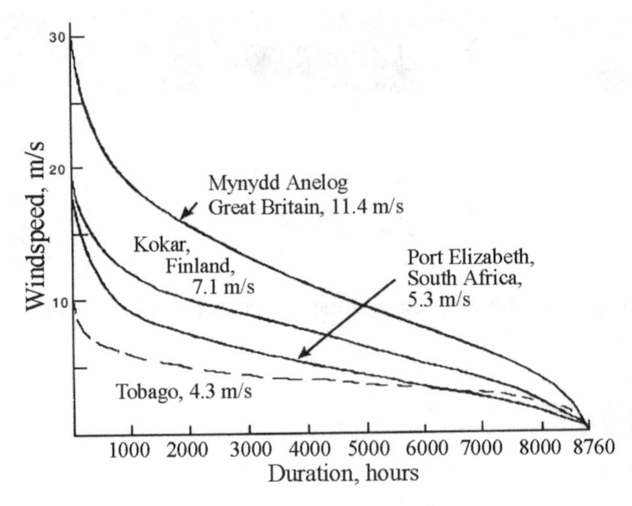

Figure 2.26 Velocity duration curve example (Rohatgi and Nelson, 1994). Reproduced by permission of Alternative Energy Institute

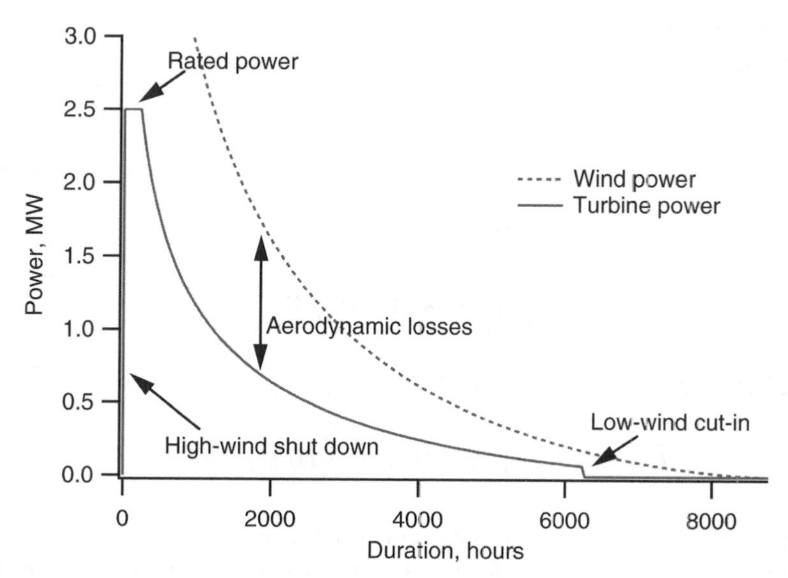

Figure 2.27 Machine productivity curve

2.4.3 Statistical analysis of wind data

Statistical analysis can be used to determine the wind energy potential of a given site and to estimate the wind energy output at this site. The development of the statistical analysis of wind data for resource estimation follows summaries of this type of analysis in several references (Justus, 1978, Johnson, 1985, and Rohatgi and Nelson, 1994). If time series measured data are available at the desired location and height, there may be little need for a data analysis in terms of probability distributions and statistical techniques. That is, the previously described analyses may be all that is needed. On the other hand, if projection of measured data from one location to another is required, or when only summary data are available, then there are distinct advantages in the use of analytical representations for the probability distribution of wind speed.

This type of analysis relies on the use of the probability density function, $p(U)$, of wind speed. This mathematical function was previously mentioned as one of the statistical variables that are used to characterize turbulence (see Section 2.3.2.2). One way to define the probability density function is that the probability of a wind speed occurring between U_a and U_b is given by:

$$p(U_a \le U \le U_b) = \int_{U_a}^{U_b} p(U)\mathrm{d}U \tag{2.4.13}$$

Also, the total area under the probability distribution curve is given by:

$$\int_0^{\infty} p(U)\,\mathrm{d}U = 1 \tag{2.4.14}$$

If $p(U)$ is known, the following parameters can be calculated:

Mean wind speed, $\bar{U}$:

$$\bar{U} = \int_0^\infty U p(U)\, \mathrm{d}U \tag{2.4.15}$$

Standard deviation of wind speed, σ_U :

$$\sigma_U = \sqrt{\int_0^\infty \left(U - \bar{U}\right)^2 p(U)\, \mathrm{d}U} \tag{2.4.16}$$

Mean available wind power density, $\bar{P}/A$

$$\bar{P}/A = (1/2)\rho \int_0^\infty U^3 p(U)\, \mathrm{d}U = (1/2)\rho \overline{U^3} \tag{2.4.17}$$

where $\overline{U^3}$ is the expected value for the cube of the wind speed.

It should be noted that the probability density function can be superimposed on a wind velocity histogram by scaling it to the area of the histogram.

Another important statistical parameter is the cumulative distribution function $F(U)$. $F(U)$ represents the time fraction or probability that the wind speed is smaller than or equal to a given wind speed, U'. That is: $F(U)$ = Probability ($U' \leq U$) where U' is a dummy variable. It can be shown that:

$$F(U) = \int_0^U p(U')\mathrm{d}U' \tag{2.4.18}$$

Also, the slope of the cumulative distribution function is equal to the probability density function, i.e.:

$$p(U) = \frac{\mathrm{d}F(U)}{\mathrm{d}U} \tag{2.4.19}$$

In general, either of two probability distributions (or probability density functions) are used in wind data analysis: (1) Rayleigh and (2) Weibull. The Rayleigh distribution uses one parameter, the mean wind speed. The Weibull distribution is based on two parameters and, thus, can better represent a wider variety of wind regimes. Both the Rayleigh and Weibull distributions are called 'skew' distributions in that they are defined only for values greater than 0.

2.4.3.1 Rayleigh distribution

This is the simplest velocity probability distribution to represent the wind resource since it requires only a knowledge of the mean wind speed, $\bar{U}$. The probability density function and the cumulative distribution function are given by:

$$p(U) = \frac{\pi}{2}\left(\frac{U}{\overline{U}^2}\right)\exp\left[-\frac{\pi}{4}\left(\frac{U}{\overline{U}}\right)^2\right] \tag{2.4.20}$$

$$F(U) = 1 - \exp\left[-\frac{\pi}{4}\left(\frac{U}{\overline{U}}\right)^2\right] \tag{2.4.21}$$

Figure 2.28 illustrates a Rayleigh probability density function for different mean wind speeds. As shown, a larger value of the mean wind speed gives a higher probability at higher wind speeds.

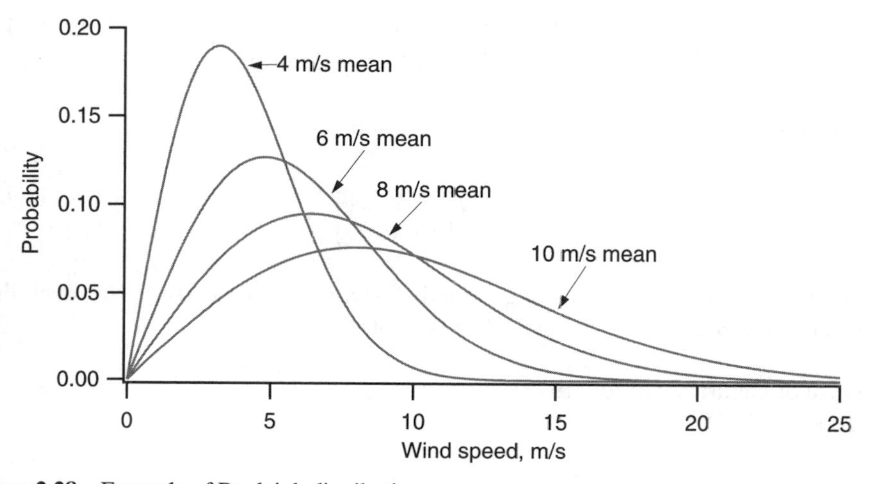

Figure 2.28 Example of Rayleigh distribution

2.4.3.2 Weibull distribution

Determination of the Weibull probability density function requires a knowledge of two parameters: k, a shape factor and c, a scale factor. Both these parameters are a function of $\overline{U}$ and σ_U. The Weibull probability density function and the cumulative distribution function are given by:

$$p(U) = \left(\frac{k}{c}\right)\left(\frac{U}{c}\right)^{k-1}\exp\left[-\left(\frac{U}{c}\right)^k\right] \tag{2.4.22}$$

$$F(U) = 1 - \exp\left[-\left(\frac{U}{c}\right)^k\right] \tag{2.4.23}$$

Examples of a Weibull probability density function, for various values of k, are given in Figure 2.29. As shown, as the value of k increases, the curve has a sharper peak, indicating that there is less wind speed variation. Methods to determine k and c from $\overline{U}$ and σ_U are presented below.

Using Equation 2.4.22 for the Weibull distribution, it is possible to calculate the average velocity as follows:

$$\overline{U} = c\Gamma\left(1+\frac{1}{k}\right) \tag{2.4.24}$$

where $\Gamma(x)$ = gamma function = $\int_0^\infty e^{-t}\ t^{x-1}\ dt$

The gamma function can be approximated by (Jamil, 1994):

$$\Gamma(x)=\left(\sqrt{2\pi\, x}\right)\left(x^{x-1}\right)\left(e^{-x}\right)\left(1+\frac{1}{12x}+\frac{1}{288x^2}-\frac{139}{51840x^3}+\ldots\right) \tag{2.4.25}$$

It can also be shown that for the Weibull distribution:

$$\sigma_U^2 = \overline{U}^2\left[\frac{\Gamma(1+2/k)}{\Gamma^2(1+1/k)}-1\right] \tag{2.4.26}$$

It is not a straightforward process to get c and k in terms of $\overline{U}$ and σ_U . However, there are a number of methods that can be used. For example:

1) Analytical or empirical (Justus, 1978)
   Use (good for $1 \leq k < 10$):

$$k = \left(\frac{\sigma_U}{\overline{U}}\right)^{-1.086} \tag{2.4.27}$$

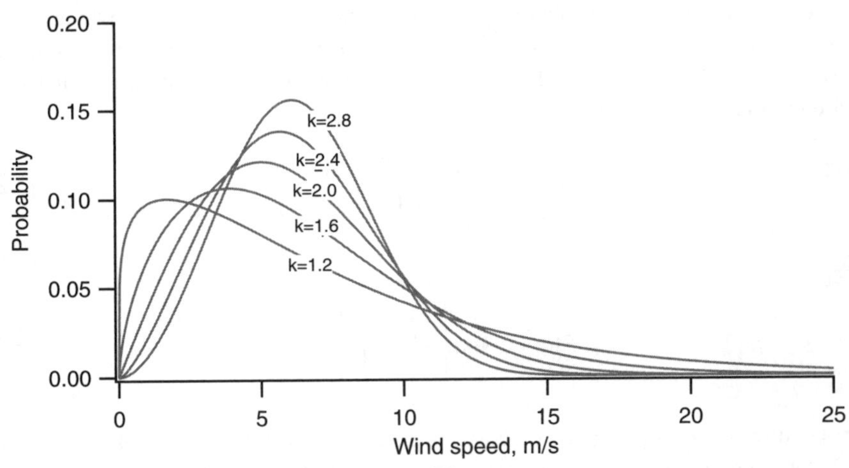

Figure 2.29 Example Weibull distributions for $\overline{U}$ = 8 m/s

Equation 2.4.24 can then be used to solve for c:

$$c = \frac{\overline{U}}{\Gamma(1+1/k)} \tag{2.4.28}$$

This method still requires use of the gamma function.

2) Empirical (Lysen, 1983)
Use Equation 2.4.27 to find k. Then, find c from the following approximation:

$$\frac{c}{\overline{U}} = \left(0.568 + 0.433/k\right)^{\frac{-1}{k}} \tag{2.4.29}$$

3) Graphical: log–log plot
Using this method, a straight line is drawn through a plot of wind speed, U, on the x-axis and $F(U)$ on the y-axis of log–log paper. The slope of the straight line gives k. Then, the intersection of a horizontal line with $F(U) = 0.632$ gives an estimate of c on the x-axis.

Based on the Weibull distribution (this assumes that c and k are known), the following other parameters of interest can be calculated.

(a) Standard deviation of wind speed.

$$\sigma_U^2 = c^2\left[\Gamma(1+2/k) - \Gamma^2(1+1/k)\right] = \overline{U}^2\left[\frac{\Gamma(1+2/k)}{\Gamma^2(1+1/k)} - 1\right] \tag{2.4.30}$$

(b) Expected value of the cube of the wind speed, $\overline{U^3}$:

$$\overline{U^3} = \int_0^\infty U^3 p(U)\mathrm{d}U = c^3\Gamma(1+3/k) \tag{2.4.31}$$

One should note that normalized variations of these two parameters depend only on the shape factor k. For example, the energy pattern factor, K_e (defined as the total amount of power available in the wind divided by the power calculated from cubing the average wind speed) is given by:

$$K_e = \frac{\overline{U^3}}{\left(\overline{U}\right)^3} = \frac{\Gamma(1+3/k)}{\Gamma^3(1+1/k)} \tag{2.4.32}$$

Some example values of some parameters of interest are given in Table 2.4.

It should also be noted that a Weibull distribution for which $k = 2$ is a special case of the Weibull distribution. It equals the Rayleigh distribution. That is, for $k = 2$ $\Gamma^2(1+1/2) = \pi/4$. One can also note that $\sigma_U / \overline{U} = 0.523$ for a Rayleigh distribution.

Table 2.4 Variation of parameters with Weibull k shape factor

k	$\sigma_U/\overline{U}$	K_e
1.2	0.837	3.99
2	0.523	1.91
3	0.363	1.40
5.0	0.229	1.15

2.5 Wind Turbine Energy Production Estimates Using Statistical Techniques

For a given wind regime probability distribution, $p(U)$, and a known machine power curve, $P_w(U)$, the average wind machine power, $\overline{P}_w$, is given by:

$$\overline{P}_w = \int_0^\infty P_w(U)\,p(U)\,\mathrm{d}U \tag{2.5.1}$$

As will be discussed in Chapter 6, it is possible to determine a machine power curve based on the power available in the wind and the rotor power coefficient, C_p. The result is the following expression for $P_w(U)$:

$$P_w(U) = \tfrac{1}{2}\rho A C_p \eta U^3 \tag{2.5.2}$$

where η is the drive train efficiency (generator power/rotor power). The rotor power coefficient is defined by:

$$C_p = \frac{Rotor\ power}{Dynamic\ power} = \frac{P_{rotor}}{\frac{1}{2}\rho A U^3} \tag{2.5.3}$$

In the next chapter, it will be shown that C_p can generally be expressed as a function of the tip speed ratio, λ, defined by:

$$\lambda = \frac{Blade\,tip\,speed}{Wind\,speed} = \frac{\Omega R}{U} \tag{2.5.4}$$

where Ω is the angular velocity of the wind rotor and R is the radius of the wind rotor.

Therefore, assuming a constant value for drive train efficiency, another expression for the average wind machine power is given by:

$$\overline{P}_w = \tfrac{1}{2}\rho A \eta \int_0^\infty C_p(\lambda) U^3 p(U)\,\mathrm{d}U \tag{2.5.5}$$

One is now in a position to use statistical methods for the estimation of the energy productivity of a specific wind turbine at a given site with a minimum of information. Two examples using Rayleigh and Weibull statistics as a basis for the analysis follow.

2.5.1 *Idealized machine productivity calculations using Rayleigh distribution*

A measure of the maximum possible average annual power from a given rotor diameter can be calculated assuming an ideal wind machine and using a Rayleigh probability density function. The analysis, based on the work of Carlin (1997), assumes the following:

- Idealized wind turbine, no losses, machine power coefficient, C_p, equal to the Betz limit ($C_{p,Betz}$ = 16/27). As will be discussed in the next chapter, the Betz limit is the theoretical maximum possible power coefficient.
- Wind speed probability distribution is given by a Rayleigh distribution.

The average wind machine power, $\overline{P}_w$, is given by Equation 2.5.5, and, for a Rayleigh distribution, is given by :

$$\overline{P}_w = \tfrac{1}{2}\rho A\eta \int_0^\infty C_p(\lambda)U^3 \left\{ \frac{2U}{U_c^2} \exp\left[-\left(\frac{U}{U_c} \right)^2 \right] \right\} \mathrm{d}U \tag{2.5.6}$$

where U_c is a characteristic wind velocity given by: $U_c = 2\overline{U}/\sqrt{\pi}$.

For an ideal machine, η = 1, and the power coefficient can be replaced with the Betz value of $C_{p,Betz}$ = 16/27, thus:

$$\overline{P}_w = \tfrac{1}{2}\rho A U_c^{\,3} C_{p,Betz} \int_0^\infty \left(\frac{U}{U_c} \right)^3 \left\{ \frac{2U}{U_c} \exp\left[-\left(\frac{U}{U_c} \right)^2 \right] \right\} \mathrm{d}U/U_c \tag{2.5.7}$$

One can now normalize the wind speed by defining a dimensionless wind speed, x, such that: $x = U/U_c$. This simplifies the previous integral as follows:

$$\overline{P}_w = \tfrac{1}{2}\rho A U_c^{\,3} C_{p,Betz} \int_0^\infty (x)^3 \left\{ 2x \exp\left[-(x)^2 \right] \right\} \mathrm{d}x \tag{2.5.8}$$

Note that the wind machine constants have been removed from the integral. The integral can now be evaluated over all wind speeds. Its value is $(3/4)\sqrt{\pi}$. Thus:

$$\overline{P}_w = \tfrac{1}{2}\rho A U_c^{\,3} (16/27)(3/4)\sqrt{\pi} \tag{2.5.9}$$

Substituting for the rotor disk area, $A = \pi D^2/4$, and for the characteristic velocity, U_c, the equation for average power is further simplified to:

$$\overline{P}_w = \rho \left(\frac{2}{3} D \right)^2 \overline{U}^3 \tag{2.5.10}$$

Carlin called this the one-two-three equation! (The density is raised to the first power).

For a numerical example, one could calculate the average annual production of an 18 meter diameter Rayleigh–Betz machine at sea level in a 6 m/s average annual wind velocity regime. For this example:

$$\overline{P}_w = \left(1.225\ \mathrm{kg/m^3}\right)\left(\frac{2}{3} \times 18\,\mathrm{m}\right)^2 (6\ \mathrm{m/s})^3 = 38.1\,\mathrm{kW}$$

Multiplication of this by 8760 hr/yr yields an expected annual energy production of 334,000 kWhr.

2.5.2 *Productivity calculations for a real wind turbine using a Weibull distribution*

Similar to the previous example, the average wind machine power is calculated using Equation 2.5.1:

$$\overline{P}_w = \int_0^\infty P_w(U)\, p(U)\,\mathrm{d}U \tag{2.5.11}$$

Based on Equation 2.4.18, it is possible to rewrite this equation using the cumulative distribution function, thus:

$$\overline{P}_w = \int_0^\infty P_w(U)\ \mathrm{d}F(U) \tag{2.5.12}$$

For a Weibull distribution, Equation 2.4.23 gives the following expression for $F(U)$:

$$F(U) = 1 - \exp\left[-\left(\frac{U}{c}\right)^k\right] \tag{2.5.13}$$

Therefore, replacing the integral in Equation 2.5.12 with a summation over N_B bins, the following expression can be used to find the average wind machine power:

$$\overline{P}_w = \sum_{j=1}^{N_B} \left\{ \exp\left[-\left(\frac{U_{j-1}}{c}\right)^k\right] - \exp\left[-\left(\frac{U_j}{c}\right)^k\right] \right\} P_w\left(\frac{U_{j-1} + U_j}{2}\right) \tag{2.5.14}$$

Note that Equation 2.5.14 is the statistical method's equivalent to Equation 2.4.11. In particular, the relative frequency, f/N, corresponds to the term in brackets and the wind turbine power is calculated at the midpoint between U_{j-1} and U_j.

2.6 Overview of Available Resource Assessment Data

Numerous studies have been carried out in various parts of the world in order to evaluate the regionally available wind energy resource. Some of these studies (e.g., in the United States and Europe) have resulted in the completion of detailed wind atlases. The use of available wind resource data is an important part of any resource assessment or wind siting program. In evaluating available wind data, however, it is important to realize their limitations. That is, little of this type of information has been collected for the purpose of wind energy assessment, and many data collection stations were located near or in cities, in relatively flat terrain or areas with low elevation. Thus, this type of data can provide a general description of the wind resource within a large area, but typically does not provide enough information for the detailed identification of candidate sites for wind development. In recent times, however, the analysis of this type of data and other data collected with wind resource assessment in mind has resulted in the publishing of wind atlases which are designed to quantify a particular location's wind resource.

This section presents a general review of the type of wind resource assessment data that is currently available.

2.6.1 United States resource information

In the 1970s, a preliminary wind resource assessment of the United States was carried out that produced 12 regional wind energy atlases. The atlases depicted the annual and seasonal wind resource on a state and regional level. They also included the wind resource's certainty rating (an indication of the reliability of the data) and an estimate for percentage of land suitable for wind energy development based on variations in land-surface form.

These data were used to produce a general wind power potential map that gave an indication of the wind resource (in W/m^2) for all locations in the United States in one map (see Figure 2.30). Almost as soon as these results were published, it was realized that these resource maps were not adequate and an intensive program was initiated by the Pacific Northwest Laboratories (PNL) to better characterize the wind energy potential in the United States. This work resulted in the publication of a new wind energy resource atlas.

The 1987 wind atlas integrated the pre-1979 wind measurements with topography and land form characteristics to determine US wind resource estimates. Data from approximately 270 post-1979 sites, including nearly 200 that were instrumented specifically for wind resource estimation purposes, were used to verify, or update the original wind resource values. The updated wind resource values are depicted on gridded maps, 1/4° latitude by 1/3° longitude resolution (about 120 km^2), on both a national scale as well as a state-by-state basis.

In the 1987 atlas, the magnitude of the wind resource is expressed in terms of seven wind power classes, rather than as a function of wind speed. The wind power classes range from Class 1 (for winds containing the least energy) to Class 7 (for winds containing the most energy). Each class represents a range of mean power density (W/m^2) or equivalent mean wind speed at specified heights above the ground. Table 2.5 shows the wind power classes in terms of their mean wind power density and the mean wind speed at 10 m (33 ft), 30 m (98 ft), and 50 m (164 ft) above the ground. Note that the 30 and 50 m heights correspond to the range of hub heights of many wind turbines then operating or under development. This table was constructed using the following assumptions:

- Vertical extrapolation of wind power density and wind speed were based on the 1/7 power law.
- Mean wind speed was estimated assuming a Rayleigh distribution of wind speeds and standard sea-level air density.

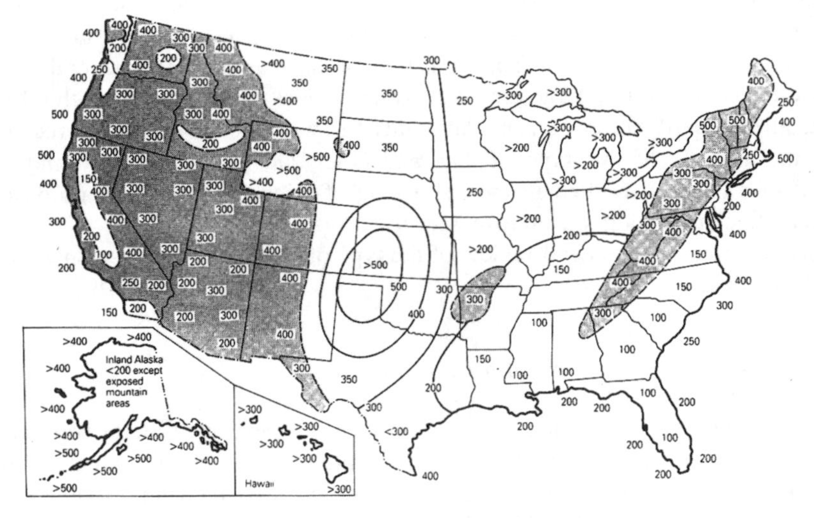

Figure 2.30 Initial wind power potential map of the United States (Elliot, 1977)

Table 2.5 Wind power density classes

	10 m (33 ft)		30 m (98 ft)		50 m (164 ft)	
Wind power class	Power Density W/m^2	Speed m/s (mph)	Power Density W/m^2	Speed m/s (mph)	Power Density W/m^2	Speed m/s (mph)
1	0–100	0–4.4 (0–9.8)	0–160	0–5.1 (0–11.4)	0–200	0–5.6 (0–12.5)
2	100–150	4.4–5.1 (9.8–11.5)	160–240	5.1–5.8 (11.4–13.2)	200–300	5.6–6.4 (12.5–14.3)
3	150–200	5.1–5.6 (11.5–12.5)	240–320	5.8–6.5 (13.2–14.6)	300–400	6.4–7.0 (14.3–15.7)
4	200–250	5.6–6.0 (12.5–13.4)	320–400	6.5–7.0 (14.6–15.7)	400–500	7.0–7.5 (15.7–16.8)
5	250–300	6.0–6.4 (13.4–14.3)	400–480	7.0–7.4 (15.7–16.6)	500–600	7.5–8.0 (16.8–17.8)
6	300–400	6.4–7.0 (14.3–15.7)	480–640	7.4–8.2 (16.6–18.3)	600–800	8.0–8.8 (17.8–19.7)
7	400–1000	7.0–9.4 (15.7–21.1)	640–1600	8.2–11.0 (18.3–24.7)	800–2000	8.8–11.9 (19.7–26.6)

Areas designated as class 4 or greater are generally considered to be suitable for most wind turbine applications. class 3 areas are suitable for wind energy development using tall (e.g., 50 m hub height) turbines. class 2 areas are marginal and class 1 areas are unsuitable for wind energy development. It should be noted that the results of this categorization also indicate the certainty of the wind resource based on data reliability and their areal distribution. They do not account, however, for the variability in mean wind speed on a local scale; rather, they indicate regions where a high wind resource is likely to exist. For example, in regions with complex terrain, a class 2 area could contain specific sites with a higher wind resource. Correspondingly, one cannot assume that every part of a class 6 region experiences consistently high winds.

Certainty ratings assigned to each grid cell indicate the degree of confidence in the resource estimate. The degree of certainty depends on the following:

- Abundance and quality of wind data
- Complexity of terrain
- Geographical variability of wind resource

The highest degree of certainty (rating 4) was assigned to grid cells with relatively simple terrain for which there were abundant historical data. Regions with complex terrain or for which little data was available received a low certainty rating of 1. An example of the type of data summary contained in this atlas is given in Figure 2.31, which presents the average wind power (by class) for the state of Maine.

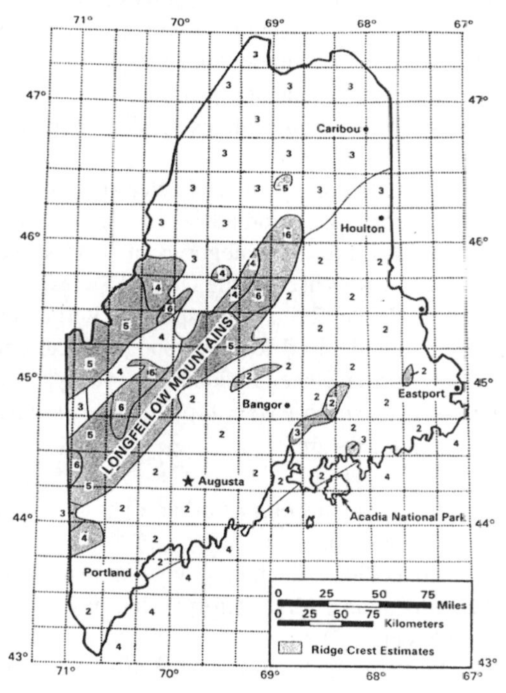

Figure 2.31 Average annual wind power for Maine (Elliot et al., 1987)

One should refer to the work of Elliot et al. (1991) for an example of how the data contained in this wind atlas can be used. In this report, based on information contained in the atlas, they prepared a detailed estimate of land areas with various levels of wind energy resource and resultant wind energy potential for each state in the United States.

2.6.2 European resource information

The wind energy resources vary considerably over Europe and are influenced by:

- Large temperature differences between the polar air in the north and subtropical air in the south
- The distribution of land and sea with the Atlantic Ocean to the west, Asia to the east, and the Mediterranean Sea and Africa to the south
- Major orographic barriers such as the Alps, the Pyrenees, and the Scandinavian mountain chain

In order to characterize these resources, the European Community (Troen and Petersen, 1989) developed a detailed wind resource atlas for Europe. The atlas is divided into the following three parts:

- Wind resource. This section, via graphs and tables, gives an overall view of the wind climate and magnitude and distribution of wind resources in the European Union
- Determining the wind resource. This section provides information for regional wind resource assessments. It also gives a methodology for the local estimation of the mean power produced by a specific wind turbine at a specific site
- The models and the analysis. This section contains the documentation part of the atlas

It should be noted that the methodology for characterizing for the wind resource was different than that used by PNL for the US wind atlases. As shown in Table 2.6, the maps of the European wind atlas are divided into five categories of wind speed, indicated by colors on the maps, instead of the seven classes used in the United States.

Additionally, as shown in Table 2.6, wind speed criteria for each category are subdivided into estimates of the wind speed for five topographical conditions. Note that in the wind atlas, the different terrains have been divided into four types, each characterized by its roughness elements. Furthermore, each terrain type may be associated with a roughness class. The five topographical conditions are:

1. Sheltered terrain. This includes such terrain as urban district, forest, and farm land with many windbreaks (roughness class 3).
2. Open plain. This is described as flat land with a few windbreaks (roughness class 1).
3. Sea coast. This describes a location with a uniform wind direction and land surface with a few wind breaks (roughness class 3).
4. Open sea. This condition defines a sea location 10 km offshore (roughness class 0).
5. Hills and ridges. These are characterized by a height of 400 m and a base diameter of 4 km.

Table 2.6 European Wind Atlas (Troen and Petersen, 1989) classification of wind resource at 50 m

Map	Sheltered terrain		Open plain		Sea coast		Open sea		Hills and ridges	
Color	m/s	W/m^2	m/s	W/m^2	m/s	W/m^2	m/s	W/m^2	m/s	W/m^2
Blue	0–3.5	<50	0–4.5	<100	0–5.0	<150	0–5.5	0–200	0–7.0	0–400
Green	3.5–	50–	4.5–	100–	5.0–	150–	5.5–	200–	7.0–	400–
	4.5	100	5.5	200	6.0	250	7.0	400	8.5	700
Orange	4.5–	100–	5.5–	200–	6.0–	250–	7.0–	400–	8.5–	700–
	5.0	150	6.5	300	7.0	400	8.0	600	10.0	1200
Red	5.0–	150–	6.5–	300–	7.0–	400–	8.0–	600–	10.0–	1200–
	6.0	250	7.5	500	8.5	700	9.0	800	11.5	1800
Purple	>6.0	250	7.5	>500	>8.5	700	9.0	800	11.5	>1800

2.6.3 Wind resource information for other parts of the world

There are numerous technical publications that summarize wind resource information for other parts of the work. Today, however, there is no one publication, or wind atlas, that summarizes all of this work. In 1981 the US Department of Energy's (DOE's) Pacific Northwest Laboratory (PNL) created a world resource map based on ship data, national weather data, and terrain (Cherry et al., 1981). At the present time, wind resource maps for different countries and regions of the world are generally not based on as good data as those in the United States and Europe.

Some examples of the type of wind resource information available for international resources are given in Rohatgi and Nelson (1994). For example, Figure 2.32 gives their wind resource map for Brazil, showing lines of average annual wind speed in m/s.

Figure 2.32 Annual average wind speeds for Brazil (m/s) (Rohatgi and Nelson, 1994). Reproduced by permission of Alternative Energy Institute

In recent times, US based organizations such as the National Renewable Energy Laboratory, PNL, Sandia National Laboratory, the US DOE, the Agency for International Development, and the American Wind Energy Association, and numerous European research and development agencies have provided technical assistance for wind resource assessment in developing countries. These have included Mexico, Indonesia, the Caribbean Islands, the former Soviet Union, Brazil, Chile, and Argentina. Resource assessments in these countries have focused on the development of rural wind power applications.

As more countries measure the wind resource for determining energy or power potential, the resource maps will become more detailed. It is expected that they will enable wind energy researchers to be better able to predict locations suitable for wind power sites.

2.7 Wind Measurements and Instrumentation

So far in this chapter it has been assumed that one has perfectly reliable meteorological wind speed data, for a location close to the desired location. The use of this meteorological data requires that information on site parameters (location, height, period of data collection, etc.), sensors (characteristics and calibration), and the type of data recorded has to be obtained. In most wind energy applications this is not the case, however, and measurements must be made specifically for determining the wind resource at the candidate location.

It is important to note that there are three types of instrument systems used for wind measurements:

- Instruments used by national meteorological services
- Instruments designed specifically for measuring and characterizing the wind resource
- Instruments specially designed for high sampling rates for determining gust, turbulence, and inflow wind information for analyzing wind turbine response

For each wind energy application, the type and amount of instrumentation required varies widely. For example, this can vary from a simple system just containing one wind speed anemometer/recorder to a very complex system designed to characterize the turbulence at a particular site. Figure 2.33 shows an example of the latter type of system developed by PNL. This system consisted of two towers and eight anemometers, with data sampled at a rate of 5 Hz.

2.7.1 Overview

Instrumentation for wind energy applications is a most important subject and has been discussed in detail by numerous authors. They include the early work of Golding (1977) and Putnam (1948) as well as in the more current references of Johnson (1985), Freris (1990) and Rohatgi and Nelson (1994), and the siting handbook of Hiester and Pennell (1981). In addition, it should be pointed out that the performance test codes for wind turbines of the American Society of Mechanical Engineers (ASME, 1989) and the measurement standards of the American Wind Energy Association (AWEA, 1986) also contain much useful information on wind instrumentation equipment and procedures.

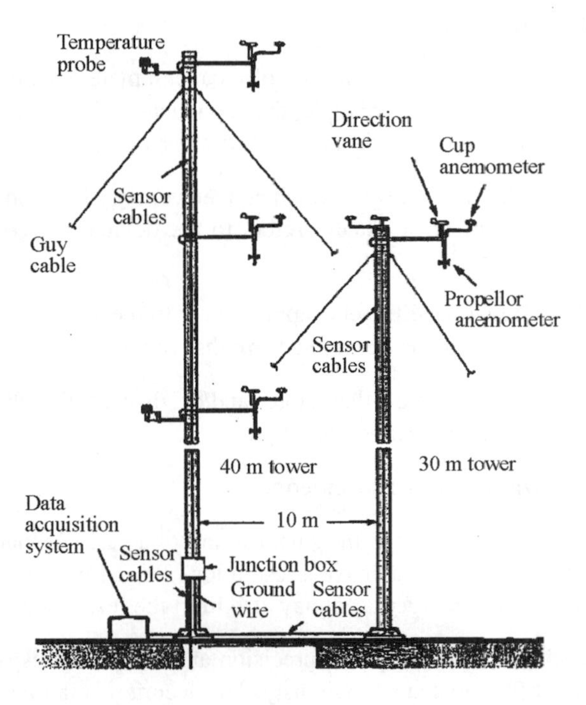

Figure 2.33 Turbulence characterization system of Pacific Northwest Laboratories (Wendell et al., 1991)

Wind energy applications use the following types of meteorological instruments:

- Anemometers to measure wind velocity
- Wind vanes to measure wind direction
- Thermometers to measure the ambient air temperature
- Barometers to measure the air pressure

In this section the discussion will be limited to the first two types of instruments. For more detail on the use of the third and fourth instrument type, one should refer to the wind resource assessment handbook of Bailey et al. (1996). Furthermore, wind instrumentation systems consist of three major components: sensors, signal conditioners, and recorders. In the following review, these components will be discussed in more detail.

2.7.2 General characteristics of instruments

Before discussing instrumentation systems, it is important to review some basics of measurement systems. The important parameters and concepts of instrumentation and measurement systems are reviewed below. This review is divided into three sections:

- System components
- Characterization of measurements
- Instrument characteristics

2.7.2.1 System components

Sensors A sensor is a device, such as a cup of an anemometer or a hot wire, which reacts to changes in the environment. For example, the cup reacts to the force of the wind, whereas the hot wire reacts to the wind flow via a temperature response.

Transducers A transducer is a device which converts energy from one form to another. In the case of wind measurement, it usually refers to the device that converts a mechanical motion to an electrical signal.

Signal conditioner Signal conditioners supply power to the sensor when required, receive the signal from the sensor, and convert it to a form that can be used by a recorder or display.

Recorder Recorders are devices that store and/or display the data obtained by the sensor/transducer/signal conditioner combination.

2.7.2.2 Characterization of measurements

Resolution Resolution is defined as the smallest unit of a variable that is detectable by the sensor. As an example, a sensor may have a resolution of ± 0.1 m/s or ± 1 m/s depending on the instrument. The type of recorder used may also limit the resolution.

Accuracy and precision Accuracy and precision are two measures of instrument system performance that are often treated ambiguously. The accuracy of an instrument refers to the mean difference between the output of the instrument and the true value of the measured variable. Precision refers to the dispersion about that mean. For example, an instrument may produce the same measured value every time, but that value may be 50% off. Thus, that system has high precision, but low accuracy. Another instrument measuring a variable may produce measurements with no mean error, but the dispersion of a single measurement may vary widely about the mean. This instrument has high accuracy, but low precision. In general, for wind measurement systems, the precision is usually high so that accuracy is the principal concern.

Error Error is the difference between the indication and the true value of the measured signal, e.g., for an anemometer, ± 1 m/s.

Reliability The reliability of an instrument is a measure of the probability that it will continue to perform within specified limits of error for a specified time under specified conditions. The best indicator of reliability is the past performance of similar instruments. In general, simple and rugged instruments with fewer parts are more reliable than those with a large number of parts.

Repeatability The repeatability of an instrument is the closeness of agreement among a number of consecutive measurements of output for the same input value, provided the measurements are made under the same conditions.

Reproducibility The closeness of agreement among measurements of the same quantity where the individual measurements are made under different conditions defines measurement reproducibility.

2.7.2.3 Instrument characteristics

Time constant The period required for a sensor to respond to 63.2% (1–1/e) of a stepwise change in an input signal defines its time constant.

Distance constant The distance constant is the length of fluid flow past a sensor required to cause it to respond to 63.2% of a step change in speed. It is calculated by multiplying the sensor time constant times the average wind speed. Standard cup anemometers can have distance constants as high as 10 m, depending on their size and weight. Small, lightweight cup anemometers, used for turbulence measurements, have distance constants between 1.5 to 3 m. For lightweight propeller anemometers, the distance constant is close to 1 m.

Response time The response time is the time required for an instrument to register a designated percentage (usually 90% or 95%) of a step change in the variable being measured.

Sampling rate Sampling rate is the frequency (Hz) at which the signal is sampled. It can be a function of the data collection system.

Resolution Resolution is defined as the smallest change in a variable that causes a detectable change in the indication of the instrument.

Sensitivity The sensitivity of an instrument is the ratio of the full-scale output of an instrument to the full-scale input value.

2.7.3 Wind speed measuring instrumentation

The sensors of wind measuring instrumentation can be classified according to their principle of operation via the following (ASME, 1989):

- Momentum transfer - cups, propellers, and pressure plates
- Pressure on stationary sensors - pitot tubes and drag spheres
- Heat transfer - hot wires and hot films
- Doppler effects - acoustics and laser
- Special methods - ion displacement, vortex shedding, etc.

Despite the number of potential instruments available for wind speed measurements, in most wind energy applications four different systems have been used. As discussed below, they include:

- Cup anemometers
- Propeller anemometers
- Kite anemometers
- Acoustic Doppler sensors (SODAR)

2.7.3.1 Cup anemometers

Cup anemometers use their rotation, which varies in proportion to the wind speed, to generate a signal. Today's most common designs feature three cups mounted on a small shaft. The rate of rotation of the cups can be measured by:

- Mechanical counters registering number of rotations
- Electrical or electronic voltage changes (AC or DC)
- A photoelectric switch

The mechanical type anemometers indicate the wind flow in distance. The mean wind speed is obtained by dividing the wind flow by time (this type is also called a wind-run anemometer). For remote sites, this type of anemometer has the advantage of not requiring a power source. Some of the earliest types of mechanical anemometers also drove a pen recorder directly. However, these systems were expensive and difficult to maintain.

An electronic cup anemometer gives a measurement of instantaneous wind speeds. The lower end of the rotating spindle is connected to a miniature AC or DC generator and the analog output is converted to wind speed via a variety of methods. The photoelectric switch type has a disc containing up to 120 slots and a photocell. The periodic passage of the slots produces pulses during each revolution of the cup.

The response and accuracy of a cup anemometer are determined by its weight, physical dimensions, and internal friction. By changing any of these parameters, the response of the instrument will vary. If turbulence measurements are desired, small, lightweight, low-friction sensors should be used. Typically, the most responsive cups have a distance constant of about 1 m. Where turbulence data are not required, the cups can be larger and heavier, with distance constants from 2 to 5 m. This limits the maximum usable data sampling rate to no greater than once every few seconds. Typical accuracy values (based on wind tunnel tests) for cup anemometers are about ±2%.

Numerous environmental problems can plague cup anemometers and reduce their reliability. These include icing or blowing dust. Dust can lodge in the bearings, causing an increase in friction and wear and reducing anemometer wind speed readings. If an anemometer ices up, its rotation will slow, or completely stop, causing erroneous wind speed signals, until the sensor thaws completely. Heated cup anemometers can be used, but they require a significant source of power. Because of these problems, the assurance of reliability for cup anemometers depends on calibration and service visits. The frequency of these visits depends on the site environment and the value of the data.

A widely used anemometer in the wind industry is the Maximum cup anemometer. This sensor is about 15 cm in diameter (see Figure 2.34).

This anemometer has a generator that provides a sine wave voltage output. It has a Teflon® sleeve bushing bearing system that is not supposed to be affected by dust, water, or lack of lubrication. The frequency of the sine wave is related to the wind speed. Special anemometers based on this design (16 pole magnet) can be used for some turbulence measurements with a 1 Hz sampling rate.

Figure 2.34 Maximum anemometer

2.7.3.2 Propeller anemometers

Propeller anemometers use the wind blowing into a propeller to turn a shaft that drives an AC or DC (most common) generator, or a light chopper to produce a pulse signal. The designs used for wind energy applications have a fast response and behave linearly in changing wind speeds. In a typical horizontal configuration, the propeller is kept facing the wind by a tail-vane, which also can be used as a direction indicator. The accuracy of this design is about ± 2%, similar to the cup anemometer. The propeller is usually made of polystyrene foam or polypropylene. The problems of reliability of propeller anemometers are similar to those discussed for cup anemometers.

When mounted on a fixed vertical arm, the propeller anemometer is especially suited for measuring the vertical wind component. A configuration for measuring three components of wind velocity is shown in Figure 2.35. The propeller anemometer responds primarily to wind parallel to its axis, and the wind perpendicular to the axis has no effect.

Figure 2.35 Propeller type anemometer for measuring three wind velocity components

2.7.3.3 Kite anemometers

In the past kites have been used in applications where wind measurements were desired at heights greater than conventional meteorological towers (see Hiester and Pennel, 1981). One of the most popular kite systems was the TALA (Tethered Aerodynamic Lifting Anemometer) kite. It used the tension on the kite line as an indicator of wind speed. They have been used in the preliminary assessment of wind power sites by numerous researchers, and, when used in groups, can be used to measure the wind shear profile at a site. Researchers have attempted to measure turbulence with TALA kites, but they have had limited success. Another use of TALA kites was in identifying areas of high turbulence (as a function of height) in complex terrain. A limitation of this type of device is the small amount of data produced, especially in comparison to how labor intensive they are.

2.7.3.4 Acoustic Doppler sensors (SODAR)

The acoustic doppler sensor system (or SODAR, standing for sonic detection and ranging) is based on the principle of acoustic backscattering. That is, an acoustic pulse transmitted into the air experiences backscattering from small temperature inhomogeneities (of a size on the order of the wave length). The travel time between emission and reception determines the height the signal represents. In addition, the Doppler shift in the frequency is proportional to the wind speed along the beam axis.

SODAR is classified as a remote sensing system, since it can make measurements without placing an active sensor at the point of measurement. Since such systems do not need tall (and expensive) towers for their use, the potential advantages of their use are obvious.

SODARs have recently been used for both onshore (Maeda et al., 1999) and offshore (Coelingh et al., 1999) wind siting studies. There has been a great deal of development on these devices over the last several years, and they are now commercially available from a number of sources.

2.7.4 Wind direction instrumentation

Wind direction is normally measured via the use of a wind vane. A conventional wind vane consists of a broad tail that the wind keeps on the downwind side of a rotating vertical shaft, and a counterweight at the upwind end to provide balance at the junction of the vane and shaft. Friction at the shaft is reduced with bearings, and so the vane requires a minimum force to initiate movement. For example, the usual threshold of this force occurs at wind speeds on the order of 1 m/s. Also, it is normal to damp the motion of the vane in order to prevent rapid changes of direction.

Wind vanes usually produce signals by contact closures or by potentiometers. Details of the circuitry required for these designs, and the overall design considerations (i.e., turning moment analysis) of such devices are given by Johnson (1985). The accuracy obtained from potentiometers is higher than that from contact closures, but the potentiometer-based wind vanes usually cost more. As with cup and propeller anemometers, environmental problems (blowing dust, salt, and ice) affect the reliability of wind vanes.

2.7.5 *Instrumentation towers*

Since it is desirable to collect wind data at the hub height of turbines, one needs to use towers that can reach from a minimum of 20 m up to about 150 m. Instrumentation towers come in many styles: self supporting, lattice or tubular towers, guyed lattice towers, and tilt-up guyed towers. Sometimes communications towers are available near the site under consideration. In most cases, however, towers must be installed specifically for wind measurement systems.

More details of this subject are included in the wind resource assessment handbook of Bailey et al. (1996). As noted here, guyed tilt-up towers that can be erected from the ground are the type most commonly used today. These towers have been designed specifically for wind measurements and they are lightweight and can be moved easily. They require small foundations and can usually be installed in less than a day.

2.7.6 *Data recording systems*

In the development of a wind measurement program one must select some type of data recording system in order to display, record, and analyze the data obtained from the sensors and transducers. The types of displays used for wind instruments are either of the analog type (meters) or of the digital type (LED, LCD) and supply one with current information. Typical displays use dials, lights, and digital counters. Recorders can provide past information, and also may provide current information. The recorders used in wind instrumentation systems generally fall into four classes:

- Counters
- Strip charts
- Magnetic tape
- Solid-state devices

The simplest recorder is the single counter or accumulator. A device of this type only records the total amount of wind passing the sensor, like the odometer of a car. In order to calculate wind speed one must know the total elapsed time. Some recorders combine a number of accumulators. For example, a recorder might include 10 accumulators, each associated with a given wind speed, resulting in a wind speed frequency distribution (a direct use of the previously discussed method of bins).

Strip chart recorders were a standard means of recording wind data for many years. This method of data recording, however, is less common today. It was first replaced by magnetic tape recorders, which in turn have been replaced by solid-state devices. Solid-state recorders can accomplish a large amount of the data analysis before storing the data.

In general, the favored method to handle the large amount of data needed for complete analysis is the use of data loggers or data acquisition using personal computers. A number of data logging systems are available on the market that record wind speed and direction averages and standard deviations, as well as maximum wind speed during the averaging interval. These systems often record the data on removable storage cards. Some allow the data to be downloaded via modem.

The choices of methods and data recording systems are large, and each has its own advantages and disadvantages. The particular situation will define the data requirements, which, in turn, will dictate the choice of recording methods. As shown in Table 2.7, the AWEA Standard for meteorological measurements at a potential wind site (AWEA, 1986) defines three classes of wind measurement systems.

More details of recording system classification schemes are given by Hiester and Pennell (1981) and in Rohatgi and Nelson (1994).

Table 2.7 American Wind Energy Association description of major classes of wind measurement systems

Class	Storage capability	Recording medium	Primary application	Comments
I	None	Manual records	Real time, instantaneous data	Low-cost equipment; human factor could introduce bias error
II	Single register	Counter or electronic	Weekly or monthly averages	Minimum system for average speed or annual energy
III	Multiple register sequential and processed	Strip chart; magnetic tape; solid state	Summarized bin data; detailed statistical data analysis	Raw data; some internal processing; data storage dependent on processing and logging systems

2.7.7 *Wind data analysis*

The data produced by a wind monitoring system can be analyzed in a number of ways. These may include, but are not limited to:

- Average horizontal wind speeds over specified time intervals
- Variations in the horizontal wind speed over the sampling intervals (standard deviation, turbulence intensity, maximums)
- Average horizontal wind direction
- Variations in the horizontal wind direction over the sampling intervals (standard deviation)
- Speed and direction distributions
- Persistence
- Determining gust parameters
- Statistical analysis, including autocorrelation, power spectral density, length and time scales, and spatial and time correlations with nearby measurements.
- Steady and fluctuating u, v, w wind components
- Diurnal, seasonal, annual, interannual and directional variations of any of the above parameters

Some mention has been made of each of these measures of wind data, except for persistence. Persistence is the duration of the wind speed within a given wind speed range. For example, histograms of the frequency of continuous periods of winds between the cut-in and cut-out wind speeds would provide information on the expected length of periods of continuous turbine operation.

A wind rose is a diagram showing the temporal distribution of wind direction and azimuthal distribution of wind speed at a given location. A wind rose (an example is shown in Figure 2.36) is a convenient tool for displaying anemometer data (wind speed and direction) for siting analysis. This figure illustrates the most common form, which consists of several equally spaced concentric circles with 16 equally spaced radial lines (each represents a compass point). The line length is proportional to the frequency of the wind from the compass point, with the circles forming a scale. The frequency of calm conditions is indicated in the center. The longest lines identify the prevailing wind directions. Wind roses generally are used to represent annual, seasonal, or monthly data.

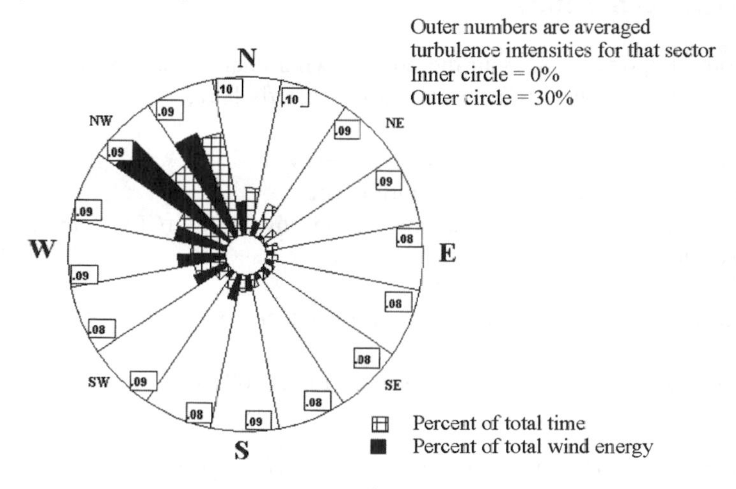

Figure 2.36 Example of a wind rose diagram

2.7.8 *Overview of a wind monitoring program*

One is now in a position to apply the fundamentals of wind characteristics, resource measurement and evaluation, and wind energy measurement systems. This section presents an outline of the procedure for conducting a successful wind monitoring program.

In the United States much work has been carried out on this, and the researchers and wind engineers involved in this subject have carefully documented their work. More specifically, under the auspices of the US Department of Energy and their subcontractor, the National Renewable Energy Laboratory (NREL), a detailed handbook on this subject was prepared by AWS Scientific (Bailey et al., 1996). The handbook, which has been designed for use in wind energy training seminars, contains ten chapters and an appendix. Following the approach of this handbook a wind assessment and monitoring program includes the following components:

- Review of guiding principles of a wind resource assessment program
- Determination of costs and labor requirements for a wind monitoring program
- Siting of monitoring systems
- Determination of measurement parameters
- Selection of monitoring station instrumentation
- Installation of monitoring systems
- Station operation and maintenance
- Data collection and handling
- Data validation, processing, and reporting

2.8 Advanced Topics

There are some important topics in the area of wind characterization that are beyond the scope of this chapter, but they will be summarized briefly here.

- Application of stochastic processes to wind energy
- Analysis and characterization of wind turbulence
- Use of numerical or computational fluid dynamic (CFD) models for flow characterization
- Micrositing
- Advanced statistically based resource assessment techniques

A short description of each of these advanced topics follows.

2.8.1 Application of stochastic processes to wind energy

In Section 2.4.3, deterministic statistical functions such as the Weibull or Rayleigh distribution were used to characterize wind speed variation. The variation in wind speed is a random process, however, and it is not possible to predict what will happen in the future, even with extensive wind speed data at the same site. Thus, one must find the chance, or probability, that the wind speed will be within certain limits. This type of variation is called a stochastic, probabilistic, or a random process.

Stochastic models are based on the concept that turbulence is made up of sinusoidal waves, or eddies with periods and random amplitudes. These types of models may also use probability distribution or other statistical parameters. Stochastic analysis can be invaluable in developing models of the wind inflow, because it facilitates the following critical engineering analysis or design tasks:

- Representation of many data points taken in field tests
- Evaluation of fatigue loads
- Comparison of model field data with historical field data

More details of these types of models as applied to wind energy may be found in Spera (1994) and Rohatgi and Nelson (1994). For information on the analytical methods for stochastic process analysis see Bendat and Piersol (1993).

2.8.2 *Analysis and characterization of wind turbulence*

A knowledge of the fundamentals of turbulence is important because turbulence causes random, fluctuating loads and power output, and stresses over the whole turbine and tower structure. It is important to consider turbulence for the following purposes:

- Maximum load prediction
- Structural excitation
- Fatigue
- Control
- Power quality

Turbulence is a complex subject, considered in many advanced fluid mechanics textbooks. There has been considerable progress made in the application of turbulence research to wind energy applications. See, for example, the discussions of this subject in the texts of Spera (1994) and Rohatgi and Nelson (1994).

To illustrate the instrumentation required for a detailed turbulence study, Figure 2.37 shows the turbulence characterization scheme used by NREL for their 'combined experiment' test program. This system consisted of a plane array of 13 anemometers used to measure the wind inflow. The system could collect a large amount of data in a very short time.

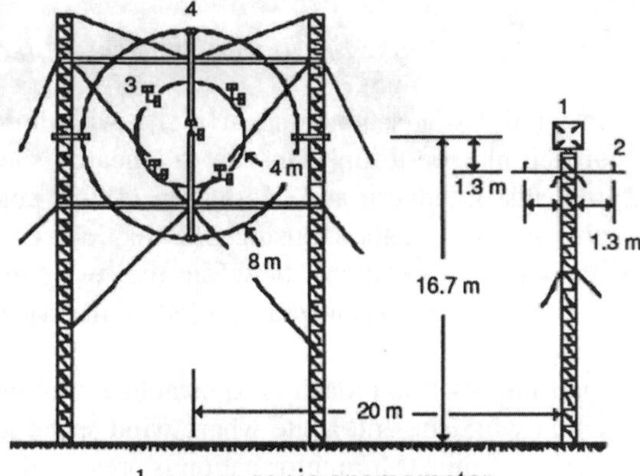

1 *u-v-w* sonic anemometer
2 Hot film anemometer
3 *u-v-w* Gill anemometer
4 Propellor vane anemometer

Figure 2.37 National Renewable Energy Laboratory combined experiment anemometry system (Butterfield, 1989)

2.8.3 Use of numerical or computational fluid dynamic models for flow characterization

The progress in computational or numerical modeling of complex flow fields has spread to the field of wind energy. For example, since many potential locations for wind turbines involve siting in complex terrain, it is useful to have analytical tools that can characterize the wind fields in these locations. One of the tools used here is modeling the flow by numerical models.

Today, the use of computational fluid dynamic (CFD) modeling is one of the most rapidly expanding areas in fluid mechanics (see Anderson, 1995). The rapid progress in CFD models and the ability to analyze complex flows are expected to expand with the ever increasing power of digital computers, and the associated graphical routines.

2.8.4 Micrositing

Micrositing is defined as a resource assessment tool used to determine the exact position of one or more wind turbines on a parcel of land to maximize the power production. An objective of micrositing is to locate the wind turbines in the wind farm to maximize annual energy production, or yield the largest financial return for the owners of the wind farm. Chapter 8 discusses this subject.

Effective micrositing depends on a combination of detailed wind resource information for the specific site and, generally, the use of CFD models to predict the detailed flow field in the wind farm (including machine wake effects). The output is then combined in another model that gives a prediction of the energy output of the wind farm. Some micrositing models are even able to give the optimized location for wind turbine placement. Some examples of micrositing models are summarized by Rohatgi and Nelson (1994).

2.8.5 Advanced statistically-based resource assessment techniques

For an estimate of the wind resource potential at a site with little or no wind resource measurement, one approach may be to link this site to a nearby site that has long duration wind resource measurements. Landberg and Mortensen (1993) note that this link can be accomplished using either physical methods (using CFD models) or via the use of statistical methods (based on statistical correlations between the two time series of data). A statistically based technique that has been widely used is the Measure–Correlate–Predict (MCP) approach.

The basic idea behind the MCP approach is to establish relations between wind speed and direction at a potential wind site and a site where wind speed and direction have been measured over a long period of time. More information is presented in Chapter 8.

2.9 References

Anderson, J. D. (1995) *Computational Fluid Dynamics: The Basics with Applications*, McGraw-Hill, New York.

ASME (1989) Performance Test Code for Wind Turbines. ASME/ANSI PTC 42-1988, American Society of Mechanical Engineers, New York.

Aspliden, C. I., Elliot, D. L., Wendell, L. L. (1986) Resource Assessment Methods, Siting, and Performance Evaluation, in *Physical Climatology for Solar and Wind Energy* (eds. R. Guzzi and C. G. Justus) World Scientific, New Jersey.

Avallone E. A., Baumeister III, T. (eds.) (1978) *Mark's Standard Handbook for Mechanical Engineers*, McGraw-Hill, New York, NY.

AWEA (1986) Standard Procedures for Meteorological Measurements at a Potential Wind Site. AWEA Standard 8.1, American Wind Energy Association, Washington, DC.

AWEA (1988) Standard Performance Testing of Wind Energy Conversion Systems. AWEA Standard 1.1, American Wind Energy Association, Washington, DC.

Bailey, B. H., McDonald, S. L., Bernadett, D. W., Markus, M. J., Elsholtz, K. V. (1996) Wind Resource Assessment Handbook. AWS Scientific Report (NREL Subcontract No. TAT-5-15283-01).

Balmer, R. T. (1990) *Thermodynamics*, West Publishing, St. Paul, MN.

Bendat, J. S., Piersol, A. G. (1993) *Engineering Applications of Correlation and Spectral Analysis*, Wiley, New York.

Butterfield, C. P., (1989) Aerodynamic Pressure and Flow-Visualization Measurement from a Rotating Wind Turbine Blade., Proc. 8th ASME Wind Energy Symposium, 245-256.

Carlin, P. W. (1997) Analytic Expressions for Maximum Wind Turbine Average Power in a Rayleigh Wind Regime. Proc. 1997 ASME/AIAA Wind Symposium, 255–263.

Cherry, N. J., Elliot, D. L., Aspliden, C. I. (1981) World-Wide Wind Resource Assessment, Proceedings AWEA Wind Workshop V, American Wind Energy Association, Washington, DC.

Coelingh, J. P., Folkerts, J., van Zuylen, E. J., Wiegerink, G. (1999) Using SODAR measurements in the POWER project, Proc. 1999 21st BWEA Wind Energy Conference, 283–287.

Counihan, J. (1975) Adiabatic Atmospheric Boundary Layers: A Review and Analysis of Data Collected from the Period 1880- 1972. *Atmospheric Environment*, 9, 871–905.

Eldridge, F. R. (1980) *Wind Machines, 2nd edn*, Van Nostrand Reinhold, New York.

Elliot, D. L. (1977) Adjustment and Analysis of Data for Regional Wind energy Assessments, Workshop on Wind Climate, Ashville, NC.

Elliot, D.L., Holladay, C.G. Barchet, W.R., Foote, H.P., Sandusky, W.F. (1987) Wind Energy Resource Atlas of the United States. Pacific Northwest Laboratories Report DOE/CH10094-4, NTIS.

Elliot, D. L., Wendell L.L., Gower G.L. (1991) An Assessment of the Available Windy Land Area and Wind Energy Potential in the Contiguous United States. Pacific National Laboratories Report PNL-7789, NTIS.

Freris, L. L. (1990) *Wind Energy Conversion Systems*, Prentice Hall, London.

Golding, E. W. (1977) *The Generation of Electricity by Wind Power*, E&FN Spon, London.

Gustavson, M. R. (1979) Limits to Wind Power Utilization. *Science*, 204, 6 April, 13–18.

Hiester, T. R., Pennell, W. T. (1981) The Meteorological Aspects of Siting Large Wind Turbines. Pacific Northwest Laboratories Report PNL- 2522, NTIS.

International Electrotechnical Commission (1998) Wind Turbine Power Performance Measurements: IEC 1400-12, Geneva.

Jamil, M. (1994) Wind Power Statistics and Evaluation of Wind Energy Density, *Wind Engineering*, 18, No.5, 227–240.

Johnson, G. L. (1985) *Wind Energy Systems*, Prentice Hall, Englewood Cliffs, NJ.

Justus, C. G. (1978) *Winds and Wind System Performance*, Franklin Institute Press, Philadelphia, PA.

Landberg, L., Mortensen, N. G. (1993) A Comparison of Physical and Statistical Methods for Estimating the Wind Resource at a Site. Proc. 15th British Wind Energy Association Conference, 119–125.

Lysen, E. H. (1983) *Introduction to Wind Energy*, SWD Publication SWD 82-1, The Netherlands.

Maeda, T., Yokata, T., Shimizu, Y., Maniwa, Y., Hyodo, H., Mori, T. (1999) Measurement of Atmospheric Boundary Layer for Siting Wind Farms. Proc. 1999 EWEC, 1228-1231.

Putnam, P. C. (1948) *Power from the Wind*, Van Nostrand Reinhold, New York.

Rohatgi, J. S., Nelson, V. (1994) *Wind Characteristics: An Analysis for the Generation of Wind Power*. Alternative Energy Institute, Canyon, TX.

Spera, D. A. (ed.) (1994) *Wind Turbine Technology: Fundamental Concepts of Wind Turbine Engineering*, ASME Press, New York.

Troen, I., Petersen, E. L. (1989) *European Wind Atlas*, Riso National Laboratory, Denmark.

Wegley, H. L., Ramsdell J. V., Orgill, M. M. and Drake, R. L. (1980) A Siting Handbook for Small Wind Energy Conversion Systems, Battelle Pacific Northwest Lab., PNL-2521, Rev. 1, NTIS.

Wendell, L. L., Morris, V. R., Tomich S. D., Gower, G. L. (1991) Turbulence Characterization for Wind Energy Development. Proc. Windpower '91. AWEA, 254–265.

World Energy Council (1993) *Renewable Energy Resources: Opportunities and Constraints 1990–2020*. World Energy Council, London.

Wortman, A. J. (1982) *Introduction to Wind Turbine Engineering*, Butterworth, Boston, MA.

3

Aerodynamics of Wind Turbines

3.1 General Overview

Wind turbine power production depends on the interaction between the rotor and the wind. As discussed in Chapter 2, the wind may be considered to be a combination of the mean wind and turbulent fluctuations about that mean flow. Experience has shown that the major aspects of wind turbine performance (mean power output and mean loads) are determined by the aerodynamic forces generated by the mean wind. Periodic aerodynamic forces caused by wind shear, off-axis winds and rotor rotation and randomly fluctuating forces induced by turbulence and dynamic effects are the source of fatigue loads and are a factor in the peak loads experienced by a wind turbine. These are, of course, important, but can only be understood once the aerodynamics of steady state operation have been understood. Accordingly, this chapter focuses primarily on steady state aerodynamics. An overview of the complex phenomena of unsteady aerodynamics is presented at the end of the chapter.

Practical horizontal axis wind turbine designs use airfoils to transform the kinetic energy in the wind into useful energy. The material in this chapter provides the background to enable the reader to understand power production with the use of airfoils, to calculate an optimum blade shape for the start of a blade design and to analyse the aerodynamic performance of a rotor with a known blade shape and airfoil characteristics. A number of authors have derived methods for predicting the steady state performance of wind turbine rotors. The classical analysis of the wind turbine was originally developed by Betz and Glauert (Glauert, 1935) in the 1930's. Subsequently, the theory was expanded and adapted for solution by digital computers (see Wilson and Lissaman, 1974, Wilson et al., 1976 and de Vries, 1979). In all of these methods, momentum theory and blade element theory are combined into a strip theory that enables calculation of the performance characteristics of an annular section of the rotor. The characteristics for the entire rotor are then obtained by integrating, or summing, the values obtained for each of the annular sections. This approach is the one used in this chapter.

The chapter starts with the analysis of an idealized wind turbine rotor. The discussion introduces important concepts and illustrates the general behavior of wind turbine rotors and the airflow around wind turbine rotors. The analyses are also used to determine theoretical performance limits for wind turbines.

General aerodynamic concepts and the operation of airfoils are then introduced. This information is then used to consider the advantages of using airfoils for power production over other approaches.

The majority of the chapter details the classical analytical approach for the analysis of horizontal axis wind turbines, as well as some applications and examples of its use. First the details of momentum theory and blade element theory are developed and used to calculate the optimum blade shape for simplified, ideal operating conditions. The results illustrate the derivation of the general blade shape used in wind turbines. The combination of the two approaches, called strip theory or blade element momentum (BEM) theory, is then used to outline a procedure for the aerodynamic design and performance analysis of a wind turbine rotor. Aerodynamic losses and off-design performance are discussed and a starting optimum blade design for a more realistic flow field is developed. Finally, a simplified design procedure is presented that can be used for quick analyses.

The last two sections of the chapter discuss limitations on the maximum theoretical performance of a wind turbine and introduce advanced topics. These advanced topics include the effects of non-ideal steady state aerodynamics, turbine wakes and their effects on turbine operation, unsteady aerodynamics, computer codes for analyzing rotor performance and other theoretical approaches to rotor performance analysis.

Every attempt has been made to make the material in this chapter accessible to readers without a fluid dynamics background. Nevertheless, it would be helpful to be familiar with a variety of concepts including Bernoulli's equation, streamlines, control volume analyses and the concepts of laminar and turbulent flow. The material does require an understanding of basic physics.

3.2 One-Dimensional Momentum Theory and the Betz Limit

A simple model, generally attributed to Betz (1926), can be used to determine the power from an ideal turbine rotor, the thrust of the wind on the ideal rotor and the effect of the rotor operation on the local wind field. This simple model is based on a linear momentum theory developed over 100 years ago to predict the performance of ship propellers.

The analysis assumes a control volume, in which the control volume boundaries are the surface of a stream tube and two cross-sections of the stream tube (see Figure 3.1). The only flow is across the ends of the stream tube. The turbine is represented by a uniform "actuator disk" which creates a discontinuity of pressure in the stream tube of air flowing through it. Note that this analysis is not limited to any particular type of wind turbine.

This analysis uses the following assumptions:

- Homogenous, incompressible, steady state fluid flow
- No frictional drag
- An infinite number of blades
- Uniform thrust over the disk or rotor area
- A nonrotating wake
- The static pressure far upstream and far downstream of the rotor is equal to the undisturbed ambient static pressure

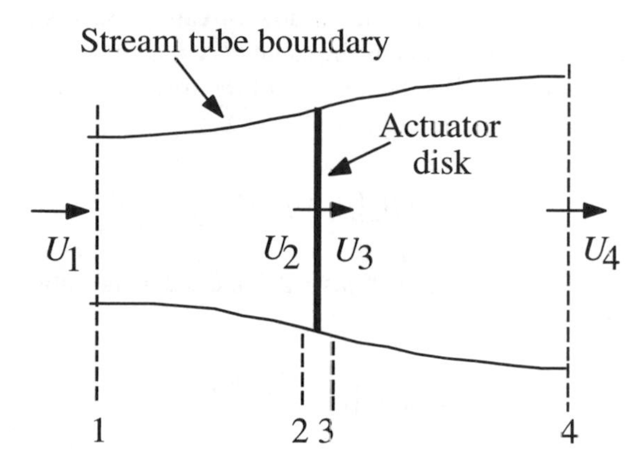

Figure 3.1 Actuator disk model of a wind turbine; U, mean air velocity; 1, 2, 3 and 4 indicate locations

Applying the conservation of linear momentum to the control volume enclosing the whole system, one can find the net force on the contents of the control volume. That force is equal and opposite to the thrust, T, which is the force of the wind on the wind turbine. From the conservation of linear momentum for a one-dimensional, incompressible, time-invariant flow, the thrust is equal and opposite to the change in momentum of air stream:

$$T = U_1(\rho A U)_1 - U_4(\rho A U)_4 \tag{3.2.1}$$

where ρ is the air density, A is the cross sectional area, U is the air velocity and the subscripts indicate values at numbered cross sections in Figure 3.1.

For steady state flow, $(\rho A U)_1 = (\rho A U)_4 = \dot{m}$, where $\dot{m}$ is the mass flow rate. Therefore:

$$T = \dot{m}(U_1 - U_4) \tag{3.2.2}$$

The thrust is positive so the velocity behind the rotor, U_4, is less than the free stream velocity, U_1. No work is done on either side of the turbine rotor. Thus the Bernoulli function can be used in the two control volumes on either side of the actuator disk. In the stream tube upstream of the disk:

$$p_1 + \tfrac{1}{2}\rho U_1^2 = p_2 + \tfrac{1}{2}\rho U_2^2 \tag{3.2.3}$$

In the stream tube downstream of the disk:

$$p_3 + \tfrac{1}{2}\rho U_3^2 = p_4 + \tfrac{1}{2}\rho U_4^2 \tag{3.2.4}$$

where it is assumed that the far upstream and far downstream pressures are equal ($p_1 = p_4$) and that the velocity across the disk remains the same ($U_2 = U_3$).

The thrust can also be expressed as the net sum of the forces on each side of the actuator disc:

$$T = A_2 (p_2 - p_3) \tag{3.2.5}$$

If one solves for ($p_2 - p_3$) using Equations 3.2.3 and 3.2.4 and substitutes that into Equation 3.2.5, one obtains:

$$T = \tfrac{1}{2} \rho A_2 \left(U_1^{\,2} - U_4^{\,2} \right) \tag{3.2.6}$$

Equating the thrust values from Equations 3.2.2 and 3.2.6 and recognizing that the mass flow rate is $A_2 U_2$, one obtains:

$$U_2 = \frac{U_1 + U_4}{2} \tag{3.2.7}$$

Thus, the wind velocity at the rotor plane, using this simple model, is the average of the upstream and downstream wind speeds.

If one defines the axial induction factor, a, as the fractional decrease in wind velocity between the free stream and the rotor plane, then

$$a = \frac{U_1 - U_2}{U_1} \tag{3.2.8}$$

$$U_2 = U_1 (1 - a) \tag{3.2.9}$$

and

$$U_4 = U_1 (1 - 2a) \tag{3.2.10}$$

The quantity, $U_1 a$, is often referred to as the induced velocity at the rotor, in which case velocity of the wind at the rotor is a combination of the free stream velocity and the induced wind velocity. As the axial induction factor increases from 0, the wind speed behind the rotor slows more and more. If a = 1/2, the wind has slowed to zero velocity behind the rotor and the simple theory is no longer applicable.

The power out, P , is equal to the thrust times the velocity at the disk:

$$P = \tfrac{1}{2} \rho A_2 \left(U_1^{\,2} - U_4^{\,2} \right) U_2 = \tfrac{1}{2} \rho A_2 U_2 (U_1 + U_4)(U_1 - U_4) \tag{3.2.11}$$

Substituting for U_2 and U_4 from Equations 3.2.9 and 3.2.10 gives

$$P = \tfrac{1}{2}\rho A U^3 4a(1-a)^2 \tag{3.2.12}$$

where the control volume area at the rotor, A_2, is replaced with A, the rotor area, and the free stream velocity U_1 is replaced by U.

Wind turbine rotor performance is usually characterized by its power coefficient, C_P:

$$C_P = \frac{P}{\frac{1}{2}\rho U^3 A} = \frac{\text{Rotor power}}{\text{Power in the wind}} \tag{3.2.13}$$

The non-dimensional power coefficient represents the fraction of the power in the wind that is extracted by the rotor. From Equation 3.2.12, the power coefficient is:

$$C_P = 4a(1-a)^2 \tag{3.2.14}$$

The maximum C_P is determined by taking the derivative of the power coefficient (Equation 3.2.14) with respect to a and setting it equal to zero, yielding a = 1/3. Thus:

$$C_{P,\max} = 16/27 = 0.5926 \tag{3.2.15}$$

when a = 1/3. For this case, the flow through the disk corresponds to a stream tube with an upstream cross-sectional area of 2/3 the disk area that expands to twice the disk area downstream. This result indicates that, if an ideal rotor were designed and operated such that the wind speed at the rotor were 2/3 of the free stream wind speed, then it would be operating at the point of maximum power production. Furthermore, given the basic laws of physics, this is the maximum power possible.

From Equations 3.2.6, 3.2.9 and 3.2.10, the axial thrust on the disk is:

$$T = \tfrac{1}{2}\rho A U_1^2 \left[4a(1-a)\right] \tag{3.2.16}$$

Similarly to the power, the thrust on a wind turbine can be characterized by a non-dimensional thrust coefficient:

$$C_T = \frac{T}{\frac{1}{2}\rho U^2 A} = \frac{\text{Thrust force}}{\text{Dynamic force}} \tag{3.2.17}$$

From Equation 3.2.16, the thrust coefficient for an ideal wind turbine is equal to $4a(1-a)$. C_T has a maximum of 1.0 when a = 0.5 and the downstream velocity is zero. At maximum power output (a = 1/3), C_T has a value of 8/9. A graph of the power and thrust coefficients for an ideal Betz turbine and the non-dimensionalized downstream wind speed are illustrated in Figure 3.2.

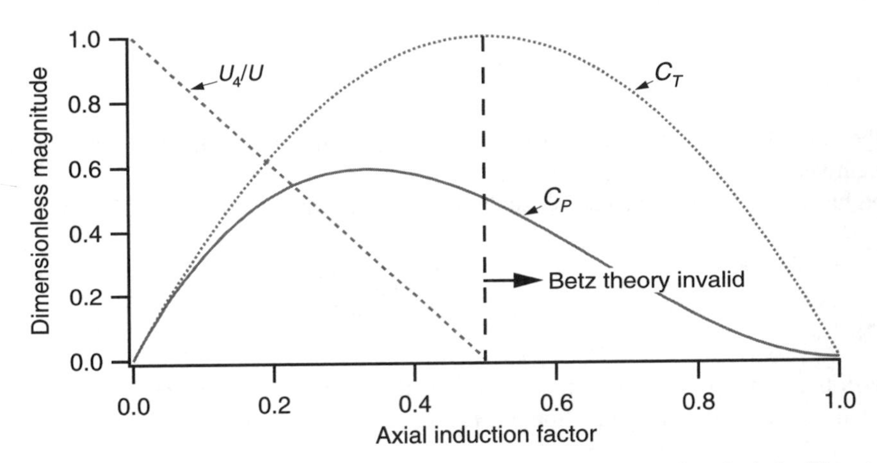

Figure 3.2 Operating parameters for a Betz turbine; U , velocity of undisturbed air; U_4 , air velocity behind the rotor,; C_P , power coefficient; C_T , thrust coefficient

As mentioned above, this idealized model is not valid for axial induction factors greater than 0.5. In practice (Wilson et al., 1976), as the axial induction factor approaches and exceeds 0.5, complicated flow patterns that are not represented in this simple model result in thrust coefficients that can go as high as 2.0. The details of wind turbine operation at these high axial induction factors appear in Section 3.7.

The Betz limit, $C_{P,\max} = 16/27$, is the maximum theoretically possible rotor power coefficient. In practice three effects lead to a decrease in the maximum achievable power coefficient:

- Rotation of the wake behind the rotor
- Finite number of blades and associated tip losses
- Non-zero aerodynamic drag

Note that the overall turbine efficiency is a function of both the rotor power coefficient and the mechanical (including electrical) efficiency of the wind turbine:

$$\eta_{overall} = \frac{P_{out}}{\frac{1}{2}\rho A U^3} = \eta_{mech} C_p \tag{3.2.18}$$

Thus:

$$P_{out} = \tfrac{1}{2}\rho A U^3 \left(\eta_{mech} C_P\right) \tag{3.2.19}$$

3.3 Ideal Horizontal Axis Wind Turbine with Wake Rotation

In the previous analysis using linear momentum theory, it was assumed that no rotation was imparted to the flow. The previous analysis can be extended to the case where the rotating rotor generates angular momentum, which can be related to rotor torque. In the case of a rotating wind turbine rotor, the flow behind the rotor rotates in the opposite direction to the rotor, in reaction to the torque exerted by the flow on the rotor. An annular stream tube model of this flow, illustrating the rotation of the wake, is shown in Figure 3.3.

The generation of rotational kinetic energy in the wake results in less energy extraction by the rotor than would be expected without wake rotation. In general, the extra kinetic energy in the wind turbine wake will be higher if the generated torque is higher. Thus, as will be shown here, slow running wind turbines (with a low rotational speed and a high torque) experience more wake rotation losses than high-speed wind machines with low torque.

Figure 3.4 gives a schematic of the parameters involved in this analysis. Subscripts denote values at the cross sections identified by numbers. If it is assumed that the angular velocity imparted to the flow stream, ω , is small compared to the angular velocity, Ω , of the wind turbine rotor, then it can also be assumed that the pressure in the far wake is equal to the pressure in the free stream (see Wilson et al., 1976). The analysis that follows is based on the use of an annular stream tube with a radius r and a thickness $\mathrm{d}r$, resulting in a cross-sectional area equal to $2\pi r \mathrm{d}r$ (see Figure 3.4). The pressure, wake rotation and induction factors are all assumed to be a function of radius.

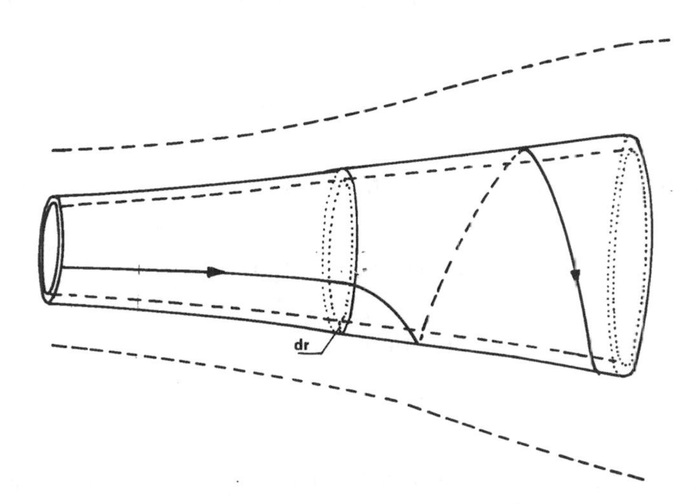

Figure 3.3 Stream tube model of flow behind rotating wind turbine blade. Picture of stream tube with wake rotation, from *Introduction to Wind Energy*, by E. H. Lysen, published by SWD (Steering Committee Wind Energy Developing Countries), Amersfoort, the Netherlands, 1982. Reproduced by permission of the author.

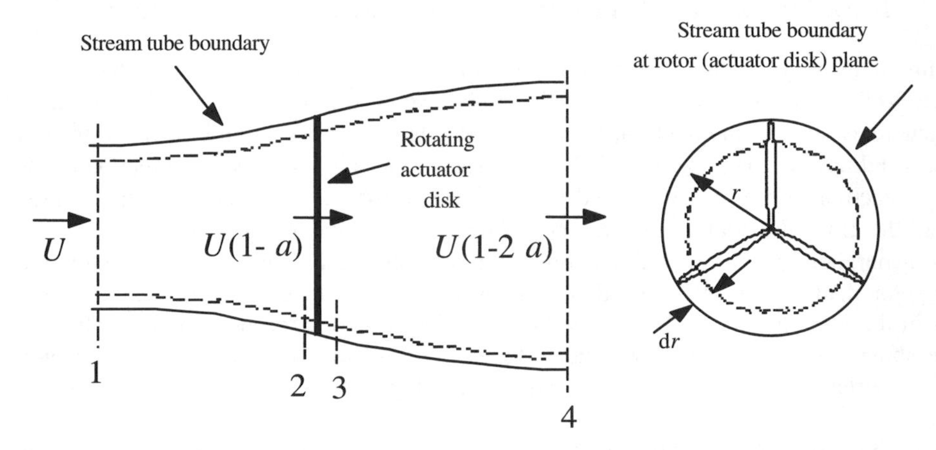

Figure 3.4 Geometry for rotor analysis; U, velocity of undisturbed air; a, induction factor; r, radius

If one uses a control volume that moves with the angular velocity of the blades, the energy equation can be applied in the sections before and after the blades to derive an expression for the pressure difference across the blades (see Glauert, 1935, for the derivation). Note that across the flow disk, the angular velocity of the air relative to the blade increases from Ω to $\Omega + \omega$, while the axial component of the velocity remains constant. The results are:

$$p_2 - p_3 = \rho(\Omega + \tfrac{1}{2}\omega)\omega r^2 \tag{3.3.1}$$

The resulting thrust on an annular element, dT, is:

$$dT = (p_2 - p_3)dA = \left[\rho(\Omega + \tfrac{1}{2}\omega)\omega r^2\right]2\pi r dr \tag{3.3.2}$$

The angular induction factor, a', is defined as:

$$a' = \omega / 2\Omega \tag{3.3.3}$$

Note that when wake rotation is included in the analysis, the induced velocity at the rotor consists of not only the axial component, Ua, but also a component in the rotor plane, $r\Omega a'$.

The expression for the thrust becomes:

$$dT = 4a'(1 + a')\tfrac{1}{2}\rho\Omega^2 r^2 2\pi r dr \tag{3.3.4}$$

Following the previous linear momentum analysis, the thrust on an annular cross-section can also be determined by the following expression that uses the axial induction factor, a, (note that U_1, the free stream velocity, is designated by U in this analysis):

$$dT = 4a(1-a)\tfrac{1}{2}\rho U^2 2\pi r dr \quad (3.3.5)$$

Equating the two expressions for thrust gives:

$$\frac{a(1-a)}{a'(1+a')} = \frac{\Omega^2 r^2}{U^2} = \lambda_r^2 \quad (3.3.6)$$

where λ_r is the local speed ratio (see below). This result will be used later in the analysis.

The tip speed ratio, λ, defined as the ratio of the blade tip speed to the free stream wind speed is given by:

$$\lambda = \Omega R / U \quad (3.3.7)$$

The tip speed ratio often occurs in the aerodynamic equations for the rotor. The local speed ratio is the ratio of the rotor speed at some intermediate radius to the wind speed:

$$\lambda_r = \Omega r / U = \lambda r / R \quad (3.3.8)$$

Next, one can derive an expression for the torque on the rotor by applying the conservation of angular momentum. For this situation, the torque exerted on the rotor, Q, must equal the change in angular momentum of the wake. On an incremental annular area element this gives:

$$dQ = d\dot{m}(\omega r)(r) = (\rho U_2 2\pi r dr)(\omega r)(r) \quad (3.3.9)$$

Since $U_2 = U(1-a)$ and $a' = \omega / 2\Omega$, this expression reduces to

$$dQ = 4a'(1-a)\tfrac{1}{2}\rho U \Omega \; r^2 2\pi \; r dr \quad (3.3.10)$$

The power generated at each element, dP, is given by:

$$dP = \Omega dQ \quad (3.3.11)$$

Substituting for dQ in this expression and using the definition of the local speed ratio, λ_r, (Equation 3.3.9), the expression for the power generated at each element becomes:

$$\mathrm{d}P = \tfrac{1}{2}\rho A U^3 \left[\frac{8}{\lambda^2} a'(1-a)\lambda_r^3 \, \mathrm{d}\lambda_r \right] \tag{3.3.12}$$

It can be seen that the power from any annular ring is a function of the axial and angular induction factors and the tip speed ratio. The axial and angular induction factors determine the magnitude and direction of the airflow at the rotor plane. The local speed ratio is a function of the tip speed ratio and radius.

The incremental contribution to the power coefficient, $\mathrm{d}C_P$, from each annular ring is given by:

$$\mathrm{d}C_P = \frac{\mathrm{d}P}{\tfrac{1}{2}\rho A U^3} \tag{3.3.13}$$

Thus

$$C_P = \frac{8}{\lambda^2}\int_0^{\lambda} a'(1-a)\lambda_r^3 \, \mathrm{d}\lambda_r \tag{3.3.14}$$

In order to integrate this expression, one needs to relate the variables a, a', and λ_r, (see Glauert 1948, Sengupta and Verma, 1992). Solving Equation 3.3.6 to express a' in terms of a, one gets:

$$a' = -\tfrac{1}{2} + \tfrac{1}{2}\sqrt{\left[1 + \frac{4}{\lambda_r^2} a(1-a)\right]} \tag{3.3.15}$$

The aerodynamic conditions for the maximum possible power production occur when the term $a'(1-a)$ in Equation 3.3.14 is at its greatest value. Substituting the value for a' from Equation 3.3.15 into $a'(1-a)$ and setting the derivative with respect to a equal to zero yields:

$$\lambda_r^2 = \frac{(1-a)(4a-1)^2}{1-3a} \tag{3.3.16}$$

This equation defines the axial induction factor for maximum power as a function of the local tip speed ratio in each annular ring. Substituting into Equation 3.3.6, one finds that that, for maximum power in each annular ring:

$$a' = \frac{1-3a}{4a-1} \tag{3.3.17}$$

If Equation 3.3.16 is differentiated with respect to a, one obtains a relationship between $\mathrm{d}\lambda_r$ and $\mathrm{d}a$ at those conditions that result in maximum power production:

$$2\lambda_r \mathrm{d}\lambda_r = \left[6(4a-1)(1-2a)^2/(1-3a)^2\right]\mathrm{d}a \tag{3.3.18}$$

Now, substituting the Equations 3.3.16–3.3.18 into the expression for the power coefficient (Equation 3.3.14) gives:

$$C_{P,\max} = \frac{24}{\lambda^2}\int_{a_1}^{a_2}\left[\frac{(1-a)(1-2a)(1-4a)}{(1-3a)}\right]^2 \mathrm{d}a \tag{3.3.19}$$

Here the lower limit of integration, a_1, corresponds to axial induction factor for $\lambda_r = 0$ and the upper limit, a_2, corresponds to the axial induction factor at $\lambda_r = \lambda$. Also, from Equation 3.3.16:

$$\lambda^2 = (1-a_2)(1-4a_2)^2/(1-3a_2) \tag{3.3.20}$$

Note that from Equation 3.3.16, $a_1 = 0.25$ gives λ_r a value of zero.

Equation 3.3.20 can be solved for the values of a_2 that correspond to operation at tip speed ratios of interest. Note also from Equation 3.3.20, $a_2 = 1/3$ is the upper limit of the axial induction factor, a, giving an infinitely large tip speed ratio.

The definite integral can be evaluated by changing variables: substituting x for $(1-3a)$ in Equation 3.3.19. The result is (see Eggleston and Stoddard, 1987):

$$C_{P,\max} = \frac{8}{729\lambda^2}\left\{\frac{64}{5}x^5 + 72x^4 + 124x^3 + 38x^2 - 63x - 12[\ln(x)] - 4x^{-1}\right\}_{x=(1-3a_2)}^{x=0.25} \tag{3.3.21}$$

Table 3.1 presents a summary of numerical values for $C_{P,\max}$ as a function of λ, with corresponding values for the axial induction factor at the tip, a_2.

The results of this analysis are graphically represented in Figure 3.5, which also shows the Betz limit of the ideal turbine based on the previous linear momentum analysis. The results show that, the higher the tip speed ratio, the greater the maximum theoretical C_P.

These equations can be used to look at the operation of an ideal wind turbine, assuming wake rotation. For example, Figure 3.6 shows the axial and angular induction factors for a turbine with a tip speed ratio of 7.5. It can be seen that the axial induction factors are close to the ideal of 1/3 until one gets near the hub. Angular induction factors are close to zero in the outer parts of the rotor, but increase significantly near the hub.

Table 3.1 Power coefficient, $C_{P,\max}$, as a function of tip speed ratio, λ; a_2 = axial induction factor when the tip speed ratio equals the local speed ratio

λ	a_2	$C_{P,\max}$
0.5	0.2983	0.289
1.0	0.3170	0.416
1.5	0.3245	0.477
2.0	0.3279	0.511
2.5	0.3297	0.533
5.0	0.3324	0.570
7.5	0.3329	0.581
10.0	0.3330	0.585

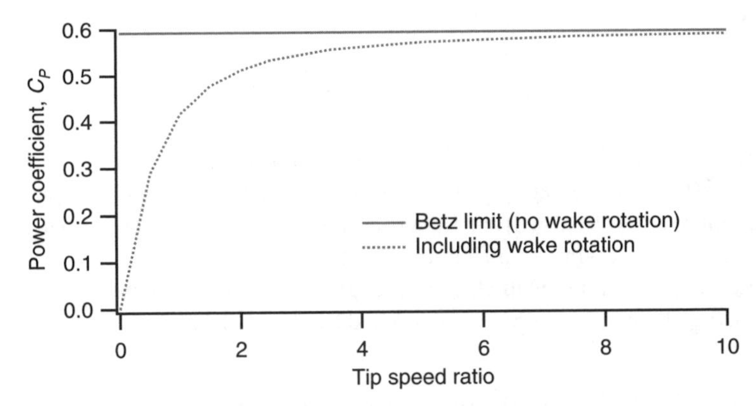

Figure 3.5 Theoretical maximum power coefficient as a function of tip speed ratio for an ideal horizontal axis wind turbine, with and without wake rotation

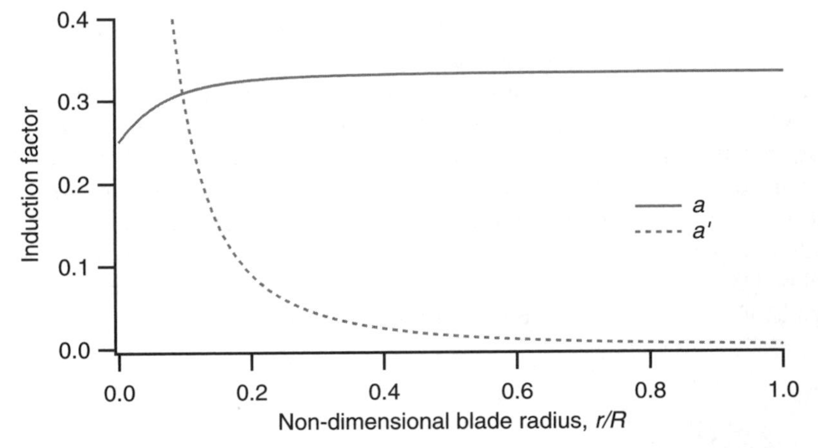

Figure 3.6 Induction factors for an ideal wind turbine with wake rotation; tip speed ratio, λ = 7.5; a, axial induction factor; a', angular induction factor; r, radius; R rotor radius

In the previous two sections, basic physics has been used to determine the nature of the airflow around a wind turbine and theoretical limits on the maximum power that can be extracted from the wind. The rest of the chapter explains how airfoils can be used to approach this theoretically achievable power extraction.

3.4 Airfoils and General Concepts of Aerodynamics

Wind turbine blades use airfoils to develop mechanical power. The cross-sections of wind turbine blades have the shape of airfoils. The width and length of the blade are functions of the desired aerodynamic performance, the maximum desired rotor power, the assumed airfoil properties and strength considerations. Before the details of wind turbine power production are explained, aerodynamic concepts related to airfoils need to be reviewed.

3.4.1 Airfoil terminology

A number of terms are used to characterize an airfoil, as shown in Figure 3.7. The mean camber line is the locus of points halfway between the upper and lower surfaces of the airfoil. The most forward and rearward points of the mean camber line are on the leading and trailing edges, respectively. The straight line connecting the leading and trailing edges is the chord line of the airfoil, and the distance from the leading to the trailing edge measured along the chord line is designated as the chord, c, of the airfoil. The camber is the distance between the mean camber line and the chord line, measured perpendicular to the chord line. The thickness is the distance between the upper and lower surfaces, also measured perpendicular to the chord line. Finally, the angle of attack, α, is defined as the angle between the relative wind and the chord line. Not shown in the figure is the span of the airfoil, which is the length of the airfoil perpendicular to its cross-section. The geometric parameters that have an effect on the aerodynamic performance of an airfoil include: the leading edge radius, mean camber line, maximum thickness and thickness distribution of the profile and the trailing edge angle.

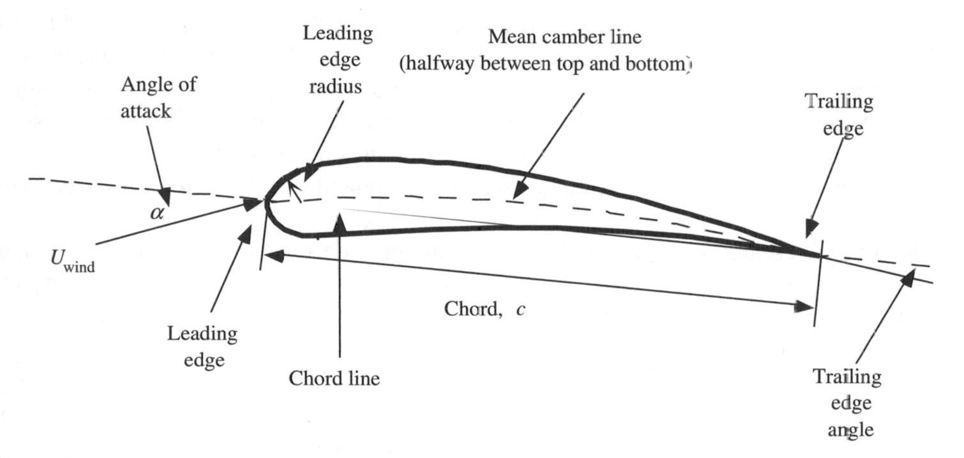

Figure 3.7 Airfoil nomenclature

There are many types of airfoils (see Abbott and Von Doenhoff, 1959, Althaus and Wortmann, 1981, Althaus, 1996, and Tangler, 1987). A few examples of ones that have been used in wind turbine designs are shown in Figure 3.8. The NACA 0012 is a 12% thick symmetric airfoil. The NACA 63(2)-215 is a 15% thick airfoil with a slight camber, and the LS(1)-0417 is a 17% thick airfoil with a larger camber.

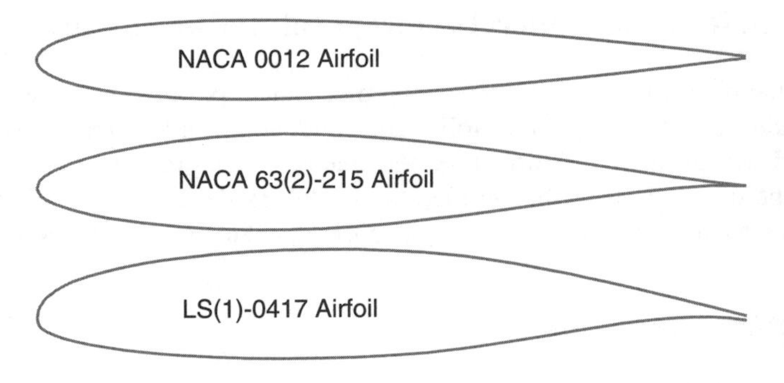

Figure 3.8 Sample airfoils

3.4.2 Lift, drag and non-dimensional parameters

Airflow over an airfoil produces a distribution of forces over the airfoil surface. The flow velocity over airfoils increases over the convex surface resulting in lower average pressure on the 'suction' side of the airfoil compared with the concave or 'pressure' side of the airfoil. Meanwhile, viscous friction between the air and the airfoil surface slows the airflow to some extent next to the surface.

As shown in Figure 3.9, the resultant of all of these pressure and friction forces is usually resolved into two forces and a moment that act along the chord at a distance of $c/4$ from the leading edge (at the 'quarter chord'):

- Lift force – defined to be perpendicular to direction of the oncoming airflow. The lift force is a consequence of the unequal pressure on the upper and lower airfoil surfaces
- Drag force – defined to be parallel to the direction of oncoming airflow. The drag force is due both to viscous friction forces at the surface of the airfoil and to unequal pressure on the airfoil surfaces facing toward and away from the oncoming flow
- Pitching moment – defined to be about an axis perpendicular to the airfoil cross-section

Theory and research have shown that many flow problems can be characterized by non-dimensional parameters. The most important non-dimensional parameter for defining the characteristics of fluid flow conditions is the Reynolds number. The Reynolds number, *Re*, is defined by:

$$Re = \frac{UL}{\nu} = \frac{\rho UL}{\mu} = \frac{\text{Inertial force}}{\text{Viscous force}} \tag{3.4.1}$$

where, ρ is the fluid density, μ is fluid viscosity, $\nu = \mu/\rho$ is the kinematic viscosity and U and L are a velocity and length that characterize the scale of the flow. These might be the free stream velocity and the chord length on an airfoil.

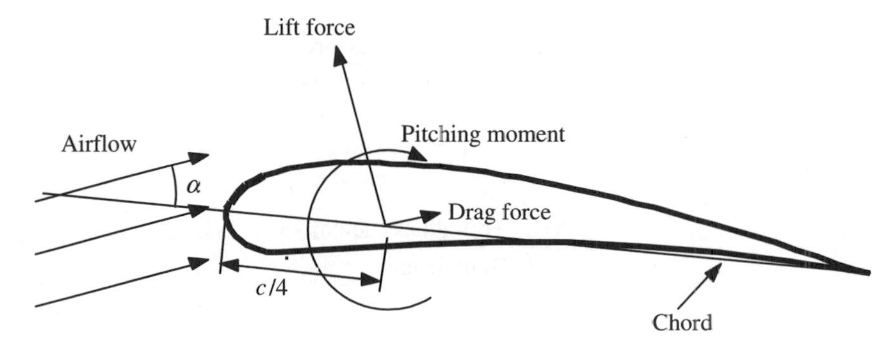

Figure 3.9 Drag and lift forces on stationary airfoil; α, angle of attack; c, chord

Force and moment coefficients, which are a function of Reynolds number, can be defined for two- or three-dimensional objects. Force and moment coefficients for flow around two-dimensional objects are usually designated with a lower case subscript, as in C_d for the two-dimensional drag coefficient. In that case, the forces measured are forces per unit span. Lift and drag coefficients that are measured for flow around three-dimensional objects are usually designated with an upper case subscript, as in C_D. Rotor design usually uses two-dimensional coefficients, determined for a range of angles of attack and Reynolds numbers, in wind tunnel tests. The two-dimensional lift coefficient is defined as:

$$C_l = \frac{L/l}{\frac{1}{2}\rho U^2 c} = \frac{\text{Lift force / unit length}}{\text{Dynamic force / unit length}} \tag{3.4.2}$$

The two-dimensional drag coefficient is defined as:

$$C_d = \frac{D/l}{\frac{1}{2}\rho U^2 c} = \frac{\text{Drag force / unit length}}{\text{Dynamic force / unit length}} \tag{3.4.3}$$

and the pitching moment coefficient is:

$$C_m = \frac{M}{\frac{1}{2}\rho U^2 Ac} = \frac{\text{Pitching moment}}{\text{Dynamic moment}} \tag{3.4.4}$$

where: ρ is the density of air, U is the velocity of undisturbed airflow, A is the projected airfoil area (cord $\times$ span), c is the airfoil chord length and l is the airfoil span.

Other dimensionless coefficients that are important for the analysis and design of wind turbines include the power and thrust coefficients and the tip speed ratio, mentioned above, the pressure coefficient, which is used to analyse airfoil flow:

$$C_p = \frac{p - p_\infty}{\frac{1}{2}\rho U^2} = \frac{\text{Static pressure}}{\text{Dynamic pressure}} \tag{3.4.5}$$

and surface roughness ratio:

$$\frac{\varepsilon}{L} = \frac{\text{Surface roughness height}}{\text{Body length}} \tag{3.4.6}$$

3.4.3 *Airfoil behavior*

It is useful to consider the behavior of a symmetric airfoil as a starting point for looking at airfoils for wind turbines. It can be shown (Currie, 1974) that, under ideal conditions, the theoretical lift coefficient of a flat plate is:

$$C_l = 2\pi \sin(\alpha) \tag{3.4.7}$$

and that, under similar ideal conditions, symmetric airfoils of finite thickness have similar theoretical lift coefficients. This would mean that lift coefficients would increase with increasing angles of attack and continue to increase until the angle of attack reaches 90 degrees. The behavior of real symmetric airfoils does indeed approximate this theoretical behavior at low angles of attack. For example, typical lift and drag coefficients for a symmetric airfoil, the NACA 0012 airfoil, the profile of which is shown in Figure 3.8, are shown in Figure 3.10 as a function of angle of attack and Reynolds number. The lift coefficient for a flat plate under ideal conditions is also shown for comparison.

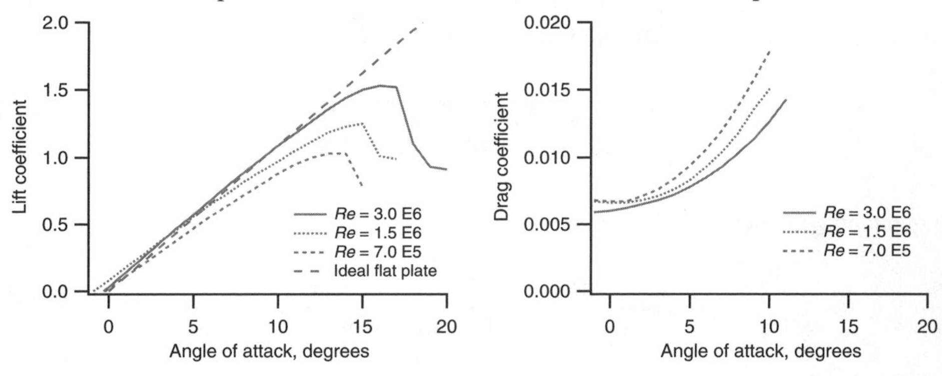

Figure 3.10 Lift and drag coefficients for the NACA 0012 symmetric airfoil (Miley, 1982); *Re*, Reynolds number

Note that, in spite of the very good correlation at low angles of attack, there are significant differences between actual airfoil operation and the theoretical performance at higher angles of attack. The differences are due primarily to the assumption, in the theoretical estimate of the lift coefficient, that air has no viscosity. Surface friction due to viscosity slows the airflow next to the airfoil surface, resulting in a separation of the flow from the surface at higher angles of attack and a rapid decrease in lift. This condition is referred to as stall and is discussed more below.

Airfoils for horizontal axis wind turbines (HAWTs) often are designed to be used at low angles of attack, where lift coefficients are fairly high and drag coefficients are fairly low. The lift coefficient of this symmetric airfoil is about zero at an angle of attack of zero and increases to over 1.0 before decreasing at higher angles of attack. The drag coefficient is usually much lower than the lift coefficient at low angles of attack. It increases at higher angles of attack.

Note, also, that there are significant differences in airfoil behavior at different Reynolds numbers. Rotor designers must make sure that appropriate Reynolds number data are available for the detailed analysis of a wind rotor system.

The lift coefficient at low angles of attack can be increased and drag can often be decreased by using a cambered airfoil (Eggleston and Stoddard, 1987). For example, the DU-93-W-210 airfoil is used in some European wind turbines. Its cross-sectional profile is shown in Figure 3.11. The lift, drag, and pitching moment coefficients for this same airfoil are shown in Figures 3.12 and 3.13 for a Reynolds number of 3 million.

In a manner similar to the behavior of the symmetric airfoil, the lift coefficient for the DU-93-W-210 airfoil increases to about 1.35 but then decreases as the angle of attack increases. Similarly, the drag coefficient starts out very low, but increases at about the same angle of attack that the lift coefficient decreases. This behavior is common to most airfoils. This cambered airfoil also has a nonzero lift coefficient at an angle of attack of zero.

Airfoil behavior can be categorized into three flow regimes: the attached flow regime, the high lift/stall development regime, and the flat plate/fully stalled regime (Spera, 1994). These flow regimes are described below and can be seen in the lift curves above and in Figure 3.14. Figure 3.14 shows the lift and drag coefficients for the S809 airfoil, which has also been used in wind turbines.

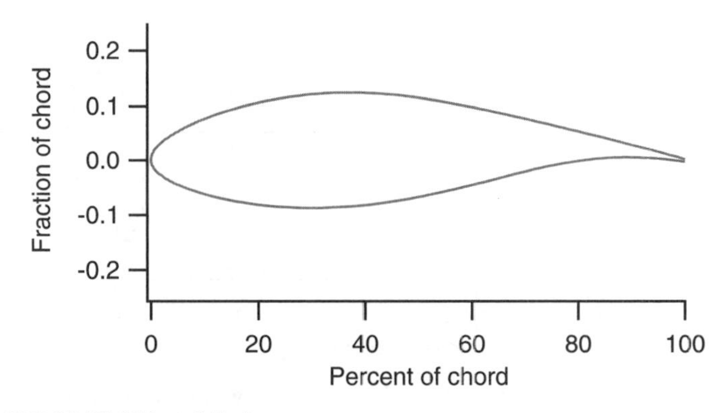

Figure 3.11 DU-93-W-210 airfoil shape

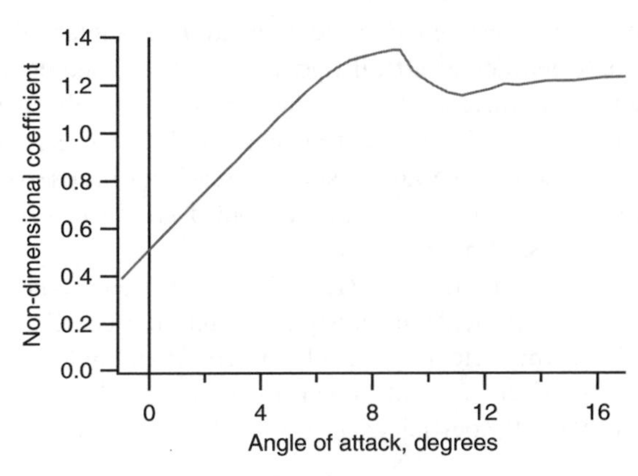

Figure 3.12 Lift coefficients for the DU-93-W-210 airfoil

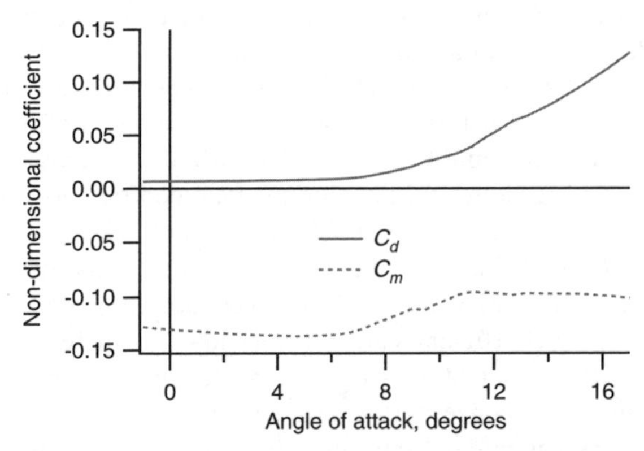

Figure 3.13 Drag and pitching moment coefficients for the DU-93-W-210 airfoil; C_d and C_m, respectively

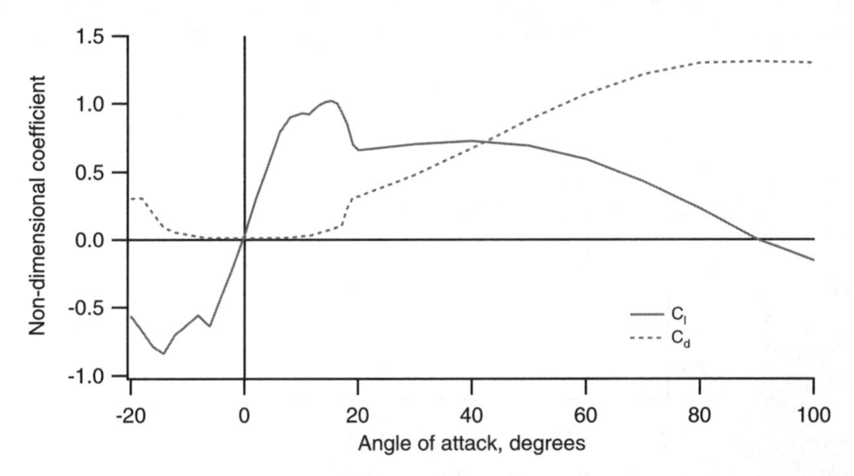

Figure 3.14 Lift and drag coefficients, C_l and C , respectively, for the S809 airfoil; Reynolds Number Re = 75,000,000.

3.4.3.1 Attached flow regime

At low angles of attack (up to about 7 degrees for the DU-93-W-210 airfoil), the flow is attached to the upper surface of the airfoil. In this attached flow regime, lift increases with the angle of attack and drag is relatively low.

3.4.3.2 High lift/stall development regime.

In the high lift/stall development regime (from about 7 to 11 degrees for the DU-93-W-210 airfoil), the lift coefficient peaks as the airfoil becomes increasingly stalled. Stall occurs when the angle of attack exceeds a certain critical value (say 10 to 16 degrees, depending on the Reynolds number) and separation of the boundary layer on the upper surface takes place, as shown in Figure 3.15. This causes a wake to form above the airfoil, which reduces lift and increases drag.

This condition can occur at certain blade locations or conditions of wind turbine operation. It is sometimes used to limit wind turbine power in high winds. For example, many wind turbine designs using fixed pitch blades rely on power regulation control via aerodynamic stall of the blades. That is, as wind speed increases, stall progresses outboard along the span of the blade (toward the tip) causing decreased lift and increased drag. In a well designed, stall regulated machine (see Chapter 7), this results in nearly constant power output as wind speeds increase.

3.4.3.3 Flat plate/fully stalled regime

In the flat plate/fully stalled regime, at larger angles of attack up to 90 degrees, the airfoil acts increasingly like a simple flat plate with approximately equal lift and drag coefficients at an angle of attack of 45 degrees and zero lift at 90 degrees.

3.4.4 Modelling of post-stall airfoil characteristics

Measured wind turbine airfoil data are used to design wind turbine blades. Wind turbine blades may often operate in the stalled region of operation, but data at high angles of attack are sometimes unavailable. Because of the similarity of stalled behavior to flat plate behavior, models have been developed to model lift and drag coefficients for stalled operation. Information on modelling post stall behavior of wind turbine airfoils can be found in Viterna and Corrigan (1981). Summaries of the Viterna and Corrigan model can be found in Spera (1994) and Eggleston and Stoddard (1987).

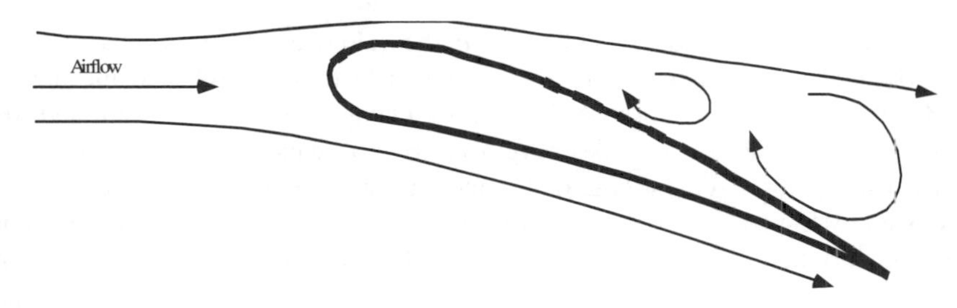

Figure 3.15 Illustration of airfoil stall

3.4.5 Airfoils for wind turbines

Modern HAWT blades have been designed using airfoil 'families' (Hansen and Butterfield, 1993). That is, the blade tip is designed using a thin airfoil, for high lift to drag ratio, and the root region is designed using a thick version of the same airfoil for structural support. Typical Reynolds numbers found in wind turbine operation are in the range of 500,000 and 10 million. A catalogue of airfoil data at these 'low' Reynolds numbers was compiled by Miley (1982). Generally, in the 1970s and early 1980s, wind turbine designers felt that minor differences in airfoil performance characteristics were far less important than optimising blade twist and taper. For this reason, little attention was paid to the task of airfoil selection. Thus, airfoils that were in use by the helicopter industry were chosen because the helicopter was viewed as a similar application. Aviation airfoils such as the NACA 44xx and NACA 230xx (Abbott and Von Doenhoff, 1959) were popular airfoil choices because they had high maximum lift coefficients, low pitching moment, and low minimum drag.

The NACA classification has 4, 5, and 6 series wing sections. For wind turbines, 4 digit series are generally used, for example: NACA 4415. The first integer indicates the maximum value of the mean camber line ordinate in percent of the chord. The second integer indicates the distance from the leading edge to maximum camber in tenths of the chord. The last two integers indicate the maximum section thickness in percent of the chord.

In the early 1980s wind turbine designers became aware of airfoils such as the NASA LS(1) MOD, and this airfoil was chosen by U.S. and British designers for its reduced sensitivity to leading edge roughness, compared to the NACA 44xx and NACA 230xx series airfoils (Tangler et al., 1990). Danish wind turbine designers began to use the NACA 63(2)-xx instead of the NACA 44xx airfoils for the same reasons.

The experience gained from operating these traditional airfoils has highlighted the shortcomings of such airfoils for wind turbine applications. Specifically, stall controlled HAWTs commonly produced too much power in high winds. This caused generator damage. Stall controlled turbines were operating with some part of the blade in deep stall for more than 50% of the life of the machine. Peak power and peak blade loads all occurred while the turbine was operating with most of the blade stalled, and predicted loads were only 50% to 70% of the measured loads. Designers began to realize that a better understanding of airfoil stall performance was important. In addition, leading edge roughness affected rotor performance. For example, with the early airfoil designs, when the blades accumulate insects and dirt along the leading edge, power output can drop as much as 40% of its clean value. Even the LS(1) MOD airfoils, which were designed to tolerate surface roughness, experienced a loss of power in the field once blades became soiled.

As a consequence of these experiences, airfoil selection criteria, and the designs for wind turbine airfoils and blades, have had to change to achieve high and reliable performance. New airfoil design codes have been used by wind energy engineers to design airfoils specifically for HAWTs. One of the most used codes in wind energy engineering was developed by Eppler and Somers (1980). This code combines a variety of techniques to optimize boundary layer characteristics and airfoil shapes to achieve specified performance criteria.

Using the Eppler code, researchers at the National Renewable Energy Laboratory (Spera, 1994) have developed 'special purpose families' of airfoils for three different classes of wind turbines (the SERI designated classification of airfoils). As reported by Tangler et al. (1990) these S-Series airfoils have been tested on 8 m long blades and have been shown to be relatively insensitive to leading edge surface roughness and contributed to increased annual energy production by allowing a larger rotor diameter without increased peak power. These airfoils are now used on some commercial wind turbines.

3.4.6 Lift versus drag machines

Wind energy converters that have been built over the centuries can be divided into lift machines and drag machines. Lift machines use lift forces to generate power. Drag machines use drag forces. The horizontal axis wind turbines that are the primary topic of this book (and almost all modern wind turbines) are lift machines, but many useful drag machines have been developed. The advantages of lift over drag machines are described in this section through the use of a few simple examples.

A simple drag machine, shown in Figure 3.16, was used in the Middle East over a thousand years ago (Gasch, 1996). It includes a vertical axis rotor consisting of flat surfaces in which half of the rotor is shielded from the wind. The simplified model on the right in Figure 3.16 is used to analyse the performance of this drag machine.

Figure 3.16 Simple drag machine and model; U, velocity of the undisturbed airflow; Ω, angular velocity of wind turbine rotor; r, radius

The drag force, F_D, is a function of the relative wind velocity at the rotor surface (the difference between the wind speed, U, and the speed of the surface is Ωr):

$$F_D = C_D\left[\tfrac{1}{2}\rho(U-\Omega r)^2 A\right] \tag{3.4.8}$$

where A is the drag surface area and where the three-dimensional drag coefficient, C_D, for a square plate is assumed to be 1.1.

The rotor power is the product of the drag force and rotational speed of the rotor surfaces:

$$P = C_D\left[\tfrac{1}{2}\rho A(U-\Omega r)^2\right]\Omega r = \left(\rho A U^3\right)\left[\tfrac{1}{2}C_D\lambda(1-\lambda)^2\right] \tag{3.4.9}$$

The power coefficient, shown in Figure 3.17, is a function of λ, the ratio of the surface velocity to the wind speed, and is based on an assumed total machine area of 2A:

$$C_P = \left[\tfrac{1}{2} C_D \lambda (1-\lambda)^2\right] \tag{3.4.10}$$

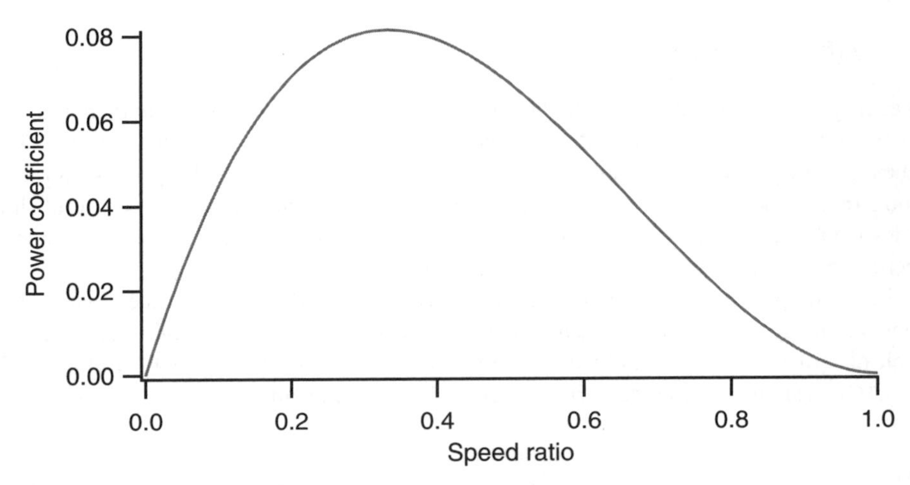

Figure 3.17 Power coefficient of flat plate drag machine

The power coefficient is zero at speed ratios of zero (no motion) and 1.0 (the speed at which the surface moves at the wind speed and experiences no drag force). The peak power coefficient of 0.08 occurs as a speed ratio of 1/3. This power coefficient is significantly lower than the Betz limit of 0.593. This example also illustrates one of the primary disadvantages of a pure drag machine: the rotor surface cannot move faster than the wind speed. Thus, the wind velocity relative to the power producing surfaces of the machine, U_{rel}, is limited to the free stream velocity:

$$U_{rel} = U(1-\lambda) \qquad \lambda < 1 \tag{3.4.11}$$

The forces in lift machines are also a function of the relative wind velocity and of the lift coefficient:

$$F_L = C_L \left(\tfrac{1}{2} \rho A U_{rel}^2\right) \tag{3.4.12}$$

The maximum lift and drag coefficients of airfoils are of similar magnitude. One significant difference in the performance between lift and drag machines is the much higher relative wind velocities that can be achieved with lift machines. Relative velocities are always greater than the free stream wind speed, sometimes by an order of magnitude. As illustrated in Figure 3.18, the relative wind velocity at the airfoil of a lift machine is:

$$U_{rel} = \sqrt{U^2 + (\Omega r)^2} = U\sqrt{1+\lambda^2} \tag{3.4.13}$$

With speed ratios of up to 10, and forces that are a function of the square of the relative speed, it can be seen that the forces that can be developed by a lift machine are significantly greater than those achievable with a drag machine with the same surface area. The larger forces allow for much greater power coefficients.

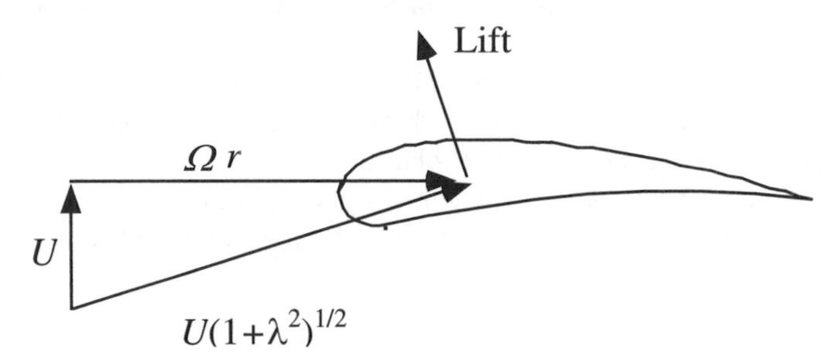

Figure 3.18 Relative velocity of a lift machine; for notation, see Figure 3.16

It should be pointed out that some drag-based machines, such as the Savonius rotor, may achieve maximum power coefficients of greater than 0.2 and may have tip speed ratios greater than 1.0. This is primarily due to the lift developed when the rotor surfaces turn out of the wind as the rotor rotates (Wilson et al., 1976). Thus, the Savonius rotor and many other drag-based devices may also experience some lift forces.

3.5 Momentum Theory and Blade Element Theory

3.5.1 Overview

The calculation of rotor performance and aerodynamically effective blade shapes is presented in this and the following sections. The analysis builds on the foundation presented in the previous sections. A wind turbine rotor consists of airfoils that generate lift by virtue of the pressure difference across the airfoil, producing the same step change in pressure seen in the actuator disc type of analysis. In Sections 3.2 and 3.3, the flow field around a wind turbine rotor, represented by an actuator disc, was determined using the conservation of linear and angular momentum. That flow field, characterized by axial and angular induction factors that are a function of the rotor power extraction and thrust, will be used to define the airflow at the rotor airfoils. The geometry of the rotor and the lift and drag characteristics of the rotor airfoils, described in Section 3.4, can then be used to determine either the rotor shape, if certain performance parameters are known, or rotor performance, if the blade shape has already been defined.

The analysis here uses momentum theory and blade element theory. Momentum theory refers to a control volume analysis of the forces at the blade based on the conservation of linear and angular momentum. Blade element theory refers to an analysis of forces at a section of the blade, as a function of blade geometry. The results of these approaches can be combined into what is known as strip theory or blade element momentum (BEM) theory. This theory can be used to relate blade shape to the rotor's ability to extract power from the wind. The analysis in this and the following sections covers:

- Momentum and blade element theory
- The simplest 'optimum' blade design with an infinite number of blades and no wake rotation
- Performance characteristics (forces, rotor airflow characteristics, power coefficient) for a general blade design of known chord and twist distribution, including wake rotation, drag, and losses due to a finite number of blades
- A simple 'optimum' blade design including wake rotation and an infinite number of blades. This blade design can be used as the start for a general blade design analysis

3.5.2 Momentum theory

The forces on a wind turbine blade and flow conditions at the blades can be derived by considering conservation of momentum since force is the rate of change of momentum. The necessary equations have already been developed in the derivation of the performance of an ideal wind turbine including wake rotation. The present analysis is based on the annular control volume shown in Figure 3.4. In this analysis, the axial and angular induction factors are assumed to be functions of the radius, r.

The result, from Section 3.3, of applying the conservation of linear momentum to the control volume of radius r and thickness $\mathrm{d}r$ (Equation 3.3.5) is an expression for the differential contribution to the thrust:

$$\mathrm{d}T = \rho U^2 4a(1-a)\pi r \mathrm{d}r \tag{3.5.1}$$

Similarly, from the conservation of angular momentum equation, Equation 3.3.10, the differential torque, Q, imparted to the blades (and equally, but oppositely, to the air) is:

$$\mathrm{d}Q = 4a'(1-a)\rho U \pi r^3 \Omega \, \mathrm{d}r \tag{3.5.2}$$

Thus, from momentum theory one gets two equations, Equations 3.5.1 and 3.5.2, that define the thrust and torque on an annular section of the rotor as a function of the axial and angular induction factors (i.e. of the flow conditions).

3.5.3 Blade element theory

The forces on the blades of a wind turbine can also be expressed as a function of lift and drag coefficients and the angle of attack. As shown in Figure 3.19, for this analysis, the blade is assumed to be divided into *N* sections (or elements). Furthermore, the following assumptions are made:

- There is no aerodynamic interaction between elements
- The forces on the blades are determined solely by the lift and drag characteristics of the airfoil shape of the blades

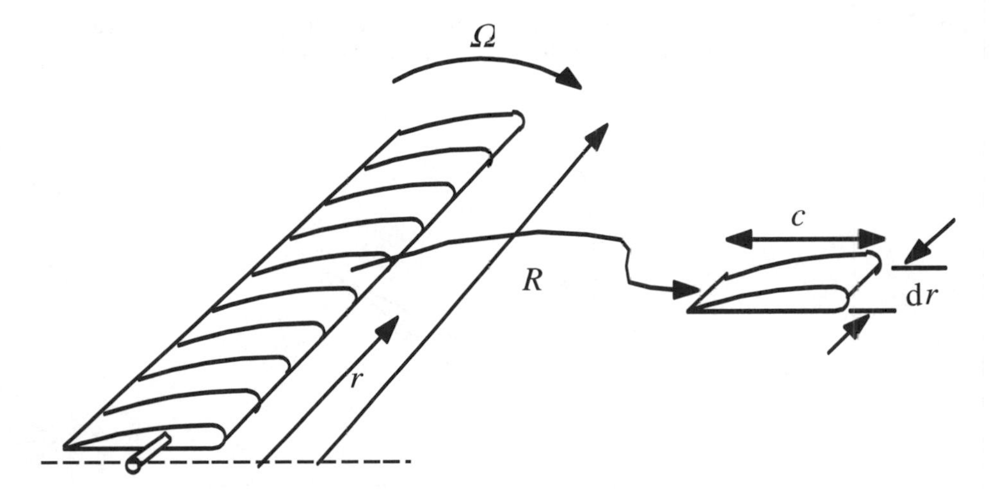

Figure 3.19 Schematic of blade elements; c, airfoil chord length; $\mathrm{d}r$, radial length of element; r, radius; R, rotor radius; Ω, angular velocity of rotor

In analysing the forces on the blade section, it must noted that the lift and drag forces are perpendicular and parallel, respectively, to an effective, or relative, wind. The relative wind is the vector sum of the wind velocity at the rotor, $U(1-a)$, and the wind velocity due to rotation of the blade. This rotational component is the vector sum of the blade section velocity, Ωr, and the induced angular velocity at the blades from conservation of angular momentum, $\omega r/2$, or

$$\Omega r+(\omega/2)\,\mathrm{r}=\Omega r+\Omega a' r=\Omega r\,(1+a') \tag{3.5.3}$$

The overall flow situation is shown in Figure 3.20 and the relationships of the various forces, angles, and velocities at the blade, looking down from the blade tip, is shown in Figure 3.21.

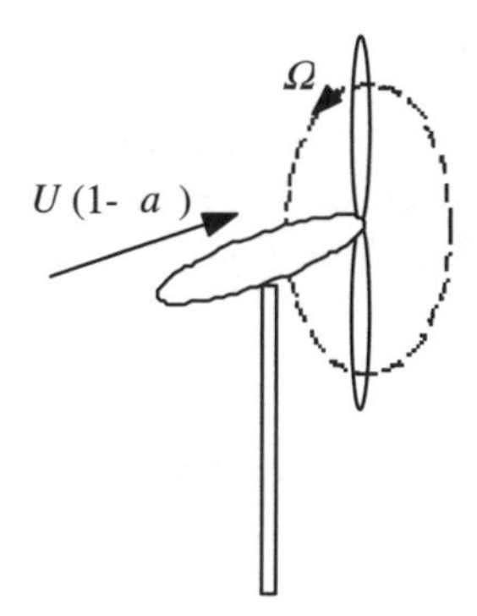

Figure 3.20 Overall geometry for a downwind horizontal axis wind turbine analysis; a, axial induction factor; U, velocity of undisturbed flow; Ω, angular velocity of rotor

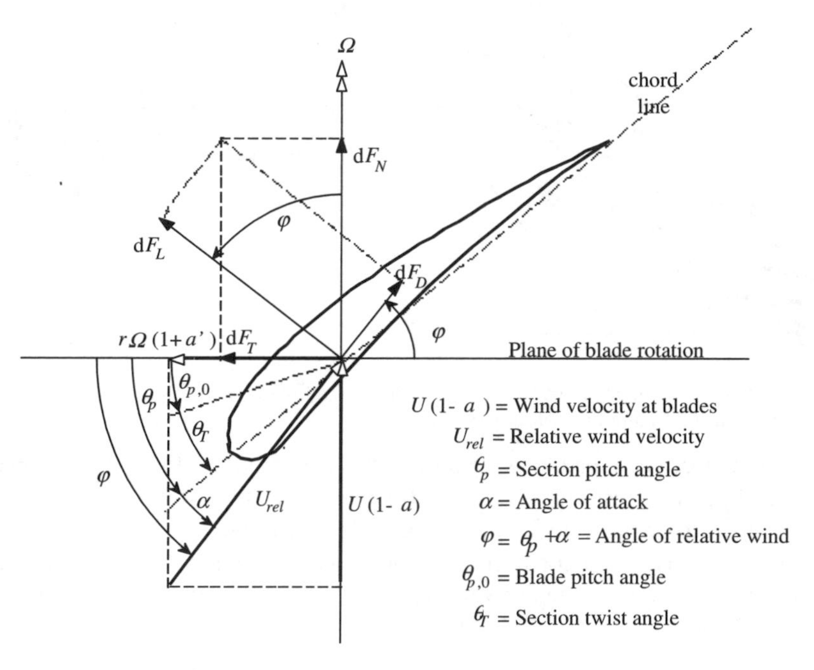

Figure 3.21 Blade geometry for analysis of a horizontal axis wind turbine; for definition of variables, see text

Here θ_p is the section pitch angle, which is the angle between the chord line and the plane of rotation, $\theta_{p,0}$ is the blade pitch angle at the tip, θ_T is the blade twist angle, α is the angle of attack (the angle between the chord line and the relative wind), φ is the angle of relative wind, dF_L is the incremental lift force, dF_D is the incremental drag force, dF_N is the incremental force normal to the plane of rotation (this contributes to thrust), and dF_T is the incremental force tangential to the circle swept by the rotor. This is the force creating useful torque. Finally, U_{rel} is the relative wind velocity.

Note also that, here, the blade twist angle, θ_T, is defined relative to the blade tip (it could be defined otherwise). Therefore:

$$\theta_T = \theta_p - \theta_{p,0} \tag{3.5.4}$$

where $\theta_{p,0}$ is the blade pitch angle at the tip. The twist angle is, of course, a function of the blade geometry, whereas θ_p changes if the position of the blade, $\theta_{p,0}$, is changed. Note, also, that the angle of the relative wind is the sum of the section pitch angle and the angle of attack:

$$\varphi = \theta_p + \alpha \tag{3.5.5}$$

From the figure, one can determine the following relationships:

$$\tan\varphi = \frac{U(1-a)}{\Omega r(1+a')} = \frac{1-a}{(1+a')\lambda_r} \tag{3.5.6}$$

$$U_{rel} = U(1-a)/\sin\varphi \tag{3.5.7}$$

$$\mathrm{d}F_L = C_l \tfrac{1}{2}\rho U_{rel}^2 c\,\mathrm{d}r \tag{3.5.8}$$

$$\mathrm{d}F_D = C_d \tfrac{1}{2}\rho U_{rel}^2 c\,\mathrm{d}r \tag{3.5.9}$$

$$\mathrm{d}F_N = \mathrm{d}F_L \cos\varphi + \mathrm{d}F_D \sin\varphi \tag{3.5.10}$$

$$\mathrm{d}F_T = \mathrm{d}F_L \sin\varphi - \mathrm{d}F_D \cos\varphi \tag{3.5.11}$$

If the rotor has B blades, the total normal force on the section at a distance, r, from the centre is:

$$\mathrm{d}F_N = B\tfrac{1}{2}\rho U_{rel}^2 (C_l \cos\varphi + C_d \sin\varphi) c\mathrm{d}r \tag{3.5.12}$$

The differential torque due to the tangential force operating at a distance, r, from the center is given by:

$$dQ = B\, r\, dF_T \tag{3.5.13}$$

so

$$dQ = B \tfrac{1}{2} \rho U_{rel}^2 (C_l \sin\varphi - C_d \cos\varphi) c r dr \tag{3.5.14}$$

Note that the effect of drag is to decrease torque and hence power, but to increase the thrust loading.

Thus, from blade element theory, one also obtains two equations (Equations 3.5.12 and 3.5.14) that define the normal force (thrust) and tangential force (torque) on the annular rotor section as a function of the flow angles at the blades and airfoil characteristics. These equations will be used below, with additional assumptions or equations, to determine ideal blade shapes for optimum performance and to determine rotor performance for any arbitrary blade shape.

3.6 Blade Shape for Ideal Rotor without Wake Rotation

As mentioned above, one can combine the momentum theory relations with those from blade element theory to relate blade shape to performance. Because the algebra can get complex, a simple, but useful example will be presented here to illustrate the method.

In the first example of this chapter, the maximum possible power coefficient from a wind turbine, assuming no wake rotation or drag, was determined to occur with an axial induction factor of 1/3. If the same simplifying assumptions are applied to the equations of momentum and blade element theory, the analysis becomes simple enough that an ideal blade shape can be determined. The blade shape approximates one that would provide maximum power at the design tip speed ratio of a real wind turbine.

In this analysis, the following assumptions will be made:

- There is no wake rotation; thus $a' = 0$
- There is no drag; thus $C_d = 0$.
- There are no losses from a finite number of blades
- For the Betz optimum rotor, $a = 1/3$ in each annular stream tube

First, a design tip speed ratio, λ, the desired number of blades, B, the radius, R, and an airfoil with known lift and drag coefficients as a function of angle of attack need to be chosen. An angle of attack (and, thus, a lift coefficient at which the airfoil will operate) is also chosen. This angle of attack should be selected where C_d / C_l is minimum in order to most closely approximate the assumption that $C_d = 0$. These choices allow the twist and chord distribution of a blade that would provide Betz limit power production (given the input assumptions) to be determined. With the assumption that $a = 1/3$, one gets from momentum theory (Equation 3.5.1):

$$dT = \rho U^2 4\left(\tfrac{1}{3}\right)\left(1 - \tfrac{1}{3}\right)\pi r\, dr = \rho U^2\, \tfrac{8}{9} \pi r\, dr \tag{3.6.1}$$

and from blade element theory (Equation 3.5.12, with $C_d = 0$):

$$dF_N = B\tfrac{1}{2}\rho U_{rel}^2 (C_l \cos\varphi) c dr \tag{3.6.2}$$

A third equation, Equation 3.5.7, can be used to express U_{rel} in terms of other known variables:

$$U_{rel} = U(1-a)/\sin\varphi = \frac{2U}{3\ \sin\varphi} \tag{3.6.3}$$

BEM theory or strip theory refers to the determination of wind turbine blade performance by combining the equations of momentum theory and blade element theory. In this case, equating Equations 3.6.1 and 3.6.2 and using Equation 3.6.3, yields:

$$\frac{C_l B c}{4\pi r} = \tan\varphi \sin\varphi \tag{3.6.4}$$

A fourth equation, Equation 3.5.6, which relates a, a' and φ based on geometrical considerations, can be used to solve for the blade shape. Equation 3.5.6, with $a' = 0$ and $a = 1/3$, becomes

$$\tan\varphi = \frac{2}{3\lambda_r} \tag{3.6.5}$$

Therefore

$$\frac{C_l B c}{4\pi r} = \left(\frac{2}{3\lambda_r}\right)\sin\varphi \tag{3.6.6}$$

Rearranging, and noting that $\lambda_r = \lambda(r/R)$, one can determine the angle of the relative wind and the chord of the blade for each section of the ideal rotor:

$$\varphi = \tan^{-1}\left(\frac{2}{3\lambda_r}\right) \tag{3.6.7}$$

$$c = \frac{8\pi r \sin\varphi}{3BC_l\lambda_r} \tag{3.6.8}$$

These relations can be used to find the chord and twist distribution of the Betz optimum blade. As an example, suppose: $\lambda = 7$, $R = 5$ m, the airfoil has a lift coefficient of $C_l = 1$, C_d/C_l has a minimum at $\alpha = 7^\circ$ and, finally, that there are three blades, so $B = 3$. Then, from Equations 3.6.7 and 3.6.8 we get the results shown in Table 3.2. In this process, Equations 3.5.4 and 3.5.5 are also used to relate the various blade angles to each other (see Figure 3.21). The twist angle is assumed to start at 0 at the tip. The chord and twist of this blade are illustrated in Figures 3.22 and 3.23.

It can be seen that blades designed for optimum power production have an increasingly large chord and twist angle as one gets closer to the blade root. One consideration in blade design is the cost and difficulty of fabricating the blade. An optimum blade would be very difficult to manufacture at a reasonable cost, but the design provides insight into the blade shape that might be desired for a wind turbine.

Table 3.2 Twist and chord distribution for a Betz optimum blade; r/R, fraction of rotor radius

r/R	Chord, m	Twist Angle (deg)	Angle of Rel. Wind (deg)	Section Pitch (deg)
0.1	1.375	38.2	43.6	36.6
0.2	0.858	20.0	25.5	18.5
0.3	0.604	12.2	17.6	10.6
0.4	0.462	8.0	13.4	6.4
0.5	0.373	5.3	10.8	3.8
0.6	0.313	3.6	9.0	2.0
0.7	0.269	2.3	7.7	0.7
0.8	0.236	1.3	6.8	-0.2
0.9	0.210	0.6	6.0	-1.0
1	0.189	0	5.4	-1.6

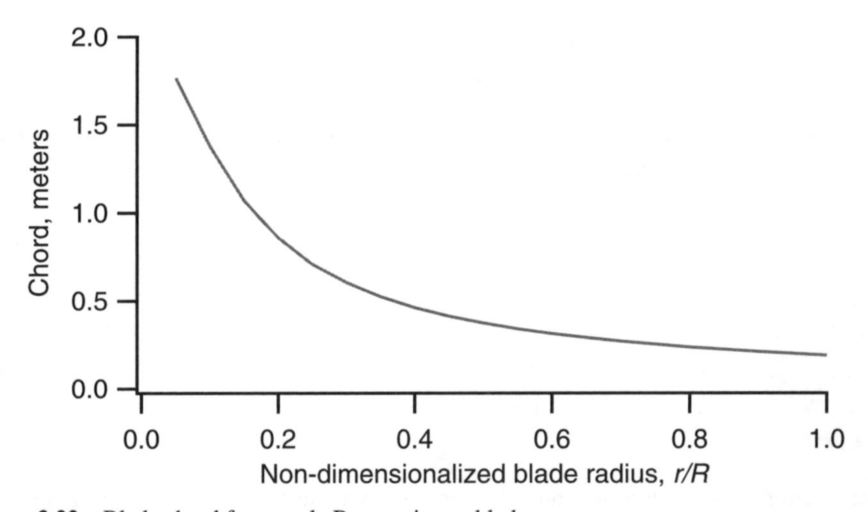

Figure 3.22 Blade chord for sample Betz optimum blade

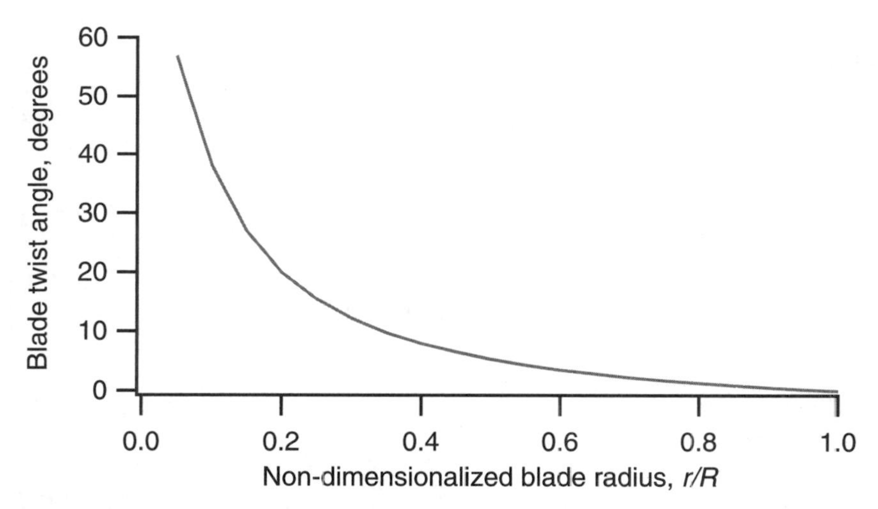

Figure 3.23 Blade twist angle for sample Betz optimum blade

3.7 General Rotor Blade Shape Performance Prediction

In general, a rotor is not of the optimum shape because of fabrication difficulties. Furthermore, when an 'optimum' blade is run at a different tip speed ratio than the one for which it is designed, it is no longer 'optimum'. Thus, blade shapes must be designed for easy fabrication and for overall performance over the range of wind and rotor speeds that they will encounter. In considering non-optimum blades, one generally uses an iterative approach. That is, one can assume a blade shape and predict its performance, try another shape and repeat the prediction until a suitable blade has been chosen.

So far, the blade shape for an ideal rotor without wake rotation has been considered. In this section, the analysis of arbitrary blade shapes is considered. The analysis includes wake rotation, drag, losses from a finite number of blades and off-design performance. In subsequent sections these methods will be used to determine an optimum blade shape, including wake rotation, and as part of a complete rotor design procedure.

3.7.1 Strip theory for a generalized rotor, including wake rotation

The analysis of a blade that includes wake rotation builds on the analysis used in the previous section. Here we also consider the non-linear range of the lift coefficient versus angle of attack curve, i.e. stall. The analysis starts with the four equations derived from momentum and blade element theories. In this analysis, it is assumed that the chord and twist distributions of the blade are known. The angle of attack is not known, but additional relationships can be used to solve for the angle of attack and performance of the blade.

The forces and moments derived from momentum theory and blade element theory must be equal. Equating these, one can derive the flow conditions for a turbine design.

3.7.1.1 Momentum theory

From axial momentum:

$$dT = \rho U^2 4a(1-a)\pi r \, dr \tag{3.5.1}$$

From angular momentum:

$$dQ = 4a'(1-a)\rho U \pi r^3 \Omega \, dr \tag{3.5.2}$$

3.7.1.2 Blade element theory

From blade element theory:

$$dF_N = B \tfrac{1}{2} \rho U_{rel}^2 (C_l \cos\varphi + C_d \sin\varphi) c dr \tag{3.5.12}$$

$$dQ = B \tfrac{1}{2} \rho U_{rel}^2 (C_l \sin\varphi - C_d \cos\varphi) c r dr \tag{3.5.14}$$

where the thrust, dT, is the same force as the normal force, dF_N. The relative velocity can be expressed as a function of the free stream wind using Equation 3.5.7. Thus, Equations 3.5.12 and 3.5.14 from blade element theory can be written as:

$$dF_N = \sigma' \pi \rho \frac{U^2(1-a)^2}{\sin^2\varphi} (C_l \cos\varphi + C_d \sin\varphi) r \, dr \tag{3.7.1}$$

$$dQ = \sigma' \pi \rho \frac{U^2(1-a)^2}{\sin^2\varphi} (C_l \sin\varphi - C_d \cos\varphi) r^2 dr \tag{3.7.2}$$

where σ' is the local solidity, defined by:

$$\sigma' = Bc/2\pi r \tag{3.7.3}$$

3.7.1.3 Blade element momentum theory

In the calculation of induction factors, a and a', accepted practice is to set C_d equal to zero (see Wilson and Lissaman, 1974). For airfoils with low drag coefficients, this simplification introduces negligible errors. So, when the torque equations from momentum and blade element theory are equated (Equations 3.5.2 and 3.7.2), with $C_d = 0$, one gets:

$$a'/(1-a)=\sigma' C_l/(4\lambda_r \sin\varphi) \tag{3.7.4}$$

By equating the normal force equations from momentum and blade element theory (Equations 3.5.1 and 3.7.1), one obtains:

$$a/(1-a)=\sigma' C_l \cos\varphi/(4\sin^2\varphi) \tag{3.7.5}$$

After some algebraic manipulation using Equation 3.5.6 (which relates a, a', • and $\bullet_r$, based on geometric considerations) and Equations 3.7.4 and 3.7.5, the following useful relationships result:

$$C_l = 4\sin\varphi \frac{(\cos\varphi - \lambda_r \sin\varphi)}{\sigma'(\sin\varphi + \lambda_r \cos\varphi)} \tag{3.7.6}$$

$$a'/(1+a')=\sigma' C_l/(4\cos\varphi) \tag{3.7.7}$$

Other useful relationships that may be derived include:

$$a/a' = \lambda_r/\tan\varphi \tag{3.7.8}$$

$$a = 1/\left[1+4\sin^2\varphi/(\sigma' C_l \cos\varphi)\right] \tag{3.7.9}$$

$$a' = 1/[(4\cos\varphi/(\sigma' C_l))-1] \tag{3.7.10}$$

3.7.1.4 Solution methods

Two solution methods will be proposed using these equations to determine the flow conditions and forces at each blade section. The first one uses the measured airfoil characteristics and the BEM equations to solve directly for C_l and a. This method can be solved numerically, but it lends itself to a graphical solution that clearly shows the flow conditions at the blade and the existence of multiple solutions (see Section 3.7.4). The second solution is an iterative numerical approach that is most easily extended for flow conditions with large axial induction factors.

Method 1 – Solving for C_l and α Since $\varphi = \alpha + \theta_p$, for a given blade geometry and operating conditions, there are two unknowns in Equation 3.7.6, C_l and α at each section. In order to find these values, one can use the empirical C_l vs. α curves for the chosen airfoil (see de Vries, 1979). One then finds the C_l and α from the empirical data that satisfy Equation 3.7.6. This can be done either numerically, or graphically (as shown in

Figure 3.24). Once C_l and α have been found, a' and a can be determined from any two of Equations 3.7.7 through 3.7.10. It should be verified that the axial induction factor at the intersection point of the curves is less than 0.5 to ensure that the result is valid.

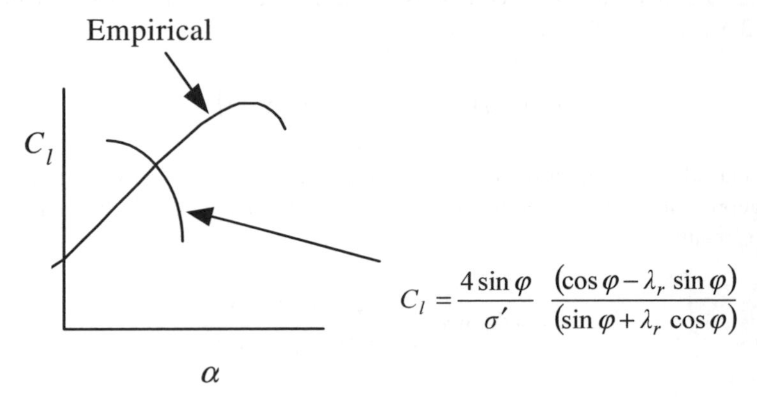

Figure 3.24 Angle of attack – graphical solution method; C_l, two-dimensional lift coefficient; α, angle of attack; λ_r, local speed ratio; φ, angle of relative wind; σ', local rotor solidity

Method 2 – Iterative solution for a and a' Another equivalent solution method starts with guesses for a and a', from which flow conditions and new inductions factors are calculated. Specifically:

1. Guess values of a and a'
2. Calculate the angle of the relative wind from Equation 3.5.6
3. Calculate the angle of attack from $\theta = \alpha + \theta_p$ and then C_l and C_d
4. Update a and a' from Equations 3.7.4 and 3.7.5 or 3.7.9 and 3.7.10

The process is then repeated until the newly calculated induction factors are within some acceptable tolerance of the previous ones. This method is especially useful for highly loaded rotor conditions, as described in Section 3.7.4.3.

3.7.2 Calculation of power coefficient

Once a has been obtained from each section, the overall rotor power coefficient may be calculated from the following equation (Wilson and Lissaman, 1974):

$$C_P = (8/\lambda^2)\int_{\lambda_h}^{\lambda} \lambda_r^3 a'(1-a)[1-(C_d/C_l)\cot\varphi]\mathrm{d}\lambda_r \tag{3.7.11}$$

where λ_h is the local speed ratio at the hub. Equivalently (de Vries, 1979):

$$C_P = (8/\lambda^2)\int_{\lambda_h}^{\lambda} \sin^2\varphi(\cos\varphi - \lambda_r \sin\varphi)(\sin\varphi + \lambda_r \cos\varphi)[1-(C_d/C_l)\cot\varphi]\lambda_r^2\mathrm{d}\lambda_r \tag{3.7.12}$$

Usually, these equations are solved numerically, as will be discussed later. Note that even though the axial induction factors were determined assuming the $C_d = 0$, the drag is included here in the power coefficient calculation.

The derivation of Equation 3.7.11 follows. The power contribution from each annulus is

$$dP = \Omega \, dQ \tag{3.7.13}$$

where Ω is the rotor rotational speed. The total power from the rotor is:

$$P = \int_{r_h}^{R} dP = \int_{r_h}^{R} \Omega \, dQ \tag{3.7.14}$$

where r_h is the rotor radius at the hub of the blade. The power coefficient, C_P, is

$$C_P = \frac{P}{P_{wind}} = \frac{\int_{r_h}^{R} \Omega \, dQ}{\frac{1}{2}\rho \pi R^2 U^3} \tag{3.7.15}$$

Using the expression for the differential torque from Equation 3.7.2 and the definition of the local tip speed ratio (Equation 3.3.8):

$$C_P = \frac{2}{\lambda^2} \int_{\lambda_h}^{\lambda} \sigma' C_l (1-a)^2 (1/\sin\varphi)[1-(C_d/C_l)\cot\varphi]\lambda_r^{\,2} d\lambda_r \tag{3.7.16}$$

where λ_h is the local tip speed ratio at the hub. From Equations 3.7.5 and 3.7.8:

$$\sigma' C_l (1-a) = 4a \sin^2\varphi / \cos\varphi \tag{3.7.17}$$

$$a \tan\varphi = a' \lambda_r \tag{3.7.18}$$

Substituting these into Equation 3.7.16, one gets the desired result, that is, Equation 3.7.11:

$$C_P = (8/\lambda^2) \int_{\lambda_h}^{\lambda} \lambda_r^{\,3} a'(1-a)[1-(C_d/C_l)\cot\varphi] \, d\lambda_r \tag{3.7.11}$$

Note that when $C_d = 0$, this equation for C_P is the same as the one derived from momentum theory, including wake rotation, Equation (3.3.14). The derivation of Equation 3.7.12 is algebraically complex and is left as an exercise for the interested reader.

3.7.3 *Tip loss: effect on power coefficient of number of blades*

Because the pressure on the suction side of a blade is lower than that on the pressure side, air tends to flow around the tip from the lower to upper surface, reducing lift and hence power production near the tip. This effect is most noticeable with fewer, wider blades.

A number of methods have been suggested for including the effect of the tip loss. The most straightforward approach to use is one developed by Prandtl (see de Vries, 1979). According to this method, a correction factor, F, must be introduced into the previously discussed equations. This correction factor is a function of the number of blades, the angle of relative wind, and the position on the blade. Based on Prandtl's method:

$$F = (2/\pi)\cos^{-1}\left[\exp\left(-\left\{\frac{(B/2)[1-(r/R)]}{(r/R)\sin\varphi}\right\}\right)\right] \tag{3.7.19}$$

where the angle resulting from the inverse cosine function is assumed to be in radians. Note, also, that F is always between 0 and 1. This tip loss correction factor characterizes the reduction in the forces at a radius r along the blade that is due to the tip loss at the end of the blade.

The tip loss correction factor affects the forces derived from momentum theory. Thus Equations 3.5.1 and 3.5.2 become:

$$dT = F\rho U^2 4a(1-a)\pi r\,\mathrm{dr} \tag{3.5.1a}$$

and

$$dQ = 4Fa'(1-a)\rho U\pi r^3\Omega\,\mathrm{dr} \tag{3.5.2a}$$

Note that in this subsection, modifications of previous equations use the original equation numbers followed by an "a" for easy comparison with the original equations.

Equations 3.5.4 through 3.5.14 are all based on the definition of the forces used in the blade element theory and remain unchanged. When the forces from momentum theory and from blade element theory are set equal, using the methods of strip theory, the derivation of the flow conditions is changed, however. Carrying the tip loss factor through the calculations, one finds these changes:

$$a'/(1-a) = \sigma' C_l / 4F\lambda_r \sin\varphi \tag{3.7.4a}$$

$$a/(1-a) = \sigma' C_l \cos\varphi / 4F\sin^2\varphi \tag{3.7.5a}$$

$$C_l = 4F \sin\varphi \frac{(\cos\varphi - \lambda_r \sin\varphi)}{\sigma' (\sin\varphi + \lambda_r \cos\varphi)} \tag{3.7.6a}$$

$$a'/(1+a') = \sigma' C_l / 4F \cos\varphi \tag{3.7.7a}$$

$$a = 1/\left[1 + 4F \sin^2\varphi / (\sigma' C_l \cos\varphi)\right] \tag{3.7.9a}$$

$$a' = 1/[(4F \cos\varphi/(\sigma' C_l)) - 1] \tag{3.7.10a}$$

and

$$U_{rel} = \frac{U(1-a)}{\sin\varphi} = \frac{U}{(\sigma' C_l / 4F)\cot\varphi + \sin\varphi} \tag{3.7.20}$$

Note that Equation 3.7.8 remains unchanged. The power coefficient can be calculated from

$$C_P = 8/\lambda^2 \int_{\lambda_h}^{\lambda} F \lambda_r^3 a'(1-a)[1 - (C_d/C_l)\cot\theta] d\lambda_r \tag{3.7.11a}$$

or:

$$C_P = (8/\lambda^2) \int_{\lambda_h}^{\lambda} F \sin^2\varphi (\cos\varphi - \lambda_r \sin\varphi)(\sin\varphi + \lambda_r \cos\varphi)[1 - (C_d/C_l)\cot\varphi] \lambda_r^2 d\lambda_r \tag{3.7.12a}$$

3.7.4 *Off-design performance issues*

When a section of blade has a pitch angle or flow conditions very different from the design conditions, a number of complications can affect the analysis. These include multiple solutions in the region of transition to stall and solutions for highly loaded conditions with values of the axial induction factor approaching and exceeding 0.5.

3.7.4.1 Multiple solutions to blade element momentum equations

In the stall region, as shown in Figure 3.25, there may be multiple solutions for C_l. Each of these solutions is possible. The correct solution should be that one which maintains the continuity of the angle of attack along the blade span.

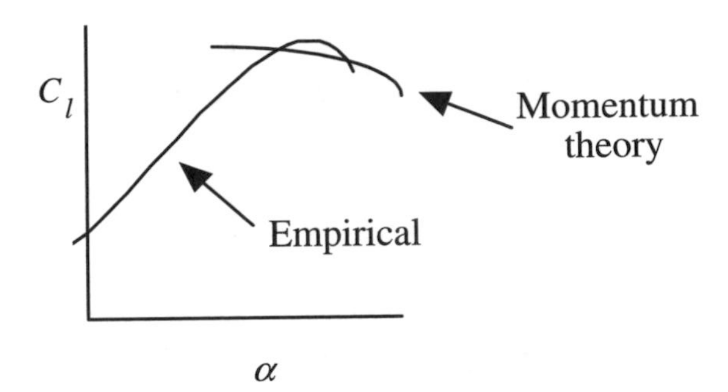

Figure 3.25 Multiple solutions ; α , angle of attack; C_l , two-dimensional lift coefficient

3.7.4.2 Wind turbine flow states

Measured wind turbine performance closely approximates the results of BEM theory at low values of the axial induction factor. Momentum theory is no longer valid at axial induction factors greater than 0.5, because the wind velocity in the far wake would be negative. In practice, as the axial induction factor increases above 0.5, the flow patterns through the wind turbine become much more complex than those predicted by momentum theory. A number of operating states for a rotor have been identified (see Eggleston and Stoddard, 1987). The operating states relevant to wind turbines are designated the windmill state and the turbulent wake state. The windmill state is the normal wind turbine operating state. The turbulent wake state occurs under operation in high winds. Figure 3.26 illustrates fits to measured thrust coefficients for these operating states. The windmill state is characterized by the flow conditions described by momentum theory for axial induction factors less than about 0.5. Above $a = 0.5$, in the turbulent wake state, measured data indicate that thrust coefficients increase up to about 2.0 at an axial induction factor of 1.0. This state is characterized by a large expansion of the slipstream, turbulence and recirculation behind the rotor. While momentum theory no longer describes the turbine behavior, empirical relationships between C_T and the axial induction factor are often used to predict wind turbine behavior.

3.7.4.3 Rotor modelling for the turbulent wake state

The rotor analysis discussed so far uses the equivalence of the thrust forces determined from momentum theory and from blade element theory to determine the angle of attack at the blade. In the turbulent wake state the thrust determined by momentum theory is no longer valid. In these cases, the previous analysis can lead to a lack of convergence to a solution or a situation in which the curve defined by Equation 3.7.6a or 3.7.6 would lie below the airfoil lift curve.

In the turbulent wake state, a solution can be found by using the empirical relationship between the axial induction factor and the thrust coefficient in conjunction with blade element theory. The empirical relationship developed by Glauert, and shown in Figure 3.26, (see Eggleston and Stoddard, 1987), including tip losses, is:

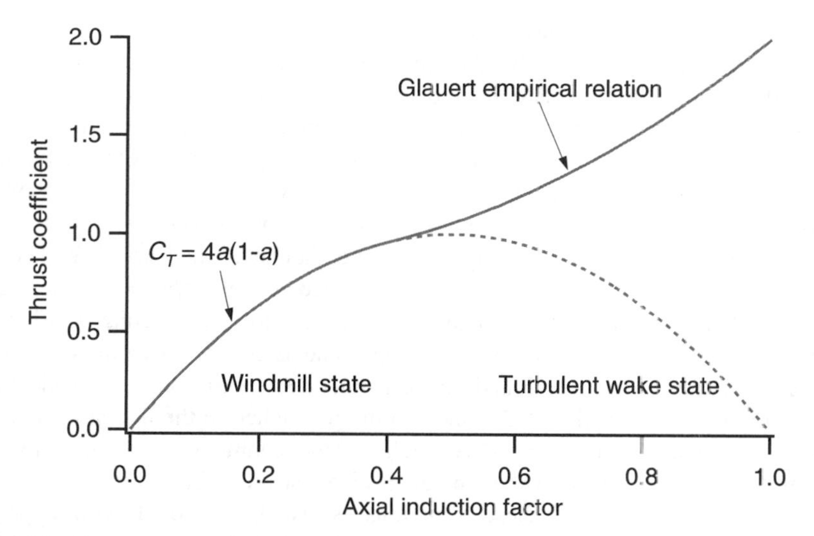

Figure 3.26 Fits to measured wind turbine thrust coefficients

$$a = (1/F)\left[0.143 + \sqrt{0.0203 - 0.6427(0.889 - C_T)}\right] \tag{3.7.21}$$

This equation is valid for $a > 0.4$ or, equivalently for $C_T > 0.96$.

The Glauert empirical relationship was determined for the overall thrust coefficient for a rotor. It is customary to assume that it applies equally to equivalent local thrust coefficients for each blade section. The local thrust coefficient, C_{T_r}, can be defined for each annular rotor section as (Wilson et al., 1976):

$$C_{T_r} = \frac{\mathrm{d}F_N}{\frac{1}{2}\rho U^2 2\pi r \mathrm{d}r} \tag{3.7.22}$$

From the equation for the normal force from blade element theory, Equation 3.7.1, the local thrust coefficient is:

$$C_{T_r} = \sigma'(1-a)^2 (C_l \cos\varphi + C_d \sin\varphi)/\sin^2\varphi \tag{3.7.23}$$

The solution procedure can then be modified to include heavily loaded turbines. The easiest procedure to use is the iterative procedure (Method 2) that starts with the selection of possible values for a and a'. Once the angle of attack and C_l and C_d have been determined, the local thrust coefficient can be calculated according to Equation 3.7.23. If $C_{T_r} < 0.96$ then the previously derived equations can be used. If $C_{T_r} > 0.96$ then the next estimate for the axial induction factor should be determined using the local thrust coefficient and Equation 3.7.21. The angular induction factor, a', can be determined from 3.7.7a.

3.7.4.4 Off-axis flows and blade coning

The analysis in this chapter assumes that the prevailing wind is uniform and aligned with the rotor axis and that the blades rotate in a plane perpendicular to the rotor axis. These assumptions are rarely the case because of wind shear, yaw error, vertical wind components, turbulence and blade coning. Wind shear will result in wind speeds across the disk that vary with height. Wind turbines often operate with a steady state or transient yaw error (misalignment of the rotor axis and the wind direction about the vertical yaw axis of the turbine). Yaw error results in a flow component perpendicular to the rotor disk. The winds at the rotor may also have a vertical component, especially at sites in complex terrain. Turbulence results in a variety of wind conditions over the rotor. The angular position of the blade in the rotor plane is called the azimuth angle and is measured from some suitable reference. Each of the effects mentioned above results in conditions at the blade varying with blade azimuth angle. Finally, blades are also often attached to the hub at a slight angle to the plane perpendicular to the rotor axis. This blade coning may be done to reduce bending moments in the blades or to keep the blades from striking the tower.

In a rotor analysis, each of these situations is usually handled with appropriate geometrical transformations. Blade coning is handled by resolving the aerodynamic forces into components that are perpendicular and parallel to the rotor plane. Off-axis flow is also resolved into the flow components that are perpendicular and parallel to the rotor plane. Rotor performance is then determined for a variety of rotor azimuth angles. The axial and in-plane components of the flow that depend on the blade position result in angles of attack and aerodynamic forces that fluctuate cyclically as the blades rotate. BEM equations that include terms for blade coning are provided by Wilson et al. (1976). Linearized methods for dealing with small off-axis flows and blade coning are discussed in Chapter 4.

3.8 Blade Shape for Optimum Rotor with Wake Rotation

The blade shape for an ideal rotor that includes the effects of wake rotation can be determined using the analysis developed for a general rotor. This optimisation includes wake rotation, but ignores drag ($C_d = 0$) and tip losses ($F = 1$). One can perform the optimisation by taking the partial derivative of that part of the integral for C_P (Equation 3.7.12) which is a function of the angle of the relative wind, φ, and setting it equal to zero, i.e.:

$$\frac{\partial}{\partial \varphi}\left[\sin^2\varphi(\cos\varphi - \lambda_r \sin\varphi)(\sin\varphi + \lambda_r \cos\varphi)\right] = 0 \tag{3.8.1}$$

This yields:

$$\lambda_r = \sin\varphi(2\cos\varphi - 1)/[(1-\cos\varphi)(2\cos\varphi + 1)] \tag{3.8.2}$$

Some more algebra reveals that:

$$\varphi = (2/3)\tan^{-1}(1/\lambda_r) \quad (3.8.3)$$

$$c = \frac{8\pi r}{BC_l}(1-\cos\varphi) \quad (3.8.4)$$

Induction factors can be calculated from:

$$a' = \frac{1-3a}{4a-1} \quad (3.3.17)$$

$$a = 1/\left[1 + 4\sin^2\varphi/(\sigma' C_l \cos\varphi)\right] \quad (3.7.9)$$

These results can be compared with the result for an ideal blade without wake rotation, for which:

$$\varphi = \tan^{-1}\left(\frac{2}{3\lambda_r}\right) \quad (3.6.7)$$

$$c = \frac{8\pi r}{BC_l}\left(\frac{\sin\varphi}{3\lambda_r}\right) \quad (3.6.8)$$

Note, that the optimum values for φ and c, including wake rotation, are often similar to, but could be significantly different from, those obtained without assuming wake rotation. Also, as before, select α where C_d/C_l is minimum.

Solidity is the ratio of the area of the blades to the swept area, thus:

$$\sigma = \frac{1}{\pi R^2}\int_{r_h}^{R} c\,\mathrm{d}r \quad (3.8.5)$$

The optimum blade rotor solidity can be found from methods discussed above. When the blade is modelled as a set of N blade sections of equal span, the solidity can be calculated from:

$$\sigma \cong \frac{B}{N\pi}\left(\sum_{i=1}^{N} c_i/R\right) \quad (3.8.6)$$

The blade shape for three sample optimum rotors, assuming wake rotation, are given in Table 3.3. Here C_{l1} is assumed to be 1.00 at the design angle of attack. In these rotors, the blade twist is directly related to the angle of the relative wind because the angle of attack is assumed to be constant (see Equations 3.5.4 and 3.5.5). Thus, changes in blade twist would mirror the changes in the angle of the relative wind shown in Table 3.3. It can be seen that the slow 12 bladed machine would have blades that had a roughly constant chord over the outer half of the blade and smaller chords closer to the hub. The blades would also have a significant twist. The two faster machines would have blades with an increasing chord as one went from the tip to the hub. The blades would also have significant twist, but much less than the 12 bladed machine. The fastest machine would have the least twist, which is a function of local speed ratio only. It would also have the smallest chord because of the low angle of the relative wind and only two blades (see Equations 3.8.3 and 3.8.4).

Table 3.3 Three optimum rotors

	$\lambda=1$ $B=12$		$\lambda=6$ $B=3$		$\lambda=10$ $B=2$	
r/R	φ	c/R	φ	c/R	φ	c/R
0.95	31	0.284	6.6	0.053	4.0	0.029
0.85	33.1	0.289	7.4	0.059	4.5	0.033
0.75	35.4	0.291	8.4	0.067	5.1	0.037
0.65	37.9	0.288	9.6	0.076	5.8	0.042
0.55	40.8	0.280	11.2	0.088	6.9	0.050
0.45	43.8	0.263	13.5	0.105	8.4	0.060
0.35	47.1	0.234	17.0	0.128	10.6	0.075
0.25	50.6	0.192	22.5	0.159	14.5	0.100
0.15	54.3	0.131	32.0	0.191	22.5	0.143
Solidity, σ		0.86		0.088		0.036

Note: B, number of blades; c, airfoil chord length; r, blade section radius; R, rotor radius; λ, tip speed ratio; φ, angle of relative wind

3.9 Generalized Rotor Design Procedure

3.9.1 Rotor design for specific conditions

The previous analysis can be used in a generalized rotor design procedure. The procedure begins with the choice of various rotor parameters and the choice of an airfoil. An initial blade shape is then determined using the optimum blade shape assuming wake rotation. The final blade shape and performance are determined iteratively considering drag, tip losses, and ease of manufacture. The steps in determining a blade design follow.

3.9.1.1 Determine basic rotor parameters

1. Begin by deciding what power, P, is needed at a particular wind velocity, U. Include the effect of a probable C_P and efficiencies, η, of various other components (e.g., gearbox, generator, pump, etc.). The radius, R, of the rotor may be estimated from:

$$P = C_P \eta \tfrac{1}{2} \rho \pi R^2 U^3 \tag{3.9.1}$$

2. According to the type of application, choose a tip speed ratio, λ. For a water pumping windmill, for which greater torque is needed, use $1 < \lambda < 3$. For electric power generation, use $4 < \lambda < 10$. The higher speed machines use less material in the blades and have smaller gearboxes, but require more sophisticated airfoils.
3. Choose a number of blades, B, from Table 3.4. Note: If fewer than three blades are selected, there are a number of structural dynamic problems that must be considered in the hub design. One solution is a teetered hub (see Chapter 6).

Table 3.4 Suggested blade number, B, for different tip speed ratios, λ

λ	B
1	8–24
2	6–12
3	3–6
4	3–4
>4	1–3

4. Select an airfoil. If $\lambda < 3$, curved plates can be used. If $\lambda > 3$, use a more aerodynamic shape.

3.9.1.2 Define the blade shape

5. Obtain and examine the empirical curves for the aerodynamic properties of the airfoil at each section (the airfoil may vary from the root to the tip), i.e. C_l vs. α, C_d vs. α. Choose the design aerodynamic conditions, $C_{l,design}$ and α_{design}, such that $C_{d,design} / C_{l,design}$ is at a minimum for each blade section.
6. Divide the blade into N elements (usually 10–20). Use the optimum rotor theory to estimate the shape of the ith blade with a midpoint radius of r_i:

$$\lambda_{r,i} = \lambda (r_i / R) \tag{3.9.2}$$

$$\varphi_i = (2/3) \tan^{-1} (1/\lambda_{r,i}) \tag{3.9.3}$$

$$c_i = \frac{8\pi r_i}{BC_{l,design,i}}(1-\cos\varphi_i) \quad (3.9.4)$$

$$\theta_{T,i} = \theta_{p,i} - \theta_{p,0} \quad (3.9.5)$$

$$\varphi_i = \theta_{p,i} + \alpha_{design,i} \quad (3.9.6)$$

7. Using the optimum blade shape as a guide, select a blade shape that promises to be a good approximation. For ease of fabrication, linear variations of chord, thickness and twist might be chosen. For example, if a_1, b_1 and a_2 are coefficients for the chosen chord and twist distributions, then the chord and twist can be expressed as:

$$c_i = a_1 r_i + b_1 \quad (3.9.7)$$

$$\theta_{T,i} = a_2(R - r_i) \quad (3.9.8)$$

3.9.1.3 Calculate rotor performance and modify blade design

8. As outlined above, one of two methods might be chosen to solve the equations for the blade performance.

Method 1 – Solving for C_l and α Find the actual angle of attack and lift coefficients for the centre of each element, using the following equations and the empirical airfoil curves:

$$C_{l,i} = 4F_i \sin\varphi_i \frac{(\cos\varphi_i - \lambda_{r,i}\sin\varphi_i)}{\sigma_i'(\sin\varphi_i + \lambda_{r,i}\cos\varphi_i)} \quad (3.9.9)$$

$$\sigma_i' = Bc_i/2\pi r_i \quad (3.9.10)$$

$$\varphi_i = \alpha_i + \theta_{T,i} + \theta_{P,0} \quad (3.9.11)$$

$$F_i = (2/\pi)\cos^{-1}\left[\exp\left(-\left\{\frac{(B/2)[1-(r_i/R)]}{(r_i/R)\sin\varphi_i}\right\}\right)\right] \quad (3.9.12)$$

The lift coefficient and angle of attack can be found by iteration or graphically. A graphical solution is illustrated in Figure 3.27. The iterative approach requires an initial estimate of the tip loss factor. To find a starting F_i, start with an estimate for the angle of the relative wind of:

$$\varphi_{i,1} = (2/3)\tan^{-1}(1/\lambda_{r,i}) \tag{3.9.13}$$

For subsequent iterations, find F_i using:

$$\varphi_{i,j+1} = \theta_{P,i} + \alpha_{i,j} \tag{3.9.14}$$

where j is the number of the iteration. Usually, few iterations are needed.

Finally, calculate the axial induction factor:

$$a_i = 1/\left[1 + 4\sin^2\varphi_i / (\sigma_i' C_{l,i}\cos\varphi_i)\right] \tag{3.9.15}$$

If a_i is greater than 0.4, use method 2.

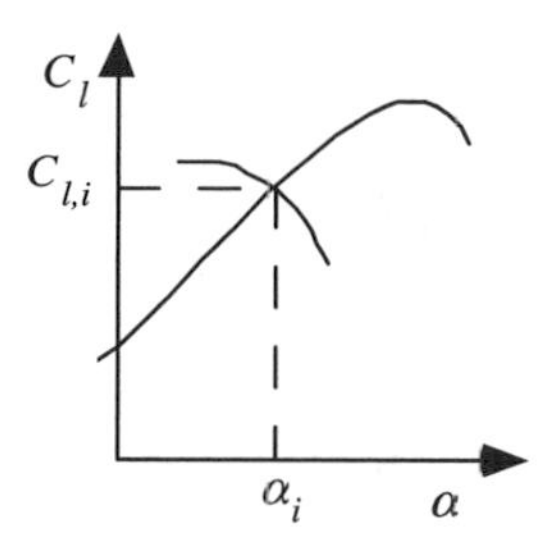

Figure 3.27 Graphical solution for angle of attack, α ; C_l, two-dimensional lift coefficient; $C_{l,i}$ and α_i, C_l and α, respectively, for blade section, i

Method 2 – Iterative solution for a and a' Iterating to find the axial and angular induction factors using method 2 requires initial guesses for their values. To find initial values, start with values from an adjacent blade section, values from the previous blade design in the iterative rotor design process or use an estimate based on the design values from the starting optimum blade design:

$$\varphi_{i,1} = (2/3)\tan^{-1}(1/\lambda_{r,i}) \tag{3.9.16}$$

$$a_{i,1} = \frac{1}{\left[1 + \dfrac{4\sin^2(\varphi_{i,1})}{\sigma'_{i,design} C_{l,design}\cos\varphi_{i,1}}\right]} \tag{3.9.17}$$

$$a'_{i,1} = \frac{1-3a_{i,1}}{(4a_{i,1})-1} \tag{3.9.18}$$

Having guesses for $a_{i,1}$ and $a'_{i,1}$, start the iterative solution procedure for the *j*th iteration. For the first iteration *j* = 1. Calculate the angle of the relative wind and the tip loss factor:

$$\tan\varphi_{i,j} = \frac{U(1-a_{i,j})}{\Omega r(1+a'_{i,j})} = \frac{1-a_{i,j}}{(1+a'_{i,j})\lambda_{r,i}} \tag{3.9.19}$$

$$F_{i,j} = (2/\pi)\cos^{-1}\left[\exp\left(-\left\{\frac{(B/2)[1-(r_i/R)]}{(r_i/R)\sin\varphi_{i,j}}\right\}\right)\right] \tag{3.9.20}$$

Determine $C_{l,i,j}$ and $C_{d,i,j}$ from the airfoil lift and drag data, using:

$$\alpha_{i,j} = \varphi_{i,j} - \theta_{p,i} \tag{3.9.21}$$

Calculate the local thrust coefficient:

$$C_{T_r,i,j} = \frac{\sigma_i'(1-a_{i,j})^2(C_{l,i,j}\cos\varphi_{i,j} + C_{d,i,j}\sin\varphi_{i,j})}{\sin^2\varphi_{i,j}} \tag{3.9.22}$$

Update a and a' for the next iteration. If $C_{T_r,i,j} < 0.96$:

$$a_{i,j+1} = \frac{1}{\left[1+\frac{4F_{i,j}\sin^2(\varphi_{i,j})}{\sigma_i' C_{l,i,j}\cos\varphi_{i,j}}\right]} \tag{3.9.23}$$

If $C_{T_r,i,j} > 0.96$:

$$a_{i,j} = (1/F_{i,j})\left[0.143 + \sqrt{0.0203 - 0.6427(0.889 - C_{T_r,i,j})}\right] \tag{3.9.24}$$

$$a'_{i,j+1} = \frac{1}{\frac{4F_{i,j}\cos\varphi_{i,j}}{\sigma' C_{l,i,j}} - 1} \tag{3.9.25}$$

If the newest induction factors are within an acceptable tolerance of the previous guesses, then the other performance parameters can be calculated. If not, then the procedure starts again at Equation 3.9.19 with $j = j+1$.

9. Having solved the equations for the performance at each blade element, the power coefficient is determined using a sum approximating the integral in Equation 3.7.12a:

$$C_P = \sum_{i=1}^{N} \left(\frac{8 \Delta \lambda_r}{\lambda^2} \right) F_i \sin^2 \varphi_i (\cos \varphi_i - \lambda_{ri} \sin \varphi_i)(\sin \varphi_i + \lambda_{ri} \cos \varphi_i) \left[1 - \left(\frac{C_d}{C_l} \right) \cot \varphi_i \right] \lambda_{ri}^2 \qquad (3.9.26)$$

If the total length of the hub and blade is assumed to be divided into N equal length blade elements, then:

$$\Delta \lambda_r = \lambda_{ri} - \lambda_{r(i-1)} = \lambda / N \qquad (3.9.27)$$

$$C_P = \frac{8}{\lambda N} \sum_{i=k}^{N} F_i \sin^2 \varphi_i (\cos \varphi_i - \lambda_{ri} \sin \varphi_i)(\sin \varphi_i + \lambda_{ri} \cos \varphi_i) \left[1 - \left(\frac{C_d}{C_l} \right) \cot \varphi_i \right] \lambda_{ri}^2 \qquad (3.9.28)$$

where k is the index of the first "blade" section consisting of the actual blade airfoil.

10. Modify the design if necessary and repeat steps 8–10, in order to find the best design for the rotor, given the limitations of fabrication.

3.9.2 $C_P - \lambda$ curves

Once the blade has been designed for optimum operation at a specific design tip speed ratio, the performance of the rotor over all expected tip speed ratios needs to be determined. This can be done using the methods outlined in Section 3.7. For each tip speed ratio, the aerodynamic conditions at each blade section need to be determined. From these, the performance of the total rotor can be determined. The results are usually presented as a graph of power coefficient versus tip speed ratio, called a $C_P - \lambda$ curve, as shown in Figure 3.28.

$C_P - \lambda$ curves can be used in wind turbine design to determine the rotor power for any combination of wind and rotor speed. They provide immediate information on the maximum rotor power coefficient and optimum tip speed ratio. Care must be taken in using $C_P - \lambda$ curves. The data for such a relationship can be found from turbine tests or from modelling. In either case, the results depend on the lift and drag coefficients of the airfoils, which may vary as a function of the flow conditions. Variations in airfoil lift and drag coefficients depend on the airfoil and the Reynolds numbers being considered, but, as shown in Figure 3.10, airfoils can have remarkably different behavior when the Reynolds number changes by as little as a factor of 2.

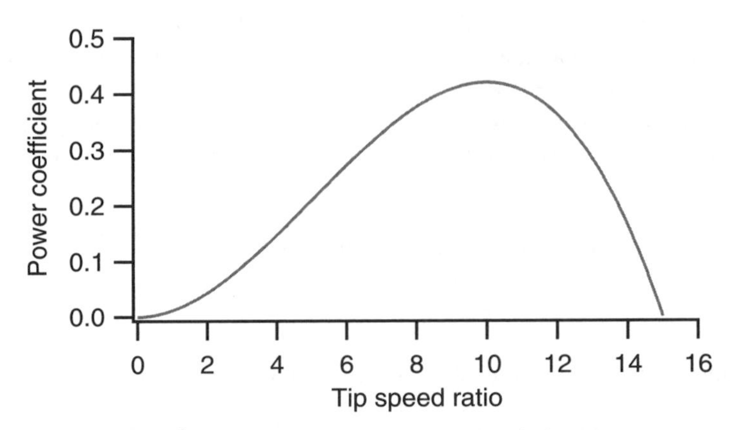

Figure 3.28 Sample $C_P - \lambda$ curve for a high tip speed ratio wind turbine

3.10 Simplified HAWT Rotor Performance Calculation Procedure

Manwell (1990) proposed a simplified method for calculating the performance of a horizontal axis wind turbine rotor that is particularly applicable for an unstalled rotor, but may also be useful under certain stall conditions. The method uses the previously discussed blade element theory and incorporates an analytical method for finding the blade angle of attack. Depending on whether tip losses are included, few or no iterations are required. The method assumes that two conditions apply:

- The airfoil section lift coefficient vs. angle of attack relation must be linear in the region of interest
- The angle of attack must be small enough that the small-angle approximations may be used

These two requirements normally apply if the section is unstalled. They may also apply under certain partially stalled conditions for moderate angles of attack if the lift curve can be linearized.

The simplified method is the same as method 1 outlined above, with the exception of a simplification for determining the angle of attack and the lift coefficient for each blade section. The essence of the simplified method is the use of an analytical (closed-form) expression for finding the angle of attack of the relative wind at each blade element. It is assumed that the lift and drag curves can be approximated by:

$$C_l = C_{l,0} + C_{l,\alpha}\alpha \tag{3.10.1}$$

$$C_d = C_{d,0} + C_{d,\alpha 1}\alpha + C_{d,\alpha 2}\alpha^2 \tag{3.10.2}$$

When the lift curve is linear and when small-angle approximations can be used, it can be shown that the angle of attack is given by:

$$\alpha = \frac{-q_2 \pm \sqrt{q_2^2 - 4q_1q_3}}{2\,q_3} \tag{3.10.3}$$

where:

$$q_1 = C_{l,0}\; d_2 - \frac{4F}{\sigma'} d_1 \sin\theta_p \tag{3.10.4}$$

$$q_2 = C_{l,\alpha} d_2 + d_1 C_{l,0} - \frac{4F}{\sigma'} (d_1 \cos\theta_\mathrm{p} - d_2 \sin\theta_\mathrm{p}) \tag{3.10.5}$$

$$q_3 = C_{l,\alpha} d_1 + \frac{4F}{\sigma'} d_2 \cos\theta_\mathrm{p} \tag{3.10.6}$$

$$d_1 = \cos\theta_\mathrm{p} - \lambda_\mathrm{r} \sin\theta_\mathrm{p} \tag{3.10.7}$$

$$d_2 = \sin\theta_\mathrm{p} + \lambda_\mathrm{r} \cos\theta_\mathrm{p} \tag{3.10.8}$$

Using this approach, the angle of attack can be calculated from Equation 3.10.3 once an initial estimate for the tip loss factor is determined. The lift and drag coefficients can then be calculated from Equation 3.10.1 and 3.10.2, using Equation 3.9.14. Iteration with a new estimate of the tip loss factor may be required.

The simplified method provides angles of attack very close to those of the more detailed method for many operating conditions. For example, results for the analysis of one blade of the University of Massachusetts WF-1 wind turbine are shown in Figure 3.29. This was a three-bladed turbine with a 10 m rotor, using near optimum tapered and twisted blades. The lift curve of the NACA 4415 airfoil was approximated by $C_l = 0.368 + 0.0942\ \alpha$. The drag coefficient equation constants were 0.00994, 0.000259, and 0.0001055. Figure 3.29 compares the results from the simplified method and the conventional strip theory method for the angle of attack for one of the blade elements. The point at which the curves cross the empirical lift line determines the angle of attack and the lift coefficient. Also plotted on Figure 3.29 is the axial induction factor, a, for the section. Note that it is the right-hand intersection point which gives a value of $a < 1/2$, as is normally the case.

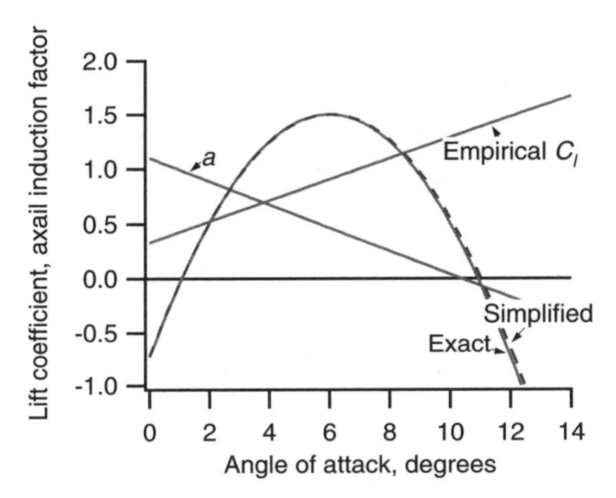

Figure 3.29 Comparison of calculation methods for one blade element; a, axial induction factor; C_l, two-dimensional lift coefficient

3.11 Effect of Drag and Blade Number on Optimum Performance

At the beginning of the chapter, the maximum theoretically possible power coefficient for wind turbines was determined as a function of tip speed ratio. As explained in this chapter, airfoil drag and tip losses that are a function of the total number of blades reduce the power coefficients of wind turbines. The maximum achievable power coefficient for turbines with an optimum blade shape but a finite number of blades and aerodynamic drag has been calculated by Wilson et al. (1976). Their fit to the data is accurate to within 0.5% for tip speed ratios from 4 to 20, lift to drag ratios (C_l/C_d) from 25 to infinity and from one to three blades (B):

$$C_{p,\max} = \left(\frac{16}{27}\right)\lambda\left[\lambda + \frac{1.32 + \left(\frac{\lambda - 8}{20}\right)^2}{B^{\frac{2}{3}}}\right]^{-1} - \frac{(0.57)\lambda^2}{\frac{C_l}{C_d}\left(\lambda + \frac{1}{2B}\right)} \tag{3.11.1}$$

Figure 3.30, based on this equation, shows the maximum achievable power coefficients for a turbine with 1, 2, and 3 optimum blades and no drag. The performance for ideal conditions (an infinite number of blades) is also shown. It can be seen that the fewer the blades the lower the possible C_p at the same tip speed ratio. Most wind turbines use two or three blades and, in general, most two-bladed wind turbines use a higher tip speed ratio than most three-bladed wind turbines. Thus, there is little practical difference in the maximum achievable C_p between typical two- and three-bladed designs, assuming no drag. The effect of the lift to drag ratio on maximum achievable power coefficients for a three-bladed rotor is shown in Figure 3.31. There is clearly a significant reduction in maximum achievable power as the airfoil drag increases. For reference, the DU-93-W-210 airfoil has a maximum

C_l/C_d ratio of 140 at an angle of attack of 6 degrees, and the 19% thick LS(1) airfoil has a maximum C_l/C_d ratio of 85 at an angle of attack of 4 degrees. It can be seen that it clearly benefits the blade designer to use airfoils with high lift to drag ratios. Practical rotor power coefficients may be further reduced as a result of non-optimum blade designs that are easier to manufacture, the lack of airfoils at the hub and aerodynamic losses at the hub end of the blade.

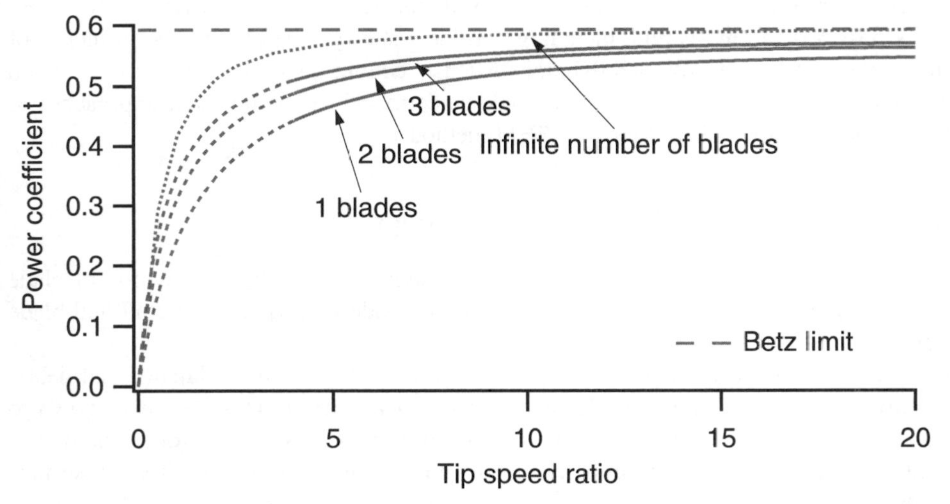

Figure 3.30 Maximum achievable power coefficients as a function of number of blades, no drag

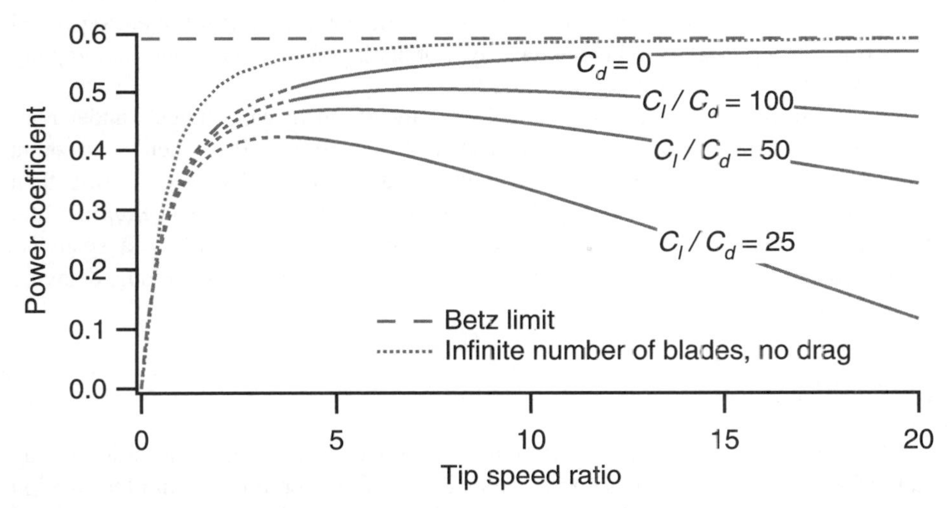

Figure 3.31 Maximum achievable power coefficients of a three-bladed optimum rotor as a function of the lift to drag ratio, C_l/C_d.

3.12 Advanced Aerodynamic Topics

As mentioned in the beginning of this chapter, the aerodynamic performance of wind turbines are primarily a function of the steady state aerodynamics that are discussed above. The analysis presented in this chapter provides a method for determining average loads on a wind turbine. There are, however, a number of important steady state and dynamic effects that cause increased loads or decreased power production from those expected with the BEM theory presented here, especially increased transient loads. An overview of those effects is provided in this section, including non-ideal steady state effects, the influence of turbine wakes and unsteady aerodynamics. This section also includes comments on computer programs that can be used for rotor performance modelling and approaches to modelling rotor aerodynamics other than BEM methods.

3.12.1 Non-ideal steady state aerodynamic issues

Steady state effects that influence wind turbine behavior include the degradation of blade performance due to surface roughness, the effects on blade performance of stall and blade rotation.

As mentioned above in Section 3.4.5, blade surfaces roughened by damage and debris can significantly increase drag and decrease the lift of an airfoil. This has been shown to decrease power production by as much as 40% on certain airfoils. The only solution is frequent blade repair and cleaning or the use of airfoils that are less sensitive to surface roughness.

Parts of a wind turbine blade may operate at times in the stall region. On stall-controlled horizontal axis rotors much of the blade may be stalled under some conditions. Stalled airfoils do not always exhibit the simple relationship between the angle of attack and aerodynamic forces that are evident in lift and drag coefficient data. The turbulent separated flow occurring during stall can induce rapidly fluctuating flow conditions and rapidly fluctuating loads on the wind turbine.

Finally, the lift and drag behavior of airfoils are measured in wind tunnels under non-rotating conditions. Investigation has shown that the same airfoils, when used on a horizontal axis wind turbine, may exhibit delayed stall and may produce more power than expected. The resulting unexpectedly high loads at high wind speeds can reduce turbine life. This behavior has been linked to spanwise pressure gradients that result in a spanwise velocity component along the blade that helps keep the flow attached to the blade, delaying stall and increasing lift.

3.12.2 Turbine wakes

Many of the primary features of the air flow in and around wind turbines are described by the results of BEM theory: the induced velocities due to power production and the rotation of the turbine wake and the expanding wake downwind of the turbine. The actual flow field, however, is much more complicated. Some of the details of flow around and behind a horizontal axis wind turbine are described here. The consequences of these flow patterns affect downwind turbines and may result in a ‘skewed wake,’ which causes increased fluctuating loads which are not predicted by BEM theory. Before considering their consequences, the details of turbine wakes are described in the following paragraph.

Wind turbine wakes are often thought of as consisting of a near wake and a far wake (Voutsinas et al., 1993). The difference between the near and far wakes is a function of the spatial distribution and intensity of turbulence in the flow field. Fluid flow modelling and experiments have shown that each blade generates a sheet of vortices that is transported through the wake by the mean axial and rotational flow in the wake. In addition to the vortex sheet from the trailing edge of the blades, vortices generated at the hub and especially strong vortices, generated at the blade tips, are also convected downstream. The tip vortices cause the tip losses mentioned in Section 3.5. All of these vortices and mechanically generated turbulence are dissipated and mixed in the near wake (within 1 to 3 rotor diameters downstream of the rotor). The stream of tip vortices from each blade merge in the near wake into a cylindrical sheet of rotating turbulence as they mix and diffuse through the flow (Sorensen and Shen, 1999). Much of the periodic nature of the flow is lost in the near wake (Ebert and Wood, 1994). Thus, turbulence and vorticity generated at the rotor are diffused in the near wake, resulting in more evenly distributed turbulence and velocity profiles in the far wake. Meanwhile, the mixing of the slower axial flow of the turbine wake with the free stream flow slowly re-energizes the flow. The vortex sheet from the tip vortices results in an annular area in the far wake of higher relative turbulence surrounding the less turbulent core of the wake. Mixing and diffusion continue in the far wake until the turbine-generated turbulence and velocity deficit with respect to the free stream flow have disappeared.

The consequences of the vorticity and turbulence in turbine wakes are increased loads and fatigue. The most obvious effect is the increased turbulence in the flow at turbines that are downwind of other turbines in a wind farm (see Chapter 8). The nature of turbine wakes also affects the loads on the turbine producing the wake. For example, tip and hub vortices reduce the energy capture from the rotor.

Another important effect occurs with off-axis winds, i.e. those whose direction is not perpendicular to the plane of the rotor. Off-axis flows, whether due to yaw error or vertical wind components, result in a skewed wake in which the wake is not symmetric with the turbine axis. Skewed wakes result in the downwind side of the rotor being closer to the wake centerline than the upwind side of the rotor. The result is higher induced velocities on the downwind side of the rotor than the upwind side. This effect has been shown to result in higher turbine forces than otherwise would be expected (Hansen, 1992). One commonly used approach to modelling the effects of a skewed wake is the Pitt and Peters model (Pitt and Peters, 1981; Goankar and Peters, 1986). The model applies a multiplicative correction factor to the axial induction factor that is a function of yaw angle, radial position and blade azimuth angle. For more information on its use in wind turbine modelling, see Hansen (1992).

3.12.3 Unsteady aerodynamic effects

There are a number of unsteady aerodynamic phenomena that have a large effect on wind turbine operation. The turbulent eddies carried along with the mean wind cause rapid changes in speed and direction over the rotor disk. These changes cause fluctuating aerodynamic forces, increased peak forces, blade vibrations, and significant material fatigue. Additionally, the transient effects of tower shadow, dynamic stall, dynamic inflow and rotational sampling (all explained below) change turbine operation in unexpected ways. Many of these effects occur at the rotational frequency of the rotor or at multiples of that frequency. Effects that occur once per revolution are often referred to as having a frequency of $1P$. Similarly, effects that occur at 3 or n times per revolution of the rotor are referred to as occurring at a frequency of $3P$ or nP.

Tower shadow refers to the wind speed deficit behind a tower caused by the tower obstruction. The blades of a downwind rotor with B blades will encounter the tower shadow once per revolution, causing a rapid drop in power and BP vibrations in the turbine structure.

Dynamic stall refers to rapid aerodynamic changes that may bring about or delay stall behavior. Rapid changes in wind speed (for example, when the blades pass through the tower shadow) cause a sudden detachment and then reattachment of airflow along the airfoil. Such effects at the blade surface cannot be predicted with steady state aerodynamics, but may affect turbine operation, not only when the blades encounter tower shadow, but also during operation in turbulent wind conditions. Dynamic stall effects occur on time scales of the order of the time for the relative wind at the blade to traverse the blade chord, approximately $c/\Omega r$. For large wind turbines, this might be on the order of 0.2 seconds at the blade root to 0.01 seconds at the blade tip (Snel and Schepers, 1991). Dynamic stall can result in high transient forces as the wind speed increases, but stall is delayed. A variety of dynamic stall models have been used in computer rotor performance codes including those of Gormont (1973) and Beddoes (Björck et al., 1999). The Gormont model, for example, modifies the angle of attack calculated with the BEM theory by adding a factor that depends on the rate of change of the angle of attack.

Dynamic inflow refers to the response of the larger flow field to turbulence and changes in rotor operation (pitch or rotor speed changes, for example). Steady state aerodynamics suggests that increased wind speed and, thus, increased power production should result in an instantaneous increase in the axial induction factor and changes in the flow field upstream and downstream of the rotor. During rapid changes in the flow and rapid changes in rotor operation, the larger field cannot respond quickly enough to instantly establish steady state conditions. Thus the aerodynamic conditions at the rotor are not necessarily the expected conditions, but an ever-changing approximation as the flow field changes. The time scale of dynamic flow effects is on the order of D/U, the ratio of the rotor diameter to the mean ambient flow velocity. This might be as much as 10 seconds (Snel and Schepers, 1991). Phenomena occurring slower than this can be considered using a steady state analysis. For more information on dynamic inflow see Snel and Schepers (1991, 1993) and Pitt and Peters (1981).

Finally, *rotational sampling* (see Connell, 1982) causes some unsteady aerodynamic effects and increases fluctuating loads on the wind turbine. These effects are all induced or complicated because the wind as seen by the rotor is constantly changing as the rotor rotates. The general flow turbulence may bring wind speed changes on a time scale of about 5 seconds. The turbulent eddies may be smaller than the rotor disk, resulting in different winds at different parts of the disk. If the blades are rotating once a second, the blades 'sample' different parts of the flow field at a much faster rate than the general changes in the wind field itself, causing rapidly changing flow at the blade.

3.12.4 Computer codes for performance and load estimation

A number of computer codes are available that can predict rotor performance and aerodynamic loads. The National Renewable Energy Laboratory (NREL) in Boulder, CO, has supported the development of a number of these codes and has made some of the codes available over the Internet. The aerodynamic performance codes that are available from NREL include:

- WT_Perf
- YawDyn and AeroDyn

WT_Perf: WT_Perf (Buhl, 2000) is a rotor performance code for horizontal axis wind turbines. The aerodynamics analysis uses momentum and strip theory to determine blade performance, including corrections for tip losses and wind shear. Three-dimensional aerodynamics calculations have also been added. The code and a users manual are available from NREL.

YawDyn: YawDyn (Hansen, 1992) is a complete aerodynamics and dynamics analysis code for constant speed horizontal axis wind turbines. The code was designed to evaluate wind turbine yaw dynamics. The aerodynamics part of the code, AeroDyn, uses a detailed implementation of strip theory with modifications to improve results for unsteady winds. More details on YawDyn are given in Section 4.4.1.

3.12.5 Other performance prediction and design methods

In this chapter, the BEM theory approach has been used to predict rotor performance. An iterative approach to blade design has also been outlined based on the analysis methods detailed in the text. There are other approaches to predicting blade performance and to designing blades that may be more applicable in some situations. Some of the disadvantages of the BEM theory include errors under conditions with large induced velocities (Glauert, 1948) or yawed flow and inability to predict delayed stall due to rotational effects.

Vortex wake methods have been used in the helicopter industry in addition to BEM methods. Vortex wake methods calculate the induced velocity field by determining the distribution of vorticity in the wake. This method is computationally intensive, but promises to have advantages for yawed flow and operation subject to three-dimensional boundary layer effects (Hansen and Butterfield, 1993).

There are also other possible theoretical approaches. Researchers at Delft University of Technology have reported initial work on a model employing asymptotic acceleration potential methods (Hansen and Butterfield, 1993). Cascade Theory, often used in turbomachinery design, has also been used to analyse wind turbine performance. Cascade theory takes into consideration aerodynamic interactions between blades. Although it is more computationally intensive, cascade theory has been shown to provide better results than BEM theory for high-solidity, low tip-speed rotors (Islam and Islam, 1994). Computational fluid dynamics (CFD), while very computationally intensive, has also been applied to wind turbine rotors (see, for example, Sorenson and Michelsen, 2002, and Duque et al., 1999).

Finally, each of these analysis methods could be used in an iterative fashion to define the final blade design for a wind turbine, but work has been also preformed to develop a computer code to approach the blade design problem from the opposite direction (Selig and Tangler, 1992). This approach allows the designer to input desired rotor performance characteristics and blade aerodynamic characteristics and the code determines the corresponding blade geometry. The code has been used successfully to design blades for a commercial wind turbine.

References

Abbott, I. A., Von Doenhoff, A.E. (1959) *Theory of Wing Sections*. Dover Publications, New York.

Althaus, D. (1996) *Airfoils and experimental results from the laminar wind tunnel of the Institute for Aerodynamik and Gasdynamik of the University of Stuttgart*. University of Stuttgart.

Althaus, D., Wortmann, F. X. (1981) *Stuttgarter Profilkatalog.*, Friedr. Vieweg and Sohn, Braunschweig/Wiesbaden.

Betz, A. (1926) *Windenergie und Ihre Ausnutzung durch Windmüllen*. Vandenhoeck and Ruprecht , Gottingen, Germany, 1926.

Björck, A., Mert, M., Madsen, H. A. (1999) Optimal Parameters for the FFA-Beddoes Dynamic Stall Model. *Proc. 1999 EWEC*, 1–5 March 1999, Nice, 125–129.

Buhl, M. L., Jr., *WT_Perf User's Guide*. National Wind Technology Center, Golden CO.

Connell, J. R. (1982) The Spectrum of Wind Speed Fluctuations Encountered by a Rotating Blade of a Wind Energy Conversion System. *Solar Energy*, 29 (5) 363–375.

Currie, I. G. (1974) *Fundamental Mechanics of Fluids*. McGraw-Hill, New York.

de Vries, O. (1979) *Fluid Dynamic Aspects of Wind Energy Conversion*. Advisory Group for Aerospace Research and Development, North Atlantic Treaty Organization, AGARD-AG-243.

Duque, P. N., van Dam, C. P., Hughes, S. C. (1999) Navier-Stokes Simulations of the NREL Combined Experiment Rotor, AIAA Paper 99-0037, *Proc. 37th AIAA Aerospace Sciences Meeting and Exhibit*, Reno, NV.

Ebert, P. R., Wood, D. H. (1994) Three dimensional measurements in the wake of a wind turbine, *Proceedings of the European Wind Energy Conference*, Thessalonika, 10–14 October, 460–464.

Eggleston, D. M., Stoddard, F. S. (1987) *Wind Turbine Engineering Design*. Van Nostrand Rheinhold, New York.

Eppler, R., and Somers, K. M. *A Computer Program for the Design and Analysis of Low-Speed Airfoils*, NASA TM-80210, Hampton, Virginia: NASA Langley Research Center.

Gasch, R. (ed.) (1996) *Windkraftanlagen*. B. G. Teubner, Stuttgart.

Glauert, H. (1935) Airplane Propellers. *Aerodynamic Theory* (ed. W. F. Durand), Div. L. Chapter XI, Springer Verlag, Berlin (reprinted by Peter Smith, Gloucester, MA, 1976).

Glauert, H. (1948) *The Elements of Aero Foil and Airscrew Theory*. Cambridge University Press, Cambridge, England.

Goankar, G. H., Peters, D. A. (1986) Effectiveness of Current Dynamic-Inflow Models in Hover and Forward Flight. *Journal of the American Helicopter Society* 31 (2).

Gormont, R. E. (1973) *A Mathematical Model of Unsteady Aerodynamics and Radial Flow for Application to Helicopter Rotors*. US. Army Air Mobility Research and Development Laboratory, Technical Report, 76–67.

Hansen, A. C. (1992) *Yaw Dynamics of Horizontal Axis Wind Turbines: Final Report.* SERI Report, Subcontract No. XL-6-05078-2, January 1992.

Hansen, A. C., Butterfield, C. P. (1993) Aerodynamics of Horizontal Axis Wind Turbines. *Annual Review of Fluid Mechanics* 25, 115–149.

Islam, M. Q., Islam, A. K. M. S. (1994) The Aerodynamic Performance of a Horizontal Axis Wind Turbine Calculated by Strip Theory and Cascade Theory. *JSME International Journal* Series B, 37, 871–877.

Manwell, J. F. (1990) A Simplified Method for Predicting the Performance of a Horizontal Axis Wind Turbine Rotor. *Proc. AWEA 1990.*

Miley, S. J. (1982) *A Catalog of Low Reynolds Number Airfoil Data for Wind Turbine Applications.* Rockwell Int., Rocky Flats Plant RFP-3387, NTIS.

Pitt, D. M., Peters, D. A. (1981) Theoretical Predictions of Dynamic Inflow Derivatives. *Vertica*, Vol. 5 (1) 21–34.

Selig, M. S., Tangler, L. T. (1992) Development and Application of a Multipoint Inverse Design Method for Horizontal Axis Wind Turbines. *Wind Engineering*, 19 (2).

Sengupta, A., Verma, M. P. (1992) An Analytical Expression for the Power Coefficient of an Ideal Horizontal-Axis Wind Turbine. *Int. J. of Energy Research*, 16, 453–456.

Snel, H., Schepers, J. G. (1991) Engineering Models for Dynamic Inflow Phenomena. *Proc. EWEC '91*, Amsterdam, Netherlands, 390–396.

Snel, H., Schepers, J. G. (1993) Investigation and Modelling of Dynamic Inflow Effects. *Proc. EWEC 1993*, 8–12 March 1993, Lübeck, Germany, 371–375.

Sorensen, J. N, and Shen, W. Z. (1999) Computation of Wind Turbine Wakes Using Combined Navier-Stokes Actuator-Line Methodology. *Proc. 1999 EWEC*, 1–5 March, 1999, Nice, 156–159.

Sorenson, N.N., Michelsen, J.A. (2002) Navier-Stokes Predictions of the NREL Phase VI Rotor in the NASA Ames 80-by-120 Wind Tunnel, *Proc. 2002 ASME Wind Energy Symposium*, 40th AIAA Aerospace Sciences Meeting and Exhibit, Reno, NV.

Spera. D. A. (ed) (1994) *Wind Turbine Technology.* American Society of Mechanical Engineers, New York.

Tangler, J. L. (1987) *Status of Special Purpose Airfoil Families.* SERI/TP-217-3264, National Renewable Energy Laboratory, Golden, CO, January.

Tangler, J., Smith, B., Jager, D., Olsen, T (1990) Atmospheric Performance of the Status of the Special–purpose SERI Thin-airfoil Family: Final Results. *Proc. EWEC Conf.* Madrid Spain, 10–14 September.

Viterna, L. A., Corrigan, R. D. (1981) Fixed Pitch Rotor Performance of Large Horizontal Axis Wind Turbines. *Proceedings, Workshop on Large Horizontal Axis Wind Turbines*, NASA CP-2230, DOE Publication CONF-810752, NASA Lewis Research Center, Cleveland OH, 69–85.

Voutsinas, S. G., Rados K. G., Zervos, A. (1993) Wake Effects in Wind Parks. A New Modelling Approach. *Proc. ECWEC*, 8–12 March 1993, Lübeck, Germany, 444–447.

Wilson, R. E., Lissaman, P. B. S. (1974) *Applied Aerodynamics of Wind Power Machine.* Oregon State University.

Wilson, R. E., Lissaman, P. B. S., Walker, S. N. (1976) *Aerodynamic Performance of Wind Turbines.* Energy Research and Development Administration, ERDA/NSF/04014-76/1.

Wilson, R. E., Walker, S. N. Heh P. (1999) *Technical and User's Manual for the FAST_AD Advanced Dynamics Code.* OSU/NREL Report 99-01, Oregon State Univeristy, Corvallis, OR.

4

Mechanics and Dynamics

4.1 General Background

The interplay of the forces from the external environment, primarily due to the wind, and the motions of the various components of the wind turbine, results not only in the desired energy production from the turbine, but also in stresses in the constituent materials. For the turbine designer, these stresses are of primary concern, because they directly affect the strength of the turbine and how long it will last.

To put it succinctly, in order to be a viable contender for providing energy, a wind turbine must:

- Produce energy
- Survive
- Be cost effective

That means that the turbine design must not only be functional in terms of extracting energy. It must also be structurally sound so that it can withstand the loads it experiences, and the costs to make it structurally sound must be commensurate with the value of the energy it produces.

The purpose of this chapter is to discuss the mechanical framework within which the turbine must be designed if it is to meet those three requirements. Conceptually, the chapter is divided into two parts. The first part provides some background on the fundamental principles relevant to wind turbine dynamics. The second part deals more directly with wind turbine motions, forces, and stresses.

The fundamentals section includes a very brief overview of the basics: statics, dynamics, and strength of materials. This overview is brief because it is assumed that the reader already has some familiarity with the concepts involved. A slightly longer discussion, however, is devoted to a few concepts of particular relevance. The fundamentals section includes more detailed discussion of two other topics: vibrations and fatigue. These topics are discussed in more detail because it is presumed that they may be less familiar to many readers. The topic of vibrations is important not only because the concepts are relevant to actual structural vibrations, but also because the concepts can be used to describe many aspects of the behavior of the wind turbine rotor. The topic of fatigue is included, because it provides an understanding of how turbine components would be expected to withstand a lifetime of continuously varying loads.

The section on wind turbine motions and forces examines the response of the wind turbine to aerodynamic forces, using progressively more detailed approaches. The first approach uses a simple steady state approach that follows directly from the discussions in Section 3.7 of rotor performance. The second approach includes the effects of the rotation of wind turbine rotor. In addition, the operating environment is not assumed to be uniform, so the effects of vertical wind shear, yaw motion, and turbine orientation are all accounted for. This approach uses a number of simplifications to make the problem solvable, resulting in solutions that are more representative than precise. Finally, some of the more detailed approaches to investigating wind turbine dynamics are discussed. These are in general more comprehensive or specialized, but are also more complex and less intuitive. For that reason, they are only discussed briefly and the reader is referred to other sources for more details.

Although the fundamentals apply to all types of wind turbine designs, the focus of this next section is exclusively on horizontal axis turbines.

4.1.1 Types of loads

In this chapter, the term 'load' refers to forces or moments that may act upon the turbine. The loads that a turbine may experience are of primary concern in assessing the turbine's structural requirements. These loadings may be divided into seven types:

- Static (non-rotating)
- Steady (rotating)
- Cyclic
- Transient
- Impulsive
- Stochastic
- Resonance-induced loads

The key characteristics of these loads and some examples for wind turbines are summarized below.

4.1.1.1 Static (non-rotating) loads

Static loads, as used in this text, refer to constant (non-time varying) loads that impinge on a non-moving structure. For example, a steady wind blowing on a stationary wind turbine would induce static loads on the various parts of the machine.

4.1.1.2 Steady (rotating) loads

Steady loads are also non-time varying loads, but the structure may be in motion. For example, a steady wind blowing on a rotating wind turbine rotor while it is generating power would induce steady loads on the blades and other parts of the machine. The calculation of these particular loads has been detailed in Chapter 3.

4.1.1.3 Cyclic loads

Cyclic loads are those which vary in a regular or periodic manner. The term applies particularly to loads which arise due to the rotation of the rotor. Cyclic loads arise as a result of such factors as the weight of the blades, wind shear, and yaw motion. Cyclic loads may also be associated with the vibration of the turbine structure or some of its components.

As noted in Section 3.12, a load which varies an integral number of times in relation to a complete revolution of the rotor is known as a 'Per rev' load, and given the symbol P. For example, a blade rotating in wind with wind shear will experience a cyclic $1P$ load. If the turbine has three blades the main shaft will experience a cyclic $3P$ load.

4.1.1.4 Transient loads

Transient loads are time-varying loads which arise in response to some temporary external event. There may be some oscillation associated with the transient response, but they eventually decay. Examples of transient loads include those in the drive train resulting from the application of a brake.

4.1.1.5 Impulsive loads

Impulsive loads are time varying loads of relatively short duration, but of perhaps significant peak magnitude. One example of an impulsive load is that experienced by a blade of a downwind rotor when it passes behind the tower (through the 'tower shadow'). Two bladed rotors are often 'teetered' or pinned at the low-speed shaft. This allows the rotor to rock back and forth, reducing bending loads on the shaft, but necessitating the use of teeter dampers. The force on a teeter damper when the normal range of teeter is exceeded is another example of an impulsive load.

4.1.1.6 Stochastic loads

Stochastic loads are time varying, as are cyclic, transient, and impulsive loads. In this case, the loading varies in a more apparently random manner. In many cases the mean value may be relatively constant, but there may be significant fluctuations from that mean. Examples of stochastic loads are those which arise in the blades when the wind is very turbulent.

4.1.1.7 Resonance-induced loads

Resonance-induced loads are cyclic loads that result from the dynamic response of some part of the wind turbine being excited at one of its natural frequencies. They may reach high magnitudes. Resonance-induced loads are to be avoided whenever possible, but may occur under unusual operating circumstances or due to poor design. While these loads are not truly a different type of load, they are mentioned separately because of their possibly serious consequences.

4.1.2 Sources of loads

Three are four primary sources of loads to consider in wind turbine design:

- Aerodynamics
- Gravity
- Dynamic interactions
- Mechanical control

The loads and motions may interact in complicated ways, but it is worthwhile considering them separately. The follow sections describe each of these loads.

4.1.2.1 Aerodynamics

The first source of wind turbine loads that one would typically consider is aerodynamics. Aerodynamics, particularly as related to power production, was described in Chapter 3. The loadings of particular concern in the structural design are those which could arise in very high winds, or those which generate fatigue damage. When a wind turbine is stationary in high winds, drag forces are the primary consideration. When the turbine is operating, it is lift forces which create the aerodynamic loadings of concern.

4.1.2.2 Gravity

Gravity is an important source of loads on the blades of large turbines, although it is less so on smaller machines. In any case, tower top weight is significant to the tower design and to the installation of the machine.

4.1.2.3 Dynamic interactions

Motion induced by aerodynamic and gravitational forces in turn induces loads in other parts of the machine. For example, virtually all horizontal axis wind turbines allow some motion about a yaw axis. When yaw motion occurs while the rotor is turning there will be induced gyroscopic forces. These forces can be substantial when the yaw rate is high.

4.1.2.4 Mechanical control

Control of the wind turbine may sometimes be a source of significant loads. For example, starting a turbine which uses an induction generator or stopping the turbine by applying a brake can generate substantial loads throughout the structure.

4.1.3 Effects of loads

The loadings experienced by a wind turbine are important in two primary areas: (i) ultimate strength and (ii) fatigue. Wind turbines may occasionally experience very high loads, and they must be able to withstand those loads. Normal operation is accompanied by widely varying loads, due to starting and stopping, yawing, and the passage of blades through continuously changing winds. These varying loads can fatigue machine components, such that a given component may eventually fail at much lower loads than it would have when new.

4.2 General Principles

This section presents an overview of some of the principles of basic mechanics and dynamics which are of particular concern in wind turbine design. The fundamentals of the mechanics of wind turbines are essentially the same as those of other similar structures. For that reason, the topics taught in engineering courses in statics, strength of materials, and dynamics are equally applicable here. Topics of particular relevance include Newton's Second Law, especially when applied in the normal-tangential coordinate system; moments of inertia; bending moments, and stresses and strains. These topics are well discussed in many physics or engineering texts, such as Pytel and Singer (1987), Merriam (1978), Beer and Johnston (1976) and Den Hartog (1961) and will not be pursued here in any detail, except where a particular example is especially relevant.

4.2.1 Selected topics from basic mechanics

There are a few topics from basic engineering mechanics which are worth singling out, because they have a particular relevance to wind energy, and may not be familiar to all readers. These are summarized briefly below.

4.2.1.1 Inertia forces

In most current teaching of dynamics, the forces that are considered are exclusively real forces. It is sometimes convenient, however, to describe certain accelerations in terms of fictitious 'inertia forces'. This is often done in the case of rotating systems, including in the analysis of wind turbine rotor dynamics. For example, the effect of normal acceleration is accounted for by the inertial centrifugal force. The effect of the inertia force, as reflected in the 'Principle of D'Alembert,' is that the sum of all forces acting on a particle, including the inertia force is zero. The method is most useful when dealing with larger rigid bodies, which can be considered to be made up of a large number of particles rigidly connected together. Specifically, the Principle of D'Alembert states that the internal forces in a rigid body having accelerated motion can be calculated by the methods of statics on that body under the influence of the external and inertia forces. Furthermore, a rigid body of any size will behave as a particle if the resultant of its external forces passes through its center of gravity.

4.2.1.2 Bending of cantilevered beams

The bending of beams is an important topic of strength of materials. A wind turbine blade is basically a cantilevered beam so the topic is of particular relevance. One simple but interesting example is a uniformly loaded cantilevered beam. For this case, the bending moment diagram, $M(x)$, is described by an inverted partial parabola:

$$M(x) = \frac{w}{2}(L - x)^2 \tag{4.2.1}$$

The maximum bending moment, M_{max}, is at the side of attachment and is given by:

$$M_{max} = \frac{w L^2}{2} = \frac{W L}{2} \tag{4.2.2}$$

where L is the length of the beam, x is the distance from the fixed end of the beam, w is the loading (force/unit length), W is the total load.

The maximum stress in the beam is also at the point of attachment and, in addition, is at the maximum distance, c, from neutral axis. For a wind turbine blade in bending, the neutral axis would be nearly the same as the chord line and c would be approximately half the airfoil thickness.

In equation form, the maximum stress σ_{max} in a beam with area moment of inertia I would be:

$$\sigma_{max} = \frac{M_{max}c}{I} \tag{4.2.3}$$

4.2.1.3 Rigid body planar rotation

Two dimensional rotation When a body, such as a wind turbine rotor, is rotating, it acquires angular momentum. Angular momentum, $\boldsymbol{H}$, is characterized by a vector whose magnitude is the product of the rotational speed Ω and the polar mass moment of inertia, J. The direction of the vector is determined by the right hand rule (whereby if the fingers of the right hand are curled in the direction of rotation, the thumb points in the appropriate direction). In equation form:

$$\boldsymbol{H} = J\,\boldsymbol{\Omega} \tag{4.2.4}$$

From basic principles, the sum of the applied moments, M, about the mass center is equal to the time rate of change of the angular momentum about the mass center. That is:

$$\sum M = \dot{\boldsymbol{H}} \tag{4.2.5}$$

In most situations of interest in wind turbine dynamics the moment of inertia can be considered constant. Therefore the magnitude of the sum of the moments is:

$$\left|\sum M\right| = J\,\dot{\Omega} = J\,\alpha \tag{4.2.6}$$

where α is the angular acceleration of the inertial mass.

In the context of this text, a continuous moment applied to a rotating body is referred to as torque, and denoted by Q. The relation between applied torques and angular acceleration, α, is analogous to that between force and linear acceleration:

$$\sum Q = J\,\alpha \tag{4.2.7}$$

When rotating at a constant speed, there is no angular acceleration or deceleration. Thus the sum of any applied torques must be zero. For example, if a wind turbine rotor is turning at a constant speed in a steady wind, the driving torque from the rotor must be equal to the generator torque plus the loss torques in the drive train.

Rotational power/energy A rotating body contains kinetic energy, E, given by:

$$E = \frac{1}{2}J\,\Omega^2 \tag{4.2.8}$$

The power P consumed or generated by a rotating body is given by the product of the torque times the rotational speed:

$$P = Q\Omega \tag{4.2.9}$$

4.2.1.4 Gears and gear ratios

Gears are frequently used to transfer power from one shaft to another, while maintaining a fixed ratio between the speeds of the shafts. While the input power in an ideal gear train remains equal to the output power, the torques and speed vary in inverse proportion to each other. In going from a smaller gear (1) to a larger one (2), the rotational speed drops, but the torque increases. In general:

$$Q_1\Omega_1 = Q_2\Omega_2 \tag{4.2.10}$$

The ratio between the speeds of two gears, Ω_1 / Ω_2, is inversely proportional to the ratio of the number of teeth on each gear, N_1 / N_2. The latter is proportional to the gear diameter. Thus:

$$\Omega_1 / \Omega_2 = N_2 / N_1 \tag{4.2.11}$$

When dealing with geared systems, consisting of shafts, inertias and gears, as shown in Figure 4.1, it is possible to refer shaft stiffness (Ks) and inertias (Js) to equivalent values on a single shaft. (Note that in the following it is assumed that the shafts themselves have no inertia.) This is done by multiplying all stiffnesses and inertias of the geared shaft by n^2 where n is the speed ratio between the two shafts. The equivalent system is shown in Figure 4.2. These relations can be derived by applying principles of kinetic energy (for inertias) and potential energy (for stiffnesses), as described in Thomson (1981). More information on gears and gear trains is provided in Chapter 6.

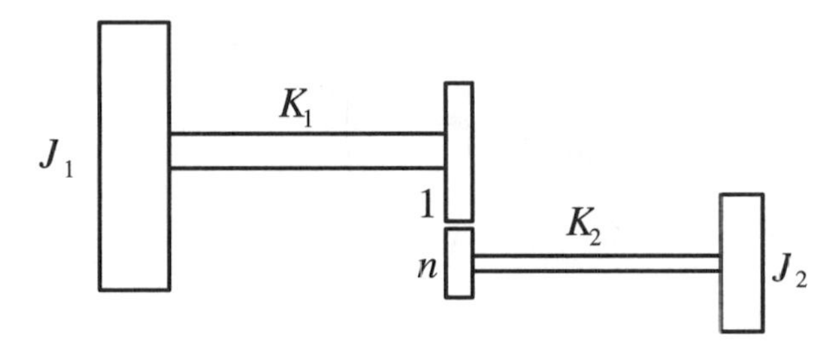

Figure 4.1 Geared system; J, inertia; K, stiffness; n, speed ratio between the shafts; subscripts 1 and 2, gear 1 and 2

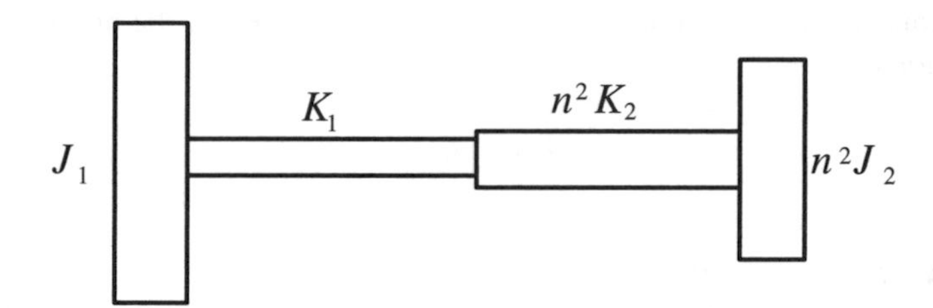

Figure 4.2 System equivalent to geared system; *J*, inertia; *K*, stiffness; *n*, speed ratio between the shafts; subscripts 1 and 2, gear 1 and 2

4.2.1.5 Gyroscopic motion

Gyroscopic motion is of particular concern in the design of wind turbines, because yawing of the turbine while the rotor is spinning may result in significant gyroscopic loads. The effects of gyroscopic motion are illustrated in the Principal Theorem of the Gyroscope, which is summarized below.

In this example, it is assumed that a rigid body, with a constant polar mass moment of inertia of *J*, is rotating with angular momentum $J\Omega$. The Principal Theorem of the Gyroscope states that if a gyroscope of angular momentum $J\Omega$ rotates with speed ω about an axis perpendicular to Ω ('precesses'), then a couple, $J\Omega\omega$, acts on the gyroscope about an axis perpendicular to both the gyroscopic axis, Ω, and the precession axis, ω. Conversely, an applied moment that is not parallel to Ω can induce precession.

Gyroscopic motion may be considered with the help of Figure 4.3. A bicycle wheel of weight *W* is shown rotating in the counter clockwise direction, supported at the end of one axle by a string. The wheel would fall down if it were not rotating. In fact, it is rotating, and rather than falling, it precesses in a horizontal plane (counterclockwise when seen from above).

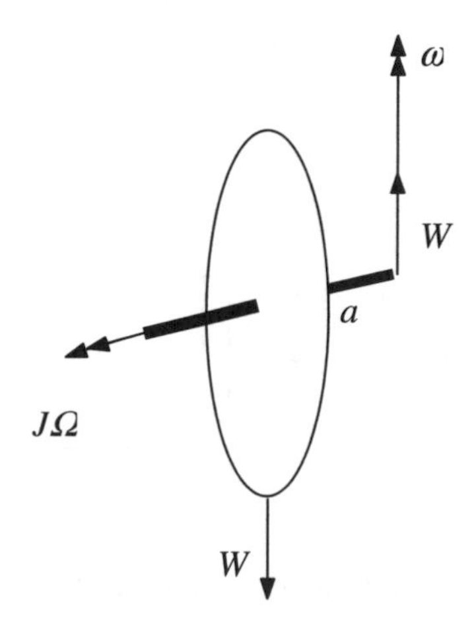

Figure 4.3 Gyroscopic motion; *a*, distance to weight; *J*, inertia; *W*, weight; ω, rate of precession; Ω, angular velocity

The moment acting on the wheel is *Wa*, so

$$Wa = J\Omega\omega \tag{4.2.12}$$

Thus the rate of precession is

$$\omega = \frac{Wa}{J\Omega} \tag{4.2.13}$$

The relative directions of rotation are related to each other by cross products and the right-hand rule as follows:

$$\sum \boldsymbol{M} = Wa = \omega \times J\Omega \tag{4.2.14}$$

where, $\boldsymbol{M}$, ω, and Ω are the moment, rate of precession and angular velocity vectors, respectively.

4.2.2 Vibrations

The term vibration refers to the limited reciprocating motion of a particle or an object in an elastic system. The motion goes to either side of an equilibrium position. Vibration is important in wind turbines, because they are partially elastic structures, and they operate in an unsteady environment that tends to result in a vibrating response. The presence of vibrations can result in deflections that must be accounted for in turbine design and can also result in the premature failure (due to fatigue) of the materials which comprise the turbine. In addition, much of wind turbine operation can be best understood in the context of vibratory motion. The following section provides an overview of those aspects of vibrations most important to wind turbine applications.

4.2.2.1 Single degree of freedom systems

Undamped vibrations The simplest vibrating system is mass m attached to a massless spring with spring constant *k*, as shown in Figure 4.4. When displaced a distance *x* and allowed to freely move, the mass will vibrate back and forth.
Applying Newton's Second Law, the governing equation is:

$$m\ddot{x} = -kx \tag{4.2.15}$$

When $t = 0$ at $x = x_0$ the solution is:

$$x = x_0 \cos(\omega_n t) \tag{4.2.16}$$

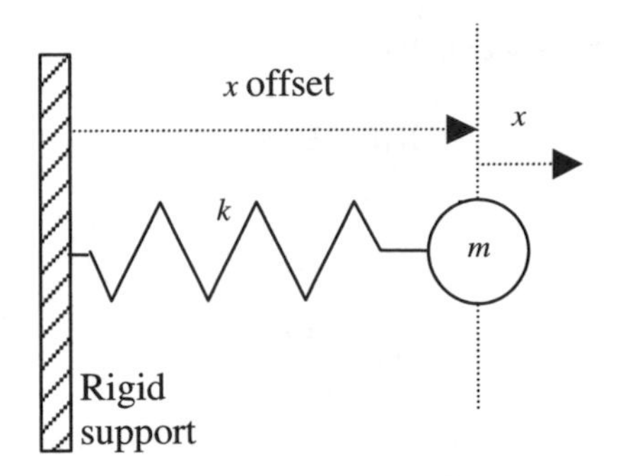

Figure 4.4 Undamped vibrating system; k, spring constant; m, mass; x, displacement

where $\omega_n = \sqrt{k/m}$ is the natural frequency of the motion. In general, the solution contains two sinusoids:

$$x = x_0 \cos(\omega_n t) + \frac{\dot{x}_0}{\omega_n} \sin(\omega_n t) \tag{4.2.17}$$

where $\dot{x}_0$ is the velocity at t = 0.

The solution can also be written in terms of a single sinusoid of amplitude C and phase angle ϕ. That is:

$$x = C \sin(\omega_n t + \phi) \tag{4.2.18}$$

The amplitude and phase angle can be expressed in terms of the other parameters as:

$$C = \sqrt{x_0^2 + \left(\frac{\dot{x}_0}{\omega_n}\right)^2} \tag{4.2.19}$$

$$\phi = \tan^{-1}\left[\frac{x_0 \omega_n}{\dot{x}_0}\right] \tag{4.2.20}$$

Damped vibrations Vibrations as described above will continue indefinitely. In all real vibrations, the motions will eventually die out. This effect can be modeled by including a viscous damping term. Damping involves a force, usually assumed to be proportional to the velocity, which opposes the motion. Then the equation of motion becomes:

$$m \ddot{x} = -c \dot{x} - k x \tag{4.2.21}$$

where c is the damping constant and k is the spring constant. Depending on the ratio of the damping and spring constant, the solution may be oscillatory ('underdamped') or non-oscillatory ('overdamped'). The limiting case between the two is 'critically damped,' in which case:

$$c = c_c = 2\sqrt{k\,m} = 2\,m\,\omega_n \tag{4.2.22}$$

For convenience a non-dimensional damping ratio $\xi = c / c_c$ is used to characterize the motion. For $\xi < 1$ the motion is underdamped; for $\xi > 1$ the motion is overdamped.

The solution for underdamped oscillation is:

$$x = Ce^{-\xi \omega_n t} \sin(\omega_d t + \phi) \tag{4.2.23}$$

where $\omega_d = \omega_n \sqrt{1 - \xi^2}$ is the natural frequency for damped oscillation. Note that the frequency of damped oscillation is slightly different from that of undamped vibration. The amplitude, C, and phase angle, ϕ, are determined from initial conditions.

Forced harmonic vibrations Consider the mass, spring, and damper discussed above. Suppose that it is driven by a sinusoidal force of magnitude F_0 and frequency ω (which is not necessarily equal to ω_n or ω_d). The equation of motion is then:

$$m\ddot{x} + c\dot{x} + kx = F_0 \sin(\omega\, t) \tag{4.2.24}$$

It may be shown that the steady state solution to this equation is:

$$x(t) = \frac{F_0}{k} \frac{\sin(\omega\, t - \phi)}{\sqrt{\left[1 - (\omega / \omega_n)^2\right]^2 + \left[2\xi(\omega / \omega_n)^2\right]}} \tag{4.2.25}$$

The phase angle in this case is:

$$\phi = \tan^{-1}\left[\frac{2\,\xi\,(\omega / \omega_n)}{1 - (\omega / \omega_n)^2}\right] \tag{4.2.26}$$

It is of particular interest to consider the non-dimensional response amplitude which is given by:

$$\frac{x\,k}{F_0} = \frac{1}{\sqrt{\left[1 - (\omega / \omega_n)^2\right]^2 + \left[2\xi(\omega / \omega_n)^2\right]}} \tag{4.2.27}$$

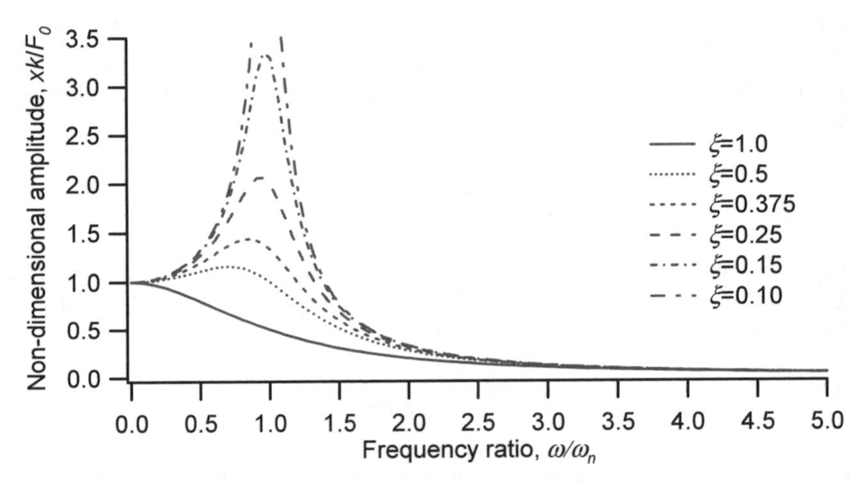

Figure 4.5 System responses to forced vibrations; ξ , non-dimensional damping ratio

As the excitation (forcing) frequency gets closer to the natural frequency, the amplitude of the response gets larger. Increasing the damping reduces the peak value, and also shifts it slightly. Furthermore, although the peak is highest when the excitation frequency equals the natural frequency (ignoring the effect of damping), there may still be a significant increase in amplitude when the excitation is close to the natural frequency, as shown in Figure 4.5.

Rotational vibration If a body with polar mass moment J is attached to a rigid support via a rotational spring with rotational stiffness k_θ, as shown in Figure 4.6, its equation of motion is given by:

$$J\ddot{\theta} = -k_\theta \theta \tag{4.2.28}$$

Solutions for rotational vibration are analogous to those for linear vibrations discussed previously. Rotational vibration is of particular importance in wind turbine design for shaft systems and drive train dynamics. It is also used in the linearized hinge–spring rotor model which is described in Section 4.3.2.

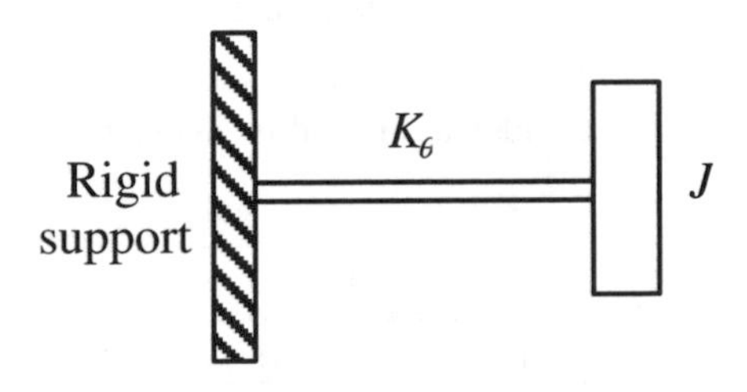

Figure 4.6 Rotational vibrating system; J, polar mass moment of inertia; k_θ , rotational stiffness of spring

4.2.2.2 Vibration of cantilevered beams

It is of interest to consider the vibration of a cantilevered beam, since many aspects of wind turbines have similarities to cantilevered beams. These include in particular the tower of the turbine, and the blades.

Modes and mode shapes Recall that a single vibrating mass on a spring oscillates with a single characteristic natural frequency. There is also only one path that the mass will take during its motion. For multiple masses the number of natural frequencies and possible paths will increase. For continuous objects there are actually an infinite number of natural frequencies. To each natural frequency there corresponds a characteristic mode shape of vibration. In practice, however, only the lowest few natural frequencies of a beam are usually important.

Vibration of uniform cantilevered beams Vibration of an ideal, uniform beam of constant cross section and material properties can be modelled by an analytic equation known as the Euler Equation for beams. This equation is particularly useful because it allows first order approximations of natural frequencies to be made very easily for many beams. The reader is referred to other sources (e.g. Thomson, 1981) for the details of the development of the Euler beam equation. This section will confine itself to presenting the equation for a cantilevered beam in enough detail for it to be applied.

The form of the Euler equation for the deflection, y_i, of a uniform cantilevered beam of length L and mode shape i is:

$$y_i = A\left\{\cosh\left(\frac{(\beta L)_i}{L}x\right)-\cos(\beta x)-\frac{\sinh(\beta L)_i-\sin(\beta L)_i}{\cosh(\beta L)_i+\cos(\beta L)_i}\left[\sinh\left(\frac{(\beta L)_i}{L}x\right)-\sin\left(\frac{(\beta L)_i}{L}x\right)\right]\right\}$$

(4.2.29)

The parameter $(\beta L)_i$ is related to the natural frequencies, ω_i, density per unit length, $\tilde{\rho}$, area moment of inertia, I, and modulus of elasticity, E, of the beam by the following equation:

$$(\beta L)_i^4 = \tilde{\rho}\omega_i^2/(EIL^4) \qquad (4.2.30)$$

From Equation 4.2.30, the natural frequencies (radians/second) are:

$$\omega_i = \frac{(\beta L)_i^2}{L^2}\sqrt{\frac{EI}{\tilde{\rho}}} \qquad (4.2.31)$$

for values of $(\beta L)_i$ that solve:

$$\cosh(\beta L)_i \cos(\beta L)_i + 1 = 0 \qquad (4.2.32)$$

Equation 4.2.29 enables us to determine the mode shapes of the vibration. Note that since A is not known it may be determined by assuming a deflection of $y_i = 1$ at the free end of the beam.

The products $(\beta L)_i^2$ are constants, and for the case of a cantilevered beam the numerical values for the first three modes are 3.52, 22.4, and 61.7. Figure 4.7 illustrates the shape of the first three vibration modes of a uniform cantelivered beam, based on equations 4.2.29 and 4.2.32.

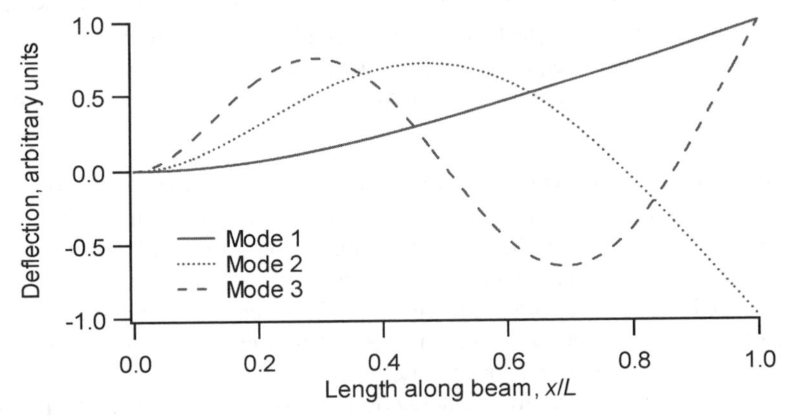

Figure 4.7 Vibration modes of uniform cantilevered beam

Vibration of general beams The previous discussion dealt with a uniform beam. This section concerns a more general case. It summarizes the use of the Myklestad method, which involves modeling the beam with a number of lumped masses connected by massless beam elements. The Myklestad method is quite versatile and can be applied to nearly any beam, but as before, the discussion will be confined to the cantilevered beam. More details may be found in Thomson (1981).

The method is illustrated in Figure 4.8. The beam is divided into n - 1 sections with the same number of masses, m_i. An additional station is at the point of attachment. All of the distances λ_i between masses are the same. In the figure it is assumed that the masses are located at the center of equal sized sections, so the distance from the mass closest to the attachment is 1/2 of the others. The flexible connections have moments of inertia I_i and modulus of elasticity E_i. The beam in this example may be rotating with a speed Ω about an axis perpendicular to the beam and passing through station 'n'.

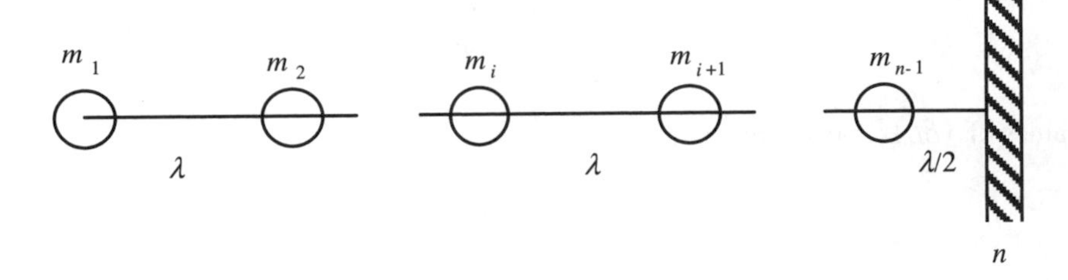

Figure 4.8 Model of cantilevered beam; m_i, mass; λ_i, distance between masses; n, number of stations

The Myklestad method involves solving a set of sequence equations. The sequence equations can be developed by considering the forces and moments acting on each of the masses and the point of attachment. Figure 4.9 illustrates the free body diagram of one section of a rotating beam. The diagram shows shear forces, S_i, inertial (centrifugal) forces, F_i, bending moments, M_i, deflections, y_i, and angular deflections, ϑ_i.

The complete equations for a rotating beam follow. The centrifugal forces at distance x_j from the attachment are:

$$F_i = \Omega^2 \sum_{j=1}^{i-1} m_j x_j \tag{4.2.33a}$$

$$F_{i+1} = F_i + \Omega^2 m_i x_i \tag{4.2.33b}$$

Using small-angle approximations, the shear forces are:

$$S_{i+1} = S_i - m_i \omega^2 y_i - F_{i+1} \theta_i \tag{4.2.34}$$

The moments are:

$$M_{i+1} = \left[M_i - S_{i+1} \left(\lambda_i - F_{i+1} \frac{\lambda_i^3}{3\, E_i I_i} \right) + \theta_i\, \lambda_i\, F_{i+1} \right] \Big/ \left(1 - F_{i+1} \frac{\lambda_i^2}{2\, E_i I_i} \right) \tag{4.2.35}$$

The angle of the beam sections from horizontal are:

$$\theta_{i+1} = \theta_i + M_{i+1} \left(\frac{\lambda_i}{E_i I_i} \right) + S_{i+1} \left(\frac{\lambda_i^2}{2 E_i I_i} \right) \tag{4.2.36}$$

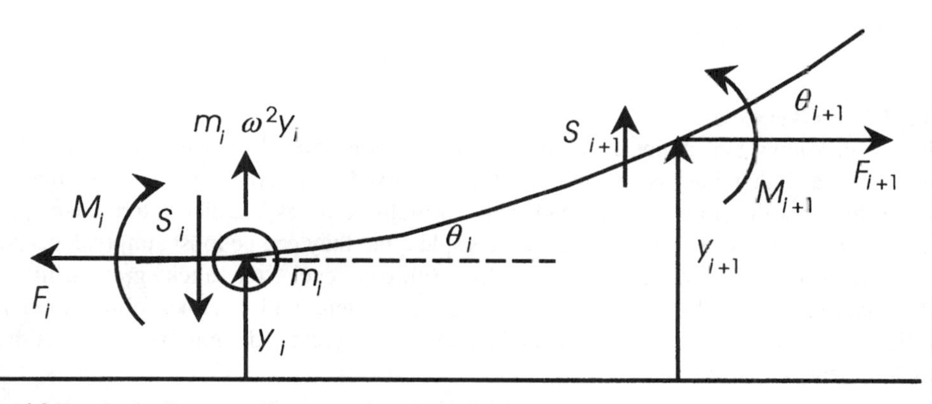

Figure 4.9 Free body diagram of beam section; for definition of variables, see text

Finally, the deflections from horizontal line, passing through the fixed end are:

$$y_{i+1} = y_i + \theta_i \lambda_i + M_{i+1}\left(\frac{\lambda_i^2}{2\,E_i I_i}\right) + S_{i+1}\left(\frac{\lambda_i^3}{3\,E_i I_i}\right) \tag{4.2.37}$$

The series of equations are solved by iteratively finding natural frequencies which result in a calculated deflection of zero at the point of attachment. The process – normally done with a computer – begins by assuming a natural frequency ω and performing a series of calculations. The calculations are repeated with a new assumption for the natural frequency until a deflection sufficiently close to zero is found. Two sets of calculations are undertaken. They start at the free end of the beam first with $y_{1,1} = 1$ and $\theta_{1,2} = 0$ and then with $y_{1,2} = 0$ and $\theta_{1,2} = 1$ where the second subscript refers to the sets of calculations. Calculations are performed sequentially for each section until section n is reached. The calculations yield $y_{n,1}$, $y_{n,2}$, $\theta_{n,1}$ and $\theta_{n,2}$. The desired deflection (which should approach zero) is:

$$y_n = y_{n,1} - y_{n,2}\left(\theta_{n,1} / \theta_{n,2}\right) \tag{4.2.38}$$

The entire process can be repeated to find additional natural frequencies. There should be as many natural frequencies as there are masses. Because inertial forces on the rotating beam effectively stiffen the beam, natural frequencies of a rotating beam will be higher than those of the same non-rotating beam.

4.2.2.3 Torsional Systems

Many wind turbine components, particularly those in the drive train, can be modelled by a series of discs connected by shafts. In these models, the discs are assumed to have inertia, but are completely rigid, whereas the shafts have stiffness, but no inertia. The natural frequencies for such systems can be determined by Holzer's Method. As with the Myklestad Method sequence equations can be used to determine each angle and torque along the shaft. The details of this method are beyond the scope of this book. The reader is referred to Thomson (1981) or similar texts for more information.

4.2.3 Fatigue

4.2.3.1 Overview

It is well known that many materials which can withstand a load when applied once, will not survive if that load is applied and then removed ('cycled') a number of times. This increasing inability to withstand loads applied multiple times is called fatigue damage. The underlying causes of fatigue damage are complex, but they can be most simply conceived of as deriving from the growth of tiny cracks. With each cycle, the cracks grow a little, until the material fails. This simple view is also consistent with another observation about fatigue: the lower the magnitude of the load cycle, the greater the number of cycles that the material can withstand.

Wind turbines, by their very nature are subject to a great number of cyclic loads. The lower bound on the number of many of the fatigue-producing stress cycles in turbine components is proportional to the number of blade revolutions over the turbine's lifetime. The total cycles, n_L, over a turbine's lifetime would be:

$$n_L = 60\, k\, n_{rotor} H_{op} Y \tag{4.2.39}$$

where k is the number of cyclic events per revolution, n_{rotor} is the rotational speed of rotor (rpm), H_{op} is the operating hours per year and Y is the years of operation.

For blade root stress cycles, k would be at least equal to 1 while, for the drive train or tower, k would be at least equal to the number of blades. A large turbine with an rpm of approximately 30 to 70 operating 4000 hours per year would experience from 10^8 to 10^9 cycles over a 20-year lifetime. This may be compared to many other manufactured items, which would be unlikely to experience more than 10^6 cycles over their lifetime.

In fact the number of cyclic events on a blade can be much more than once per revolution of the rotor. Figure 4.10 illustrates a 5 second time trace of the bending moment in the flapwise (axial) direction at the blade root of a typical three-bladed wind turbine, using arbitrary units. The rotor in this case turns at just a little more than once per second. As can be seen, the largest cycles do occur at approximately that frequency, but smaller ones occur considerably more often. In any case there are a large number of cycles, of varying frequency and magnitude, and these would result in fluctuating stresses. These all contribute to fatigue damage, and need to be considered in wind turbine design.

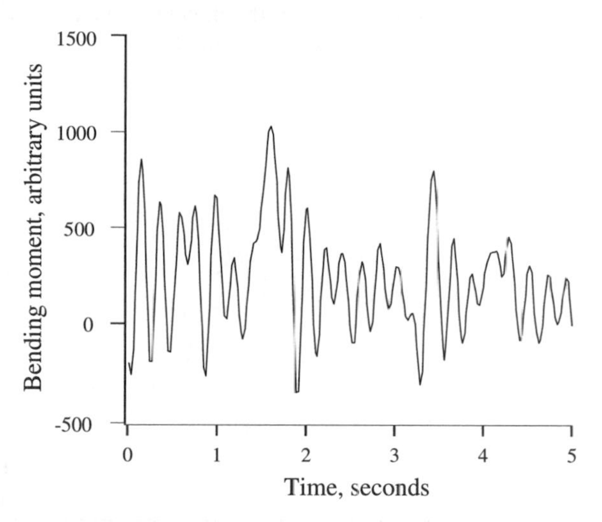

Figure 4.10 Sample root flap bending moment

4.2.3.2. Assessment of fatigue

Procedures have been developed to estimate fatigue damage. Many of these techniques were developed for analysis of metal fatigue, but they have been extended to other materials, such as composites, as well. The following sections summarize fatigue analysis methods most commonly used in wind turbine design. More detail on fatigue in general may be found in

Shigley and Mischke (1989), Spotts (1985) and Ansell (1987). Information on fatigue in wind turbines is given in Garrad and Hassan (1986) and Sutherland et al. (1995).

Fatigue life (*S–N*) curve Fatigue resistance of materials is traditionally tested by subjecting a sample to a sinusoidally applied load until failure. One type of test is a rotating beam test. A sample is mounted on a test machine and loaded to a given stress with a side load. The sample is then rotated in the machine so that the stress reverses every cycle. The test is continued until the sample fails. The first load is somewhat less than the ultimate load. The number of cycles and the load is recorded. The load is then reduced on another sample and the process is repeated. The data is summarized in a fatigue life, or '*S–N*', curve, where *S* refers to the stress (in spite of the customary use of σ to indicate stress) and *N* refers to the number of cycles to failure. Figure 4.11 illustrates a typical *S–N* curve. With tests of this type it is important to note that (i) the mean stress is zero, (ii) the stresses are fully reversing.

Most manufactured items of commercial interest, with the exception of wind turbines, do not experience more than 10 million cycles in their lifetimes. Thus, if a sample does not fail after 10^7 cycles at a particular stress (as the loads are progressively reduced), the stress is referred to as the endurance limit, σ_e. Endurance limits typically reported are in the range of 20%–50% of the ultimate stress for many materials. In reality the material may not actually have a true endurance limit, and it may be inappropriate to use that assumption in wind turbine design.

Alternating stresses with non-zero mean Alternating stresses typically do not have a zero mean. In this case they are characterized by the mean stress, σ_m, the maximum stress, σ_{max}, and the minimum stress σ_{min} as shown in Figure 4.12.

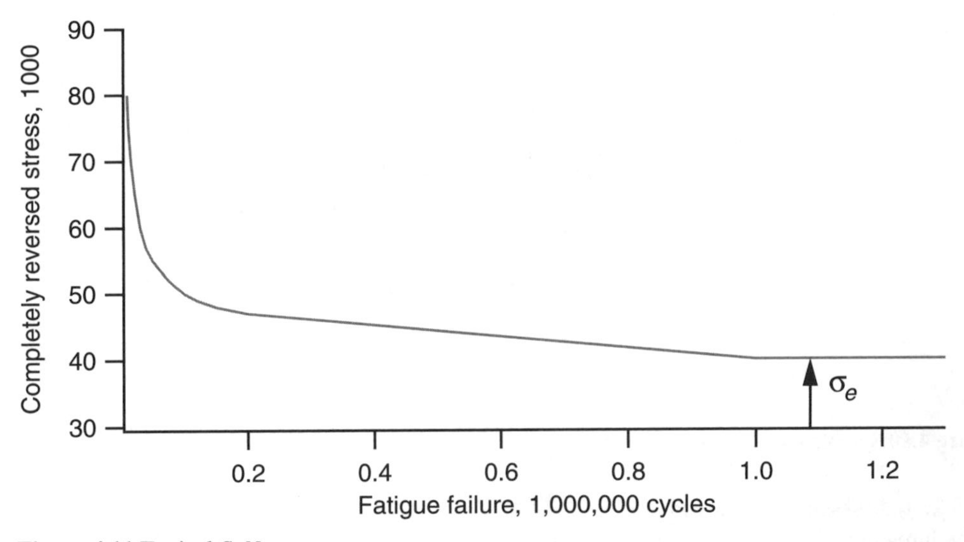

Figure 4.11 Typical *S–N* curve

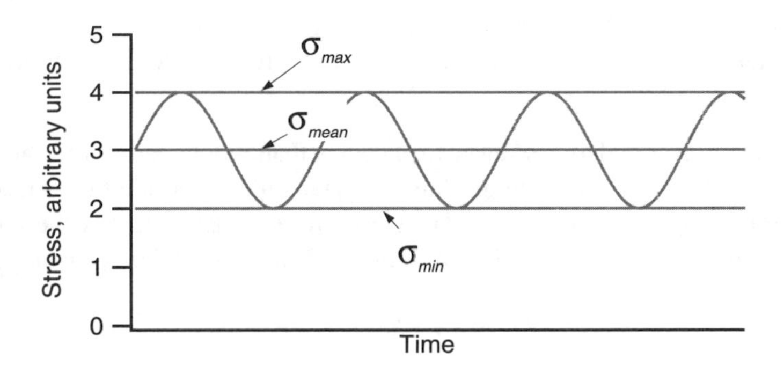

Figure 4.12 Alternating stresses with non-zero mean

The stress range, $\Delta\sigma$, is defined as the difference between the maximum and the minimum.

$$\Delta\sigma = \sigma_{\max} - \sigma_{\min} \tag{4.2.40}$$

The stress amplitude, σ_a, is half the range:

$$\sigma_a = (\sigma_{\max} - \sigma_{\min})/2 \tag{4.2.41}$$

The stress ratio, R, is the ratio between the maximum and the minimum. That is:

$$R = \sigma_{\min} / \sigma_{\max} = (\sigma_m - \sigma_a)/(\sigma_m + \sigma_a) \tag{4.2.42}$$

For alternating stress with zero mean ('completely reversing'), $R = -1$.

The fatigue life for a given alternating stress depends on R (i.e., the mean). In fact, the allowed alternating stress for a given fatigue life will decrease as the mean increases. This relationship is often approximated by Goodman's Rule:

$$\sigma_a = \sigma_e(1 - \sigma_m / \sigma_u) \tag{4.2.43}$$

where σ_a is the allowable stress amplitude for a given fatigue life and mean stress, σ_e is the zero mean ($R = -1$) alternating stress for the desired fatigue life and σ_u is the ultimate stress.

Goodman's rule can be inverted and used with non-zero mean alternating stress data to find the equivalent zero mean alternating stress (see, for example, Ansell, 1987):

$$\sigma_e = \sigma_a/(1 - \sigma_m / \sigma_u) \tag{4.2.44}$$

This form of the rule allows conventional test data to be used to predict fatigue life. For example, suppose that a component was subjected to a mean stress of 80 x 10^6 N/m^2 and an alternating stress of 20 x10^6 N/m^2, and that the ultimate stress of the material is 120 x 10^6 N/m^2. The equivalent zero mean alternating stress (to use with an *S–N* diagram) would be 60 x 10^6 N/m^2.

Damage If a component undergoes fewer load cycles than would cause it to fail, it would still suffer damage. The component might fail at a later time, after additional loads cycles are applied. To quantify this, a fractional damage term, *d*, is defined. It is the ratio of the number of cycles applied, *n*, to the number of cycles to failure, *N*, at the given amplitude.

$$d = n / N \tag{4.2.45}$$

Cumulative damage and Miner's Rule A component may experience multiple load cycles of different amplitudes, as illustrated in Figure 4.13.

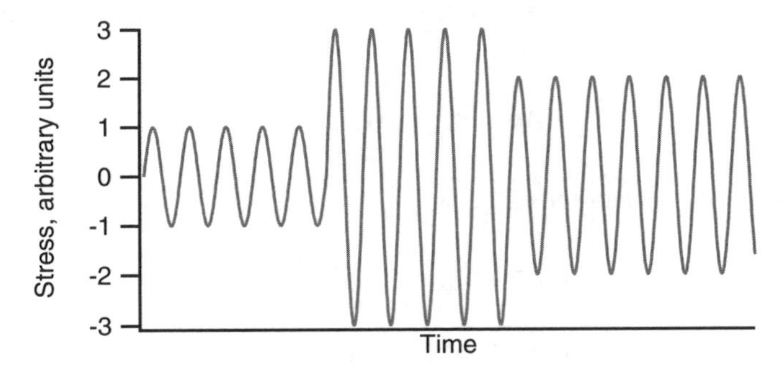

Figure 4.13 Load cycles of different amplitude

In this case, the cumulative damage, *D*, is defined by Miner's Rule to be the sum of the damages due to each of the cycles at each amplitude. The component is deemed to have failed when the total damage is equal to 1.0. In the general case of load cycles of *M* different amplitudes, the cumulative damage would be given by:

$$D = \sum n_i / N_i \leq 1 \tag{4.2.46}$$

where n_i is the number of cycles at the *i*th amplitude and N_i is the number of cycles to failure at the *i*th amplitude.

To illustrate this, suppose that for the material corresponding to Figure 4.13, the cycle life at 1 stress unit to be 20 cycles, at 4 units to be 10 cycles, and at 2.5 units to be 15 cycles. The cumulative damage due to the cycles shown would be:

$$D = (4/20) + (4/10) + (4/15) = 0.8667$$

Randomly Applied Load Cycles When loads are not applied in blocks (as was the case in Figure 4.13) but rather occur more randomly, it is difficult to identify individual load cycles.

Figure 4.14 illustrates such a situation.

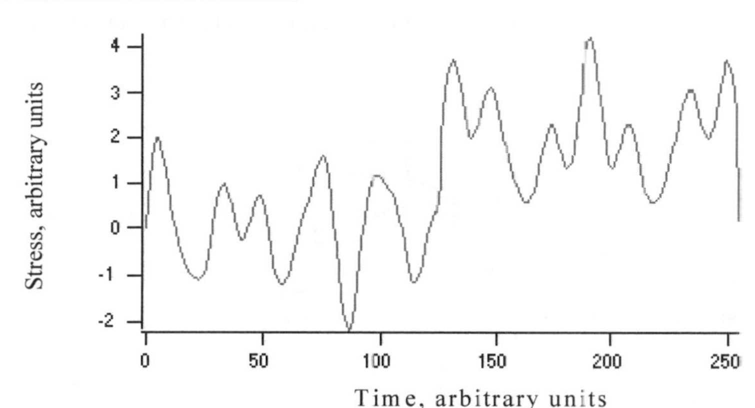

Figure 4.14 Random load cycles

A technique known as rainflow cycle counting (Downing and Socie, 1982) has been developed to identify alternating stress cycles and mean stresses from time series of randomly applied loads. Once the mean and alternating stress data have been found, they can be converted to zero mean alternating stresses and the total damage estimated using Miner's Rule, as discussed above.

The rainflow counting method is most appropriately applied to strain data, rather than stress data, and can deal with inelastic as well as elastic regions of the material. For most wind turbine applications, the material is in the elastic region, so either stress or strain data can be used. When strain is used results are ultimately converted to stresses.

The details of rainflow counting are beyond the scope of this text, but, in essence, the technique is as follows. Local highs and lows in the data are identified as 'peaks' or 'valleys'. The range between every peak and valley and between every valley and every peak are all considered to be 'half cycles'. The algorithm then pairs the half cycles to find complete cycles, and associates them with a mean.

The term 'rainflow' derives from an aspect of the algorithm, in which the completion of a cycle resembles rain water dripping from a roof (a peak) and meeting water flowing along another roof (from a valley below). In this view of the algorithm, the peak–valley history is imagined to be oriented vertically so that the 'rain' descends with increasing time. An example of a complete cycle is shown in Figure 4.15, which is derived from Figure 4.14. Here rain (indicated by heavy arrows) running down roof C–D encounters rain running down roof E–F. The cycle C–D–E is then eliminated from further consideration, but peak C is retained for use in identifying subsequent cycles. In this particular case, point F is slightly to the left of point B, so the next complete cycle is B–C–F.

4.2.3.3. Sources of wind turbine fatigue loads

The actual loads that contribute to fatigue of a wind turbine originate from a variety of sources. These include steady loads from high winds; periodic loads from wind shear, yaw error, yaw motion, and gravity; stochastic loads from turbulence; transient loads from such events as gusts, starts and stops, etc.; and resonance-induced loads from vibration of the structure. These are discussed in more detail later in this chapter and in Chapter 6.

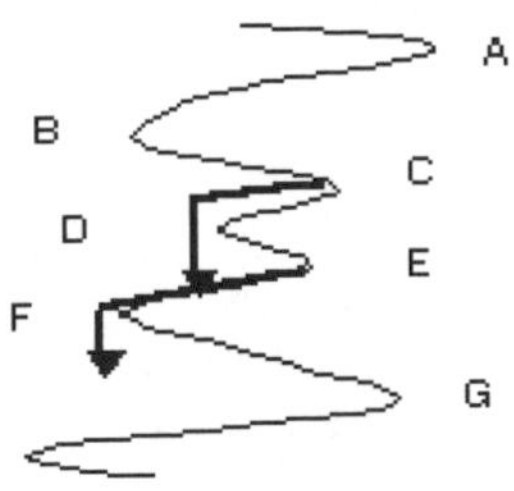

Figure 4.15 Rainflow cycle counting

4.3 Wind Turbine Rotor Dynamics

Imposed loads and dynamic interactions produce forces and motions in wind turbines which need to be understood during the design process. The effects of all of the various types of loads (static, steady, cyclic, transient, impulsive and stochastic) need to be determined. In this section two approaches for analyzing the forces and motions of a wind turbine are considered. The first approach uses a very simple ideal rigid rotor model to illustrate basic concepts about steady turbine loads. The second approach includes the development of a highly linearized dynamic model of a horizontal axis wind turbine. The model can be used to illustrate the nature of the turbine response to steady and cyclic loads. A few of the more detailed dynamic models are discussed in the final section. These can more accurately predict turbine response to stochastic or transient loads, but tend to be quite complex.

4.3.1 Loads in an ideal rotor

The most important rotor loads on a wind turbine are those associated with thrust on the blades and torque to drive the rotor. Modeling the rotor as a simple rigid, aerodynamically ideal rotor is useful to get a feeling for the steady loads on a wind turbine. As discussed in the Aerodynamics chapter (Chapter 3), the principle aerodynamic loads on an ideal rotor satisfying the Betz limit can be found fairly easily. When the rotor is operating at its ideal power coefficient, the wind speed in the wake is 1/3 of that in the free stream.

4.3.1.1 Thrust

As shown in Chapter 3, the thrust, T, may be found from the following equation.

$$T = C_T \frac{1}{2} \rho \pi R^2 U^2 \tag{4.3.1}$$

where C_T is the thrust coefficient, ρ is the density of air, R is the radius of rotor and U is the free stream velocity.

For the ideal case, $C_T = 8/9$. In terms of this simple model, then, total thrust on a given rotor varies only with the square of the wind speed.

4.3.1.2 Bending moments and stresses

Blade bending moments are usually designated as either flapwise and edgewise. Flapwise bending moments cause the blades to bend upwind or downwind. Edgewise moments are parallel to the rotor axis and give rise to the power producing torque. They are sometimes referred to as 'lead–lag.'

Axial forces and moments The flapwise bending moment at the root of an ideal blade of a turbine with multiple blades is given by the product of the thrust force per blade times 2/3 of the radius. This can be seen as follows.

Considering the rotor to be made up of a series of concentric annuli of width dr, the root flapwise bending moment M_β for a turbine with B blades is:

$$M_\beta = \frac{1}{B}\int_0^R r\left(\frac{1}{2}\rho\pi\frac{8}{9}U^2\,2r\,\mathrm{d}r\right) \tag{4.3.2}$$

Upon integrating and gathering terms the result is:

$$M_\beta = \frac{T}{B}\frac{2}{3}R \tag{4.3.3}$$

The maximum flapwise stress, $\sigma_{\beta,\max}$, due to bending at the root is given by Equation 4.2.3:

$$\sigma_{\beta,\max} = M_\beta c / I_b \tag{4.3.4}$$

where c is the distance from the flapwise neutral axis and I_b is the area moment of inertia of blade cross section at root.

It is worth noting that the term in brackets in Equation 4.3.2 is the same as the incremental normal force used in Chapter 3. The subscript, β, the angle of the blade deflection in the flap direction, is used to emphasize continuity with terms in the hinge–spring dynamic model discussed in the next section.

The shear force, S_β, in the root of the blade is simply the thrust divided by the number of blades.

$$S_\beta = T / B \tag{4.3.5}$$

In summary, for a given ideal rotor, bending forces and stresses vary with the square of the wind speed, and are independent of blade angular position (azimuth). Furthermore, rotors designed for higher tip speed ratio operation have smaller blades, so they would experience higher flapwise stresses.

Edgewise forces and moments As mentioned above, the edgewise moments give rise to the power producing torque. In terms of blade strength, edgewise moments are generally of less significance than their flapwise counterparts. The mean torque Q is the power divided by the rotational speed. For the ideal rotor, as shown in Chapter 3, torque is given by:

$$Q = \frac{P}{\Omega} = \frac{1}{\Omega}\frac{16}{27}\frac{1}{2}\rho\pi R^2 U^3 = \frac{8}{27}\rho\pi R^2 \frac{U^3}{\Omega} \tag{4.3.6}$$

In the more general case, torque can then be expressed in terms of a torque coefficient, $C_Q = C_P / \lambda$ (the power coefficient divided by the tip speed ratio), where:

$$Q = C_Q \frac{1}{2}\rho\pi R^2 U^2 \tag{4.3.7}$$

For an ideal rotor, the rotational speed varies with the wind speed, so torque varies as the square of the wind speed. Furthermore, rotors designed for higher tip speed ratio operation have lower torque coefficients, so they would experience lower torques (but not necessarily lower stresses). Again, according to this simple model there is no variation in torque with blade azimuth.

Edgewise (Lead–lag) Moment The bending moment in the edgewise direction (designated by ζ) at the root of single blade, M_ζ, is simply the torque divided by the number of blades:

$$M_\zeta = Q / B \tag{4.3.8}$$

There is no correspondingly simple relation for edgewise shear force S_ζ, but it can be found from integrating the tangential force, which is discussed in Chapter 3:

$$S_\zeta = \int_0^R \mathrm{d}F_T \tag{4.3.9}$$

4.3.2 Linearized hinge–spring blade rotor model

Actual wind turbine dynamics can get quite complicated. Loads can be variable, and the structure itself may move in ways that affect the loads. To analyze all of these interacting effects, very detailed mathematical models must be used. Nonetheless great insight can be gained by considering a simplified model of the rotor, and examining its response to simplified loads. The method described below is based on that of Eggleston and Stoddard (1987). This model provides insight not only into turbine responses to steady loads, but also to cyclic loads.

The simplified model is known as the 'linearized hinge–spring blade rotor model' or 'hinge–spring model' for short. The essence of the model is that it incorporates enough detail to be useful, but it is sufficiently simplified that analytic solutions are possible. By examining the solutions, it is possible to discern some of the most significant causes and effects of wind turbine motion. The hinge–spring model consists of four basic parts: (1) a model of each blade as a rigid body attached to a rigid hub by means of hinges and springs, (2) a linearized steady state, uniform flow aerodynamic model, (3) consideration of non-uniform flow as 'perturbations', and (4) an assumed sinusoidal form for the solutions.

4.3.2.1 Types of blade motion

The hinge–spring model allows for three directions of blade motion, and incorporates hinges and springs for all of them. The three directions of motion allowed by the hinges are: (i) flapwise, (ii) lead–lag, (iii) torsion. The springs return the blade to its 'normal' position on the hub.

As mentioned above, flapping refers to motion parallel to the axis of rotation of the rotor. For a rotor aligned with the wind, flapping would be in the direction of the wind, or opposite to it. Thrust forces in the flapping direction are of particular importance, since the largest stresses on the blades are normally due to flapwise bending.

Lead–lag motion lies in the plane of rotation. It refers to motion relative to the blade's rotational motion. In leading motion the blade will be moving faster than the overall rotational speed, and in lagging motion it will be moving slower. Lead–lag motions and forces are associated with fluctuations in torque in the main shaft and with fluctuations in power from the generator.

Torsional motion refers to motion about the pitch axis. For a fixed pitch wind turbine torsional motion is generally not of great significance. In a variable pitch wind turbine torsional motion can cause fluctuating loads in the pitch control mechanism.

In the following development, the focus will mainly be on flapping motion. The details of lead–lag and torsional motion can be found in Eggleston and Stoddard (1987).

4.3.2.2 Sources of loads

The hinge–spring model as developed below includes an analysis of how the rotor responds to six sources of loads:

- Rotor rotation
- Gravity
- Steady yaw rate
- Steady wind
- Yaw error
- Linear wind shear

These loads may be applied independently or in combination with the others. The analysis will provide a general solution for the rotor response which is a function of blade azimuth, the angular position of blade rotation. The solution will be seen to contain three terms: the first independent of azimuthal position, the second a function of the sine of azimuthal position, and the third a function of the cosine of azimuthal position. The development is broken into two parts: (i) 'free' motion, and (ii) forced motion. The free motion includes the effects of gravity and rotation. The forced motion case includes steady wind and steady yaw. Deviations from steady wind (yaw error and wind shear) are considered to be perturbations on the steady wind.

4.3.3.3 Coordinate systems for model

The following sections describe the coordinate systems to be used in all of the model analysis, develop the hinge–spring blade model, and derive the equations of motion for the hinge–spring blade for the case of flapwise motion. This section focuses on the coordinate systems used in the model.

The development is for horizontal axis wind turbines. Figure 4.16 illustrates a typical wind turbine for which the model may be applied. In the figure, and in the development which follows, the turbine is assumed to have a three–bladed, downwind rotor. In this view, the observer is looking upwind. Note, however, that the model is equally applicable to an upwind turbine and to a two-bladed turbine.

Factors affecting the free motion that are considered here include: (i) geometry, (ii) rotational speed, and (iii) blade weight. The effects of external forces on blade motion are discussed subsequently.

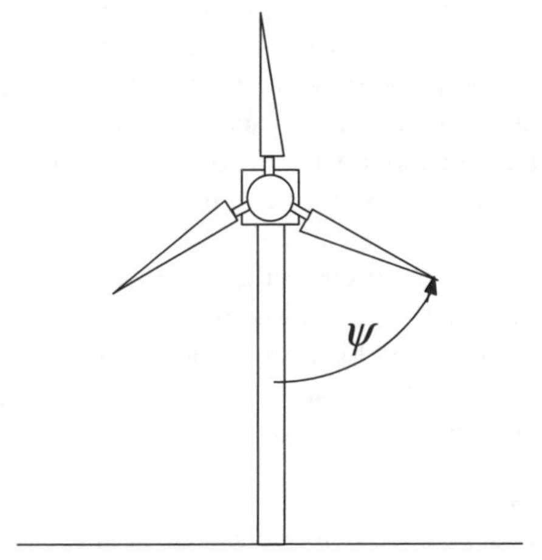

Figure 4.16 Typical turbine appropriate for hinge–spring model

The model, as developed below, embodies a number of assumptions regarding each blade. They include:

- The blade has uniform cross section
- The blade is rigid
- The blade hinge may be offset from the axis of rotation
- When rotating, the rotational speed is constant

The development of the appropriate hinge–spring stiffness and offset for the model are discussed later in this section.

Figure 4.17 illustrates the coordinate system for the model, focusing on one blade. As shown, the X', Y', Z' coordinate system is defined by the turbine itself whereas X, Y, Z are fixed to the earth. The X' axis is along the tower, the Z' axis is the axis of rotor rotation, and the Y' axis is perpendicular to both of them. The X'', Y'', Z'' axes rotate with the rotor. For the case of the blade shown, X'' is aligned with the blade, but in the plane of rotation. The blade is at an azimuth angle ψ with respect to the X'' axis. The blade itself is turned out of the plane of rotation by the flapping angle β. The figure also reflects the assumption that the direction of rotation of the rotor, as well as yaw is consistent with the right-hand rule with respect to the positive sense of the X, Y and Z axes. Specifically when looking in the downwind direction, the rotor rotates clockwise.

Figure 4.18 shows a top view of a blade which has rotated past its highest point (azimuth of π radians) and is now descending. Specifically, the view is looking along the Y'' axis.

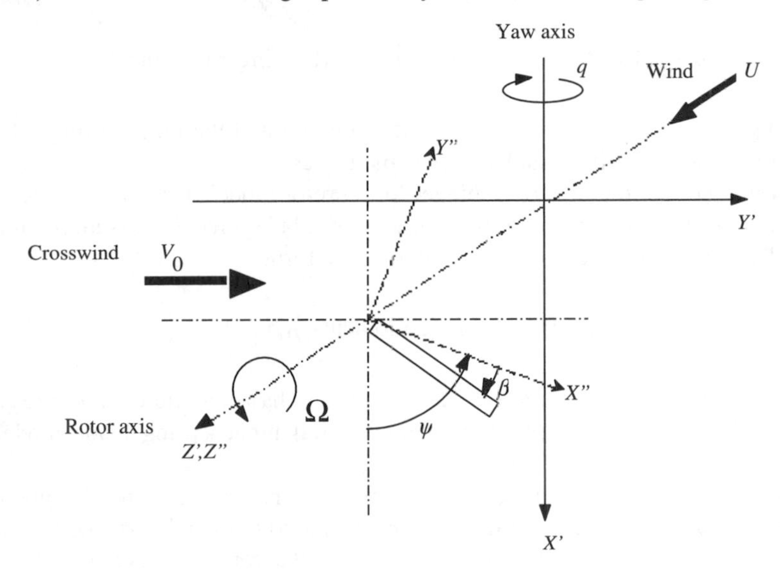

Figure 4.17 Coordinate system for hinge–spring model; q, yaw rate; U, free stream wind velocity; V_0, crosswind velocity; Ω, angular velocity (Eggleston and Stoddard, 1987). Reproduced by permission of Kluwer Academic/Plenum Publisher

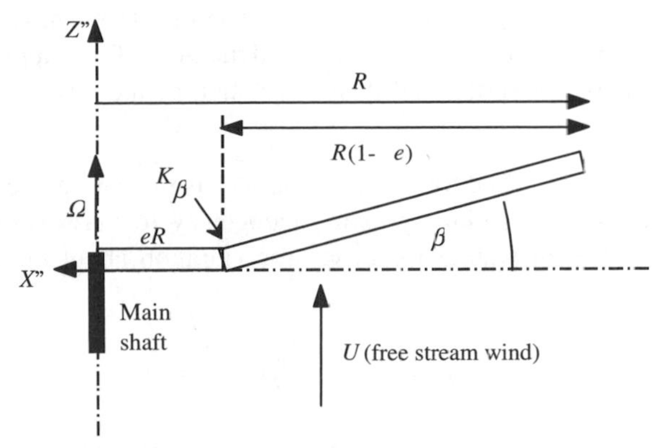

Figure 4.18 Hinge–spring model definitions, e, non-dimensional hinge offset; K_β, spring constant in flapping; R, radius of rotor; U, free stream wind velocity; Ω, angular velocity

4.3.3.4 Development of flapping blade model

This dynamic model uses the hinged and offset blade to represent a real blade. The hinge offset and spring stiffness are chosen such that the rotating hinge and spring blade has the same natural frequency and flapping inertia as the real blade. Before the details of the hinge–spring offset blade model are provided, the dynamics of a simplified hinged blade are examined. As mentioned above, the focus is on the flapping motion in order to illustrate the model.

In general the blade flapping characteristics will be described by a differential equation with constant coefficients of the form:

$$\ddot{\beta} + [f(\text{Restoring moments})]\beta = g(\text{Forcing moments})$$

The restoring moments are due to gravity, rotor rotation and the hinge spring. The forcing moments are due to yaw motion and aerodynamic forces.

The development of the linearized hinge–spring rotor model starts with the development of the equations of motion for a few simplified blade models, assuming no forcing functions. These equations of free motion all have the form:

$$\ddot{\beta} + [f(\text{Restoring moments})]\beta = 0$$

The solutions to these equations are used to illustrate characteristic dynamic responses of rotor blades. The equation of free motion for the full hinge–spring blade model is then described and derived.

The development of the full equation of motion that includes the forcing moments requires linearized models for the forcing moments due to the wind, yaw motion, yaw error and wind shear. Once these terms have been derived, the complete equation of motion can be assembled. This equation is then converted into a slightly simpler form for solution. The final form uses the derivative of the flap angle as a function of azimuth angle rather than time.

Dynamics of a simplified flapping blade model The development of the hinge–spring model is best understood by considering first the dynamics of a flapping blade with no offset. Here the effects of the spring and then of rotation of the blade with and without a spring are considered.

Spring, no rotation, no offset The first case to consider is the natural frequency (flapwise) of a non-rotating blade–hinge–spring system. Analogously to mass–spring vibration (see Section 4.2), the non-rotating natural frequency for vibration about the flapping hinge is found from:

$$\ddot{\beta} = -(K_\beta / I_b)\beta \tag{4.3.10}$$

where I_b is the blade mass moment of inertia about flapping axis and K_β is the flapping hinge spring constant.

The natural flapping frequency of the non-rotating blade, ω_{NR}, is immediately apparent:

$$\omega_{NR} = \sqrt{K_\beta / I_b} \tag{4.3.11}$$

Since one of the assumptions is that the blade has a uniform cross section, the mass moment of inertia of a blade of mass m_B (with no offset) is:

$$I_b = \int_0^R r^2 dm = \int_0^R r^2 (m_B / R) \mathrm{d} r \tag{4.3.12}$$

Thus:

$$I_b = m_B R^2 / 3 \tag{4.3.13}$$

Rotation, no spring, no offset When the blade is rotating with a hinge at the axis of rotation and has no spring, the flapwise natural frequency is the same as the speed of rotation. This can be seen as follows. The only restoring force is the centrifugal inertial force F_c. Its magnitude is proportional to the square of the speed and to the cosine of the flapping angle. The restoring component is determined by the sine of the flapping angle. Thus:

$$I_b \ddot{\beta} = \int [-r \sin(\beta)] \mathrm{d} F_c = \int_0^R [r \cos(\beta) \Omega^2][-r \sin(\beta)] \rho_{blade} \mathrm{d} r \tag{4.3.14}$$

Assuming small angles such that $\cos(\beta) \approx 1$ and $\sin(\beta) \approx \beta$:

$$\ddot{\beta} = -(\Omega^2 / I_b) \beta \int_0^R r^2 \mathrm{d} m = -\Omega^2 \beta \tag{4.3.15}$$

The solution to the above equation is obviously:

$$\omega = \Omega \tag{4.3.16}$$

Rotation, spring, no offset When a blade with no offset is hinged and also includes a spring, the natural frequency is determined by a sum of the spring solution and the rotational solution. The appropriate equation is, then:

$$\ddot{\beta} + (\Omega^2 + K_\beta / I_b) \beta = 0 \tag{4.3.17}$$

The solution to this is immediately apparent from the above discussion:

$$\omega_R^2 = \omega_{NR}^2 + \Omega^2 \tag{4.3.18}$$

where ω_R is the rotating natural frequency and is the ω_{NR} non-rotating natural frequency.

As can be seen, the rotating natural frequency is now higher than it was with the spring alone. Thus, it is said that the rotation 'stiffens' the blade.

Rotation, spring, offset In general a real blade does not behave as if it had a hinge–spring right at the axis of rotation. Thus, in general:

$$\omega_R^2 \neq \omega_{NR}^2 + \Omega^2 \quad (4.3.19)$$

To correctly model blade motions, the blade dynamics, represented by ω_R and ω_{NR}, need to be properly represented. This can be accomplished by including the hinge offset. Then all the constants of the hinge–spring model can be estimated if the natural frequency of the rotating and non-rotating real blade are known. Following the method described in Eggleston and Stoddard (1987), it may be shown that the non-dimensional offset is given by:

$$e = 2(Z-1)/[3+2(Z-1)] \quad (4.3.20)$$

where $Z = (\omega_R^2 - \omega_{NR}^2)/\Omega^2$.

The addition of the offset can be accounted for by adjusting the moment of inertia. The mass moment of inertia of the hinged blade is approximated by:

$$I_b = m_B(R^2/3)(1-e)^3 \quad (4.3.21)$$

The flapping spring constant is:

$$K_\beta = \omega_{NR}^2 I_b \quad (4.3.22)$$

Note that the rotating and non-rotating natural frequencies may be calculated by the Myklestad method as described previously or from experimental data, if they are available.

Thus, the flapping characteristics of the blade are modelled by a uniform blade with an offset hinge and spring that has one degree of freedom and that responds in the same manner to forces as the first vibration mode of the real blade. Similar constants can be determined for the lead–lag motion of the real blade. Torsion does not require an offset hinge model, but it does require a stiffness constant. With all of the constants the blade model allows for three degrees of freedom, flapping, lead–lag, and torsion.

Equations of motion of full flapping blade model (free) When gravity and an offset are included, the full flapping equation of free motion for a single blade becomes:

$$\ddot{\beta} + [\Omega^2(1+\varepsilon) + G\cos(\psi) + K_\beta / I_b]\beta = 0 \quad (4.3.23)$$

where G is the gravity term, given by $G = g\, m_B r_g / I_b$, r_g is the radial distance to the center of mass, ε is the offset term, given by $\varepsilon = 3e/[2(1-e)]$ and ψ is the azimuth angle. The derivation of the flapping equation is given in the next section.

4.3.3.5 Derivation of flapping equation of motion (free)

A flapping blade, shown in Figure 4.19, is viewed from the downwind direction (as in Figure 4.17). The blade is turned out of the plane of rotation, towards the viewer, by the flapping angle, β. The blade's long axis is also inclined at an azimuth angle ψ. Recall that zero azimuth corresponds to the blade tip pointing down. The azimuth angle increases in the direction of rotation. The flapping angle is positive in the downwind direction. Note that in the following discussion many of the inputs vary as the sine or the cosine of the azimuth angle. They are commonly referred to as 'cyclics', more specifically as sine cylics or cosine cylics. It will later be shown that these cyclical inputs can be related to cyclic turbine responses.

In Figure 4.19, two forces act along the axis of the blade on an element of the blade with mass d*m*. The centrifugal force depends on the square of the rotational speed and the distance to the axis of rotation. The gravity component, due to the weight of the blade, depends on the azimuth.

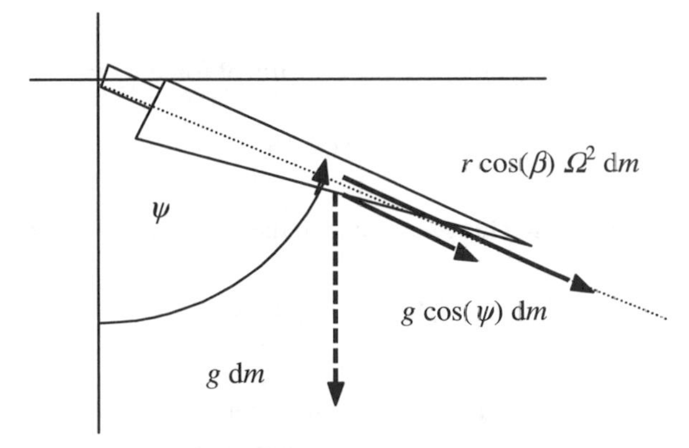

Figure 4.19 Flapping blade viewed from downwind direction; g, gravitational force; *m*, mass; *r*, radial distance from axis of rotation; β, flapping angle; Ω, rotational speed; ψ, azimuth angle

The equation of motion of a flapping blade without an offset, subject to restoring forces due to rotation, gravity, the hinge–spring and inertial forces, will be developed here. The modified equation of motion for a blade with an offset is then described. The full development of the equation with the offset included is left to the reader, but it will be seen that the two equations are very similar.

Figure 4.20 illustrates the blade as seen from above. This figure also includes the flapping spring constant and the flapping acceleration.

The following summarizes the effects of the various forces. As mentioned above, initially the effect of the blade hinge offset is ignored.

Centrifugal force The centrifugal force serves to bring the flapping blade back into the plane of rotation. As stated above, its magnitude depends on the square of the speed of rotation and is independent of blade azimuth. Centrifugal force acts on the center of mass of the blade, perpendicular to the rotor's axis of rotation.

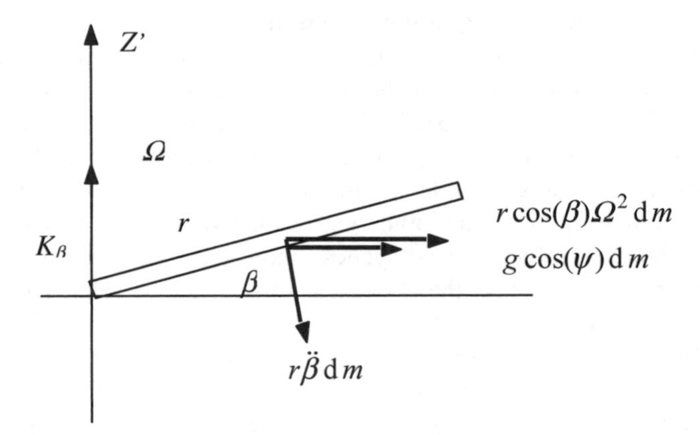

Figure 4.20 Flapping blade viewed from above (Eggleston and Stoddard, 1987). Reproduced by permission of Kluwer Academic/Plenum Publisher

As shown in the figure the magnitude of the centrifugal force F_c is:

$$F_c = r\cos(\beta)\Omega^2 \mathrm{d}\,m \tag{4.3.24}$$

The moment about the flapping hinge axis due to the centrifugal force is:

$$M_c = r\sin(\beta)[r\cos(\beta)\;\Omega^2 \mathrm{d}\,m] \tag{4.3.25}$$

Gravitational force Gravity acts downward on the center of mass of the blade. When the blade is up, gravity tends to increase the flapping angle; when down it tends to decrease it. The gravitational force is independent of rotational speed.

The magnitude of the gravitational force, F_g ,is:

$$F_g = g\cos(\psi)\mathrm{d}\,m \tag{4.3.26}$$

Since the gravitational force varies with the cosine of the azimuth, it can be said to be a 'cosine cyclic' input.

The restoring moment due to the gravitational force depends on the sine of the flapping angle. Its magnitude is:

$$M_g = r\sin(\beta)[g\cos(\psi)\mathrm{d}\,m] \tag{4.3.27}$$

Hinge–spring force The hinge–spring creates a moment, M_s , at the hinge. Its magnitude is proportional to the flapping angle. The spring tends to bring the blade back into the plane of the hinge, which in this case is the plane of rotation.

The magnitude of the spring hinge moment is:

$$M_s = K_\beta \beta \tag{4.3.28}$$

Acceleration The flapping acceleration inertial force caused by mass dm is given by the angular acceleration in the flapping direction, $\ddot{\beta}$, multiplied by its distance from the flapping axis. The moment due to this force M_f is the product of that inertia force and, again, the distance. In other words:

$$M_f = r^2 \ddot{\beta}\, \mathrm{d}m \tag{4.3.29}$$

The acceleration of mass dm is the result of the sum of the moments. In the absence of external forces, that sum is zero:

$$\sum M = M_f + M_c + M_g + M_s = 0 \tag{4.3.30}$$

Integrating over the entire blade gives:

$$\int_0^R (M_f + M_c + M_g) + M_s = 0 \tag{4.3.31}$$

Expanding the terms gives:

$$\int_0^R [r^2 \ddot{\beta} + r\cos(\beta) r\sin(\beta)\Omega^2 + r\, g\cos(\psi)\sin(\beta)]\, \mathrm{d}m + K_\beta \beta = 0 \tag{4.3.32}$$

Recall that the blade mass moment of inertia is $I_b = \int_0^R r^2 \mathrm{d}m$. The mass and distance to the center of mass, r_g, are related by:

$$\int_0^R r\, \mathrm{d}m = m_B r_g \tag{4.3.33}$$

Therefore, we have:

$$I_b \ddot{\beta} + I_b \cos(\beta)\Omega^2 \sin(\beta) + g\cos(\psi)\sin(\beta) m_B r_g + K_\beta \beta = 0 \tag{4.3.34}$$

Using the small-angle approximation for β, defining a ‘gravity term’, $G = g\, m_B r_g / I_b$, and rearranging we obtain

$$\ddot{\beta} + [\Omega^2 + G\cos(\psi) + K_\beta / I_b]\beta = 0 \tag{4.3.35}$$

Note the similarity to what has been discussed previously (Equation 4.3.23).

Recalling that the blade model assumes a uniform cross section, we can add the offset, *e*, and define an offset term, ε :

$$\varepsilon = m_B e\, r_g R / I_b = 3e / [2(1-e)] \tag{4.3.36}$$

When the offset is included in the analysis, the final flapping equation of motion becomes:

$$\ddot{\beta} + [\Omega^2(1+\varepsilon) + G\cos(\psi) + K_\beta / I_b]\beta = 0 \tag{4.3.37}$$

which was to be proved.

4.3.3.6 Equations of motion (forced)

This section expands the development of the blade equation of motions by considering the effect of forcing functions: yaw motion and wind speed.

Yaw motion Yaw motion results in gyroscopic moments acting upon the blade. By applying an analysis similar to the one given in Section 4.2.1.5, one may show that the gyroscopic moment in the flapping direction due to steady yaw motion of rate q is:

$$M_{yaw} = -2q\Omega\cos(\psi)I_b \tag{4.3.38}$$

Effects of wind: the linearized aerodynamics model The effect of wind on blade motion is incorporated via a linearized aerodynamics model. The model is similar to that developed in the aerodynamics section of this text, but it also includes some simplifying assumptions. The aerodynamic model that was developed in Chapter 3 is appropriate for performance estimates, but it is too complex to be useful in this simplified dynamic analysis. Conversely, the previous analysis only considered steady, axial flowing wind, whereas in its linearized form vertical wind shear, crosswind and yaw error can more easily be included. It should also be noted that in a truly comprehensive dynamic model, the aerodynamics model would be quite detailed. Finally, in the linearized aerodynamics model drag is ignored altogether.

The linearized aerodynamics model consists of a linearized description of the lift force on the blade and a linearized characterization of the axial and tangential components of the wind at the rotor (U_P and U_T, respectively). The lift force is a function of these two components of the wind. U_P and U_T, in turn, are functions of the mean wind speed, blade flapping speed, yaw rate, crosswind and wind shear. The equation for the flapping moment that arises from the lift force will be developed after the full aerodynamics model is explained.

The following equation is used for lift per unit length, $\tilde{L}$, on the airfoil. The derivation of this equation is developed subsequently, starting from principles presented in Chapter 3:

$$\tilde{L} \approx \frac{1}{2}\rho c\, C_{l\alpha}\left(U_P U_T - \theta_p U_T^2\right) \tag{4.3.39}$$

where c is the chord length, $C_{l\alpha}$ is the slope of lift curve, U_P is the wind velocity perpendicular (normal) to the rotor plane, U_T is the wind velocity tangential to the blade element and θ_p is the pitch angle.

The equations for the wind velocity components, which take into account axial flow, flap rate, yaw rate, yaw error and wind shear, follow. The equations for these components will also be developed subsequently. The velocity components are:

$$U_P = U(1-a) - r\,\dot{\beta} - (V_0\,\beta + q\,r)\sin(\psi) - U(r/R)K_{vs}\cos(\psi) \tag{4.3.40}$$

$$U_T = \Omega\, r - (V_0 + q\, d_{yaw})\cos(\psi) \tag{4.3.41}$$

where U is the free stream wind velocity, a is the axial induction factor, V_0 is the crosswind velocity (due to yaw error), K_{vs} is the wind shear coefficient, and d_{yaw} is the distance from rotor plane to yaw axis.

Development of linearized aerodynamics model This section develops the aerodynamics model presented above. Initially, the model is developed for a steady, axial flowing wind. Deviations from the steady wind (due to yaw error, yaw motion and wind shear) are subsequently considered as perturbations on that wind.

The development of the linearized aerodynamics model begins by reference to Figure 4.21.

In addition to the variables defined previously, the following nomenclature is used:

ϕ = Relative wind angle $= \tan^{-1}(U_P / U_T)$
α = Angle of attack
U_R = Relative wind = $\sqrt{U_P^2 + U_T^2}$

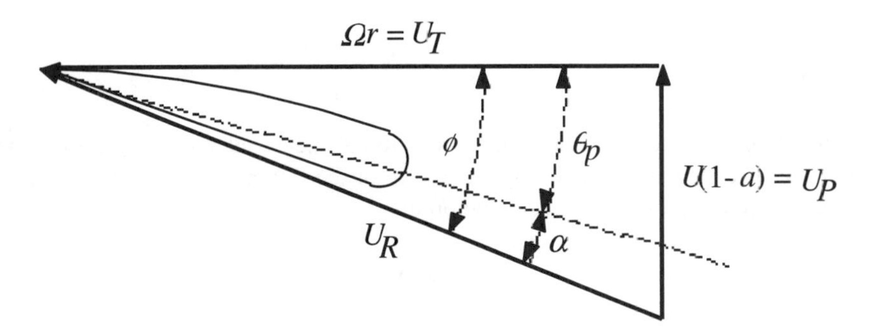

Figure 4.21 Nomenclature for linearized aerodynamics model; a, axial induction factor; r, radial distance from axis of rotation; U , relative wind velocity; U_P and U_T perpendicular and tangential components of the wind velocity respectively; Ω, angular velocity

The development is based on the following assumptions:

- The rotational speed is high relative to the wind speed. Thus $U_T >> U_P$
- The lift curve is linear and passes through zero; thus $C_l = C_{l\alpha}\alpha$, where $C_{l\alpha}$ is the slope of the lift curve
- The angle of attack is small
- Small-angle assumptions may be used as appropriate
- There is no wake rotation
- The blade has a constant cord and no twist (although the model could be expanded to include a non-constant chord and blade twist)

Axial flow Taking into account the assumption about the lift curve, and using the relations from Chapter 3, the lift per unit length on the airfoil is:

$$\tilde{L} = \frac{1}{2}\rho c\, C_l U_R^2 \approx \frac{1}{2}\rho c\, C_{l\alpha} U_R^2 \alpha \tag{4.3.42}$$

Using the small-angle approximation, $\tan^{-1}(\text{angle}) \approx \text{angle}$

$$\alpha = \phi - \theta_p = \tan^{-1}(U_P / U_T) - \theta_p \approx U_P / U_T - \theta_p \tag{4.3.43}$$

Therefore

$$\tilde{L} = \frac{1}{2}\rho\, c\, C_{l\alpha} U_R^2 [U_P / U_T - \theta_p] \tag{4.3.44}$$

Recalling that $U_R^2 = U_P^2 + U_T^2 \approx U_T^2$, one gets:

$$\tilde{L} = \frac{1}{2}\rho\, c\, C_{l\alpha} U_T^2 (U_P / U_T - \theta_p) = \frac{1}{2}\rho c\, C_{l\alpha} (U_P U_T - \theta_p U_T^2) \tag{4.3.45}$$

which was to be proved.

For the flapping blade, the tangential velocity is the usual radial position times the angular speed. It is actually reduced by the cosine of the flapping angle, but because of the small-angle approximation, that effect is ignored:

$$U_T = \Omega\, r \tag{4.3.46}$$

The perpendicular component of velocity is, as usual, the free stream wind speed less the induced axial wind speed, Ua. The flapping velocity $r\dot{\beta}$ also reduces the speed. Therefore:

$$U_P = U(1-a) - r\,\dot{\beta} \tag{4.3.47}$$

Various factors such as yaw error, crosswind, vertical wind, wind shear, etc. will affect the wind actually experienced by the blade. These are accounted for as perturbations on the main flow. These perturbations are assumed to be small, compared to the axial wind speed. They are referred to as 'deltas' in the following discussion.

Crosswind and yaw error A crosswind, V_0, is perpendicular to the axis of the rotor. It arises due to yaw error (misalignment of the rotor axis and the wind) or a sudden change in wind direction. It serves to increase the tangential velocity at the advancing blade and decrease it at the retreating blade. As shown on Figure 4.17, the crosswind is defined as going from left to right. Since the blade is turning counter clockwise, the crosswind decreases the tangential wind speed when the blade is below the hub axis and increases it when above. Thus, the change in the tangential component of the wind due to crosswind, $\Delta U_{T,crs}$ is:

$$\Delta U_{T,crs} = -V_0 \cos(\psi) \tag{4.3.48}$$

When the blade is out of the plane of rotation by the flapping angle, β, there will be a component that decreases the perpendicular wind speed at the blade, $\Delta U_{P,crs}$:

$$\Delta U_{P,crs} = -V_0 \sin(\beta)\sin(\psi) \approx -V_0\,\beta\,\sin(\psi) \tag{4.3.49}$$

Note that yaw error is predominantly a cosine cyclic disturbance, affecting the tangential velocity.

Yaw motion Yaw motion also affects blade velocities, besides inducing gyroscopic moments. Consider the blade straight up and a yaw rate of q. The blade will experience a velocity due to the yaw rotation of qd_{yaw} (where d_{yaw} is the distance from the axis of the tower to the center of the rotor). If the blade is inclined away (downwind) from the plane of rotation by the flapping angle, β, this will increase the effect by $rq\sin(\beta)$. The effect is greatest when the blade is up or down and nonexistent when the blade is horizontal.

Taking the small-angle approximation as before and recognizing that $d_{yaw}\beta + r \approx r$ and $r\beta + d_{yaw} \approx d_{yaw}$, the yaw rate deltas are:

$$\Delta U_{T,yaw} = -q\,d_{yaw}\cos(\psi) \tag{4.3.50}$$

$$\Delta U_{P,yaw} = -q\,r\sin(\psi) \tag{4.3.51}$$

Note that yaw terms are similar in form to the crosswind terms, with the tangential term varying as $\cos(\psi)$ and the perpendicular varying as $\sin(\psi)$.

Vertical wind shear In Chapter 2 of this text, vertical wind shear was modelled by the equation:

$$U_1 / U_2 = (h_1 / h_2)^{\alpha} \tag{4.3.52}$$

where α = power law exponent and the subscripts correspond to different heights

In the simplified aerodynamic model it is assumed that vertical wind shear is linear across the rotor. Consequently, the 'free stream' wind at height h is given by:

$$U_h = U[1-(r/R) K_{vs} \cos(\psi)] \tag{4.3.53}$$

where $h = 0$ when the blade tip is down, $h = 2R$ when the blade tip is up and K_{vs} is the linear wind shear constant and the subscript 'vs' indicates vertical shear.

Thus, at the center of the rotor and on the horizontal, $U_h = U$. At the top, with zero flapping, $U_h = U(1+K_{vs})$ and at the bottom $U_h = U(1-K_{vs})$.

The incremental effect of vertical shear is predominantly on the perpendicular component of wind, so the tangential wind shear delta can be assumed to equal zero:

$$\Delta U_{T,vs} = 0 \tag{4.3.54}$$

The perpendicular wind shear delta is found directly from Equation 4.3.53:

$$\Delta U_{P,vs} = -U(r/R)K_{vs} \cos(\psi) \tag{4.3.55}$$

Horizontal wind shear can be modelled analogously, but will not be discussed here.

In summary, with all angles approximated as small angles, the contributions are as shown in Table 4.1.

Table 4.1 Wind speed perturbations

Case	ΔU_P	ΔU_T
Axial Flow	$U(1-a)-r\dot{\beta}$	Ωr
Crosswind	$-V_0 \beta \sin(\psi)$	$-V_0 \cos(\psi)$
Yaw rate	$-q\, r \sin(\psi)$	$-q\, d_{yaw} \cos(\psi)$
Vertical wind shear	$-U(r/R)K_{vs} \cos(\psi)$	0

Note that vertical wind shear is exclusively a cosine perturbation. Crosswind is primarily a cosine perturbation (to the tangential velocity component). Analogously, a vertical component of the wind (upslope) should be a sine perturbation.

The total velocities in both the perpendicular and tangential directions are found by adding up all the deltas:

$$U_P = U(1-a) - r\dot{\beta} - (V_0 \beta + q\, r)\sin(\psi) - U(r/R)K_{vs} \cos(\psi) \tag{4.3.56}$$

$$U_T = \Omega\, r - \left(V_0 + q\, d_{yaw}\right)\cos(\psi) \tag{4.3.57}$$

Aerodynamic forces and moments The terms developed above can all be used to express the aerodynamic moments on the blade. It can be shown, as discussed below, that the flapping moment due to aerodynamic forces is:

$$M_\beta = \frac{1}{2}\gamma I_b \Omega^2 \left\{ \frac{\Lambda}{3} - \frac{\beta'}{4} - \frac{\theta_p}{4} - \cos(\psi)\left[\overline{V}\left(\frac{\Lambda}{2} - \frac{\beta'}{3} - \frac{2\theta_p}{3} \right) + \frac{K_{vs}\overline{U}}{4} \right] - \sin(\psi)\left[\frac{\overline{V}_0 \beta}{3} + \frac{\overline{q}}{4} \right] \right\} \tag{4.3.58}$$

where

Λ = Non-dimensional inflow, $\Lambda = U(1-a)/\Omega R$
$\overline{V}_0$ = Non-dimensional cross flow, $\overline{V}_0 = V_0/\Omega R$
$\overline{V}$ = Non-dimensional total cross flow, $\overline{V} = \left(V_0 + q\, d_{yaw}\right)/\Omega R$
β' = Azimuthal derivative of flap angle, $\beta' = \dot{\beta}/\Omega$
γ = Lock number, $\gamma = \rho C_{L\alpha} c R^4 / I_b$
$\overline{q}$ = Non-dimensional yaw rate term, $\overline{q} = q/\Omega$
$\overline{U}$ = Non-dimensional wind velocity, $\overline{U} = U/\Omega R = 1/\lambda$

Development of aerodynamic force and moment equation The components of the blade forces in the tangential (torque) and normal directions are calculated as discussed in the aerodynamics chapter (Chapter 3), except that: (i) the flapping angle is accounted for and (ii) drag is assumed to be zero. The normal force per unit length, $\tilde{F}_N$, is:

$$\tilde{F}_N = \tilde{L}\cos(\phi)\cos(\beta) \tag{4.3.59}$$

The tangential force per unit length, $\tilde{F}_T$, is

$$\tilde{F}_T = \tilde{L}\sin(\phi) \tag{4.3.60}$$

The various forces must be summed (integrated) over the blade to give the shear forces, or multiplied by the distances first and then summed (integrated) to give moments. As before, simplifications are made so that $\sin(\phi) = U_P/U_T$, $\cos(\phi) = 1$ and $\cos(\beta) = 1$.

The flapwise shear force at the root of the blade is the integral of the normal force per unit length over the length of the blade:

$$S_\beta = \int_0^R \tilde{F}_N\, \mathrm{d}r = \int_0^R \tilde{L}\cos(\phi)\cos(\beta)\,\mathrm{d}r \approx \int_0^1 \tilde{L} R\,\mathrm{d}\eta \tag{4.3.61}$$

where $\eta = r/R$.

The flapwise bending moment at the root of the blade is the integral of the normal force per unit length times the distance at which the force acts, over the length of the blade:

$$M_{\beta} = \int_0^R \tilde{F}_N r \, \mathrm{d}r = \int_0^R \tilde{L} \cos(\phi)\cos(\beta) r \, \mathrm{d}r \approx \int_0^1 \tilde{L} \, R^2 \eta \, \mathrm{d}\eta \qquad (4.3.62)$$

The flapping moment equation can be expanded, using Equation 4.3.39 to yield:

$$M_{\beta} = \int_0^1 \tilde{L} \, R^2 \eta \, \mathrm{d}\eta = \int_0^1 \left[\frac{1}{2} \rho c \, C_{l\alpha} \left(U_P U_T - \theta_p U_T^2 \right) \right] R^2 \eta \, \mathrm{d}\eta \qquad (4.3.63)$$

By making appropriate substitutions and performing the algebra one may derive Equation 4.3.58 above.

Complete equations of motion The complete flapping equation of motion is found by including the moments due to aerodynamics and yaw rate (gyroscopic effects) and performing the appropriate algebraic manipulations. It is:

$$\beta'' + \left[1 + \varepsilon + \frac{G}{\Omega^2} \cos(\psi) + \frac{K_{\beta}}{\Omega^2 I_b} \right] \beta = \frac{M_{\beta}}{\Omega^2 I_b} - 2 \bar{q} \cos(\psi) \qquad (4.3.64)$$

where $\beta'' = \ddot{\beta} / \Omega^2 =$ the azimuthal second derivative of β and where the aerodynamic forcing moment, M_{β}, is from Equation 4.3.58. Note that the equations are now expressed in terms of the azimuthal derivative, which is discussed in more detail later in this chapter. This is convenient for solving the equation.

Development of complete flapping equation of motion In this section the complete flapping equation of motion, which was presented above in Equation 4.3.64, is developed.

The original flapping equation of free motion, when the aerodynamic and gyroscopic moments are included, becomes:

$$\ddot{\beta} + \left[\Omega^2 (1 + \varepsilon) + G \cos(\psi) + \frac{K_{\beta}}{I_b} \right] \beta = \frac{M_{\beta}}{I_b} - 2 q \Omega \cos(\psi) \qquad (4.3.65)$$

Dividing by Ω^2 yields

$$\frac{\ddot{\beta}}{\Omega^2} + \left[1 + \varepsilon + \frac{G}{\Omega^2} \cos(\psi) + \frac{K_{\beta}}{\Omega^2 I_b} \right] \beta = \frac{M_{\beta}}{\Omega^2 I_b} - 2 \bar{q} \cos(\psi) \qquad (4.3.66)$$

The above flapping equation is in the time domain. Because the rotational speed is assumed to be constant, it is more interesting to express the equation as a function of angular (azimuthal) position.

The chain rule can be used to yield:

$$\dot{\beta} = \frac{\mathrm{d}\beta}{\mathrm{d}t} = \left(\frac{\mathrm{d}\beta}{\mathrm{d}\psi}\right)\left(\frac{\mathrm{d}\psi}{\mathrm{d}t}\right) = \Omega\left(\frac{\mathrm{d}\beta}{\mathrm{d}\psi}\right) = \Omega\beta' \tag{4.3.67}$$

Similarly $\ddot{\beta} = \Omega^2 \beta''$

Note that the 'dot' over the variable signifies a derivative with respect to time, whereas the prime signifies a derivative with respect to azimuth.

The flapping equation can now be rewritten as given above in Equation 4.3.64.

Arrangement of flapping equation of motion for solution Upon substituting the complete expression for the aerodynamic moment and collecting terms, one gets:

$$\beta'' + \left[\frac{\gamma}{8}\left(1 - \frac{4}{3}\overline{V}\cos(\psi)\right)\right]\beta' + \left[1 + \varepsilon + \frac{G}{\Omega^2}\cos(\psi) + \left(\frac{\gamma\overline{V}_o}{6}\right)\sin(\psi) + \frac{K_\beta}{\Omega^2 I_b}\right]\beta$$

$$= -2\,\overline{q}\cos(\psi) + \frac{\gamma}{2}\left(\frac{\Lambda}{3} - \frac{\theta_P}{4}\right) - \frac{\gamma}{8}\overline{q}\sin(\psi) - \cos(\psi)\left\{\frac{\gamma}{2}\left[\overline{V}\left(\frac{\Lambda}{2} + \frac{2\theta_P}{3}\right) + \frac{K_{vs}\overline{U}}{4}\right]\right\} \tag{4.3.68}$$

The complete flapping equation of motion can then be written in a slightly simpler form for solution:

$$\beta'' + \frac{\gamma}{8}\left[1 - \frac{4}{3}\cos(\psi)\left(\overline{V}_o + \overline{q}\overline{d}\right)\right]\beta' + \left[K + 2\,B\cos(\psi) + \frac{\gamma}{6}\overline{V}_o\sin(\psi)\right]\beta$$

$$= \frac{\gamma A}{2} - \frac{\gamma\overline{q}}{8}\sin(\psi) - \left\{2\,\overline{q} + \frac{\gamma}{2}\left[A_3\left(\overline{V}_o + \overline{q}\overline{d}\right) + \left(\frac{K_{vs}\overline{U}}{4}\right)\right]\right\}\cos(\psi) \tag{4.3.69}$$

where:

K = Flapping inertial natural frequency (includes rotation, offset, hinge spring), $K = 1 + \varepsilon + K_\beta / I_b \Omega^2$
A = Axisymmetric flow term, $= (\Lambda/3) - (\theta_p/4)$
A_3 = Axisymmetric flow term, $= (\Lambda/2) - (2\theta_P/3)$
B = Gravity term, $= G/2\Omega^2$
$\overline{d}$ = Normalized yaw moment arm, $= d_{yaw}/R$

This final equation includes all of the restoring forces and all of the forcing moments mentioned above and can now be used to determine the rotor behavior under a variety of wind and dynamic conditions.

Discussion of flapping equation A few important aspects of wind turbine dynamics are evident in the complete flapping equations of motion.

The final equation includes a damping term. The term multiplying β' depends on the Lock number, which equals 0 when aerodynamics are not involved. This means that the only damping in this model is due to aerodynamics. The Lock number, which may be thought of as the ratio of aerodynamic forces to inertial forces, is:

$$\gamma = \rho c C_{l\alpha} R^4 / I_b \tag{4.3.70}$$

If the lift curve slope is zero or negative, as in stall, the Lock number will also be zero or negative, and there would be no damping. This could be a problem for a teetered rotor with its larger range of flapping motion. It may also be a problem in rigid rotors with dynamic coupling between lead–lag and edgewise motions. In such cases negatively damped flap motions may cause edgewise vibrations.

With no crosswind or yawing the damping ratio (see Section 4.2) is:

$$\xi = \frac{\gamma}{16} \frac{1}{\omega_\beta / \Omega} \tag{4.3.71}$$

where $\omega_\beta =$ the flapping frequency. For teetered or articulated blades, $\omega_\beta \approx \Omega$, so the flapping damping ratio is approximately $\gamma / 16$. For rigid blades ω_β is on the order of 2 to 3 times higher than Ω, and the flapping damping ratio would be correspondingly smaller. So with Lock numbers ranging from 5 to 10, the damping ratio is on the order of 0.5 – 0.16. This amount of damping is enough to damp the flapping mode vibrations. While the details of lead–lag motions are not being pursued here, it should be noted that a full development of the lead–lag equation of motion would show no aerodynamic damping. The lack of damping in lead–lag can lead to blade instabilities.

Note that there is a constant term on the right-hand side of the complete flapping equation of motion ($\gamma A / 2$). This describes the blade coning, which is a constant deflection of the blades away from the plane of rotation due to the steady force of the wind. This coning will be in addition to any preconing. Preconing is sometimes incorporated in the rotor design for a number of reasons: (i) it keeps the tips away from the tower, (ii) it helps to reduce root flap bending moments on a downwind, rigid rotor, and (iii) it contributes to yaw stability.

4.3.3.7 Solutions to flapping equation of motion

The flapping equation of motion has been written in terms of constants and sines and cosines of the azimuth angle. A full solution could be written as a Fourier series, that is to say a sum of sines and cosines of the azimuth with progressively higher frequencies. The frequencies would begin with sinusoids of the azimuth, and increase by integer multiples. To a good approximation, however, the solution can also be assumed to be a sum of constants, sines and cosines of the azimuth angle. Using those assumptions, the solution to the flapping equation of motion can be expressed in terms of three constants, β_0, β_{1c}, β_{1s} and the flapping angle will be:

$$\beta \approx \beta_0 + \beta_{1c}\cos(\psi) + \beta_{1s}\sin(\psi) \qquad (4.3.72)$$

where β_0 is the coning or 'collective' response constant, β_{1c} is the cosine cyclic response constant and β_{1s} is the sine cyclic response constant

It is important to note the directional effects of the different terms in the solution equation. These are illustrated in Figure 4.22. The coning term is positive. This indicates that the blade bends away from the free stream wind, as expected. Referring to Figure 4.22, a positive cosine constant indicates that when the blade is pointing straight down, it is pushed further downwind. When pointing upwards, the blade tends to bend upwind. In either horizontal position, the cosine of the azimuth equals zero. Thus a plane determined by a path of the blade tip would tilt about a horizontal axis, upwind at the top, downwind at the bottom. The positive sine constant means that the blade which is rising tends to be bent downwind when horizontal. When descending, the blade goes upwind. Overall, the plane described by the tip would tilt to the left (in the direction of positive yaw) in the figure.

In summary:

- The constant term, β , indicates that the blade bends downwind by the same amount as it rotates (as in coning)
- The cosine term, β_{1c} , indicates that the plane of rotation tilts downwind when the blade is pointing down, upwind when pointing up
- The sine term, β_{1s} , indicates that the plane of rotation tilts downwind when the blade is rising, upwind when descending

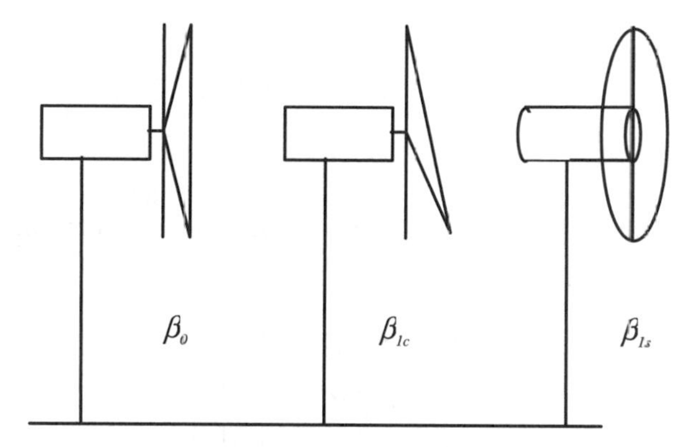

Figure 4.22 Effect of various terms in solution; β_0, collective response coefficient; β_{1c}, cosine cyclic response coefficient; β_{1c}, sine cyclic response coefficient (Eggleston and Stoddard, 1987). Reproduced by permission of Kluwer Academic/Plenum Publisher

The constants above are functions of the various parameters in the model. The coning term is related primarily to axial flow, blade weight, and the spring constant. The sine and cosine terms depend on yaw rate, wind shear, and crosswind (yaw error) as well as the same terms that affect coning.

By applying the assumption that the flapping angle can be expressed as given in Equation 4.3.72, it is possible to solve the flapping equation in closed form. This is done by taking derivatives of Equation 4.3.72, substituting the results into Equation 4.3.69, and collecting terms to match coefficients of the functions of azimuth. The result can be conveniently expressed in the following matrix equation.

$$\begin{bmatrix} K & B & -\frac{\gamma \bar{q} \bar{d}}{12} \\ 2B & K-1 & \frac{\gamma}{8} \\ \frac{\gamma \bar{U}_0}{6} & -\frac{\gamma}{8} & K-1 \end{bmatrix} \begin{bmatrix} \beta_0 \\ \beta_{1c} \\ \beta_{1s} \end{bmatrix} = \begin{bmatrix} \frac{\gamma}{2} A \\ -2\bar{q} - \frac{\gamma}{2}\left[\left(\bar{V}_0 + \bar{q}\bar{d} \right) A_3 + \left(\frac{K_{vs} \bar{U}}{4} \right) \right] \\ -\frac{\gamma}{8} \bar{q} \end{bmatrix} \tag{4.3.73}$$

Simple solutions of the flapping equation In general the useful form of the solution to the equations of motion can be found by applying Cramer's Rule to Equation 4.3.73. That approach will in fact be taken in the next section. In the meantime, it is of more interest to look at some special cases and to gain some insight into blade response to isolated inputs. It will be seen that the dynamic response of a turbine to fairly simple inputs can be complex blade motions, even without considering the effects of turbulence and non-linear aerodynamics.

Rotation only First consider the simplest case with rotation but where:

Gravity = 0
Grosswind = 0
Yaw rate = 0
Offset = 0
Hinge–spring = 0

The only non-zero terms in Equation 4.3.73 are aerodynamic and centrifugal forces:

$$\begin{bmatrix} 1 & 0 & 0 \\ 0 & 0 & \frac{\gamma}{8} \\ 0 & -\frac{\gamma}{8} & 0 \end{bmatrix} \begin{bmatrix} \beta_0 \\ \beta_{1c} \\ \beta_{1s} \end{bmatrix} = \begin{bmatrix} \frac{\gamma}{2} A \\ 0 \\ 0 \end{bmatrix} \tag{4.3.74}$$

The solution to this is immediately apparent:

$$\beta_0 = \frac{\gamma}{2} A \tag{4.3.75}$$

Thus there is only coning in this case. There is a balance between aerodynamic thrust and centrifugal force which determines the flapping angle. There is no dependence on azimuth.

Rotation + hinge–spring + offset Adding the spring and offset terms gives the same form for the solution. The flapping angle is reduced, however. The solution equation is now:

$$\begin{bmatrix} K & 0 & 0 \\ 0 & K-1 & \frac{\gamma}{8} \\ 0 & -\frac{\gamma}{8} & K-1 \end{bmatrix} \begin{bmatrix} \beta_0 \\ \beta_{1c} \\ \beta_{1s} \end{bmatrix} = \begin{bmatrix} \frac{\gamma}{2}A \\ 0 \\ 0 \end{bmatrix} \tag{4.3.76}$$

The solution is now $\beta_0 = \gamma\, A/2K$. The coning angle now results from a balance between the aerodynamic moments on the one hand and the centrifugal force and hinge–spring moments opposing them. As to be expected, the stiffer the spring, the smaller the coning angle.

Rotation + hinge–spring + offset + gravity The addition of gravity complicates the solution. As in the previous cases, there is assumed to be no yaw, no crosswind, and no wind shear. The solution matrix to the flapping equation above becomes:

$$\begin{bmatrix} K & B & 0 \\ 2B & K-1 & \frac{\gamma}{8} \\ 0 & -\frac{\gamma}{8} & K-1 \end{bmatrix} \begin{bmatrix} \beta_0 \\ \beta_{1c} \\ \beta_{1s} \end{bmatrix} = \begin{bmatrix} \frac{\gamma}{2}A \\ 0 \\ 0 \end{bmatrix} \tag{4.3.77}$$

To solve this we can use Cramer's Rule. For that we need the determinant, D, of the coefficient matrix in Equation 4.3.77. This is given in the usual way by:

$$D = K\begin{vmatrix} K-1 & \frac{\gamma}{8} \\ -\frac{\gamma}{8} & K-1 \end{vmatrix} - B\begin{vmatrix} 2B & \frac{\gamma}{8} \\ 0 & K-1 \end{vmatrix} = K\left[(K-1)^2 + \left(\frac{\gamma}{8}\right)^2\right] - 2B^2(K-1) \tag{4.3.78}$$

According to Cramer's rule, one finds the desired values by substituting the right-hand side vector into the corresponding column of the matrix, finding the determinant of the new matrix and dividing by the original determinant.

The solution, for the first time, has sines and cosines of the azimuth angle. Note the gravity term B multiplying both of the sine and cosine terms:

$$\beta_0 = \frac{\gamma}{2}\frac{A}{D}\left[(K-1)^2 + \left(\frac{\gamma}{8}\right)^2\right] \tag{4.3.79}$$

$$\beta_{1c} = -B\frac{A}{D}\gamma(K-1) \tag{4.3.80}$$

$$\beta_{1s} = -B\frac{A}{D}\left(\frac{\gamma^2}{8}\right) \tag{4.3.81}$$

Because the cosine and sine terms are negative, the rotor disc is tilted downwind and to the left.

The magnitude of the sine and cosine terms can be related by a 'cyclic sharing' term:

$$\beta_{1s} = \frac{\gamma}{8(K-1)}\beta_{1c} \tag{4.3.82}$$

The cyclic sharing term indicates the relative amount of moment backwards compared to sideways. For a rotor with independent freely hinged blades (K approaching 1) there will be mostly yawing. For a stiff-bladed machine, there will be mostly tilting. This can also be considered in terms of phase lag. Recall that gravity is a cosine input for flapping. For a stiff blade the response is mostly a cosine response, so there will be little phase lag. For a teetered rotor, the response to a cosine input is only a function of the sine of the azimuth angle. That means that the response has a $\pi/2$ (90 degree) phase lag from the disturbance.

Wind shear + hinge–spring This example ignores the gravity, yaw and crossflow terms, but includes wind shear. The equation to be solved takes the form:

$$\begin{bmatrix} K & 0 & 0 \\ 0 & K-1 & \frac{\gamma}{8} \\ 0 & -\frac{\gamma}{8} & K-1 \end{bmatrix} \begin{bmatrix} \beta_0 \\ \beta_{1c} \\ \beta_{1s} \end{bmatrix} = \begin{bmatrix} \frac{\gamma}{2}A \\ -\frac{\gamma}{8}K_{vs}\overline{U} \\ 0 \end{bmatrix} \tag{4.3.83}$$

The determinant of the coefficient matrix is simply:

$$D = K\left[(K-1)^2 + \left(\frac{\gamma^2}{8}\right)\right] \tag{4.3.84}$$

Applying Cramer's Rule again, one has:

$$\beta_0 = \frac{\gamma}{2}\frac{A}{K} \tag{4.3.85}$$

$$\beta_{1c} = -\frac{1}{D}\left[\frac{\gamma}{8}K_{vs}\bar{U}\,K\,(K-1)\right] \tag{4.3.86}$$

$$\beta_{1s} = -\frac{1}{D}\left[\frac{\gamma^2}{8}K_{vs}\bar{U}\,K\right] \tag{4.3.87}$$

Recall that wind shear is a cosine input. In response, a stiff rotor will have both cosine and sine responses. A teetered rotor with $K=1$ will only have a sine response. The turbine response to other inputs, as described below, exhibits similar combinations of sine and cosine cyclic responses.

General solution to flapping equation of motion The general solution to the flapping motion can in principal be found simply by applying Cramer's Rule to Equation 4.3.73. The resulting algebraic expressions, however, are not particularly illuminating. One approach that helps to add some clarity it to express the terms in the flapping angle as the sum of other constants, which represent the contributions of the various forcing effects:

$$\beta_{1c} = \beta_{1c,ahg} + \beta_{1c,cr} + \beta_{1c,yr} + \beta_{1c,vs} \tag{4.3.88a}$$

$$\beta_{1s} = \beta_{1s,ahg} + \beta_{1s,cr} + \beta_{1s,yr} + \beta_{1s,vs} \tag{4.3.88b}$$

where the subscripts are ahg for axial flow, hinge spring, and gravity (blade weight), cr for crosswind, vs for vertical wind shear and yr for yaw rate.

Each of these constants is also a function of the parameters in the equations of motion. Rather than expand the matrix solution by itself, we will present below the various subscripted terms which could be obtained from such a solution.

The dominant constant coning term, which includes axial flow, hinge–spring, and gravity, is:

$$\beta_0 = \frac{1}{D}\frac{\gamma A}{2}\left[(K-1)^2 + \left(\frac{\gamma}{8}\right)^2\right] \tag{4.3.89}$$

The cosine and sine terms are summarized in Table 4.2

Table 4.2 Contributions to flapping responses

*	Cosine, $\beta_{1c,*}$	Sine, $\beta_{1s,*}$
ahg	$-\frac{1}{D}\gamma B\,A(K-1)$	$-\frac{1}{D}\frac{\gamma}{8}B\,A$
cr	$\frac{\overline{V}_0}{D}\left[\left(\frac{\gamma}{8}\right)^2\left(\frac{\gamma}{2}\right)\frac{4}{3}A-\frac{\gamma\,A_3}{2}K(K-1)\right]$	$-\frac{4\overline{V}_0}{D}\left(\frac{\gamma}{8}\right)^2\left[\frac{4}{3}A(K-1)+A_3K\right]$
vs	$-\frac{K_{sh}\overline{U}}{D}\frac{\gamma}{8}K(K-1)$	$-\frac{K_{sh}\overline{U}}{D}\left(\frac{\gamma}{8}\right)^2K$
yr	$\frac{K\,\overline{q}}{D}\left[\left(\frac{\gamma}{8}\right)^2-2(K-1)-\frac{\gamma A_3}{2}(K-1)\overline{d}_{yaw}\right]$	$-\frac{K\,\overline{q}}{D}\left(\frac{\gamma}{8}\right)\left[\frac{\gamma}{2}A_3\,\overline{d}_{yaw}+K+1\right]$

The important point to note about the various subscripted constants is that they can help to illustrate the amount of the response that is due to a particular input. For example, in a particular situation, if the wind shear sine cyclic response were close in magnitude to the total sine cyclic response constant, then it would be immediately apparent that other factors were of little significance.

4.3.3.8 Blade and hub loads

The blade flapping motions, determined using Equation 4.3.73, can be used to determine blade root loads. Moments and forces on the hub and tower can then be determined from those blade forces.

For a rigid rotor turbine (with cantilevered and not teetered blades), both flapping and lead–lag moments are transmitted to the hub. Flapping is usually predominant, and in the following discussion the focus will be on the effect of those loads. In a teetered rotor, on the other hand, no flapping moments are transmitted to the hub (unless the teeter stops are hit). Only in-plane forces (in the direction of the torque) are transferred to the hub. A detailed discussion of teetered rotor response, however, is beyond the scope of this text.

Figure 4.23 illustrates the coordinate system for the forces and moments transmitted to the hub of a typical rotor.

Recall that the blade flapping angle is approximated by the sum:

$$\beta \approx \beta_0 + \beta_{1c}\cos(\psi) + \beta_{1s}\sin(\psi) \tag{4.3.90}$$

The corresponding blade root bending moment for each blade is:

$$M_\beta = K_\beta \beta \tag{4.3.91}$$

In a manner similar to the development of the flap equations, one could develop lead–lag and torsion equations and use the full set of responses to get hub loads. The terms summarized in Section 4.3.1.2 are sufficient for such an expansion of the model.

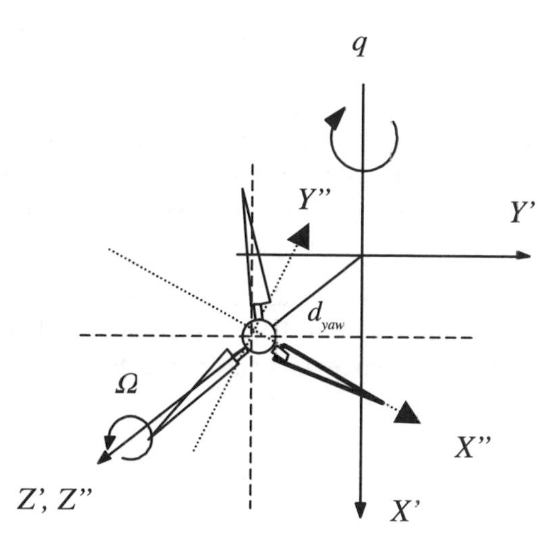

Figure 4.23 Coordinate system for hub moments and forces; d_{yaw} yaw moment arm; q, yaw rate; Ω, angular velocity

In order to find the hub loads from multiple blades, the forces and moments from all of them are determined and the effects are summed. Note that in each case the proper azimuth must be used. For example, in a rotor with three blades, azimuths should be 120 degrees apart for each of them. It should be noted that with this simplified model, if the blades are spaced symmetrically around the rotor, the cumulative effect on rotor torques would be that they would be constant throughout the rotation. Table 4.3 summarizes the key hub reactions for both teetered and cantilevered blades.

Table 4.3 Hub reactions for teetered and cantilevered blades

Hub reaction	Teetered Rotor	Cantilevered
Flapping moment	None	Full flap moment
Flapping shear	Total thrust on hinge	Thrust of each blade
Lead–lag moment	Power torque	Power torque
Lead–lag shear	Force producing torque	Force producing torque
Blade tension	Centrifugal force, weight	Centrifugal force, weight
Blade torsion	Pitching moment	1 blade pitching moment

4.3.3.9 Tower loads

Tower loads result from aerodynamic loads on the tower, the weight of the turbine and tower, and from all of the forces on the machine itself, whether steady, cyclic, or impulsive, etc.

Steady tower loads Steady tower loads include the rotor thrust, the moment from the rotor torque, the turbine weight, and hurricane loads. Hurricane loads are aerodynamic loads expected in the most extreme winds when the turbine is not operating. More details on conditions giving rise to tower loads are given in Chapter 6.

Tower vibration Tower natural frequencies (for cantilevered towers) can be calculated by methods described in Section 4.2.2, including the tower top weight. Guyed towers involve methods beyond the scope of this text. The most important consideration in tower design is to avoid natural frequencies near rotor frequencies (1*P*, 2*P* or 3*P*). A 'soft' tower is one whose fundamental natural frequency is below the blade passing frequency whereas a stiff tower has its dominant natural frequency above that frequency. Further discussion of tower vibration can be found in Chapter 6.

Dynamic tower loads Dynamic tower loads are loads on the tower resulting from the dynamic response of the wind turbine itself. For a rigid rotor, blade moments are the main source of dynamic tower loads. The three moments for each blade (flapwise, lead–lag, and torsion) are transferred to the tower coordinates as:

$$M_{X'} = M_\beta \sin(\psi) \tag{4.3.92}$$

$$M_{Y'} = -M_\beta \cos(\psi) \tag{4.3.93}$$

$$M_{Z'} = M_\zeta \tag{4.3.94}$$

where X' refers to yawing, Y' to pitching backwards, Z' to rolling of the nacelle.

For multiple blades the contribution from each blade is summed up, adjusted by the relative azimuth angle.

For a teetered rotor the flapping moment is not transferred to the hub (or the tower) unless the teeter stops are hit, and so contributes little to dynamic tower loads under normal operation.

4.3.3.10 Yaw stability

Yaw stability is an issue for free-yaw turbines. It is a complicated problem, and the simplified dynamics model is of limited utility in its analysis. Nonetheless it does provide insights into some of the basic physics. The key point is that various inputs contribute to cyclic responses. Any net sine response would result in a net torque about the yaw axis. (The cosine cyclic term would tend to rock the turbine up and down on its yaw bearing, but would not affect yaw motion.) Conversely, for the rotor to be stable under any given conditions, the sine cyclic response term must be equal to zero. The sine cyclic response is affected by gyroscopic motion, yaw error, wind shear and gravity. Referring back to the solution of the flapping equations of motion, it was pointed out that the sine cyclic term could be subdivided into the terms due to different effects (Equation 4.3.88b). The dominant contributions to the 'sine cyclic' motion are given in Table 4.2.

The first thing to note is that both vertical wind shear and the gravitational force on the bent blades tend to turn the rotor out of the wind in the same direction (in the negative yaw direction in Figure 4.23). That means that the rotor tends to experience a crosswind from the negative direction. The crosswind, due to yaw error, will tend to turn the rotor back in the other direction. For yaw stability then, if gravity, steady wind, and wind shear are the only effects considered, there will be a yaw angle such that:

$$\beta_{1s,cr} = \beta_{1s,ahg} + \beta_{1s,vs} \tag{4.3.95}$$

Using this equation, but ignoring wind shear, and assuming small-angle approximations, it can be shown that the steady state yaw error, Θ, would be approximately:

$$\Theta \approx \frac{\Omega R}{V}\left(\frac{3B}{2(2K-1)}\right) \tag{4.3.96}$$

The steady state yaw error, then, in the absence of vertical wind shear, is greater at faster rotor speeds and for softer rotors (smaller K). Vertical wind shear would increase the steady state yaw error even more. Finally, a thorough analysis of the various terms in the solution shows that a rotor with preconing tends to be more stable than one without.

The linearized hinge–spring dynamic model provides insight into yaw stability, but the subject is more complicated than a first-order model indicates. Changing angles of attack, stall, turbulence, and unsteady aerodynamic effects all influence the yaw stability of real turbines. Additional analysis of the ability of a relatively simplified dynamic model to provide insights into yaw stability can be found in Eggleston and Stoddard (1987). Experience has shown, however, that in the matter of yaw motion in particular, a more comprehensive method of analysis is required. One approach to that, which is incorporated in the computer model YawDyn, is discussed later in this chapter.

4.3.3.11 Applicability and limitations of linearized hinge–spring dynamic model

The linearized dynamic model developed above can be very useful in providing insight into wind turbine dynamics. However, there are some important aspects of rotor behavior that do not appear in the model. Actual data often exhibits oscillations which are not predicted. Referring back to Figure 4.10, we recall that the root flap bending moment depicted there shows significant higher frequency oscillations. The significance of this can be illustrated even more graphically in a power spectrum, which is a form of the power spectral density (psd).

The psd was introduced in Chapter 2 in relation to turbulent wind speed fluctuations. It can also be used to illustrate the relative amount of energy in other types of fluctuations as a function of frequency. When plotted as the psd times the frequency against frequency on a log axis, the area under the curve is proportional to the variance associated with the corresponding frequency. In this form the psd is known as the power spectrum.

Figure 4.24 shows a power spectrum (in arbitrary units) obtained from the data used in Figure 4.10. As would be expected, the $1P$ spike in the measured data (at 72 rpm or 1.2 Hz) is very strong. In addition, spikes appear at approximately $2P$ and $4P$. The other spikes are presumably due to turbulence in the wind and some natural frequencies of the blade. The important thing to note is that the linearized hinge–spring model could only reproduce (to some extent) the $1P$ response. None of the higher frequency responses would be predicted and, as can be seen from the figure, there is a significant amount of energy associated with those higher frequencies.

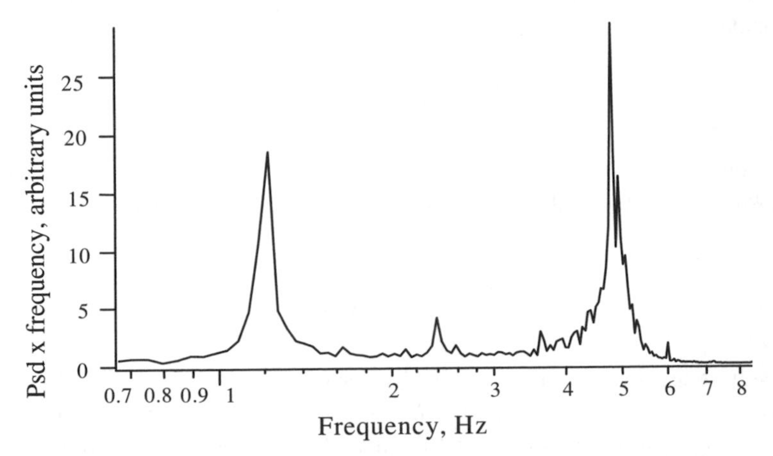

Figure 4.24 Power spectrum of root flap bending moment; psd, power spectral density

4.4 Detailed and Specialized Dynamic Models

The linearized hinge–spring dynamic model discussed in Section 4.3 has many advantages for elucidating the first-order response of a wind turbine rotor to a range of input conditions. It is less useful, however, in a number of situations where important aspects of the response are not apparent. In order to overcome this limitation, it has been necessary to develop more detailed and sometimes more specialized models. Typically, these models require numerical solutions, and are implemented in computer code. Specialized non-linear models can be used to investigate the dynamics of turbine sub-systems such as tip flaps, blades, pitch linkages, gear trains, etc. No single model has so far been able to deal with all of these situations, but an understanding of approaches to detailed modeling provides the tools for more detailed analyses. This section discusses a few of those models, with a particular focus on YawDyn (which was first introduced in Chapter 3).

4.4.1 YawDyn

YawDyn is a complete aerodynamics and dynamics analysis code for constant speed horizontal axis wind turbines (Hansen, 1992; 1996). The purpose of this model was to help in understanding some of the problems which horizontal axis wind turbines of the 1980's were experiencing due to yaw motion. The resulting model has indeed turned out to be useful for understanding yaw issues. It has also turned out to be valuable for understanding other responses.

The principles of the YawDyn model are closely related to those of the linearized hinge–spring model. There are some important differences, particularly with respect to the modelling of aerodynamics and also with regard to the method of solution.

4.4.1.1 YawDyn turbine model

The mechanical model of the turbine is basically the same as that of the linearized hinge–spring model. Blades are assumed to be rigid beams, connected to the hub by a hinge and spring. In addition, unsteady yaw motion of the entire turbine about the yaw axis is allowed. One other addition is that yaw drive train stiffness is accounted for. Another feature is that the turbines with teetered rotors may be modelled.

4.4.1.2 YawDyn aerodynamic model

The biggest differences between the linear spring model and YawDyn are in the aerodynamics. The linearized hinge–spring model uses a highly simplified aerodynamics model. It has turned out that yaw motion is greatly affected by small differences in loadings from one blade to the next, and these differences do not appear in the simplified model. So first of all, the aerodynamic model itself needed to be more accurate. In addition, it needed to account for more variation in the incoming wind (such as vertical and horizontal wind shear).

YawDyn's aerodynamic model includes the following features most of which are not in the simplified model:

- Blade element/momentum theory
- Skewed wake
- Dynamic stall
- Tower shadow
- Vertical wind shear
- Horizontal wind shear
- Vertical wind
- Tower shadow
- Unsteady inflow (turbulence)

Each of these was discussed in Chapter 3. It is also noteworthy that the aerodynamics section of YawDyn has proven sufficiently versatile that it has been separated out to function as an independent code. This piece of YawDyn, known as AeroDyn, has been used to provide the input to even more detailed dynamic models, such as ADAMS/WT (see Section 4.4.2.2).

4.4.1.3 Solutions

The linearized hinge–spring model is sufficiently simple that the equations of motion can be developed and solved with conventional algebra, although even that is a cumbersome process. The equations used in YawDyn would have been quite difficult to derive and solve by hand, so the computerized symbolic manipulation code Mathematica™ (Wolfram, 1991) was utilized. The linearized model was solved by assuming a sinusoidal solution. This yielded a convenient, closed-form result. In YawDyn it was assumed that the flap moments and yaw moments were all expressible as Fourier series, and so it was *a priori* the case that the solutions would include many harmonics. Since the inflow conditions were also assumed to vary, a closed-form solution was out of the question. Accordingly, the method of solving the equations was a numerical time step solution.

4.4.2 Other dynamics codes

A variety of other computer codes are available that can analyze rotor dynamics. The National Renewable Energy Laboratory (NREL) in Boulder, Colorado, has supported the development of a number of these codes and has made some of them available over the Internet. A brief review of two of these codes is presented below. More details of the use of the codes and their strengths and weaknesses can be found in at the NREL web site (http://www.nrel.gov). In addition to the codes available at NREL, numerous commercial codes are available that model wind turbine dynamics. The codes that are summarized here include: FAST_AD and ADAMS/WT.

4.4.2.1 FAST_AD

FAST_AD is a medium-complexity code for aerodynamic and dynamic analysis of horizontal axis wind turbines with two or three blades and a teetering or rigid hub. FAST_AD incorporates routines for simulating turbulent wind input to the rotor and routines for determining aerodynamic loads. The dynamic analysis allows for 14 degrees of freedom including multiple tower bending modes, three blade-bending modes, yaw, teeter, and drive train torsion. The code is intended to allow more degrees of freedom than YawDyn, but provide faster results than ADAMS/WT.

4.4.2.2 ADAMS/WT

ADAMS/WT (Anon., 1998) is a detailed dynamics code that can accept as input the output of the AeroDyn subroutines mentioned in Section 4.4.1.2. For the detailed analysis of larger structures appropriate input forces are required. Thus, the ability of ADAMS/WT to use available aerodynamic models for input forces makes this a powerful tool. ADAMS/WT can model the dynamics of the tower, nacelle, drive train, hub and blades. The degree of detail in each of the subsystem models is up to the user. For example, the blades and tower can be modelled as a uniform beam or a series of lumped masses with connections of different area moments of inertia and stiffnesses.

References

Anon. (1998) *ADAMS/WT 2.0 User's Guide*, Mechanical Dynamics, Inc., Mesa, AZ.

Ansell, M. P. (1987) Layman's Guide to Fatigue: The Geoff Pontin Memorial Lecture. *Proceedings of the 9th British Wind Energy Association Annual Conference*. Mechanical Engineering Publications.

Beer F. P., Johnston E. R., Jr. (1976) *Mechanics for Engineers*, 3rd Edition, McGraw Hill Book Co., New York, 1976.

Den Hartog, J. P. (1961) *Mechanics*, Dover Publications, New York.

Downing, S. D., Socie, D. F. (1982) Simple Rainflow Counting Algorithms, *International Journal of Fatigue*, January 1982, p. 31.

Eggleston, D. M., Stoddard, F. S. (1987) *Wind Turbine Engineering Design*. Van Nostrand Reinhold, New York.

Garrad, A. D., Hassan, U. (1986) The Dynamic Response of a Wind Turbine for Fatigue Life and Extreme Load Prediction. *Proceedings of the European Wind Energy Assoc. Conference (EWEC '86)*, A. Ragguzi, Bookshop for Scientific Publications, Rome.

Hansen, A. C. (1992) *Yaw Dynamics of Horizontal Axis Wind Turbines: Final Report*, National Renewable Energy Laboratory, NREL Technical Report TP 442-4822

Hansen, A. C. (1996) *User's Guide to the Wind Turbine Dynamics Computer Programs YawDyn and AeroDyn for ADAMS®, Version 9.6.* Univ. of Utah, Salt Lake City. Prepared for the National Renewable Energy Laboratory under Subcontract No. XAF-4-14076-02.

Manwell, J. F., McGowan, J. G., Abdulwahid, U., Rogers, A., McNiff, B. P. (1996) A Graphical Interface Based Model for Wind Turbine Dynamics. *Proceedings of the AWEA Annual Conference*, Denver, CO.

Merriam, J. L. (1978) *Engineering Mechanics: Vol. 1, Statics. Engineering Mechanics: Vol. 2, Dynamics.* John Wiley, New York, 1978.

Pytel, A., Singer, F. L. (1987) *Strength of Materials*, 4th Edition. Harper and Row, New York.

Shigley, R. G., Mischke, C. R. (1989) *Mechanical Engineering Design*, 5th. Edition. McGraw Hill, New York.

Spotts, M. E. (1985) *Design of Machine Elements*. Prentice-Hall, Englewood Cliffs, NJ.

Sutherland, H. J., Veers, P. S., Ashwill, T. D. (1995) *Fatigue Life Prediction for Wind Turbines: A Case Study on Loading Spectra and Parameter Sensitivity*. Standard Technical Publication 1250, American Society for Testing and Materials, 1916 Race St., Philadelphia, PA.

Thomson, W. T. (1981) *Theory of Vibrations with Applications*, 2nd Edition. Prentice-Hall, Englewood Cliffs, NJ.

Wolfram, S. (1991) *Mathematica: a System for Doing Mathematics by Computer*. Addison-Wesley, Redwood City, CA.

5

Electrical Aspects of Wind Turbines

5.1 Overview

Electricity is associated with many aspects of modern wind turbines. Most obviously, the primary function of the majority of wind turbines is the generation of electricity. A large number of topics of power systems engineering are thus directly relevant to issues associated with wind turbines. These include generation at the turbine itself as well as power transfer at the generator voltage, transforming to higher voltage, interconnection with power lines, distribution, transmission, and eventual use by the consumer. Electricity is used in the operation, monitoring, and control of most wind turbines. It is also used in site assessment and data collection and analysis. For isolated or weak grids, or systems with a large amount of wind generation, storage of electricity is an issue. Finally, lightning is a naturally occurring electrical phenomenon that may be quite significant to the design, installation and operation of wind turbines.

The principal areas in which electricity is significant to the design, installation or operation of wind turbines are summarized below in Table 5.1.

This chapter includes two main parts. First, it includes an overview of the fundamentals. Second, it presents a description of those issues related to the turbine itself, particularly generators and power converters. The interconnection of the generator to the electrical grid and issues related to the system as a whole are described in Chapter 8.

The fundamentals overview focuses on alternating current (AC). Alternating current issues include phasor notation, real and reactive power, three-phase power, fundamentals of electromagnetism and transformers. Wind turbine related issues include the common types of generators, generator starting, and synchronization, power converters, and ancillary electrical equipment.

Table 5.1 Examples of electrical issues significant to wind energy

Power generation	Generators Power electronic converters	**Storage**	Batteries Rectifiers Inverters
Interconnection and distribution	Power cables Switch gear Circuit breakers Transformers Power quality	**Lightning protection**	Grounding Lightning rods Safe paths
Control	Sensors Controller Yaw or pitch motors Solenoids	**End loads**	Lighting Heating Motors
Site monitoring	Data measurement & recording Data analysis		

5.2 Basic Concepts of Electric Power

5.2.1 Fundamentals of electricity

In this chapter it is assumed that the reader has an understanding of the basic principles of electricity, including direct current (DC) circuits. For that reason these topics will not be discussed in detail here. The reader is referred to other sources, such as Edminster (1965), for more information. Specific topics with which the reader is assumed to be familiar include:

- Voltage
- Current
- Resistance
- Resistivity
- Conductors
- Insulators
- DC circuits
- Ohm's Law
- Electrical power and energy
- Kirchhoff's laws for loops and nodes
- Capacitors
- Inductors
- Time constants of RC and RL circuits
- Series/parallel combinations of resistors

5.2.2. Alternating current

The form of electricity most commonly used in power systems is known as alternating current (AC). In this text it is assumed that the reader has some familiarity with AC circuits, so only the key points will be summarized. In AC circuits (at steady state) all voltages and currents vary in a sinusoidal manner. There is a complete sinusoidal cycle each period. The frequency, f, of the sine wave is the number of cycles per second. It is the reciprocal of the period. In the United States and much of the Western Hemisphere the standard frequency for AC is 60 cycles per second (known as "Hertz", and abbreviated Hz). In much of the rest of the world the standard frequency for AC is 50 Hz.

The instantaneous voltage, v, in an AC circuit may be described by the following equation:

$$v = V_{\max} \sin(2\pi f t + \phi) \tag{5.2.1}$$

where $V_{\max}$ is the maximum value of the voltage, t is time and ϕ is the phase angle.

The phase angle indicates the angular displacement of the sinusoid from a reference sine wave with a phase angle of zero. Phase angle is important because currents and voltages, although sinusoidal, are not necessarily in phase with each other. In analysis of AC circuits it is often useful to start by assuming that one of the sinusoids has zero phase, and then find the phase angles of the other sinusoids with respect to that reference.

An important summary measure of the voltage is its root mean square (rms) value V_{rms}:

$$V_{rms} = \sqrt{\int_{cycle} v^2 \mathrm{d}t} = V_{\max} \sqrt{\int_{cycle} \sin^2(2\,\pi\, f\, t)\mathrm{d}t} = V_{\max} \frac{\sqrt{2}}{2} \tag{5.2.2}$$

Note that the rms value of the voltage is $\sqrt{2}/2$, or about 70% of the maximum voltage for a pure sine wave. The rms voltage is often referred to as the magnitude of the voltage, so $|V| = V_{rms}$.

5.2.2.1 Capacitors in AC circuits

The current in a capacitor is proportional to the derivative of the voltage. Thus, if the voltage across a capacitor is $v = V_{\max} \sin(2\pi f t)$ then the instantaneous current, i, is:

$$i = C\frac{dv}{dt} = 2\,\pi\, f\, V_{\max} C \sin(2\,\pi\, f\, t + \pi/2) \tag{5.2.3}$$

where C is the capacitance. Equation 5.2.3 can be rewritten as:

$$i = I_{\max} \sin(2\,\pi\, f\, t + \pi/2) \tag{5.2.4}$$

where $I_{max} = V_{max}/X_C$ and $X_C = 1/(2\,\pi\, f\, C)$. The term is X_C is known as capacitive reactance. Reactance is somewhat analogous to resistance in DC circuits. Note that as the current varies in time, the sinusoid will be displaced by $\pi/2$ radians ahead of the voltage sinusoid. For that reason the current in a capacitive circuit is said to lead the voltage.

5.2.2.2 Inductors in AC circuits

The current in an inductor is proportional to the integral of the voltage. The relation between the voltage and current in an inductor can be found from:

$$i = \frac{1}{L}\int v dt = \frac{V_{max}}{2\pi f L}\sin(2\pi f t - \pi/2) \tag{5.2.5}$$

where L is the inductance. Analogously to the equation for capacitors, Equation 5.2.5 can be rewritten as:

$$i = \frac{V_{max}}{X_L}\sin(2\pi f t - \pi/2) \tag{5.2.6}$$

where X_L = inductive reactance ($X_L = 2\pi f L$). Note that in an inductive circuit the current lags the voltage.

5.2.2.3 Phasor notation

Manipulations of sines and cosines with various phase relationships can become quite complicated. Fortunately, the process can be greatly simplified, as long as frequency is constant. This is the normal case with most AC power systems. (Note that transient behavior requires a more complicated analysis method.) The procedure uses phasor notation, which is summarized below.

The use of phasors involves representing sinusoids by complex numbers. For example, the voltage in Equation 5.2.1, can be represented by a phasor:

$$\hat{V} = V_{max} e^{j\phi} = a + jb = V_{max}\angle\phi \tag{5.2.7}$$

The bold and circumflex are used to indicate a phasor. Here $j = \sqrt{-1}$, $\angle$ indicates the angle between phasor and real axis, ϕ is the phase angle and

$$a = V_{max}\cos(\phi) \tag{5.2.8}$$

$$b = V_{max}\sin(\phi) \tag{5.2.9}$$

Figure 5.1 illustrates a phasor. Note that it can be equivalently described in rectangular or polar coordinates.

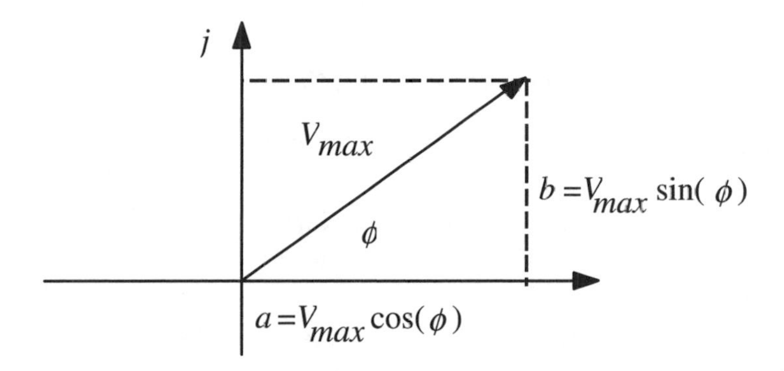

Figure 5.1 Phasor

In this method of representation the frequency is implicit. To recover the time series waveform one can use the following relation:

$$V(t) = \mathrm{Re}\left\{V_{max}\,\mathrm{e}^{j\varnothing}\;\mathrm{e}^{j2\pi f t}\right\} \tag{5.2.10}$$

where $\mathrm{Re}\{\,\}$ signifies that only the real part is to be used and, as before, $\mathrm{e}^{j\varnothing} = \cos(\phi) + j\sin(\phi)$.

In an analogous manner, current can be expressed as a phasor, $\hat{\boldsymbol{I}}$, defined by I_{max} and the phase of the current.

A few rules can be applied when using phasors. These are summarized below. Note that sometimes it is more convenient to use rectangular form; other times the polar form. Begin by defining two phasors $\hat{\boldsymbol{A}}$ and $\hat{\boldsymbol{C}}$.

$$\hat{\boldsymbol{A}} = a + jb = A_m \mathrm{e}^{j\varnothing_a} = A_m \angle \varnothing_a \tag{5.2.11}$$

$$\hat{\boldsymbol{C}} = c + jd = C_m \mathrm{e}^{j\varnothing_c} = C_m \angle \varnothing_c \tag{5.2.12}$$

where

$$\varnothing_a = \tan^{-1}(b/a) \tag{5.2.13}$$

$$\varnothing_c = \tan^{-1}(d/c) \tag{5.2.14}$$

$$A_m = \sqrt{a^2 + b^2} \tag{5.2.15}$$

$$C_m = \sqrt{c^2 + d^2} \tag{5.2.16}$$

The rules for phasor addition, multiplication, and division are:

Phasor addition:

$$\hat{\boldsymbol{A}} + \hat{\boldsymbol{C}} = (a+c) + j(b+d) \tag{5.2.17}$$

Phasor multiplication:

$$\hat{\boldsymbol{A}}\hat{\boldsymbol{C}} = A_m \angle \o_a C_m \angle \o_c = A_m C_m \angle(\o_a + \o_c) \tag{5.2.18}$$

Phasor division:

$$\frac{\hat{\boldsymbol{A}}}{\hat{\boldsymbol{C}}} = \frac{A_m \angle \o_a}{C_m \angle \o_c} = \frac{A_m}{C_m} \angle(\o_a - \o_c) \tag{5.2.19}$$

More details on phasors can be found in most texts on AC circuits, as well as in Brown and Hamilton (1984.)

5.2.2.4 Complex impedance

The AC equivalent of resistance is complex impedance, $\hat{\boldsymbol{Z}}$, which takes into account both resistance and reactance. Impedance can be used with the phasor voltage to determine phasor current and vice versa. Impedance consists of a real part (resistance) and an imaginary part (inductive or capacitive reactance.) Resistive impedance is given by, $\hat{\boldsymbol{Z}}_R = R$, where R is the resistance. Inductive and capacitive impedances are given by, respectively, $\hat{\boldsymbol{Z}}_L = j\,2\pi\, f L$ and $\hat{\boldsymbol{Z}}_C = -j/(2\pi fC)$, where f is the AC frequency in Hertz. Note that for a circuit which is completely resistive the impedance is equal to the resistance. For a circuit which is completely inductive or capacitive the impedance is equal to the reactance. Note, also, that inductive and capacitive impedances are a function of the frequency voltage fluctuations of the AC system.

The rules relating voltage, current and impedance in AC circuits are analogous to those of DC circuits.

Ohm's law:

$$\hat{\boldsymbol{V}} = \hat{\boldsymbol{I}}\hat{\boldsymbol{Z}} \tag{5.2.20}$$

Impedances in series:

$$\hat{\boldsymbol{Z}}_S = \sum_{i=1}^{N} \hat{\boldsymbol{Z}}_i \tag{5.2.21}$$

Impedances in parallel:

$$\hat{\boldsymbol{Z}}_P = 1 \Big/ \sum_{i=1}^{N} \frac{1}{\hat{\boldsymbol{Z}}_i} \tag{5.2.22}$$

where $\hat{\boldsymbol{Z}}_S$ is the effective impedance of N impedances ($\hat{\boldsymbol{Z}}_i$) in series and $\hat{\boldsymbol{Z}}_P$ is the effective impedance of N parallel impedances. Kirchhoff's Laws also apply to phasor currents and voltages in circuits with complex impedances.

5.2.2.5 Power in AC circuits

By measuring the rms voltage, V_{rms}, and rms current, I_{rms}, in an AC circuit, and multiplying them together, as would be done in DC circuit, one would obtain the apparent power, S. That is:

$$S = V_{rms} I_{rms} \tag{5.2.23}$$

Apparent power, which is measured in units of Volt-Amperes (VA), however, can be somewhat misleading. In particular, it may not correspond to the real power either consumed (in the case of a load), or produced (in the case of a generator).

Real power, P, is obtained by multiplying the apparent power by the cosine of the phase angle between the voltage and the current. It is thus given by:

$$P = V_{rms} I_{rms} \cos(\phi) \tag{5.2.24}$$

Real electrical power is measured in units of Watts.

Current that is flowing in the inductive or capacitive reactances does not result in real power, but does result in reactive power, Q. It is given by:

$$Q = V_{rms} I_{rms} \sin(\phi) \tag{5.2.25}$$

Reactive power, which is measured in units of "Volt-Amperes reactive" (VAR) is significant because it must be produced somewhere on the system. For example, currents creating the magnetic field in a generator correspond to a requirement for reactive power. Reactive current can also result in higher line losses in distribution or transmission lines, because of the resistance in the lines.

The 'power factor' of a circuit or device describes the fraction of the apparent power that is real power. Thus, power factor is simply the ratio of real to apparent power. For example, a power factor of 1 indicates that all of the power is real power. Power factor is often defined as the cosine of the phase angle between the voltage and the current, $\cos(\phi)$, This quantity is correctly called the displacement power factor. In circuits with sinusoidal currents and voltages, the two types of power factors are equivalent. In circuits with non-sinusoidal currents and voltages, displacement power factor is more appropriate.

The phase angle between current and voltage is called the power factor angle, since it is the basis for determining power factor. It is important to note that the power factor angle may be either positive or negative, corresponding to whether the current sine wave is leading the voltage sine wave or vice versa, as discussed earlier. Accordingly, if the power factor angle is positive the power factor is said to be leading. If it is negative the power factor is lagging.

An example of the waveforms relating voltage, current and apparent power is shown in Figure 5.2 for a circuit with a resistor and capacitor. For this example, the current and voltage are out of phase by 45 degrees, so the power factor is 0.707. The current sine wave precedes the voltage wave, so the power factor is leading.

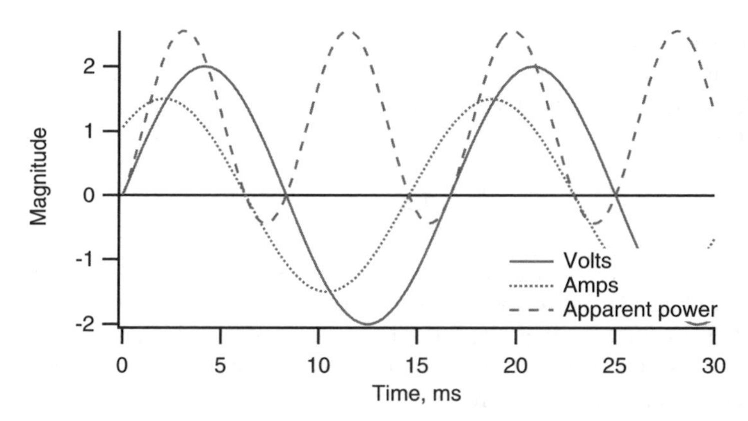

Figure 5.2 AC voltage, *v*, current, *i*, and apparent power, *vi*, in a circuit with a resistor and capacitor

A simple example of the use of phasors for calculations in an AC circuit is the following. Consider a simple circuit with an AC voltage source, a resistor, an inductor, and a capacitor all connected in series in a single loop. The resistance of the resistor is 4 Ω, the reactances of the inductor and capacitor are $j3\ \Omega$ and $-j6\ \Omega$, respectively. The voltage is $100\angle 0$. The problem is to find the current and the power dissipated in the resistor.

Solution: The total impedance, $\hat{\boldsymbol{Z}}$, is $\hat{\boldsymbol{Z}} = \hat{\boldsymbol{Z}}_R + \hat{\boldsymbol{Z}}_L + \hat{\boldsymbol{Z}}_C$ = 4 + $j3$ - $j6$ = 4 - $j3$ = $5\angle -36.9°$. The current is then $\hat{\boldsymbol{I}} = \hat{\boldsymbol{V}} / \hat{\boldsymbol{Z}}$ = $20\angle 36.9$. The power can be found from either $P = \left|\hat{\boldsymbol{I}}\right|^2 R = 20^2 4 = 1600$ W or $P = \left|\hat{\boldsymbol{I}}\right|\left|\hat{\boldsymbol{V}}\right| \cos(36.9) = 1600$ W. Note the use of the absolute value for the equivalent rms.

5.2.2.6 Three phase AC power

Power generation and large electrical loads commonly operate on a three-phase power system. A three-phase power system is one in which the voltages supplying the loads all have a fixed phase difference from each other of 120 degrees ($2\pi / 3$ radians). Individual three-phase transformers, generators or motors all have their windings arranged in one of two ways. These are: (i) Y (or wye) and (ii) Δ (delta), as illustrated in Figures 5.3 and 5.4. The appearance of the windings is responsible for the names. Note that the Y system has four wires (one of which is the neutral), whereas the Δ system has three wires.

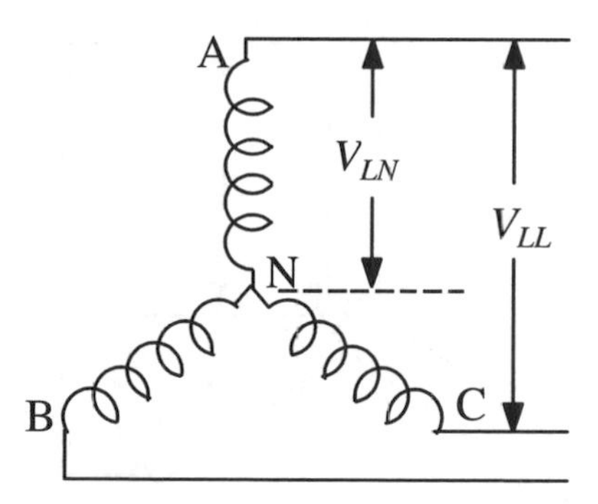

Figure 5.3 Y connected coils; V_{LN} and V_{LL}, line-to-neutral and line-to-line voltage, respectively.

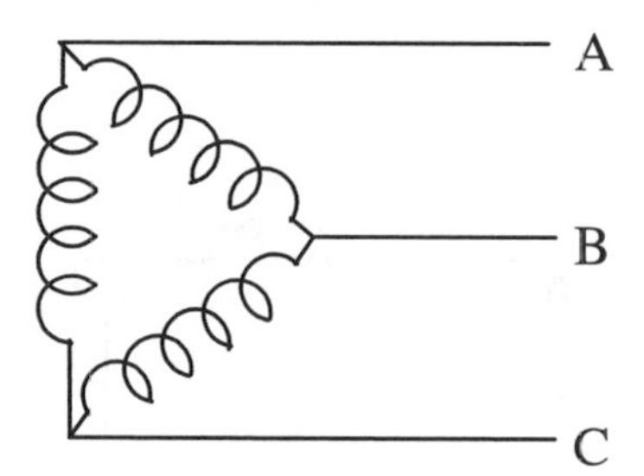

Figure 5.4 Delta connected coils

Loads in a three-phase system are, ideally, balanced. That means the impedances are all equal in each phase. If that is the case, and assuming that the voltages are of equal magnitude, then the currents are equal to each other but are out of phase from one another by 120 degrees. Voltages in three-phase systems may be line-to-neutral, V_{LN}, or line-to-line voltages, V_{LL}. They may also be described as line voltages (V_{LL}) or phase voltages (voltages across loads or coils). Currents in each conductor, outside the terminals of a load, are referred to as line currents. Currents through a load are referred to as load or phase currents. In general, in a balanced Y connected load, the line currents and phase currents are equal, the neutral current is zero, and the line-to-line voltage, V_{LL}, is $\sqrt{3}$ times the line-to-neutral voltage, V_{LN}. In a balanced delta connected load, the line voltages and phase voltages are equal, whereas the line current is $\sqrt{3}$ times the phase current. Figure 5.5 illustrates Y connected three-phase loads, assumed to be balanced and all of impedance $\hat{\boldsymbol{Z}}$.

If a three-phase system is known to be balanced, it may be characterized by a single phase equivalent circuit. The method assumes a Y connected load, in which each impedance is equal to $\hat{\boldsymbol{Z}}$. (A delta connected load could be used by applying an appropriate Y–Δ transformation to the impedances, giving $\hat{\boldsymbol{Z}}_Y = \hat{\boldsymbol{Z}}_\Delta / 3$.) The one-line equivalent circuit is one phase of a four-wire, three-phase Y connected circuit, except that the voltage used is the line-to-neutral voltage, with an assumed initial phase angle of zero. The one-line equivalent circuit is illustrated in Figure 5.6.

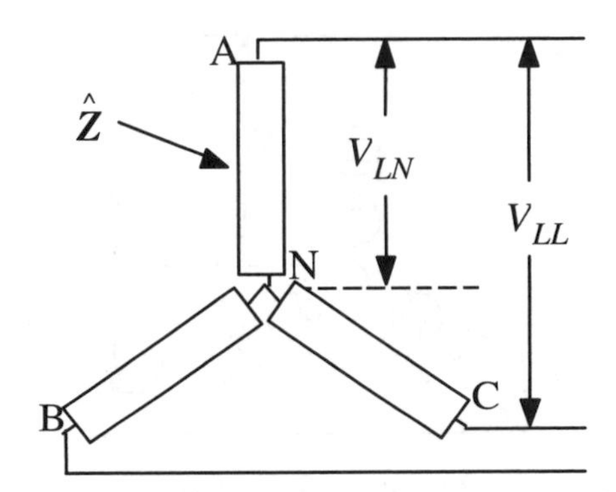

Figure 5.5 Y connected loads; V_{LN} and V_{LL}, line-to-neutral and line-to-line voltage, respectively; $\hat{Z}$, impedance

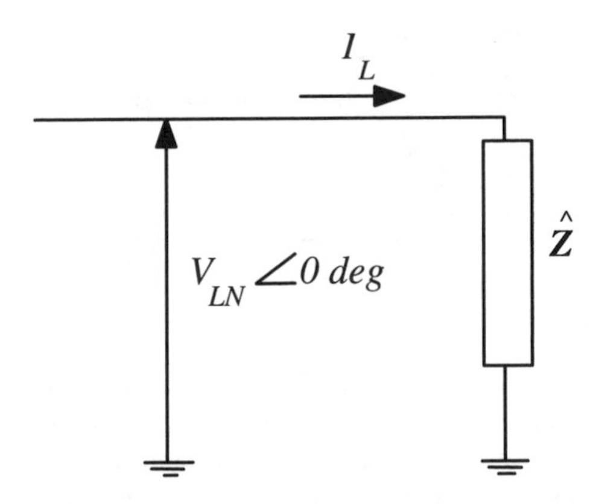

Figure 5.6 One line equivalent circuit; I_L, line current in three-phase system; V_{LN}, line-to-neutral voltage; $\hat{Z}$, impedance

More details on three-phase circuits can be found in most texts on electrical power engineering, including Brown and Hamilton (1984).

Power in three phase loads It is most convenient to be able to determine power in a three-phase system in terms of easily measurable quantities. These would normally be the line-to-line voltage difference and the line currents. In a balanced delta connected load, the power in each phase is one third of the total power. The real power in one phase, P_1, using line-to-line voltage and phase current, I_P, would be:

$$P_1 = V_{LL} I_P \cos(\phi) \tag{5.2.26}$$

Since the line current is given by $I_L = \sqrt{3} I_P$, and there are three phases, the total real power is:

$$P = \sqrt{3} V_{LL} I_L \cos(\phi) \tag{5.2.27}$$

Similarly, the total three-phase apparent power and reactive power are:

$$S = \sqrt{3} V_{LL} I_L \tag{5.2.28}$$

$$Q = \sqrt{3} V_{LL} I_L \sin(\phi) \tag{5.2.29}$$

The above relations for three-phase power also hold for balanced Y connected loads. Calculation of power in unbalanced loads is beyond the scope of this text. The interested reader should refer to any book on power system engineering for more information.

5.2.2.7 Voltage levels

One of the major advantages of AC power is that the voltage level may be readily changed by the use of power transformers. Power may be used conveniently and safely at relatively low voltage, but transformed to a much higher level for transmission or distribution. To a close approximation, power is conserved during transforming, so that when the voltage is raised currents are lowered. This serves to reduce losses in transmission or distribution lines, allowing much smaller and less expensive conductors.

Wind turbines typically produce power at 480 V (in the United States) or 690 V (in Europe). Wind turbines are often connected to distribution lines with voltages in the range of 10 kV to 69 kV. See Chapter 8 for more information on electric grids and interconnection to the grid.

5.2.3 *Fundamentals of electromagnetism*

The fundamental principles governing transformers and electrical machinery, in addition to those of electricity, which were summarized above, are those of the physics of electromagnetism. As with electricity it is assumed that the reader is familiar with the basic concepts of electromagnetism. These principles are summarized below. More details can be found in most physics or electrical machinery texts. By way of a quick overview, it may be noted that the magnetic field intensity in an electromagnet is a function of the current. The forces due to magnetic fields are a function of the magnetic flux density. This depends on the materials within the magnetic field as well as the intensity of the magnetic field.

5.2.3.1 Ampere's law

Current flowing in a conductor induces a magnetic field of intensity **H** in the vicinity of the conductor. This is described by Ampere's Law:

$$\oint \mathbf{H} \bullet \mathrm{d}\ell = I \tag{5.2.30}$$

which relates the current in the conductor, I, to the line integral of the magnetic field intensity along a path, ℓ, around the conductor.

5.2.3.2 Flux density and magnetic flux

The magnetic flux density, **B** (Wb/m^2), is related to the magnetic field intensity by the permeability, μ, of the material in which the field is occurring:

$$\boldsymbol{B} = \mu \boldsymbol{H} \tag{5.2.31}$$

where $\mu = \mu_0 \mu_r$ is the permeability (Wb/A-m) which can be expressed as the product of two terms: μ_0, the permeability of free space, $4\pi \times 10^{-7}$ Wb/A-m and μ_r, the dimensionless relative permeability of the material.

The permeability of non-magnetic materials is close to that of free space and thus the relative permeability, μ_r, is close to 1.0. The relative permeability of ferromagnetic materials is very high, in the range of 10^3 to 10^5. Consequently ferromagnetic materials are used in the cores of windings in transformers and electrical machinery in order to create strong magnetic fields.

These laws can be used to analyze the magnetic field in a coil of wire. Current flowing in a wire coil will create a magnetic field whose strength is proportional to the current and the number of turns, N, in the coil. The simplest case is a solenoid, which is a long wire wound in a close packed helix. The direction of the field is parallel to the axis of the solenoid. Using Ampere's Law, the magnitude of the flux density inside a solenoid of length L can be determined:

$$B = \mu I \frac{N}{L} \tag{5.2.32}$$

The flux density is relatively constant across the cross-section of the interior of the coil.

Magnetic flux Φ (Wb) is the integral of the product of the magnetic field flux density and the cross sectional area A through which it is directed:

$$\Phi = \int \boldsymbol{B} \bullet \mathrm{d}A \tag{5.2.33}$$

Note that the integral takes into account, via the dot product, the directions of the area through which the flux density is directed as well as that of the flux density itself. For example, magnetic flux inside a coil is proportional to the magnetic field strength and the cross sectional area, A, of the coil:

$$\Phi = BA \tag{5.2.34}$$

5.2.3.3 Faraday's Law

A changing magnetic field will induce an electromotive force (EMF, or voltage) E in a conductor within the field. This is described by Faraday's Law of Induction:

$$E = -\frac{\mathrm{d}\Phi}{\mathrm{d}t} \tag{5.2.35}$$

Note the minus sign in the above equation. This reflects the observation that the induced current flows in a direction such that it opposes the change that produced it (Lenz's Law). Note also, that in this text, the symbol E is used to indicate induced voltages, while V is used for voltages at the terminals of a device.

As a result of Faraday's Law, a coil in a changing magnetic field will have an EMF induced in it that is proportional to the number of turns:

$$E = -\frac{\mathrm{d}(N\Phi)}{\mathrm{d}t} \tag{5.2.36}$$

The term $\lambda = N\Phi$ is often referred to as the flux linkages in the device.

5.2.3.4 Induced force

A current flowing in conductor in the presence of magnetic field will result in an induced force acting on the conductor. This is the fundamental property of motors. Correspondingly, a conductor which is forced to move through a magnetic field will have a current induced in it. This is the fundamental property of generators. In either case the force d$\boldsymbol{F}$ in a conductor of incremental length $\mathrm{d}\ell$ (a vector), the current and the magnetic field d$\boldsymbol{B}$ are related by the following vector equation:

$$\mathrm{d}\boldsymbol{F} = I\,\mathrm{d}\ell \times \mathrm{d}\boldsymbol{B} \tag{5.2.37}$$

Note the cross product ($\times$) in Equation 5.2.37. This indicates that the conductor is at right angles to the field when the force is greatest. The force is also in a direction perpendicular to both the field and the conductor.

5.2.3.5 Reluctance

It is useful to consider some simple magnetic circuits. For example, suppose that a toroidal core, as shown in Figure 5.7, is wound with N turns of wire and that there is current, $i(t)$, flowing in the wire.

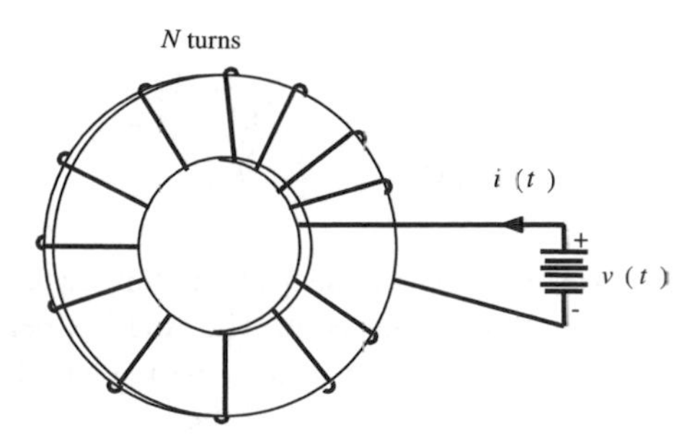

Figure 5.7 Simple magnetic circuit; $i(t)$, instantaneous current; $v(t)$, instantaneous voltage

Ampere's Law can be applied using a circular path of integration inside the core. The result is that the field intensity inside the core is

$$H_c = \frac{Ni}{2\pi r} \tag{5.2.38}$$

where r = radial distance from center of toroid. Here the quantity Ni is sometimes referred to as the magnetomotive force (MMF) that drives the flux.
The field intensity outside the core is equal to zero (ignoring the field at the wires, which are assumed to be of small diameter compared to that of the toroid). Evaluating the field intensity at the core midpoint and assuming that it is constant across the cross-section of the core results in:

$$H_c = \frac{Ni}{2\pi(r_i + r_o)/2} = \frac{Ni}{\ell_c} \tag{5.2.39}$$

where ℓ_c = length of the core at its midpoint.

The magnetic flux in the core is found by using Equations 5.2.39, 5.2.31 and 5.2.33:

$$\Phi = \int \boldsymbol{B} \bullet \mathrm{d}\boldsymbol{A} = \frac{\mu Ni A_c}{\ell_c} = \frac{Ni}{\ell_c / \mu A_c} \tag{5.2.40}$$

where A_c is the cross-sectional area of the core. The ratio of MMF to flux is known as reluctance, Re($A - t/Wb$) and can be thought of as resistance to the generation of magnetic flux by the MMF. In the above equation the reluctance of the core is $\mathrm{Re}_c = \ell_c / \mu A_c$.

Consider next a similar magnetic device with two gaps of width g in the core (see Figure 5.8). To a close approximation, it can be assumed that the magnetic flux remains in the core and in the air gap. In this case, the magnetic flux in the core and in the two gaps must be the same. From Ampere's Law, the magnetic flux is:

$$\Phi = \frac{Ni}{\dfrac{\ell_c}{\mu_r \mu_0 A_c} + \dfrac{2g}{\mu_0 A_g}} \tag{5.2.41}$$

where A_g is the area of the cross-section of the gap. In this case, the area of the cross-section of the gap is the same as that of the core. The relative permeability of magnetic materials is often about 10^4 greater than that of the air gap. Thus the reluctance of the air gap will be much greater than that of the core. Even if the gaps are 1/100 of the core length, they contribute significantly to the reduction of the magnetic flux in the device. In electrical machines magnetic flux flows across air gaps from stationary to rotating parts of the system. Normally it is important to keep these gaps as narrow as possible.

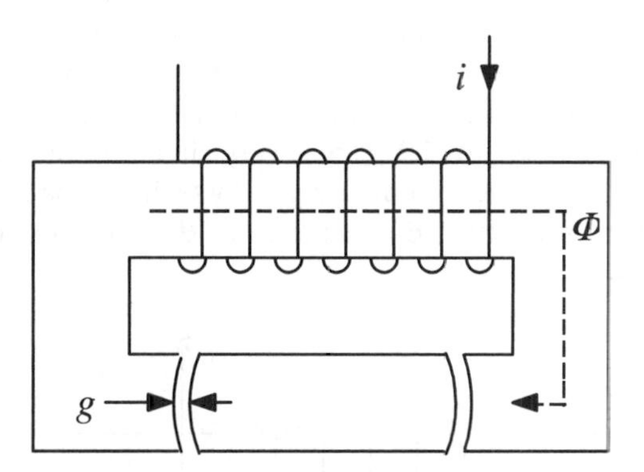

Figure 5.8 Simple magnetic device; *g*, width of air gap; *i*, current, Φ , magnetic flux. From *Electric Machines*, 1st edition, by Sarma © 1985. Reprinted with permission of Brooks/Cole, an imprint of the Wadsworth Group, a division of Thomson Learning. Fax 800 730-2215

5.2.3.6 Energy storage in magnetic fields

Energy is stored in the magnetic fields in an electromechanical device. Energy conversion in these devices involves the interchange of electrical and mechanical energy through changes in the stored energy in magnetic fields. While a full discussion of energy storage in magnetic fields is beyond the scope of this text, the use of the concept of energy storage to determine the torque in a simple device will be used to further the students' understanding of electric machines.

Determining the torque in an electric machine using Equation 5.2.37 is difficult unless the flux density and geometry of the system is known. For complicated systems, magnetic torques can be determined by an energy balance that can be expressed as:

$$Q_e = -\frac{\partial E_m(i,\theta)}{\partial \theta} + i\frac{\partial \lambda(i,\theta)}{\partial \theta} \tag{5.2.42}$$

where Q_e is the electrical torque, 6 is the angle of rotation of the device, E_m is the stored energy in the magnetic fields and λ is the flux linkages (defined above). The first term on the right of Equation 5.2.42 describes the torques developed from the change in energy stored in the magnetic fields as the rotor rotates. The second term on the right describes torques that are a function of changes in the electrical energy flowing through the system as the rotor position changes.

In order to use Equation 5.4.42, one needs to be able to express the energy stored in a magnetic field. Typically, the energy stored in the air gap field of an electric machine is much greater than that stored in the magnetic material. It can often be assumed that all of the energy in the magnetic fields is in the air gap. The energy per unit volume stored in the magnetic field, e_m, in an air gap is:

$$e_m = \tfrac{1}{2}\frac{B_g^{\,2}}{\mu_0} = \tfrac{1}{2}\frac{\Phi^2}{A_g^{\,2}\mu_0} = \tfrac{1}{2}\mu_0 H_g^{\,2} \tag{5.2.43}$$

where a subscript g indicates the quantity relates to the air gap. The total energy in the air gap field is the energy per unit volume multiplied by the volume of the air gap fields.

The use of Equation 5.2.42 can be illustrated with an example (Sarma,1985). Consider the magnetic device shown in Figure 5.9. This is very similar to the device shown above in Figure 5.8, except the material between the air gaps, which will be referred to as the rotor, can rotate. When the rotor is not aligned with the core (θ = 0), a torque is developed which attempts to realign the two.

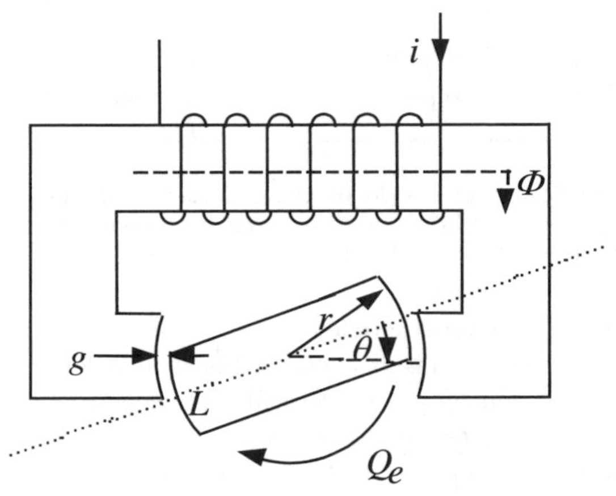

Figure 5.9 Simple magnetic torque device; g, width of air gap; i, current, Φ, magnetic flux; L, length of the face of the poles, Q_e, electrical torque; r, radius; θ, rotation angle. From *Electric Machines*, 1st edition, by Sarma © 1985. Reprinted with permission of Brooks/Cole, an imprint of the Wadsworth Group, a division of Thomson Learning. Fax 800 730-2215

A number of simplifying assumptions will be made in the analysis:

- The cross-sectional area of the air gap field is assumed to be the same as that of the core. In reality, the cross-sectional area of the field in the air gap will be larger than the core cross-sectional area
- There is no flux leakage
- All of the energy is stored in the fields in the air gap
- There are no losses in the system
- The reluctance of the air gap is much larger than that of the core

From Equation 5.4.41, the flux is a function of the reluctances in the system. The air gap reluctance is a function of the air gap length and cross-sectional area, which increases from 0 to A_c as θ increases from 0 to L/r, where L is the length of the face of each of the poles. The flux is then:

$$\Phi = \frac{Ni}{\dfrac{\ell_c}{\mu_r \mu_0 A_c} + \dfrac{2g}{\mu_0 A_g (r\theta / L)}} \tag{5.2.44}$$

Recognizing that the air gap reluctance is significantly greater than the core reluctance, the magnetic flux can be approximated as:

$$\Phi \approx \frac{Ni\mu_0 A_g}{2g}\left(\frac{r\theta}{L}\right) \tag{5.2.45}$$

The energy stored in the field is the product of the energy per unit volume (Equation 5.4.43). The volume of the field, which includes the two air gaps, increases from 0 to $2gA_g$ as • increases from 0 to *L/r*. Thus, the energy stored in the field is:

$$E_m = \frac{1}{2}\frac{\Phi^2}{A_g{}^2\mu_0}\left(2gA_g\frac{r\theta}{L}\right) \tag{5.2.46}$$

Using Equation 5.4.45 for the flux, one gets:

$$E_m \approx \frac{1}{2}\frac{N^2 i^2 \mu_0}{4g^2}\left(\frac{r\theta}{L}\right)^2\left(2gA_g\frac{r\theta}{L}\right) \tag{5.2.47}$$

Using the equation for the torque, Equation 5.4.42, and Equations 5.2.47 and 5.2.45 and the definition of flux linkages, after simplification, one gets:

$$Q_e = \frac{N^2 i^2 \mu_0}{4g^2}\left(\frac{2gA_g r}{L}\right)\left(1-\frac{3}{2}\left(\frac{r\theta}{L}\right)^2\right) \tag{5.2.48}$$

In this formulation, the torque is high when the rotor is rotated out of alignment ($6 = 0$) and decreases as the rotor approaches alignment with the core. Close to an angle of *L/r*, when the rotor poles are aligned with the core, the calculated torque becomes negative. This behavior results from the many simplifications used in the analysis. Nevertheless, the analysis does demonstrate the approximate behavior of a rotor with distinct poles in a magnetic field. The larger the current and greater the number of turns, the greater the torque. The larger the air gap, the greater the reluctance and the lower the torque.

5.2.3.7 Additional considerations

In practical electromagnetic devices, additional considerations affect machine performance, including leakage and eddy current losses, and the non-linear effects of saturation and hysteresis. Magnetic fields can never be restricted to exactly the regions where they can do useful work. Because of this there are invariably losses. These include leakage losses in transformers and electrical machinery. The effect of the leakage losses is to decrease the magnetic field from what would be expected from the current in the ideal case, or conversely to require additional current to obtain a given magnetic field. Eddy currents are secondary and, generally undesired, currents induced in parts of a circuit experiencing an alternating flux. These contribute to energy losses.

Ferromagnetic materials, such as are used in electrical machinery, often have non-linear properties. For example, the magnetic flux density, B, is not always proportional to field intensity, H, especially at higher intensities. At some point B ceases to increase even though H is increasing. This is called saturation. The properties of magnetic materials are typically shown in magnetization curves. An example of such a curve is shown in Figure 5.10. Another non-linear phenomenon that affects electric machine design is known as hysteresis. This describes a common situation in which the material becomes partially magnetized, so B does not vary with H when H is decreasing in the same way that it does when H is increasing.

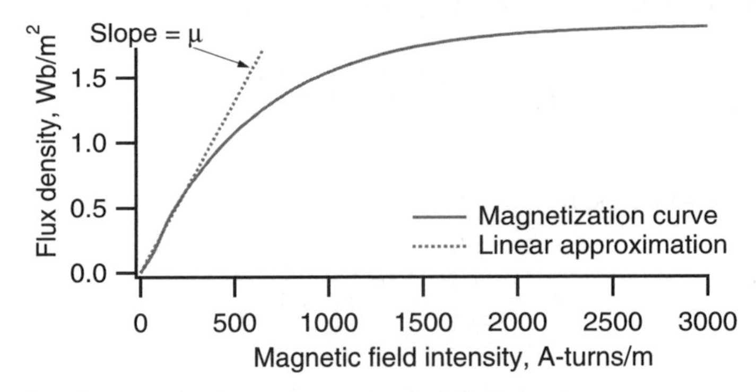

Figure 5.10 Sample magnetization curve; μ , permeability

5.3 Power Transformers

Power transformers are important components in any AC power system. Most wind turbine installations include at least one transformer for converting the generated power to the voltage of the local electrical network to which the turbine is connected. In addition, other transformers may be used to obtain voltages of the appropriate level for various ancillary equipment at the site (lights, monitoring and control systems, tools, compressors, etc.) Transformers are rated in terms of their apparent power (kVA). Distribution transformers are typically in the 5–50 kVA range, and may well be larger, depending on the application. Substation transformers are typically between from 1,000 kVA and 60,000 kVA.

A transformer is a device which has two or more coils, coupled by a mutual magnetic flux. Transformers are usually comprised of multiple turns of wire, wrapped around a laminated metal core. In the most common situation the transformer has two windings, one known as the primary, the other as the secondary. The wire is normally of copper, and is sized so there will be minimal resistance. The core consists of laminated sheets of metal separated by insulation so that there will be a minimum of eddy currents circulating in the core.

The operating principals of transformers are based on Faraday's Law of Induction (Section 5.2.3.3). An ideal transformer is one which has: (i) no losses in the windings, (ii) no losses in the core, and (iii) no flux leakage. An electrical circuit diagram of an ideal transformer is illustrated in Figure 5.11.

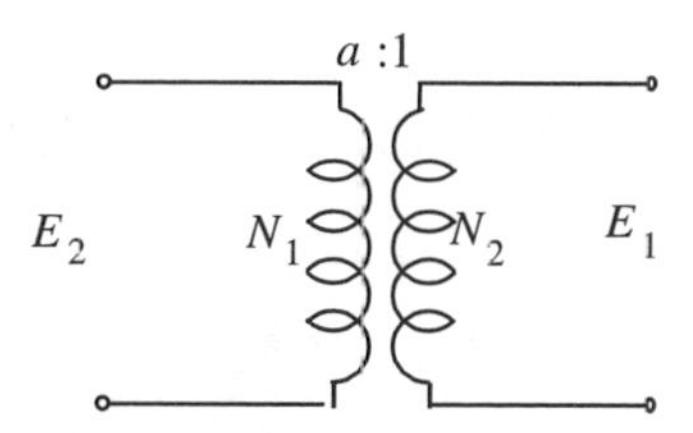

Figure 5.11 Ideal transformer; *a*, turns ratio; *E*, induced voltage; *N*, number of turns; subscripts 1 and 2 refer to the primary and secondary windings, respectively

Assume that E_1 is applied to the primary of an ideal transformer with N_1 coils on the primary and N_2 coils on the secondary. The ratio between the voltages across the primary and the secondary is equal to the ratio of the number of turns:

$$E_1 / E_2 = N_1 / N_2 = a \tag{5.3.1}$$

The parameter *a* is known as the turns ratio of the transformer.

The primary and secondary currents are inversely proportional to the number of turns (as they must be to keep the power or the product *IV* constant):

$$I_2 / I_1 = N_1 / N_2 = a \tag{5.3.2}$$

Real, or non-ideal, transformers do have losses in the core and windings, as well as leakage of flux. A non-ideal transformer can be represented by an equivalent circuit as shown in Figure 5.12:

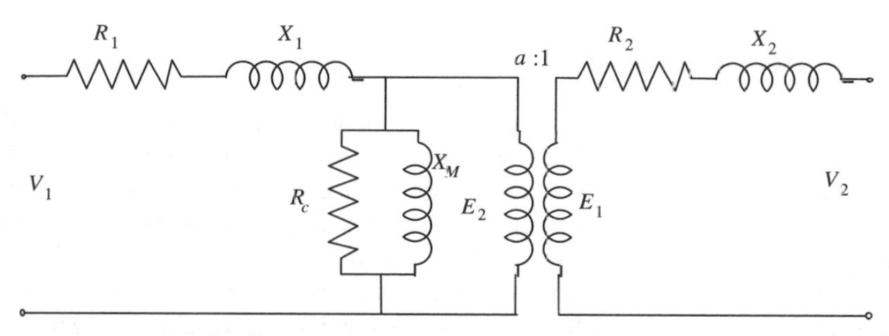

Figure 5.12 Non-ideal transformer; for the notation, see the text

In Figure 5.12 *R* refers to resistances, *X* to reactances, 1 and 2 to the primary and secondary coils, respectively. R_1 and R_2 represent the resistance of the primary and secondary windings. X_1 and X_2 represent the leakage inductances of the two windings. The subscript *M* refers to magnetizing inductance and the subscript *c* to core resistance. *V* refers to terminal voltages and E_1 and E_2 are the induced voltages at primary and secondary whose ratio is the turns ratio.

Parameters on either side of the coils may be referred to (or viewed from) one side. Figure 5.13 illustrates the equivalent circuit of the transformer when referred to the primary side.

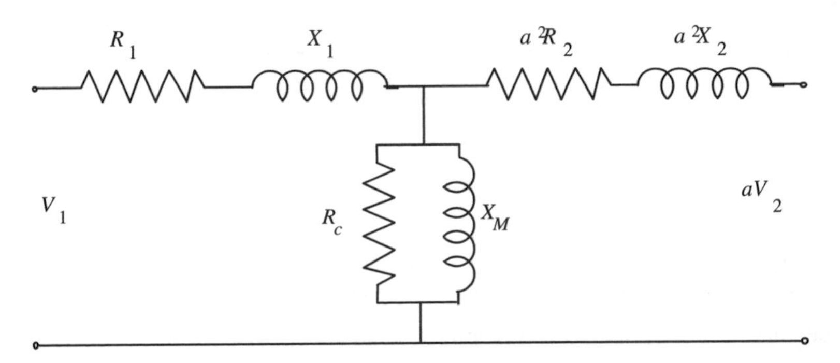

Figure 5. 13 Non-ideal transformer, referred to primary winding; for notation, see text

A transformer will draw current whether or not there is a load on it. There will be losses associated with the current, and the power factor will invariably be lagging. The magnitude of the losses and the power factor can be estimated if the resistances and reactances in Figure 5.13 are known. These parameters can be calculated by the use of two tests: (i) measurement of the voltage, current and power at no load (open circuit on one of the coils) and (ii) measurement of voltage, current and power with one of the coils short circuited. The latter test will be at reduced voltage to prevent burning out the transformer. Most texts on electrical machinery describe these tests in more detail. For example, see Nasar and Unnewehr (1979.)

It is worth noting here that the equivalent circuit of the transformer is similar in many ways to that of induction machines, which are discussed in Section 5.4.4 and which are used as generators in many wind turbines.

5.4 Electrical Machines

Generators convert mechanical power to electrical power; motors convert electrical power to mechanical power. Both generators and motors are frequently referred to as electrical machines, because they can usually be run as one or the other. The electrical machines most commonly encountered in wind turbines are those acting as generators. The two most common types are induction generators and synchronous generators. In addition, some smaller turbines use DC generators. The following section discusses the principles of electrical machines in general and then focuses on induction and synchronous generators.

5.4.1 Simple electrical machines

Many of the important characteristics of most electrical machines are evident in the operation of the simplest electrical machine, such as is shown in Figure 5.14.

In this simple electrical machine, the two magnetic poles (or pair of poles) create a field. The loop of wire is the armature. The armature can rotate, and it is assumed that there are brushes and slip rings or a commutator present to allow current to pass from a stationary frame of reference to the rotating one. (A commutator is a device which can change the direction of an electrical current. Commutators are used in DC generators to change what would otherwise be AC to DC.) If a current is flowing in the armature, a force acts on the wire. The force on the left side is down, and on the right side is up. The forces then create a torque, causing the machine to act as a motor. In this machine the torque will be a maximum when the armature loop is horizontal, and a minimum (of zero) when the loop is vertical.

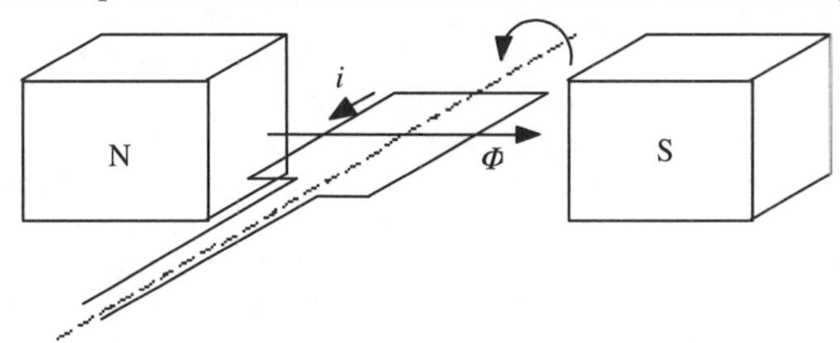

Figure 5.14 Simple electrical machine; *i*, current; Φ , magnetic flux; N, North magnetic pole; S, South magnetic pole

Conversely, if there is initially no current in the wire, but if the armature loop is rotated through the field, a voltage will be generated in accordance with Faraday's Law. If the loop is part of a complete circuit, a current will then flow. In this case the machine is acting as a generator. In general the directions of current or voltage, velocity, field direction, and force are specified by cross-product relations.

When slip rings are used there are two metal rings mounted to the shaft of the armature with one ring connected to one end of the armature coil and the other ring connected to the other coil. Brushes on the slip rings allow the current to be directed to a load. As the armature rotates the direction of the voltage will depend on the position of the wire in the magnetic field. In fact the voltage will vary sinusoidally if the armature rotates at fixed speed. In this mode, this simple machine acts as an AC generator. Similarly in the motoring mode, the force (and thus torque) reverses itself sinusoidally during a revolution.

A simple commutator for this machine would have two segments, each spanning 180 degrees on the armature. Brushes would contact one segment at a time, but segment–brush pairing would reverse itself once during each revolution. The induced voltage would then consist of a sequence of half sine waves, all of the same sign. In the motoring mode the torque would always be in the same direction. The commutator principle is the basis of conventional DC motors and generators.

Real electrical machines are similar in many ways to this simple one, but there are also some major differences:

- Except in machines with fields supplied by permanent magnets, the fields are normally produced electrically.
- The fields are most often on the rotating part of the machine (the rotor), while the armature is then on the stationary part (the stator).
- There is also a magnetic field produced by the armature which interacts with the rotor's field. The resultant magnetic field is often of primary concern in analyzing the performance of an electrical machine.

5.4.2 Rotating magnetic fields

By suitable arrangement of windings in an electrical machine it is possible to establish a rotating magnetic field, even if the windings are stationary. This property forms an important basis of the design of most AC electrical machines. In particular, it is the interaction of the stator's rotating magnetic field with the rotor's magnetic field which determines the operating characteristics of the machine.

The principle of rotating fields can be developed in a number of ways, but the key points to note are that: (1) the coils in the stator are 120 degrees ($2\pi/3$ radians) apart, (2) the magnitude of each field varies sinusoidally, with the current in each phase differing from the others by 120 degrees, and (3) the windings are such that the distribution of each field is sinusoidal. The resultant magnetic field, H, expressed in phasor form in terms of the three individual magnetic fields H_i is:

$$\boldsymbol{H} = \boldsymbol{H}_1\angle 0 + \boldsymbol{H}_2\angle 2\pi/3 + \boldsymbol{H}_3\angle 4\pi/3 \quad (5.4.1)$$

Substituting in sinusoids for the currents, and introducing an arbitrary constant C to signify that the field results from the currents we have:

$$\boldsymbol{H} = C[\cos(2\pi\, f\, t)\angle 0 + \cos(2\pi\, f\, t + 2\pi/3)\angle 2\pi/3 + \cos(2\pi\, f\, t + 4\pi/3)\angle 4\pi/3] \quad (5.4.2)$$

After performing the algebra, we obtain the interesting result that the magnitude of H is constant, and its angular position is $2\pi ft$ radians. The latter result implies immediately that the field is rotating at a constant speed of f revolutions per second, which is the same as the electrical system frequency. A graphical illustration of this result can be found from links on the web site of the Renewable Energy Research Laboratory at the University of Massachusetts (http://www.ecs.umass.edu/mie/labs/rerl/index.html).

The above discussion implicitly involved a pair of magnetic poles per phase. It is quite possible to arrange windings so as to develop an arbitrary number of pole pairs per phase. By increasing the number of poles, the resultant rotating magnetic field will rotate more slowly. At no load, the rotor of an electrical machine will rotate at the same speed as the rotating magnetic field, called the synchronous speed. In general, the synchronous speed is:

$$n = \frac{60 f}{P/2} \quad (5.4.3)$$

where n is the synchronous speed in rpm, f is the frequency of AC electrical supply in Hz and P is the number of poles.

The above equation implies, for example, that any two-pole AC machine, connected to a 60 Hz electrical network would turn with no load at 3600 rpm, a four-pole machine would turn at 1800 rpm, a six-pole machine at 1200 rpm, etc. It is worth noting here that most wind turbine generators are four-pole machines, thus having a synchronous speed of 1800 rpm when connected to a 60 Hz power system. In a 50 Hz system such generators would turn at 1500 rpm.

5.4.3 Synchronous machines

5.4.3.1 Overview of synchronous machines

Synchronous machines are used as generators in large central station power plants. In wind turbine applications they are used occasionally on large grid-connected turbines, or in conjunction with power electronic converters in variable speed wind turbines (see Section 5.5). A type of synchronous machine using permanent magnets is also used in some stand alone wind turbines (see Section 5.4.6). In this case the output is often rectified to DC before the power is delivered to the end load. Finally, synchronous machines may be used as a means of voltage control and a source of reactive power in autonomous AC networks. In this case they are known as synchronous condensers.

In its most common form the synchronous machine consists of: (1) a magnetic field on the rotor that rotates with the rotor and (2) a stationary armature containing multiple windings. The field on the rotor is created electromagnetically by a DC current (referred to as excitation) in the field windings. The DC field current is normally provided by a small DC generator mounted on the rotor shaft of the synchronous machine. This small generator is known as the 'exciter', since it provides excitation to the field. The exciter has its field stationary and its output is on the synchronous machine's rotor. The output of the exciter is rectified to DC right on the rotor and fed directly into the synchronous machine's field windings. Alternatively, the field current may be conveyed to the synchronous machine's rotor via slip rings and brushes. In either case the synchronous machine's rotor field current is controlled externally.

A simple view of a synchronous machine can give some insight into how it works. Assume, as in Section 5.4.2, that a rotating magnetic field has been set up in the stationary windings. Assume also that there is a second field on the rotor. These two fields generate a resultant field that is the sum of the two fields. If the rotor is rotating at synchronous speed, then there is no relative motion between the any of these rotating fields. If the fields are aligned, then there is no force acting upon them that could change the alignment. Next suppose that the rotor's field is displaced somewhat from that of the stator, causing a force, and hence an electrical torque, which tends to align the fields. If an external torque is continually applied to the rotor, it could balance the electrical torque. There would then be a constant angle between the fields of the stator and the rotor. There would also be a constant angle between the rotor field and the resultant field, which is known as the power angle, and it is given the symbol delta (δ). It can be thought of as a spring, since, other things being equal, the power angle increases with torque. It is important to note that as long as $\delta > 0$ the machine is a generator. If input torque drops the power angle may become negative and the machine will act as a motor. Detailed discussion of those relations is outside the scope of this text, however, and is not needed to understand the operation of a synchronous machine for the purposes of interest here. A full development is given in most electrical machinery texts. See, for example, Brown and Hamilton (1984) or Nasar and Unnewehr (1979).

5.4.3.2 Theory of synchronous machine operation

The following presents an overview of the operation of synchronous machines, based on an equivalent circuit which can be derived for it, and two electrical angles. The latter are the power factor angle, ϕ, and the power angle, δ.

The current in the synchronous machine's field windings I_f induces a magnetic flux. The flux, Φ, depends on the material and the number of turns in the winding, as explained in Section 5.2.3, but to a first approximation the flux is proportional to the current.

$$\Phi = k_1 I_f \tag{5.4.4}$$

where k_1 = constant of proportionality. The voltage produced in the stationary armature, *E*, is proportional to: (1) the magnetic flux and (2) the speed of rotation, *n*:

$$E = k_2 n \Phi \tag{5.4.5}$$

where k_2 is another constant of proportionality. This voltage causes currents to flow in the armature windings.

The generator armature winding is an inductor, which can be represented by a reactance, called the synchronous reactance, X_s, and a small resistance, R_s. The reactance has a constant value when the rotational speed (and hence grid frequency) is constant. Recall that $X_s = 2\pi f L$ where L = inductance. The synchronous impedance is:

$$\hat{\mathbf{Z}}_s = R_s + jX_s \tag{5.4.6}$$

The resistance is usually small compared to the reactance, so the impedance is often approximated by considering only the reactance.

An equivalent circuit may be developed which can be used to facilitate analysis of the machine's operation. The equivalent circuit for a synchronous machine is shown in Figure 5.15. For completeness, the resistance, R_s, is included, even though it is often ignored in analyses.

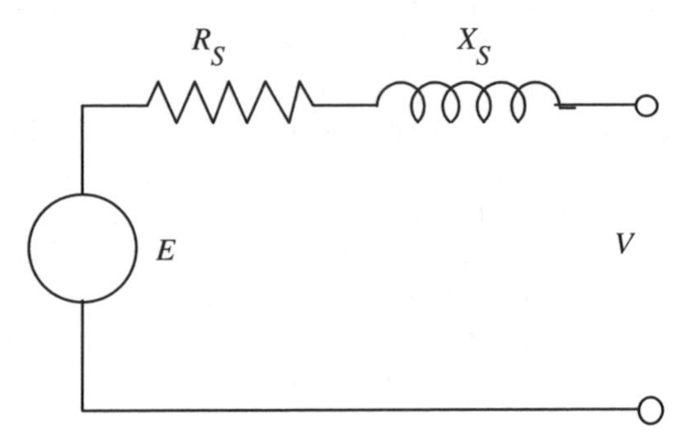

Figure 5.15 Equivalent circuit of a synchronous machine; *E*, voltage produced in stationary armature; R_s, resistance; *V*, terminal voltage; X_s, synchronous reactance

The equivalent circuit can be used to develop phasor relations, such as are shown in Figures 5.16 and 5.17. These figures illustrate the phasor relations between the field induced voltage (*E*), the terminal voltage (*V*), and the armature current I_a for a synchronous machine with a lagging or leading power factor. The figures may be derived by first assuming a terminal voltage, with reference angle of zero. With a known apparent power and power factor (lagging or leading), the magnitude and angle of the current may be found. Using the equivalent circuit, the magnitude and angle of the field voltage may then be determined. The equation corresponding to the equivalent circuit, ignoring the resistance, is:

$$\hat{\boldsymbol{E}} = \hat{\boldsymbol{V}} + jX_s\hat{\boldsymbol{I}}_{\mathrm{a}} \qquad (5.4.7)$$

The power angle, δ, shown in these Figures 5.16 and 5.17 is, by definition, the angle between the field voltage and the terminal voltage. As described above, it also is the angle between the rotor and the resultant fields.

The things to note are: (1) the armature current in Figure 5.16 is lagging the terminal voltage, indicating lagging power factor; (2) the armature current in Figure 5.17 is leading the terminal voltage, indicating leading power factor; and (3) the field induced voltage leads the terminal voltage, giving a positive power angle, as it should with the machine in a generating mode.

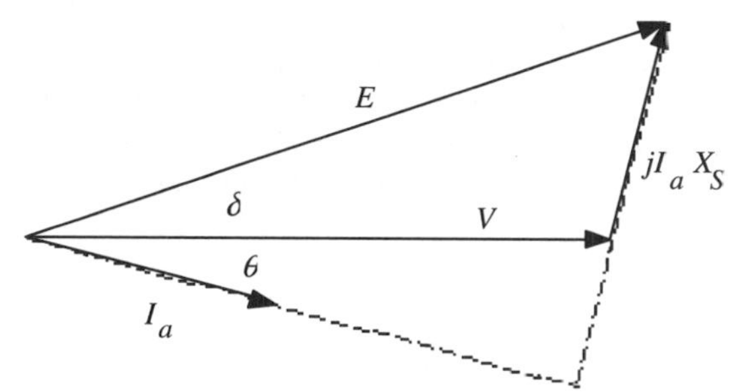

Figure 5.16 Phasor diagram for synchronous generator, lagging power factor; *E*, field induced voltage; I_a, armature current; *j*, $\sqrt{-1}$; *V*, terminal voltage; X_s , synchronous reactance; δ , power angle; 6 , power factor angle

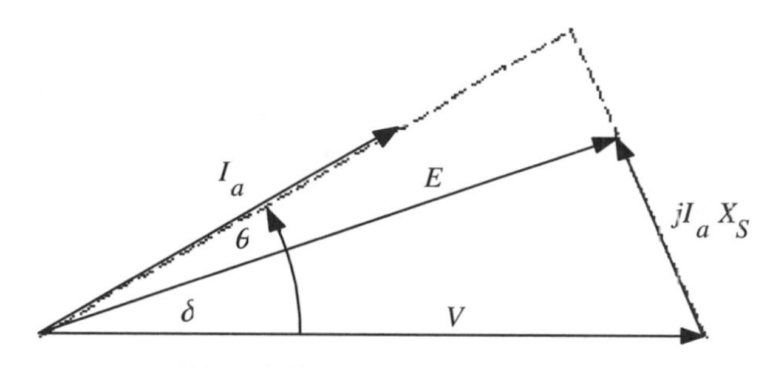

Figure 5.17 Phasor diagram for synchronous generator, leading power factor; *E*, field induced voltage; *I* , armature current; *j*, $\sqrt{-1}$; *V*, terminal voltage; X_s , synchronous reactance; δ , power angle; 6 , power factor angle

By performing the phasor multiplication with reference to Figure 5.16 or 5.17, as explained in Section 5.2.3, one can find that the real power, P, is:

$$P = \frac{|\hat{\boldsymbol{E}}||\hat{\boldsymbol{V}}|}{X_S}\sin(\delta) \tag{5.4.8}$$

Similarly, the reactive power, Q, out of the generator is:

$$Q = \frac{|\hat{\boldsymbol{E}}||\hat{\boldsymbol{V}}|\cos(\delta) - |\hat{\boldsymbol{V}}|^2}{X_S} \tag{5.4.9}$$

It is worth emphasizing that, in a grid connected application with a constant terminal voltage (controlled by other generators), a synchronous machine may serve as a source of reactive power, which may be required by loads on the system. Changing the field current will change the field induced voltage, E, while the power stays constant. For any given power level a plot of armature current vs. field current will have a minimum at unity power factor. An example of this is shown in Figure 5.18. This example is for a generator with a line-to-neutral terminal voltage of 2.4 kV. In the generator mode higher field current (and therefore higher field voltage) will result in lagging power factor; lower field (lower field voltage) current will result in leading power factor. In practice, the field current is generally used to regulate the generator's terminal voltage. A voltage regulator connected to the synchronous generator automatically adjusts the field current so as to keep the terminal voltage constant.

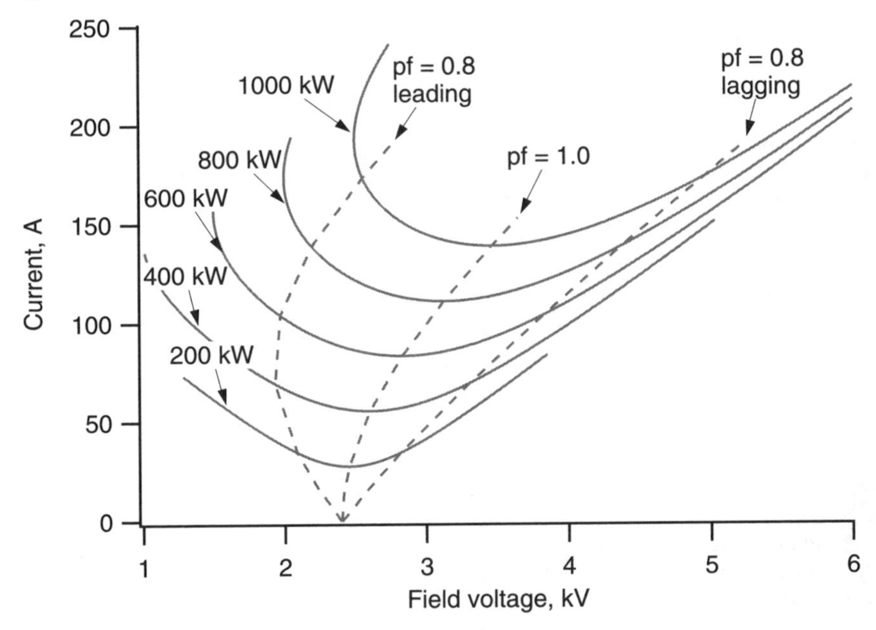

Figure 5.18 Synchronous machine armature current vs. field current in generator mode

An additional consideration relates to the details of the synchronous machine's windings. Most synchronous generators have salient poles, though there are some with round rotors. The terms direct-axis and quadrature-axis are associated with salient pole machines. These terms are discussed in detail in electrical machinery texts such as Brown and Hamilton (1984).

5.4.3.3 Starting synchronous machines

Synchronous machines are not intrinsically self-starting. In some applications the machine is brought up to speed by an external prime mover and then synchronized to the electrical network. For other applications, a self-starting capability is required. In this case the rotor is built with 'damper bars' embedded in it. These bars allow the machine to start like an induction machine does (as described in the next section.) During operation the damper bars also help to damp oscillations in the machine's rotor.

Regardless of how a synchronous machine is brought up to operating speed, particular attention must be given to synchronizing the generator with the network to which it is to be connected. A very precise match is required between the angular position of the rotor and the electrical angle of the AC power at the instant of connection. Historically, synchronization was done manually with the help of flashing lights, but it is now done with electronic controls.

Wind turbines with synchronous generators are normally started by the wind (unlike many turbines with induction generators, which can be motored up to speed). When the turbine is to be connected to an AC network which is already energized, active speed control of the turbine may be needed as part of the synchronizing process. In some isolated electrical grids, the AC power is supplied by a synchronous generator on either a diesel generator or a wind turbine, but not both. This obviates the need for a synchronizer.

5.4.4 Induction machines

5.4.4.1 Overview of induction machines

Induction machines (also known as asynchronous machines) are commonly used for motors in most industrial and commercial applications. It has long been known that induction machines could be used as generators, but they were seldom employed that way until the advent of distributed generation in the mid 1970s. Induction machines are now the most common type of generator on wind turbines, and they are used for other distributed generation (hydroelectric, engine driven) as well.

Induction machines are popular because (1) they have a simple, rugged construction, (2) they are relatively inexpensive, and (3) they may be connected and disconnected from the grid relatively simply.

The stator on an induction machine consists of multiple windings, similar to that of a synchronous machine. The rotor in the most common type of induction machine has no windings. Rather it has conducting bars, embedded in a solid, laminated core. The bars make the rotor resemble a squirrel cage. For this reason, machines of this type are commonly called squirrel cage machines.

Some induction machines do have windings on the rotor. These are known as wound rotor machines. These machines are sometimes used in variable speed wind turbines. They are more expensive and less rugged than those with squirrel cage rotors. Depending on how wound rotor machines are used they may also be referred to as doubly fed. This is because power may be sent to or taken from the rotor, as well as from the stator.

Induction machines require an external source of reactive power. They also require an external constant frequency source to control the speed of rotation. For these reasons, they are most commonly connected to a larger electrical network. In these networks synchronous generators connected to prime movers with speed governors ultimately set the grid frequency and supply the required reactive power.

When operated as a generator, the induction machine can be connected to the network and brought up to operating speed as a motor, or it can be accelerated by the prime mover, and then connected to the network. There are issues to be considered in either case. Some of these are discussed later in this section.

Induction machines often operate with a poor power factor. To improve power factor, capacitors are frequently connected to the machine at or near the point of connection to the electrical network. Care must be taken in sizing the capacitors when the machine is operated as a generator. In particular it must not be possible for the generator to be 'self-excited' if connection to the grid is lost due to a fault.

Induction machines can be used as generators in small electrical networks or even in isolated applications. In these cases, special measures must sometimes be taken for them to operate properly. The measures involve reactive power supply, maintaining frequency stability, and bringing a stationary machine up to operating speed. Some of these measures are described in Chapter 8.

5.4.4.2 Theory of induction machine operation

The theory of operation of a squirrel cage induction machine can be summarized as follows:

- The stator has windings arranged such that the phase displaced currents produce a rotating magnetic field in the stator (as explained in Section 5.4.2)
- The rotating field rotates at exactly synchronous speed (e.g. 1800 rpm for a four-pole generator in a 60 Hz electrical network)
- The rotor turns at a speed slightly different than synchronous speed (so that there is relative motion between the rotor and fields on the stator)
- The rotating magnetic field induces currents and hence a magnetic field in the rotor due to the difference in speed of the rotor and the magnetic field.
- The interaction of the rotor's induced field and the stator's field causes elevated voltage at the terminals (in the generator mode) and current to flow from the machine.

One parameter of particular importance in characterizing induction machines is the slip, s. Slip is the ratio of the difference between synchronous speed n_s and rotor operating speed n, and synchronous speed:

$$s = \frac{n_s - n}{n_s} \tag{5.4.10}$$

When slip is positive, the machine is a motor; when negative it is a generator. Slip is often expressed as a percent. Typical values of slip at rated conditions are on the order of 2%.

Most characteristics of interest can be described in terms of the equivalent circuit shown in Figure 5.19. The equivalent circuit and the relations used with it are derived in most electrical machinery texts. They will be presented, but will not be derived here.

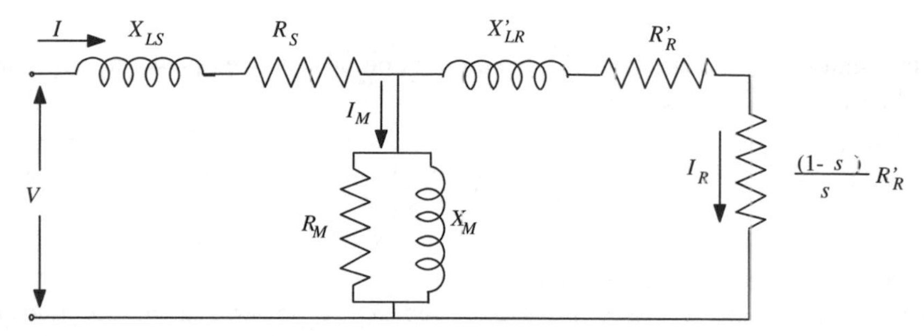

Figure 5.19 Induction machine equivalent circuit; for notation, see text

Here V is the terminal voltage, I is the stator current, I_M is the magnetizing current, I_R is the rotor current, X_{LS} is the stator leakage inductive reactance, R_S is the stator resistance, X'_{LR} is the rotor leakage inductive reactance (referred to stator), R'_R is the rotor resistance (referred to stator), X_M is the magnetizing reactance and R_M is the resistance in parallel with the mutual inductance.

A few items of particular note are:

- X_M is always much larger than X_{LS} or X'_{LR}
- The term $\frac{1-s}{s}R'_R$ is essentially a variable resistance. For a motor it is positive; for a generator it is negative
- R_M is a large resistance and is often ignored

The values of the various resistances and reactances can be derived from tests. These tests are:

- Locked rotor test (current, voltage, and power measured at approximately rated current and reduced voltage)
- No load current and voltage test (current, voltage, and power measured at no load)
- Mechanical tests to quantify windage losses and friction losses (described below)

Not all of the power converted in an induction machine is useful power. There are some losses. The primary losses are (i) mechanical losses due to windage and friction, (ii) resistive and magnetic losses in the rotor, and (iii) resistive and magnetic losses in the stator. Windage losses are those associated with drag on the rotor from air friction. Friction losses are primarily in the bearings. More information on losses may be found in most texts on electrical machinery (see Brown and Hamilton, 1984).

In the generator mode, mechanical power input to the machine, P_{in}, that is available to produce electricity is reduced by mechanical losses, $P_{mechloss}$. The mechanical power available to be converted at the generator's rotor, P_m, is

$$P_m = -(P_{in} - P_{mechloss}) \tag{5.4.11}$$

(Note: the minus sign is consistent with the convention of generated power as negative). In electrical terms this converted power is:

$$P_m = I_R^2 R'_R \frac{1-s}{s} \tag{5.4.12}$$

Recall that slip is negative for generation.

Electrical and magnetic losses in the rotor reduce the power that may be transferred from the rotor across the air gap to the stator. The power that is transferred, P_g, is:

$$P_g = I_R^2 R'_R \frac{1-s}{s} + I_R^2 R'_R \tag{5.4.13}$$

(Note: $I_R^2 R'_R$ is the electrical power loss in the rotor, making P_g less negative than P_m)
Thus:

$$P_g = \frac{P_m}{1-s} \tag{5.4.14}$$

The power lost in the stator is:

$$P_{loss} = I_s^2 R_S \tag{5.4.15}$$

The power delivered (negative) at the terminals of the generator, P_{out}, is:

$$P_{out} = P_g + P_{loss} \tag{5.4.16}$$

The overall efficiency (in the generator mode), η_{gen}, is:

$$\eta_{gen} = -\frac{P_{out}}{P_{in}} \tag{5.4.17}$$

The power factor, *PF*, (which is lagging during monitoring and leading during generating), is the ratio of the real power to apparent power:

$$PF = \frac{-P_{out}}{VI} \tag{5.4.18}$$

The previous equations show the relations between real power, reactive power and currents, but still do not allow complete calculations in terms of machine parameters. This can be done most conveniently by first simplifying the equivalent circuit in Figure 5.19, as shown in Figure 5.20.

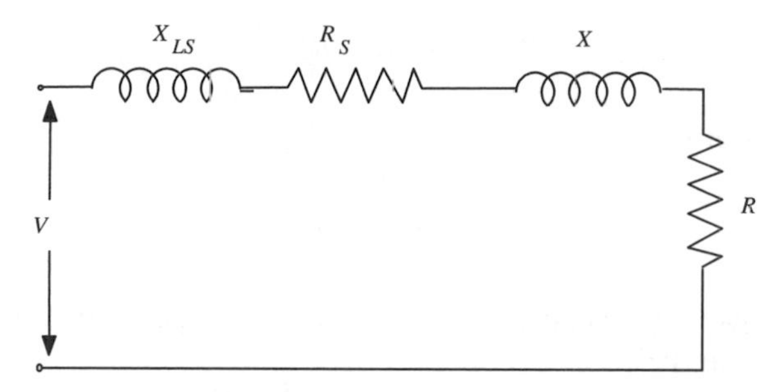

Figure 5.20 Induction machine equivalent circuit; R, resistance; X_{LS}, stator leakage inductive reactance, R_S, stator resistance; X, reactance; V, terminal voltage

If the parameters in the detailed equivalent circuit are known, then the resistance R and reactance X in this simplified model can be found as follows. From Figure 5.19, ignoring R_M, and using Equations 5.2.21 and 5.2.22 for series and parallel impedances, we have:

$$R + jX = jX_M\left(jX'_{LR} + \frac{R'_R}{s}\right) \Big/ \left[\frac{R'_R}{s} + j(X_M + X'_{LR})\right] \tag{5.4.19}$$

This gives for the resistance, R:

$$R = X_M^2 \frac{R'_R}{s} \Big/ \left[(R'_R/s)^2 + (X'_{LR} + X_M)^2\right] \tag{5.4.20}$$

The reactance, X is then:

$$X = X_M\left[(R'_R/s)^2 + X'_{LR}(X'_{LR} + X_M)\right] \Big/ \left[(R'_R/s)^2 + (X'_{LR} + X_M)^2\right] \tag{5.4.21}$$

The power converted in R will be $I_S^2 R$ where $\hat{\boldsymbol{I}}_S$ is the current (a phasor) in the loop. The total impedance is:

$$\hat{\boldsymbol{Z}} = (R + R_s) + j(X + X_{LS}) \tag{5.4.22}$$

The phasor current is:

$$\hat{\boldsymbol{I}}_S = \hat{\boldsymbol{V}} / \hat{\boldsymbol{Z}} \tag{5.4.23}$$

The mechanical power converted per phase is then:

$$P_m = (1-s) I_S^2 R \tag{5.4.24}$$

The real power generated per phase is:

$$P_{out} = I_S^2 (R_S + R) \tag{5.4.25}$$

The total reactive power is:

$$Q = I_S^2 (X_{LS} + X) \tag{5.4.26}$$

The mechanical torque, Q_m, applied to the machine (as a generator) is the input power divided by the rotational speed.

$$Q_m = P_{in} / n \tag{5.4.27}$$

Note that in this chapter Q without a subscript refers to reactive power. Q with a subscript is used for torque to maintain consistency with other chapters and conventional engineering nomenclature. The mechanical torque is, to a very good approximation, linearly related to slip over the range of slips that are generally encountered in practice.

Figures 5.21 and 5.22 illustrate the results of applying the equivalent circuit to find properties of a three-phase induction machine. Figure 5.21 shows power, current and torque for operating speeds ranging from 0 to twice synchronous speed. Between standstill and 1800 rpm, the machine is motoring; above that it is generating. As a generator in a wind turbine, the machine would never operate above approximately 3% above synchronous speed, but it would operate at speeds under 1800 while starting. Note that the peak current during startup is over 730 A, more than 5 times the rated value of 140 A. Peak torque is about two and half times rated (504 Nm). The zero speed starting torque is approximately equal to the rated value at standstill. Peak terminal power is approximately 3 times rated value (100 kW).

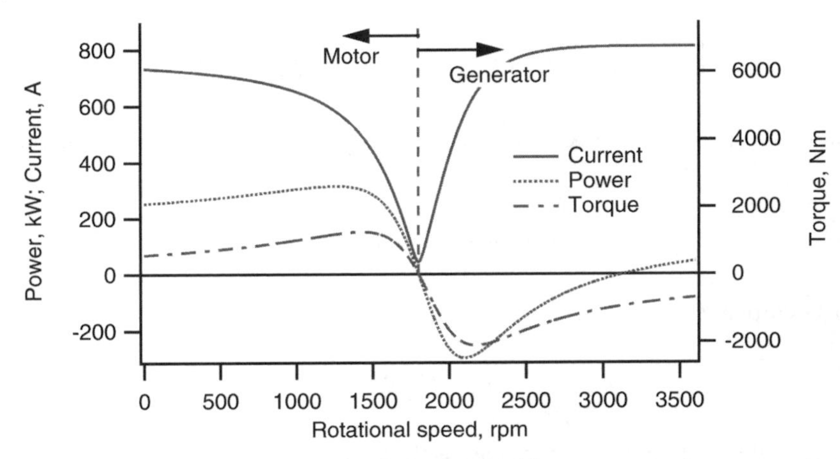

Figure 5.21 Power, current and torque of an induction machine

Figure 5.22 shows efficiency and power factor from startup to 2000 rpm. The machine has roughly the same efficiency and power factor when motoring during normal operation as when generating, but both decrease to zero at no load.

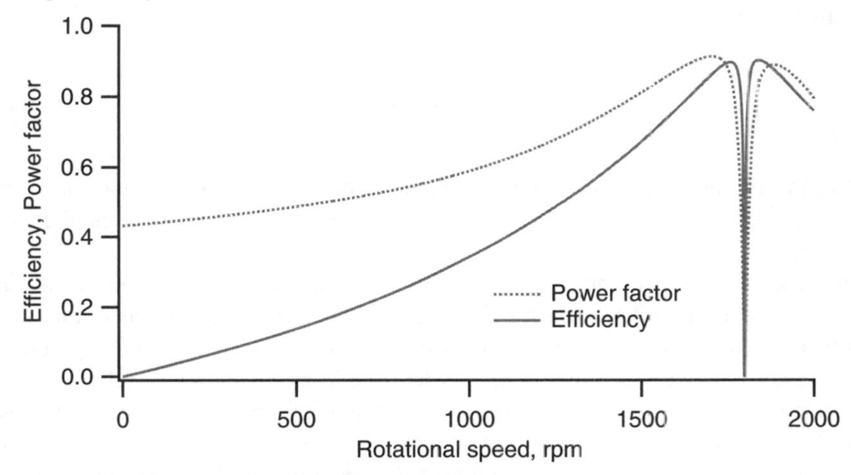

Figure 5.22 Efficiency and power factor of an induction machine

5.4.4.3 Starting wind turbines with induction generators

There are two basic methods of starting a wind turbine with an induction generator:

- Using the wind turbine rotor to bring the generator rotor up to operating speed, and then connecting the generator to the grid
- Connecting the generator to the grid and using it as a motor to bring the wind turbine rotor up to speed

When the first method is used, the wind turbine rotor must obviously be self-starting. This method is common with pitch controlled wind turbines, which are normally self-starting. Monitoring of the generator speed is required so that it may be connected when the speed is as close to synchronous speed as possible.

The second method is commonly used with stall controlled wind turbines. In this case, the control system must monitor the wind speed and decide when the wind is in the appropriate range for running the turbine. The generator may then be connected directly 'across the line' to the electrical grid, and it will start as a motor. In practice, however, across the line is not a desirable method of starting. It is preferable to use some method of voltage reduction or current limiting during starting. Options for doing this are discussed below in Section 5.6.3. As the speed of the wind turbine rotor increases, the aerodynamics become more favorable. Wind induced torque will impel the generator rotor to run at a speed slightly greater than synchronous, as determined by the torque–slip relation described in Section 5.4.4.2.

5.4.4.4 Induction machine dynamic analysis

When a constant torque is applied to the rotor of an induction machine it will operate at a fixed slip. If the applied torque is varying, then the speed of the rotor will vary as well. The relationship can be described by:

$$J\frac{\mathrm{d}\omega_r}{\mathrm{d}t} = Q_e - Q_r \tag{5.4.28}$$

where J is the moment of inertia of the generator rotor, ω_r is the angular speed of the generator rotor (rad/s), Q_e is the electrical torque and Q_r is the torque applied to the generator rotor.

When the applied torque varies slowly relative to the electrical grid frequency, a quasi steady state approach may be taken for the analysis. That is, the electrical torque may be assumed to be a function of slip as described in Equation 5.4.27 and preceding equations. The quasi steady state approach may normally be used in assessing wind turbine dynamics. This is because the frequency of fluctuations in the wind induced torque and those of mechanical oscillations are generally much less than the grid frequency. The induction generator equation used in this way was applied, for example, in the dynamic wind turbine drive train model, DrvTrnVB. This is described in Manwell et al. (1996).

It is worth noting that induction machines are somewhat "softer" in their dynamic response to changing conditions than are synchronous machines. This is because induction machines undergo a small but significant speed change (slip) as the torque in or out changes. Synchronous machines, as indicated previously, operate at constant speed, with only the power angle changing as the torque varies. Synchronous machines thus have a very 'stiff' response to fluctuating conditions.

5.4.4.5 Off-design operation of induction machines

Induction machines are designed for operation at a particular operating point. This operating point is normally the rated power at a particular frequency and voltage. In wind turbine applications, there are a number of situations when the machine may run at off-design conditions. Four of these situations are:

- Starting (mentioned above)
- Operation below rated power
- Variable speed operation
- Operation in the presence of harmonics

Operation below rated power, but at rated frequency and voltage, is a very common occurrence. It generally presents few problems. Efficiency and power factor are normally both lower under such conditions, however. Operating behavior below rated power may be examined through application of the induction machine equivalent circuit and the associated equations.

There are a number of presumed benefits of running a wind turbine rotor at variable speed. A wind turbine with an induction generator can be run at variable speed if an electronic power converter of appropriate design is included in the system between the generator and the rest of the electrical network. Such converters work by varying the frequency of the AC supply at the terminals of the generator. These converters must also vary the applied voltage. This is because an induction machine performs best when the ratio between the frequency and voltage ('Volts to Hertz ratio') of the supply is constant (or nearly so.) When that ratio deviates from the design value, a number of problems can occur. Currents may be higher, for example, resulting in higher losses and possible damage to the generator windings. Figure 5.23, for example, compares currents for an induction machine operated at its rated frequency and half that value, but both at the same voltage. The machine in this case is the same one used in the previous example (Figure 5.21).

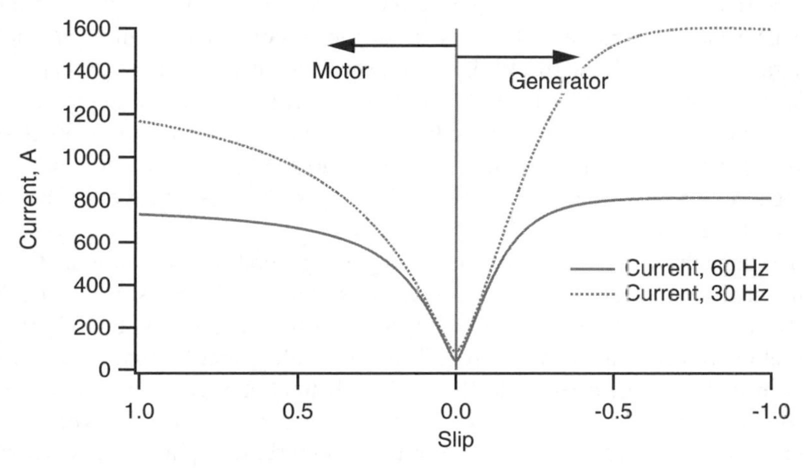

Figure 5.23 Current at different frequencies

Operation in the presence of harmonics can occur if there is a power electronic converter of significant size on the system to which the machine is connected. This would be the case for a variable speed wind turbine, and could also be the case in isolated electrical networks (see Chapter 8). Harmonics are AC voltages or currents whose frequency is an integer multiple of the fundamental grid frequency. The source of harmonics is discussed below in Section 5.5.4. Harmonics may cause bearing and electrical insulation damage and may interfere with electrical control or data signals as well.

5.4.5 DC generators

An historically important type of electrical machine for wind turbine applications is the shunt wound DC generator. These were once used commonly in smaller, battery charging wind turbines. In these generators the field is on the stator and the armature is on the rotor. The electric field is created by currents passing through the field winding which is in parallel ('shunt') with the armature windings. A commutator on the rotor in effect rectifies the generated power to DC. The full generated current must be passed out through the commutator and brushes.

In these generators, the field current, and hence magnetic field (up to a point), increases with operating speed. The armature voltage and electrical torque also increases with speed.

The actual speed of the turbine is determined by a balance between the torque from the turbine rotor and the electrical torque.

DC generators of this type are seldom used today because of high costs and maintenance requirements. The latter are associated particularly with the brushes. More details on generators of this type may be found in Johnson (1985).

5.4.6 *Permanent magnet generators*

A type of electrical machine that is being used more frequently in wind turbine applications is the permanent magnet generator. This is now the generator of choice in most small wind turbine generators, up to at least 10 kW. In these generators, permanent magnets provide the magnetic field, so there is no need for field windings or supply of current to the field. In one example, the magnets are integrated directly into a cylindrical cast aluminum rotor. The power is taken from a stationary armature, so there is no need for commutator, slip rings, or brushes. Because the construction of the machine is so simple, the permanent magnet generator is quite rugged.

The operating principles of permanent magnet generators are similar to that of synchronous machines, except that these machines are run asynchronously. That is, they are not generally connected directly to the AC network. The power produced by the generator is initially variable voltage and frequency AC. This AC is often rectified immediately to DC. The DC power is then either directed to DC loads or battery storage, or else it is inverted to AC with a fixed frequency and voltage. See Section 5.5 for discussion of power conversion. Development of a permanent magnet generator for wind turbine applications is described by Fuchs et al. (1992).

5.4.7 *Other electrical machines*

There are at least two other types of generators that may be considered for wind turbine applications: (i) direct drive generators and (ii) variable reluctance generators.

Direct drive generators are essentially synchronous machines of special design. The main difference from standard machines is that they are built with a sufficient number of poles that the generator rotor can turn at the same speed as the wind turbine rotor. This eliminates the need for a gearbox. Because of the large number of poles, the diameter of the generator is relatively large. Direct drive generators on wind turbines are frequently used in conjunction with power electronic converters. This provides some leeway in the voltage and frequency requirements of the generator itself.

Variable reluctance generators employ a rotor with distinct poles (without windings) separated from each other, such as in Figure 5.9. As the rotor turns the reluctance of the magnetic circuit linking the stator and rotor changes. The changing reluctance varies the resultant magnetic field and induces currents in the armature. A variable reluctance generator thus does not require field excitation. The variable reluctance generators currently being developed are intended for use with power electronic converters. Variable reluctance generators need little maintenance, due to their simple construction. There are no variable reluctance generators currently being used in commercial wind turbines, but research towards that end is underway. See Torrey and Childs (1993) for a description of some of that research.

5.4.8 Generator mechanical design

There are a number of issues to consider regarding the mechanical design of a generator. The rotor shaft and main bearings are designed according to the basic principles to be discussed in Chapter 6. The stator housing of the generator is normally of steel. Commercial generator housings come in standard frame sizes. Windings of the armature (and field when applicable) are of copper wire, laid into slots. The wire is not only insulated, but additional insulation is added to protect the windings from the environment and to stabilize them. Different types of insulation may be specified depending on the application.

The exterior of the generator is intended to protect the interior from condensation, rain, dust, blowing sand, etc. Two designs are commonly used: (i) open drip proof and (ii) totally enclosed, fan cooled (TEFC). The open drip proof design has been used on many wind turbines, because it is less expensive than other options, and it has been assumed that the nacelle would be sufficient to protect the generator from the environment. In many situations, however, it appears that the additional protection provided by a TEFC design may be worth the cost.

A schematic of a typical induction machine is illustrated in Figure 5.24.

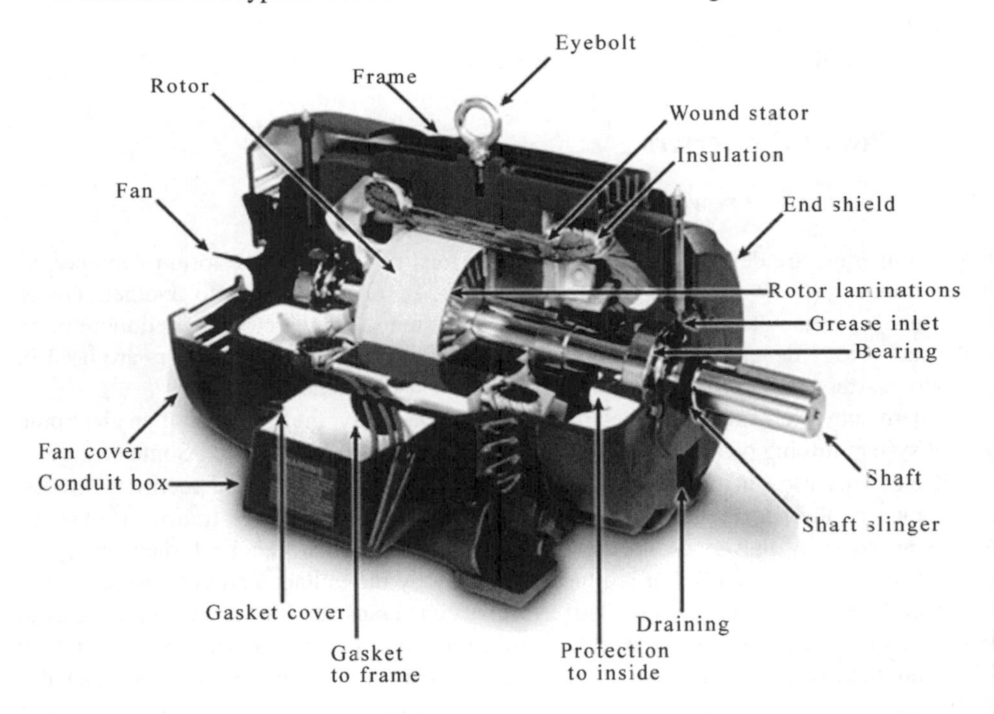

Figure 5.24 Construction of typical three phase induction machine (Rockwell International Corp.)

5.4.9 Generator specification

Wind turbine designers are not, in general, generator designers. They either select commercially available electrical machines, with perhaps some minor modifications, or they specify the general requirements of the machine to be specially designed. The basic characteristics of the important generator types have been discussed previously. The following is a summary list of the key considerations from the point of view of the wind turbine designer:

- Operating speed
- Efficiency at full load and part load
- Power factor and source of reactive power (induction machines)
- Voltage regulation (synchronous machines)
- Method of starting
- Starting current (induction machines)
- Synchronizing (synchronous machines)
- Frame size and generator weight
- Type of insulation
- Protection from environment
- Ability to withstand fluctuating torques
- Heat removal
- Feasibility of using multiple generators
- Operation with high electrical noise on conductors

5.5 Power Converters

5.5.1 Overview of power converters

Power converters are devices used to change electrical power from one form to another, as in AC to DC, DC to AC, one voltage to another, or one frequency to another. Power converters have many applications in wind energy systems. They are being used more often as the technology develops and as costs drop. For example, power converters are used in generator starters, variable speed wind turbines, and in isolated networks.

Modern converters are power electronic devices. Basically, these consist of an electronic control system turning on and off electronic switches, often called 'valves.' Some of the key circuit elements used in the inverters include diodes, silicon-controlled rectifiers (SCRs, also known as thyristors), gate turn off thyristors (GTOs), and power transistors. Diodes behave as one-way valves. SCRs are essentially diodes which can be turned on by an external pulse (at the 'gate'), but are turned off only by the voltage across them reversing. GTOs are SCRs which may be turned off as well as on. Transistors require the gate signal to be continuously applied to stay on. The overall function of power transistors is similar to GTOs, but the firing circuitry is simpler. The term 'power transistor,' as used here, includes Darlingtons, power MOSFETS and insulated gate bipolar transistors (IGBTs). The present trend it towards increasing use of IGBTs. Figure 5.25 shows the symbols used in this chapter for the most important power converter circuit elements.

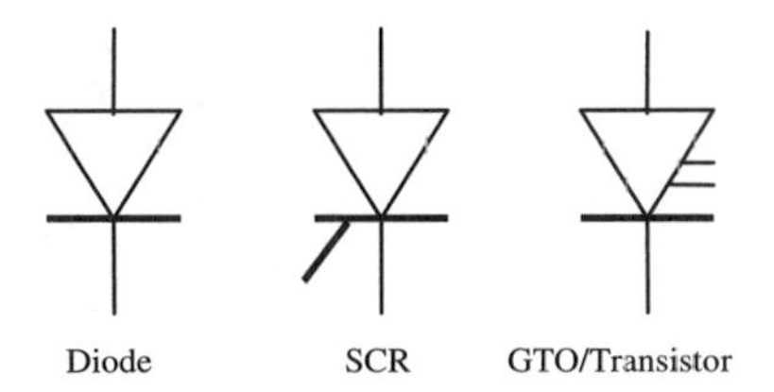

Figure 5.25 Converter circuit elements; SCR, silicon-controlled rectifier; GTO, gate turn off thyristor

5.5.2 *Rectifiers*

Rectifiers are devices which convert AC into DC. They may be used in: (1) battery charging wind systems or (2) as part of a variable speed wind power system.

The simplest type of rectifier utilizes a diode bridge circuit to convert the AC to fluctuating DC. An example of such a rectifier is shown in Figure 5.26. In this rectifier, the input is three phase AC power; the output is DC.

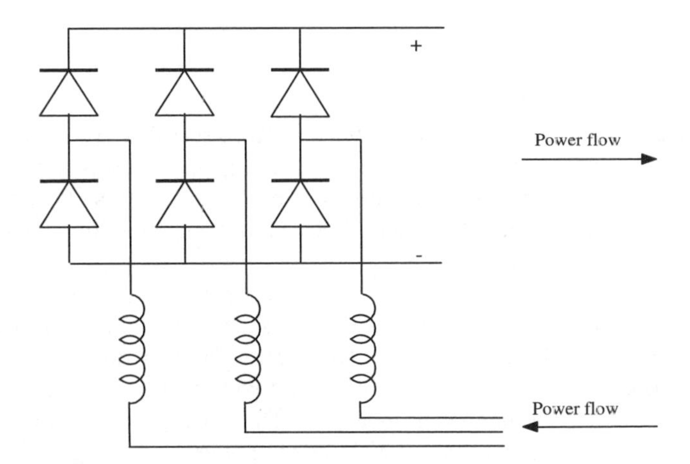

Figure 5.26 Diode bridge rectifier for three phase supply

Figure 5.27 illustrates the DC voltage that would be produced from a three phase, 480 V supply using the type of rectifier shown in Figure 5.26. Some filtering may be done (as with the inductors shown in the figure) to remove some of the fluctuations. See Section 5.5.4 for a discussion of filtering.

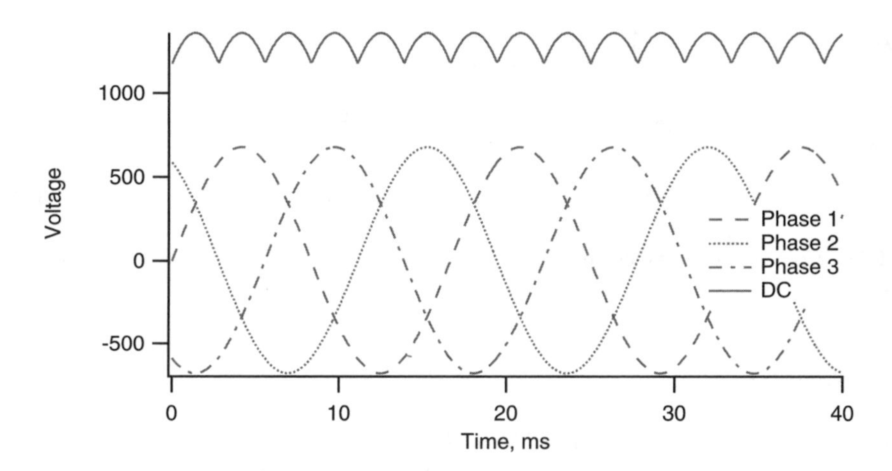

Figure 5.27 DC voltage from three phase rectifier

5.5.3 Inverters

5.5.3.1 Overview of inverters

In order to convert DC to AC, as from a battery or from rectified AC in a variable speed wind turbine, an inverter is used. Historically motor generator sets have been used to convert DC into AC. These are AC generators driven by DC motors. This method is very reliable, but is also expensive and inefficient. Because of their reliability, however, they are still used in some demanding situations

At the present time most inverters are of the electronic type. An electronic inverter typically consists of circuit elements that switch high currents and control circuitry that coordinates the switching of those elements. The control circuitry determines many aspects of the successful operation of the inverter. There are two basic types of electronic inverters: line-commutated and self-commutated inverters. The term commutation refers to the switching of current flow from one part of a circuit to another.

Inverters that are connected to an AC grid and that take their switching signal from the grid are known by the rather generic name of line-commutated inverters. Figure 5.28 illustrates an SCR bridge circuit, such as is used in a simple three-phase line-commutated inverter. The circuit is similar to the three-phase bridge rectifier shown above, but in this case the timing of the switching of the circuit elements is externally controlled and the current flows from the DC supply to the three-phase AC lines.

Self-commutated inverters do not need to be connected to an AC grid. Thus they can be used for autonomous applications. They tend to be more expensive than line-commutated inverters.

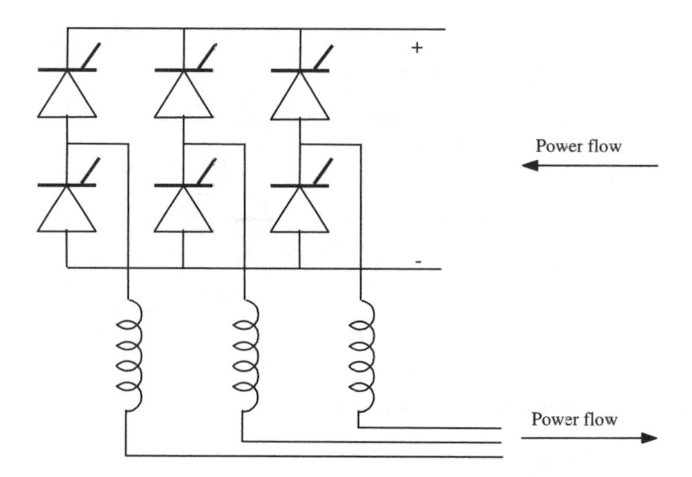

Figure 5.28 Line commutated silicon-controlled rectifier (SCR) inverter

The actual circuitry of inverters may be of a variety of designs, but inverters fall into one of two main categories: (i) voltage source inverters and (ii) current source inverters. In current source inverters, the current from the DC source is held constant, regardless of the load. They are typically used to supply high power factor loads where the impedance is constant or decreasing at harmonic frequencies. Overall efficiencies are good (around 96%), but the control circuitry is relatively complex. Voltage source inverters operate from a constant voltage DC power source. They are the type most commonly used to date in wind energy applications. (Note that most of the devices described here can operate as rectifiers or inverters, so the term converter is also appropriate.)

5.5.3.2 Voltage source inverters

Within the voltage source inverter category are two main types of interest: (1) six-pulse inverters and (2) pulse width modulation (PWM) inverters.

The simplest self-commutated voltage source inverter, referred to here as the 'six-pulse' inverter, involves the switching on and off of a DC source through different elements at specific time intervals. The switched elements are normally GTOs or power transistors, but SCRs with turn-off circuitry could be used as well. The circuit combines the resulting pulse into a staircase-like signal, which approximates a sinusoid. Figure 5.29 illustrates the main elements in such an inverter. Once again, the circuit has six sets of switching elements, a common feature of three-phase inverter and rectifier circuits, but in this case both switching on and off of the switches can be externally controlled.

If the valves are switched on one sixth of a cycle apart, in sequence according to the numbers shown on the figure, and if they are allowed to remain on for one third of a cycle, an output voltage of a step-like form will appear between any two phases of the three-phase terminals, A, B and C. A few cycles of such a waveform (with a 60 Hz fundamental frequency) are illustrated in Figure 5.30.

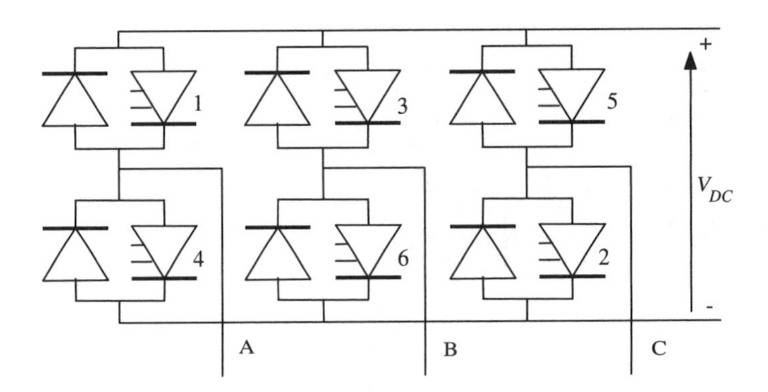

Figure 5.29 Voltage source inverter circuit elements

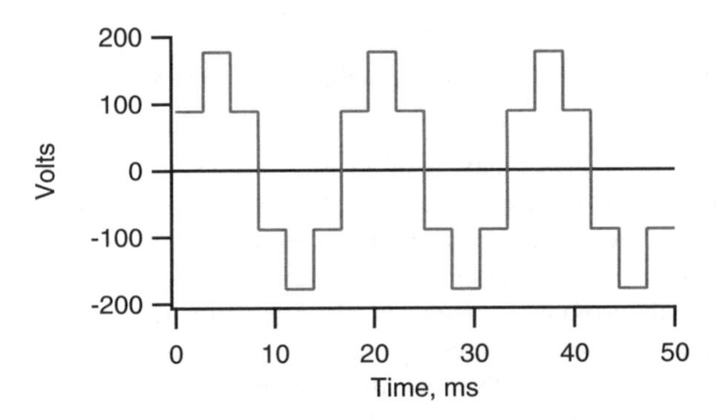

Figure 5.30 Self-commutated inverter voltage waveform

It is apparent that the voltage is periodic, but that it also differs significantly from a pure sine wave. The difference can be described by the presence of harmonic frequencies resulting from the switching scheme. These harmonics arise because of the nature of the switching. Some type of filtering is normally needed to reduce the effect of these harmonics. Harmonics and filtering are discussed in more detail below in Section 5.5.4.

In pulse width modulation (PWM) an AC signal is synthesized by the high-frequency switching on and off of the supply voltage to create pulses of a fixed height. The duration ('width') of the pulse may vary. Many pulses will be used in each half wave of the desired output. Switching frequencies on the order of 8 to 20 kHz may be used. The rate of switching is limited by the losses that occur during the switching process. Even with such losses, inverter efficiencies can be 94%. PWM inverters normally use power transistors (IGBTs) or GTOs as the switching elements.

Figure 5.31 below illustrates the principle behind one method of obtaining pulses of the appropriate width. In this method a reference sine wave of the desired frequency is compared with a high-frequency offset triangle wave. Whenever the triangle wave becomes less than the sine wave, the transistor is turned on. When the triangle next becomes greater than the sine wave, the transistor is turned off. An equivalent approach is taken during the second half of the cycle. Figure 5.31 includes two complete reference sine waves, but note that the triangle wave is of a much lower frequency than would be used in a real application. The pulse train corresponding to Figure 5.31 is shown in Figure 5.32. It is apparent that the pulses (in the absolute sense) are widest near the peaks, so the average magnitudes of the voltages are greatest there, as they should be. The voltage wave is still not a pure sine wave, but it does contain few low-frequency harmonics, so filtering is easier. Other methods are also used for generating pulses of the appropriate width. Some of these are discussed in Thorborg (1988) and Bradley (1987).

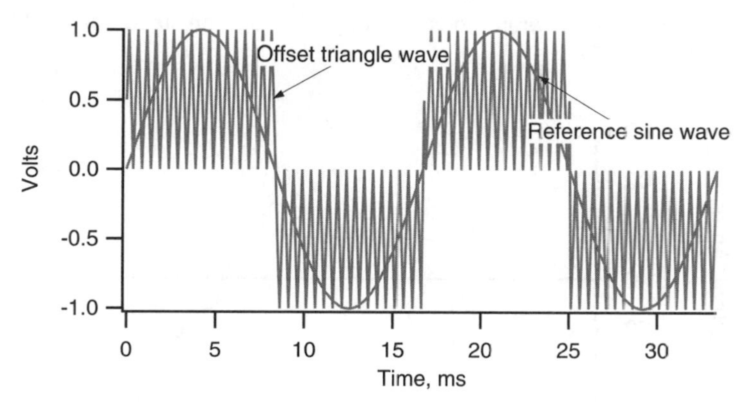

Figure 5.31 Pulse width modulation (PWM) control waves

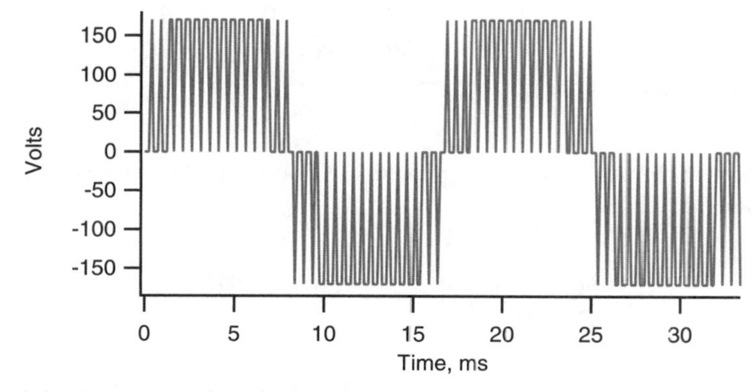

Figure 5.32 Pulse width modulation (PWM) voltage pulse train

5.5.4 Harmonics

Harmonics are AC voltages or currents whose frequency is an integer multiple of the fundamental grid frequency. Harmonic distortion refers to the effect on the fundamental waveform of non-sinusoidal or higher frequency voltage or current waveforms resulting from the operation of electrical equipment using solid state switches. Harmonic distortion is caused primarily by inverters, industrial motor drives, electronic appliances, light dimmers, fluorescent light ballasts and personal computers. It can cause overheating of transformers and motor windings, resulting in premature failure of the winding insulation. The heating caused by resistance in the windings and eddy currents in the magnetic cores is a function of the square of the current. Thus, small increases in current can have a large effect on the operating temperature of a motor or transformer. A typical example of harmonic distortion of a current waveform is illustrated in Figure 5.33.

Harmonic distortion is often characterized by a measure called total harmonic distortion (THD), which is the ratio of the total energy in the waveform at all of the harmonic frequencies to the energy in the waveform at the basic system frequency. The higher the THD, the worse the waveform. More on calculating THD and on standards for THD can be found in Chapter 8.

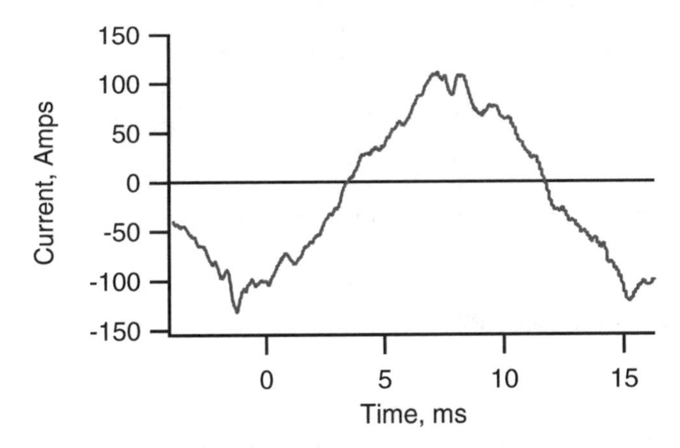

Figure 5.33 Example of harmonic distortion

Any periodic waveform $v(t)$ can be expressed as a Fourier series of sines and cosines, as given in Equation 5.5.1:

$$v(t)=\frac{a_0}{2}+\sum_{n=1}^{\infty}\left\{a_n\cos\left(\frac{n\pi t}{L}\right)+b_n\sin\left(\frac{n\pi t}{L}\right)\right\} \tag{5.5.1}$$

where n is the harmonic number, L is half the period of the fundamental frequency and

$$a_n=\frac{1}{L}\int_0^{2L} v(t)\cos\left(\frac{n\pi t}{L}\right)\mathrm{d}t \tag{5.5.2}$$

$$b_n=\frac{1}{L}\int_0^{2L} v(t)\sin\left(\frac{n\pi t}{L}\right)\mathrm{d}t \tag{5.5.3}$$

The fundamental frequency corresponds to $n = 1$. Higher order harmonic frequencies are those for which $n > 1$. For normal AC voltages and currents there is no DC component, so in general $a_0 = 0$.

Figure 5.34 shows an approximation to the converter voltage output shown originally in Figure 5.30. The approximation is done with 15 frequencies in the Fourier series. Note that if more terms were added the mathematical approximation would be more accurate and the ripples would decrease. Since the voltage as shown is an odd function, that is f(-x) = -f(x)), only sine terms are non-zero.

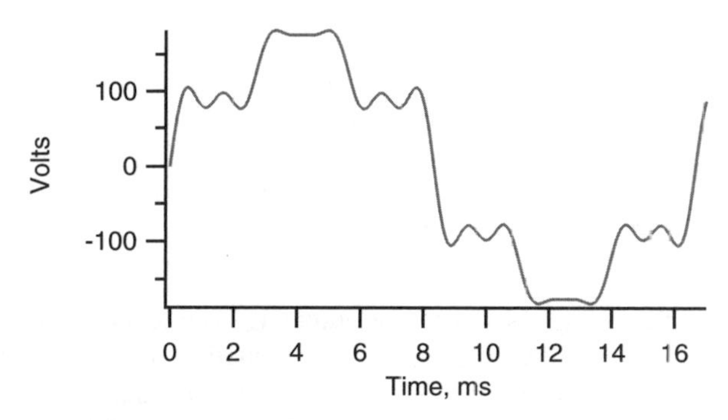

Figure 5.34 Fourier series of voltage; fifteen terms

The relative magnitudes of the coefficients of the various harmonic voltages are shown in Figure 5.35. (It can be shown in general for this particular wave that all higher harmonics are zero except those for which $n = 6k \pm 1$ where $k = 0, 1, 2, 3$, etc. and $n > 0$.)

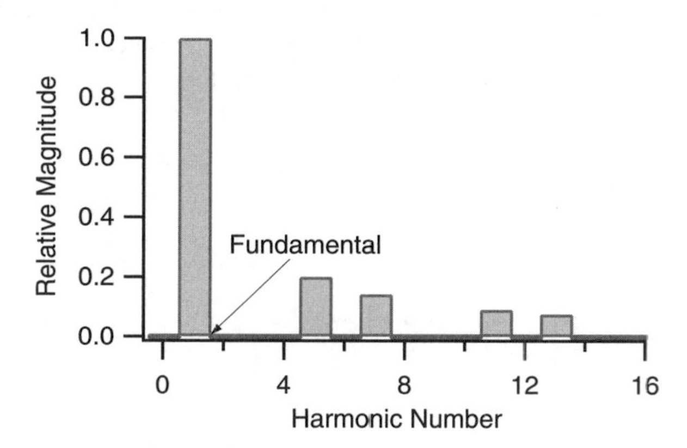

Figure 5.35 Relative harmonic content of voltage

5.5.4.1 Harmonic filters

Since the voltage and current waveforms from power electronic converters are never pure sine waves, electrical filters are frequently used. These improve the waveform and reduce the adverse effects of harmonics. Harmonic filters of a variety of types may be employed, depending on the situation. The general form of an AC voltage filter includes a series impedance and a parallel impedance, as illustrated in Figure 5.36

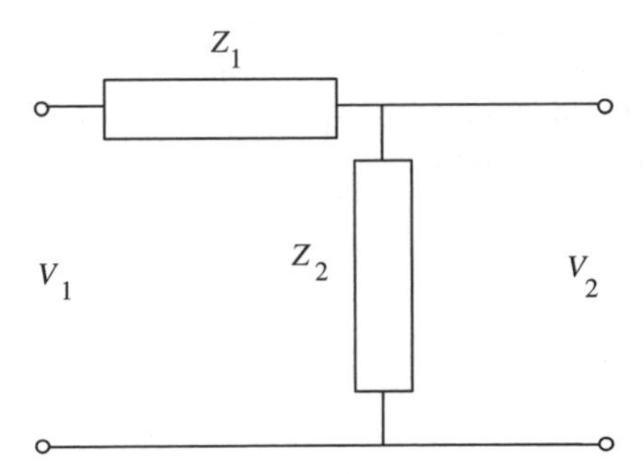

Figure 5.36 AC voltage filter; V_1. AC voltage at filter input; V_2. AC voltage at filter output; Z_1 resonant filter series impedance; Z_2 resonant filter parallel impedance

In Figure 5.36 the input voltage is V_1 and the output V_2. The ideal voltage filter results in a low reduction of the fundamental voltage and a high reduction of all the harmonics.

An example of an AC voltage filter is the series–parallel resonance filter. It consists of one inductor and capacitor in series with the input voltage and another inductor and capacitor in parallel, as shown in Figure 5.37.

As discussed in most basic electrical engineering texts, a resonance condition exists when inductors and capacitors have a particular relation to each other. The effect of resonance in a capacitor and inductor in series is, for example, to greatly increase the voltage across the capacitor relative to the total voltage across the two. Resonance in a filter helps to make the higher-order harmonics small relative to the fundamental. The resonance condition requires, for the capacitor and inductor in series, that:

$$L_1 2\pi f = \frac{1}{C_1 2\pi f} (= X') \tag{5.5.4}$$

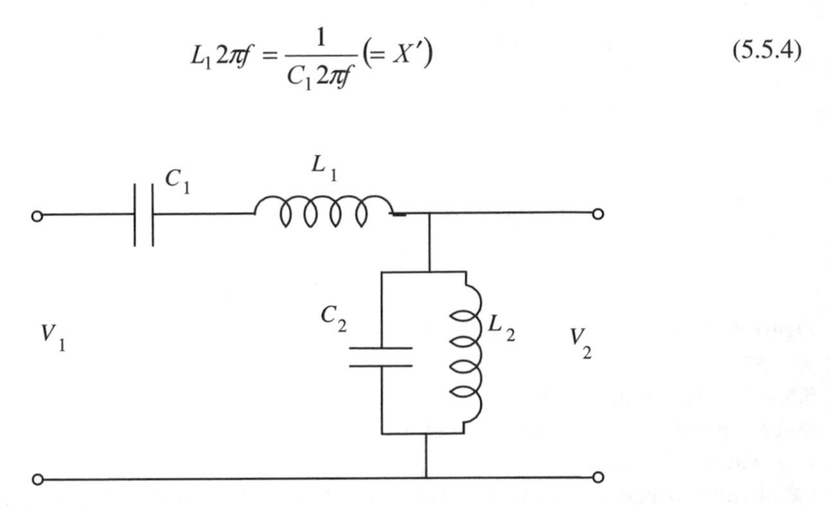

Figure 5.37 Series parallel resonance filter; C_1 and L_1. series capacitance and inductance, respectively; C_2 and L_2. parallel capacitance and inductance, respectively; V_1 and V_2 input and output voltages, respectively

where X' is the reactance of either the capacitor or the inductor and f is the frequency of the fundamental. For the capacitor and inductor in parallel, resonance implies that:

$$C_2 2\pi f = \frac{1}{L_2 2\pi f} (= Y'') \tag{5.5.5}$$

where Y'' = the reciprocal of the reactance ('admittance') of either the capacitor or the inductor.

For this particular filter, the output voltage harmonics will be reduced relative to the input by the following scale factor:

$$f(n) = \left| \frac{V_n}{V_1} \right| = \left| \frac{1}{1 - [n - (1/n)]^2 X'Y''} \right| \tag{5.5.6}$$

From Equation 5.5.6, f(1) = 1 and $f(n)$ approaches $1/n^2$ for high values of n. For certain values of n, the denominator of Equation 5.5.6 goes to zero, indicating a resonant frequency. For example, there is a resonant frequency for a value of $n > 1$ at:

$$n = \tfrac{1}{2} \left[\sqrt{\frac{1}{X'Y''}} + \sqrt{\left(\frac{1}{X'Y''} + 4 \right)} \right] \tag{5.5.7}$$

Near this resonant frequency, input harmonics are amplified. Thus, this resonant frequency should be chosen lower than the lowest occurring input harmonic voltage.

Further discussion of filters is outside the scope of this text. It is important to realize, however, that sizing of the inductors and capacitors in a filter is related to the harmonics to be filtered. Higher-frequency harmonics can be filtered with smaller components. As far as filtering is concerned, then, the higher the switching rate in a PWM inverter, for example, the better. This is because higher switching rates can reduce the lower-frequency harmonics, but increase higher ones. More information on filters is given in Thorborg (1988).

5.6 Ancillary electrical equipment

There is a variety of ancillary electrical equipment associated with a wind turbine installation. It normally includes both high-voltage (generator voltage) and low-voltage items. Figure 5.38 illustrates the main high-voltage components for a typical installation. Dotted lines indicate items that are often not included. These items are discussed briefly below.

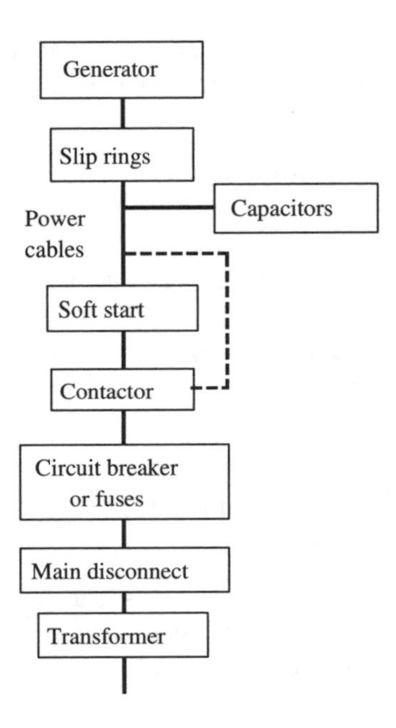

Figure 5.38 Wind turbine high-voltage equipment

5.6.1 Power cables

Power must be transferred from the generator down the tower to electrical switch gear at the base. This is done via power cables. Three-phase generators have four conductors, including the ground or neutral. Conductors are normally of copper, and they are sized to minimize voltage drop and power losses.

In most larger wind turbines, the conductors are continuous from the generator down the tower to the main contactor. In order that the cables not be wrapped up and damaged as the turbine yaws, a substantial amount of slack is left in them so that they 'droop' as they hang down the tower. The power cables are thus often referred to as droop cables. The slack is taken up as the turbine yaws and then released as it yaws back the other way. With sufficient slack, the cables seldom or never wrap up tight in most sites. When they do wrap up too far, however, they must be unwrapped. This may be done manually, after first disconnecting them, or by using a yaw drive.

5.6.2 Slip rings

Some turbines, particularly smaller ones, use discontinuous cables. One set of cables is connected to the generator. Another set goes down the tower. Slip rings and brushes are used to transfer power from one set to the other. In a typical application the slip rings are mounted on a cylinder attached to the bottom of the main frame of the turbine. The axis of the cylinder lies on the yaw axis, so the cylinder rotates as the turbine yaws. The brushes are mounted on the tower in such a way that they contact the slip rings regardless of the orientation of the turbine.

Slip rings are not commonly used on larger machines, since they become quite expensive as the current carrying capacity increases. In addition, maintenance is required as the brushes wear.

5.6.3 Soft start

As indicated in Section 5.4.4, induction generators will draw much more current while starting across the line than they would produce when running. Starting in this way has numerous disadvantages. High currents can result in early failure of the generator windings, and can result in voltage drops for loads nearby on the electrical network. Rapid acceleration of the entire wind turbine drive train can result in fatigue damage. In isolated grids with a limited supply of reactive power, it may not be possible to start a large induction machine at all.

Due to the high currents that accompany across the line starting of induction machines, most wind turbines employ some form of soft start device. These can take a variety of forms. In general, they are a type of power electronic converter that, at the very least, provides a reduced current to the generator.

5.6.4 Contactors

The main contactor is a switch that connects the generator cables to the rest of the electrical network. When a soft start is employed, the main contactor may be integrated with the soft start or it may be a separate item. In the latter case power may be directed through the main contactor only after the generator has been brought up to operating speed. At this point the soft start would be simultaneously switched out of the circuit.

5.6.5 Circuit breakers or fuses

Somewhere in the circuit between the generator and the electrical grid are circuit breakers or fuses. These are intended to open the circuit if the current gets too high, presumably as a result of a fault or short circuit. Circuit breakers can be reset after the fault is corrected. Fuses need to be replaced.

5.6.6 Main disconnect

A main disconnect switch is usually provided between the electrical grid and the entire wind turbine electrical system. This switch is normally left closed, but can be opened if any work is being done on the electrical equipment of the turbine. The main disconnect would need to be open if any work were to be done on the main contactor and would in any case provide an additional measure of safety during any electrical servicing.

5.6.7 Power factor correction capacitors

Power factor correction capacitors are frequently employed to improve the power factor of the generator when viewed from the utility. These are connected as close to the generator as is convenient, but typically at the base of the tower or in a nearby control house.

5.6.8 Turbine Electrical Loads

There may a number of electrical loads associated with the operation of wind turbines. These could include actuators, hydraulic motors, pitch motors, yaw drives, air compressors, control computers, etc. Such loads typically require 120 V or 240 V. Since the generator voltage is normally higher than that, a low-voltage supply needs to be provided by the utility, or step down transformers need to be obtained.

References

Bradley, D. A. (1987) *Power Electronics*, Van Nostrand Reinhold (UK), Wokingam, England.

Brown, D. R., Hamilton, E. P. (1984) *Electromechanical Engineering Conversion*, Macmillan Publishing, New York.

Edminister, J. A. (1965) *Electrical Circuits*. Shaum's Outline Series in Engineering, McGraw Hill Book Co., New York.

Fuchs, E. F., Erickson, R. W., Fardou, A. A. (1992) Permanent Magnet Machines for Operation with Large Speed Variations, *Proc. of the 1992 American Wind Energy Association Annual Conference*, American Wind Energy Association, Washington DC.

Johnson, Gary L. (1985) *Wind Energy Systems*, Prentice-Hall, Inc., Englewood Cliffs, NJ.

Manwell, J. F., McGowan, J. G., Adulwahid, U., Rogers, A., McNiff, B. P. (1996) A Graphical Interface Based Model for Wind Turbine Drive Train Dynamics, *Proc. of the 1996 American Wind Energy Association Annual Conference*, American Wind Energy Association, Washington DC.

Nasar, S. A., Unnewehr, L. E. (1979) *Electromechanics and Electric Machines*, John Wiley, New York.

Sarma, M.S. (1985) *Electric Machines, Steady-State Theory and Dynamic Performance*, West Publishing Company, St. Paul, MN.

Thorborg, K. (1988) *Power Electronics*, Prentice-Hall, Englewood Cliffs, NJ.

Torrey, D. A., Childs, S. E. (1993) Development of a Variable Reluctance Wind Generator, *Proc. of the 1993 American Wind Energy Association Annual Conference*, American Wind Energy Association, Washington DC.

6

Wind Turbine Design

6.1 Overview

6.1.1 Overview of design chapter

Wind turbine design involves a great many considerations, ranging from the very general to the very detailed. The approach taken in this chapter is to begin with the general and then move to the details. The chapter begins with an overview of the design process and then presents a more in depth examination of the various steps involved. This is then followed by a review of the basic wind turbine topologies. The detailed approach to design begins with a step back to the basics. A review of two topics fundamental to wind turbine design is presented: (1) material properties, and (2) machine elements. There is then a discussion of the design loads which the turbine must withstand. This is followed by a detailed look at each of the key subsystems and components that form the turbine. Then, an overview is given of some of the analysis tools available to assist in detailed evaluation of a specific design. Finally, a method to predict a new turbine's curve is presented.

6.1.2 Overview of design issues

The process of designing a wind turbine involves the conceptual assembling of a large number of mechanical and electrical components into a machine which can convert the varying power in the wind into a useful form. This process is subject to a number of constraints, but the fundamental ones involve the potential economic viability of the design. Ideally, the wind turbine should be able to produce power at a cost lower than its competitors, which are typically petroleum derived fuels, natural gas, nuclear power, or other renewables. At the present state of the technology, this is often a difficult requirement, so sometimes incentives are provided by governments to make up the difference. Even in this case, it is a fundamental design goal to keep the cost of energy lower than it would be from a turbine of a different design.

The cost of energy from a wind turbine is a function of many factors, but the primary ones are the cost of the turbine itself and its annual energy productivity. In addition to the first cost of the turbine, other costs (as discussed in more detail in Chapter 9) include installation, operation and maintenance. These will be influenced by the turbine design and must be considered during the design process. The productivity of the turbine is a function both of the turbine's design and the wind resource. The designer cannot control the resource, but must consider how best to utilize it. Other factors that affect the cost of energy, such as loan interest rates, discount rates, etc. tend to be of secondary importance and are largely outside the purview of the designer.

The constraint of minimizing cost of energy has far-reaching implications. It impels the designer to minimize the cost of the individual components, which in turn pushes him or her to consider the use of inexpensive materials. The impetus is also there to keep the weight of the components as low as possible. On the other hand, the turbine design must be strong enough to survive any likely extreme events, and to operate reliably and with a minimum of repairs over a long period of time.

Wind turbine components, because they are kept small, tend to experience relatively high stresses. By the nature of the turbine's operation, the stresses also tend to be highly variable. Varying stresses result in fatigue damage. This eventually leads to either failure of the component, or the need for replacement.

The need to balance the cost of the wind turbine with the requirement that the turbine have a long, fatigue resistant life should be the fundamental concern of the designer.

6.2 Design Procedure

There are a number of approaches that can be taken towards wind turbine design, and there are many issues that must be considered. This section outlines the steps in one approach. The following sections provide more details on those steps.

The key design steps include the following:

1. Determine application
2. Review previous experience
3. Select topology
4. Preliminary loads estimate
5. Develop tentative design
6. Predict performance
7. Evaluate design
8. Estimate costs and cost of energy
9. Refine design
10. Build prototype
11. Test prototype
12. Design production machine

Steps 1 through 7 are the subjects of this chapter. Turbine cost and cost of energy estimates (Step 8) can be made using methods discussed in Chapter 9. Steps 9–13 are beyond the scope of this text, but they are based on the principles outlined here.

6.2.1 *Determine application*

The first step in designing a wind turbine is to determine the application. Wind turbines for producing bulk power for supply to large utility networks, for example, will have a different design than will turbines intended for operation in remote communities.

The application will be a major factor in choosing the size of the turbine, the type of generator it has, the method of control, and how it is to be installed and operated. For example, wind turbines for utility power will tend to be as large as practical. At the present time, such turbines have power ratings in the range of 500 to 1500 kW, with rotor diameters in the range of 38 m (125 ft) to 61 m (200 ft). Such machines are often installed in clusters or wind farms, and may be able to utilize fairly developed infrastructure for installation, operation and maintenance.

Turbines for use by utility customers, or for use in remote communities, tend to be smaller, typically in the 10 to 200 kW range. Ease of installation and maintenance and simplicity in construction are important design considerations for these turbines.

6.2.2 *Review previous experience*

The next step in the design process should be a review of previous experience. This review should consider, in particular, wind turbines built for similar applications. A wide variety of wind turbines have been conceptualized. Many have been built and tested, at least to some degree. Lessons learned from those experiences should help guide the designer and narrow the options.

A general lesson that has been learned from every successful project is that the turbine must be designed in such a way that operation, maintenance, and servicing can be done in a safe and straightforward way.

6.2.3 *Select topology*

There are a wide variety of possible overall layouts or 'topologies' for a wind turbine. Most of these relate to the rotor. The most important choices are listed below. These choices are discussed in more detail in Section 6.3.

- Rotor axis orientation: horizontal or vertical
- Power control: stall, variable pitch, controllable aerodynamic surfaces and yaw control
- Rotor position: upwind of tower or downwind of tower
- Yaw control: driven yaw, free yaw or fixed yaw
- Rotor speed: constant or variable
- Design tip speed ratio and solidity
- Type of hub: rigid, teetering, hinged blades or gimballed
- Number of blades

6.2.4 *Preliminary loads estimate*

Early in the design process it is necessary to make a preliminary estimate of the loads that the turbine must be able to withstand. These loads will serve as inputs to the design of the individual components. Estimation of loads at this stage may involve the use of scaling of loads from turbines of similar design, 'rules of thumb', or simple computer analysis tools. These estimates are improved throughout the design phase as the details of the design are specified. At this stage it is important to keep in mind all the loads that the final turbine will need to be able to withstand. This process can be facilitated by referring to recommended design standards.

6.2.5 *Develop tentative design*

Once the overall layout has been chosen and the loads approximated, a preliminary design may be developed. The design may be considered to consist of a number of subsystems. These subsystems, together with some of their principal components, are listed below. Each of these subsystems is discussed in more detail in Section 6.7.

- Rotor (blades, hub, aerodynamic control surfaces)
- Drive train (shafts, couplings, gearbox, mechanical brakes, generator)
- Nacelle and main frame
- Yaw system
- Tower, foundation and erection

There are also a number of general considerations, which may apply to the entire turbine. Some of these include:

- Fabrication methods
- Ease of maintenance
- Aesthetics
- Noise
- Other environmental conditions

6.2.6 *Predict performance*

Early in the design process it is also necessary to predict the performance (power curve) of the turbine. This will be primarily a function of the rotor design, but will also be affected by the type of generator, efficiency of the drive train, the method of operation (constant speed or variable speed) and choices made in the control system design. Power curve prediction is discussed in Section 6.9.

6.2.7 Evaluate design

The preliminary design must be evaluated for its ability to withstand the loading the turbine may reasonably be expected to encounter. It goes almost without saying that the wind turbine must be able to easily withstand any loads likely to be encountered during normal operation. In addition, the turbine must be able to withstand extreme loads that may only occur infrequently, as well as to hold up to cumulative, fatigue-induced damage. Fatigue damage arises from varying stress levels, which may occur in a periodic manner proportional to rotor speed, a stochastic ('random') manner, or as result of transient loads.

The categories of loads the wind turbine must withstand, as described in Chapter 4, include:

- Static loads (not associated with rotation)
- Steady loads (associated with rotation, such as centrifugal force)
- Cyclic loads (due to wind shear, blade weight, yaw motion)
- Impulsive (short duration loads, such as blades passing through tower shadow)
- Stochastic loads (due to turbulence)
- Transient loads (due to starting and stopping)
- Resonance induced-loads (due to excitations near the natural frequency of the structure)

The turbine must be able to withstand these loads under all plausible conditions, both normal operation and extreme events. These conditions will be discussed in more detail in Section 6.6

The loads of primary concern are those in the rotor, especially at the blade roots, but any loads at the rotor also propagate through the rest of the structure. Therefore, the loading at each component must also be carefully assessed.

Analysis of wind turbine loads and their effects is typically carried out with the use of computer based analysis codes. In doing so, reference is normally made to accepted practices or design standards. The principles underlying the analysis of wind turbine loads were discussed in detail in Chapter 4. A more in depth discussion of wind turbine loads as related to design in given in Section 6.6.

6.2.8 Estimate costs and cost of energy

An important part of the design process is the estimation of the cost of energy from the wind turbine. The key factors in the cost of energy are the cost of the turbine itself and its productivity. It is therefore necessary to be able to predict the cost of the machine, both in the prototype stage, but most importantly in production. Wind turbine components are typically a mix of commercially available items and specially designed and fabricated items. The commercially available items will typically have prices that will be lowered only slightly when bought in volume for mass production. Special items will often be quite expensive in the prototype level, because of the design work and the effort involved in building just one or a few of the items. In mass production, however, the price for the component should drop so as to be close to that of commercial items of similar material, complexity and size.

Wind turbine costs and cost of energy calculations are discussed in more detail in Chapter 9.

6.2.9 *Refine design*

When the preliminary design has been analysed for its ability to withstand loads, its performance capability has been predicted, and the eventual cost of energy has been estimated, it is normal that some areas for refinement will have been identified. At this point another iteration on the design is made. The revised design is analyzed in a similar manner to the process summarized above. This design, or perhaps a subsequent one if there are more iterations, will be used in the construction of a prototype.

6.2.10 *Build prototype*

Once the prototype design has been completed, a prototype should be constructed. The prototype may be used to verify the assumptions in the design, test any new concepts, and ensure that the turbine can be fabricated, installed, and operated as expected. Normally the turbine will be very similar to the expected production version, although there may be provision for testing and instrumentation options which the production machine would not need.

6.2.11 *Test prototype*

After the prototype has been built and installed, it is subjected to a wide range of field tests. Power is measured and a power curve developed to verify the performance predictions. Strain gauges are applied to critical components. Actual loads are measured and compared to the predicted values.

6.2.12 *Design production machine*

The final step is the design of the production machine. The design of this machine should be very close to the prototype. It may have some differences, however. Some of these may be improvements, the need for which was identified during testing of the prototype. Others may have to do with lowering the cost for mass production. For example, a weldment may be appropriate in the prototype stage, but a casting may be a better choice for mass production.

6.3 Wind Turbine Topologies

This section provides a summary of some of the key issues relating to the most commonly encountered choices in the overall topology of modern wind turbines. The purpose of this section is not to advocate a particular design philosophy, but to provide an overview of what must be considered. It should be noted that there are in the wind energy community strong proponents of particular aspects of design, such as rotor orientation, number of blades, etc. A good overview of some issues of design philosophy are given by Doerner (1998) on his Internet site.

One of the general topics of great interest at the present time is how light a turbine can be and still survive the desired amount of time. Some of the issues in this regard are discussed by Geraets et al. (1997).

6.3.1 Rotor axis orientation: horizontal or vertical

The most fundamental decision in the design of a wind turbine is probably the orientation of the rotor axis. In most modern wind turbines the rotor axis is horizontal (parallel to the ground), or nearly so. The turbine is then referred to as a 'horizontal axis wind turbine' (HAWT), as discussed in Chapter 1. There are a number of reasons for that trend; some are more obvious than others. Two of the main advantages of horizontal axis rotors are the following:

1. The rotor solidity of a HAWT (and hence total blade mass relative to swept area) is lower when the rotor axis is horizontal (at a given design tip speed ratio). This tends to keep costs lower on a per kW basis.
2. The average height of the rotor swept area can be higher above the ground. This tends to increase productivity on a per kW basis.

The major advantage of a vertical axis rotor (resulting in a 'vertical axis wind turbine' or VAWT) is that there is no need for a yaw system. That is, the rotor can accept wind from any direction. Another advantage is that in most vertical axis wind turbines, the blades can have a constant chord and no twist. These characteristics should enable the blades to be manufactured relatively simply (e.g. by aluminum pultrusion) and thus cheaply. A third advantage is that much of the drive train (gearbox, generator, brake) can be located on a stationary tower, relatively close to the ground.

In spite of some promising advantages of the vertical axis rotor, the design has not met with widespread acceptance. Many machines that were built in the 1970s and 1980s suffered fatigue damage of the blades, especially at connection points to the rest of the rotor. This was an outcome of the cyclic aerodynamic stresses on the blades as they rotate and the fatigue properties of the aluminum from which the blades were commonly fabricated.

Incompatibilities between structure and control have also caused problems. From a structural viewpoint, the Darrieus troposkein ('skipping rope') shape has appeared most desirable (in comparison with the straight bladed design). This is because the blade is not subject to any radial bending moments, but only tension. On the other hand, it is very difficult to incorporate aerodynamic control, such as variable pitch or aerodynamic brakes, on a blade of this type. For this reason, stall control is the primary means of limiting power in high winds. Owing to the aerodynamics of the stall-controlled vertical axis rotor, the rated wind speed tends to be relatively high. This results in the need for drive train components to be larger than they might otherwise be, and for overall capacity factors of the wind turbine to be relatively low.

In summary, a horizontal axis is probably preferable. There are enough advantages, however, to the vertical axis rotor that it may be worth considering for some applications. In this case, however, the designer should have a clear understanding of what the limitations are, and should also have some plausible options in mind for addressing those limitations.

Because of the predominance of horizontal axis wind turbines presently in use or under development, the remainder of this chapter, unless otherwise specified, applies primarily to wind turbines of that type.

6.3.2 Rotor power control: stall, pitch, yaw and aerodynamic surfaces

There are a number of options for controlling power aerodynamically. The selection of which of these is used will influence the overall design in a variety of ways. The following presents a brief summary of the options, focusing on those aspects that affect the overall design of the turbine. Details on control issues are discussed in Chapter 7.

Stall control takes advantage of reduced aerodynamic lift at high angles of attack to reduce torque at high wind speeds. For stall to function, the rotor speed must be separately controlled, most commonly by an induction generator (see Chapter 5) connected directly to the electrical grid. Blades in stall-controlled machines are fastened rigidly to the rest of the hub, resulting in a simple connection. The nature of stall control, however, is such that maximum power is reached at a relatively high wind speed. The drive train must be designed to accommodate the torques encountered under those conditions, even though such winds may be relatively infrequent. Stall-controlled machines invariably incorporate separate braking systems to ensure that the turbine can be shut down under all eventualities

Variable pitch machines have blades which can be rotated about their long axis, changing the blades' pitch angle. Changing pitch also changes the angle of attack of the relative wind and the amount of torque produced. Variable pitch provides more control options than does stall control. On the other hand the hub is more complicated, because pitch bearings need to be incorporated. In addition, some form of pitch actuation system must also be included. In some wind turbines, only the outer part of blades may be pitched. This is known as partial span pitch control.

Some wind turbines utilize aerodynamic surfaces on the blades to control or modify power. These surfaces can take a variety of forms, but in any case the blades must be designed to hold them, and means must be provided to operate them. In most cases aerodynamic surfaces are used for braking the turbine. In some cases, specifically when using ailerons (see Chapter 7), the surfaces may also provide a fine-tuning effect.

Another option for controlling power is yaw control. In this arrangement, the rotor is turned away from the wind, reducing power. This method of control requires a robust yaw system. The hub must be able to withstand gyroscopic loads due to yawing motion, but can otherwise be relatively simple.

6.3.3 Rotor position: upwind of tower or downwind of tower

The rotor in a horizontal axis turbine may be either upwind or downwind of the tower. A downwind rotor allows the turbine to have free yaw, which is simpler to implement than active yaw. Another advantage of the downwind configuration is that it is easier to take advantage of centrifugal forces to reduce the blade root flap bending moments. This is because the blades are normally coned downwind, so centrifugal moments tend to counteract moments due to thrust. On the other hand, the tower produces a wake in the downwind direction, and the blades must pass through that wake every revolution. This wake is a source of periodic loads, which may result in fatigue damage to the blades and may impose a ripple on the electrical power produced. Blade passage through the wake is also a source of noise. The effects of the wake (known as 'tower shadow') may to some extent be reduced by utilizing a tower design which provides minimal obstruction to the flow.

6.3.4 Yaw control: free or active

All horizontal axis wind turbines must provide some means to orient the machine as the wind direction changes. In downwind machines yaw motion has historically been free. The turbine follows the wind like a weather vane. For free yaw to work effectively, the blades are typically coned a few degrees in the downwind direction. Free yaw machines sometimes incorporate yaw dampers to limit the yaw rate and thus gyroscopic loads in the blades.

Upwind turbines normally have some type of active yaw control. This usually includes a yaw motor, gears, and a brake to keep the turbine stationary in yaw when it is properly aligned. Towers supporting turbines with active yaw must be capable of resisting the torsional loads that will result from use of the yaw system.

6.3.5 Rotational speed: constant or variable

Most rotors on grid-connected wind turbines operate at a nearly constant rotational speed, determined by the electrical generator and the gearbox. In some turbines, however, the rotor speed is allowed to vary. The choice of whether the rotor speed is fixed or variable may have some impact on the overall design, although generally in a secondary way. For example, nearly all modern variable speed wind turbines incorporate power electronic converters to ensure that the resulting electric power is of the desired form. The presence of such a converter introduces some flexibility in the choice of the generator. Using a low-speed generator can eliminate the need for a gearbox and have a dramatic effect on the layout of the entire machine. The possible effects of electrical noise due to the power electronics in a variable speed turbine must also be taken into account in the detailed design.

6.3.6 Design tip speed ratio and solidity

The design tip speed ratio of a rotor is that tip speed ratio where the power coefficient is a maximum. Selection of this value will have a major impact on the design of the entire turbine. First of all, there is a direct relation between the design tip speed ratio and the rotor's solidity (the area of the blades relative to the swept area of the rotor), as discussed in Chapter 3. A high-speed rotor will have less blade area than the rotor of a slower machine. For a constant number of blades, the chord and thickness will decrease as the solidity decreases. Owing to structural limitations, there is a lower limit to how thin the blades may be. Thus, as the solidity decreases, the number of blades usually decreases as well.

There are a number of incentives for using higher tip speed ratios. First of all, reducing the number of blades or their weight reduces the cost. Second, higher rotational speeds imply lower torques for a given power level. This should allow the balance of the drive train to be relatively light. However, there are some drawbacks to high tip ratios as well. For one thing, high-speed rotors tend to be noisier than are slower ones (see Chapter 10).

6.3.7 Hub: rigid, teetering, hinged blades or gimballed

The hub design of a horizontal axis wind turbine is an important constituent of the overall layout. The main options are rigid, teetering, or hinged. Most wind turbines employ rigid rotors. This means that the blades cannot move in the flapwise and edgewise directions. The term 'rigid rotor' does include those with variable pitch blades, however.

The rotors in two-bladed turbines are usually teetering. That means the hub is mounted on bearings, and can teeter back and forth, in and out of the plane of rotation. The blades in turn are rigidly connected to the hub, so during teetering one blade moves in the upwind direction, while the other moves downwind. An advantage of teetering rotors is that the bending moments in the blades can be very low during normal operation.

Some two-bladed wind turbines use hinges on the hub. The hinges allow the blades to move into and out of the plane of rotation independently of each other. Since the blade weights do not balance each other, however, other provisions must be made to keep them in the proper position when the turbine is not running, or is being stopped or started.

One design variant is known as a 'gimballed turbine'. It uses a rigid hub, but the entire turbine is mounted on horizontal bearings so that the machine can tilt up or down from horizontal. This motion can help to relieve imbalances in aerodynamic forces.

6.3.8 Rigidity: flexible or stiff

Turbines with lower design tip speed ratios and higher solidities tend to be relatively stiff. Lighter, faster turbines are more flexible. Flexibility may have some advantages in relieving stresses, but blade motions may also be more unpredictable. Most obviously, a flexible blade in an upwind turbine may be far from the tower when unloaded, but could conceivably hit it in high winds. Flexible components such as blades or towers may have natural frequencies near the operating speed of the turbine. This is something to be avoided. Flexible blades may also experience 'flutter' motion, which is a form of unstable and undesirable operation.

6.3.9 Number of blades

Most modern wind turbines used for generating electricity have three blades, although some have two or even one. Three blades have the particular advantage that the polar moment of inertia with respect to yawing is constant, and is independent of the azimuthal position of the rotor. This characteristic contributes to relatively smooth operation even while yawing.

A two-bladed rotor, however, has a lower moment of inertia when the blades are vertical than when they are horizontal. This 'imbalance' is one of the reasons that most two-bladed wind turbines use a teetering rotor. Using more than three blades could also result in a rotor with a moment of inertia independent of position, but more than three blades are seldom used. This is primarily because of the higher costs that would be associated with the additional blades.

A key consideration in selecting the number of blades is that the stress in the blade root increases with the number of blades for a turbine of a given solidity. Thus all other things being equal, increasing the design tip speed ratio entails decreasing the number of blades.

A few single-bladed turbines have been built in the last twenty years. The presumed advantage is that the turbine can run at a relatively high tip speed ratio, and that the cost should be lower because of the need for only one blade. However, a counterweight must be provided to balance the weight of the single blade. The aesthetic factor of the appearance of imbalance is another consideration.

6.3.10 Tower structure

The tower of a wind turbine serves to elevate the main part of the machine up into the air. For a horizontal axis machine the tower must be at least high enough to keep the blade tips from touching the ground as they rotate. In practice, towers are usually much higher than that. Winds are nearly always much stronger as elevation above ground increases, and they are less turbulent. All other things being equal, the tower should be as high as practical. Choice of tower height is based on an economic tradeoff of increased energy capture versus increased cost.

The principal options in towers are tubular, pipe-type structures or trusses (typically bolted). One of the primary considerations is the overall tower stiffness, which also has a direct effect on its natural frequency. As mentioned in Chapter 4, stiff towers are those whose fundamental natural frequency is higher than that of the blade passing frequency (rotor's rotational speed times the number of blades). They have the advantage of being relatively insensitive to motions of the turbine itself, but, being heavy, they are also costly. Soft towers are those whose fundamental natural frequency is lower than the blade passing frequency. A further distinction is commonly made: a soft tower's natural frequency is above the rotor frequency as well as being below the blade passing frequency. A soft–soft tower is one whose natural frequency is below both the rotor frequency and blade passing frequency. These towers are generally less expensive than stiffer ones, since they are lighter. However, particularly careful analysis of the entire system is required to ensure that no resonances are excited by any motions in the rest of the turbine.

Other factors in tower selection include presumed mode of erection and aesthetics. If a turbine is to erected by tilting it up, there is a benefit to keeping the tower as light as possible. If a crane is going to used, attention must be given to the sizes of cranes expected to be available. If the tower is going to incorporate a lifting capability, which would obviate the need for a crane, planning for that would be needed early in the design process. In terms of aesthetics, it should be noted that preference seems to lie with tubular designs. It should also be noted that tubular towers appear to be preferable for minimizing impact on avian populations (see Chapter 10.)

6.3.11 Design constraints

There are inevitably a number of other factors that will influence the general design of a wind turbine. Some of these include the expected wind regime, general climate, site accessibility, and availability of expertise and equipment for installation and operation.

6.3.11.1 Climatic factors affecting design

Turbines designed for more energetic or turbulent sites need to be stronger than those in more conventional sites. Expected conditions at such sites must be considered if the turbines are to meet international standards. This topic is discussed in more detail in Section 6.6

General climate can affect turbine design in a number of ways. For example, turbines for use in hot climates may need provisions for extra cooling, whereas turbines for cold climates may require heaters, special lubricants, or even different structural materials. Turbines intended for use in marine climates would need protection from salt, and should be built of corrosion-resistant materials wherever possible.

6.3.11.2 Site-specific factors affecting design

Turbines which are intended for relatively inaccessible sites have their designs constrained in a number of ways. For example, they might need to be self-erecting. Difficulty in transport could also limit the size or weight of any one component.

Limited availability of expertise and equipment for installation and operation would be of particular importance for machines intended to operate singly or in small groups. This would be particularly important for applications in remote areas or developing countries. In this case it would be especially important to keep the machine simple, modular and designed to require only commonly available mechanical skills, tools and equipment.

6.3.11.3 Environmental factors affecting design

Wind turbine proponents inevitably extol the environmental benefits that accrue to society through the use of wind generated electricity. However, there will always be some impacts on the immediate environment where the turbine may be installed, and not all of these may be appreciated by the neighbors. Careful design, however, can minimize many of the adverse effects. Four of the most commonly noted environmental impacts of wind turbines are noise, visual appearance, effects on birds and electromagnetic interference. Some of the key issues affecting overall wind turbine design are summarized here. More details are provided in Chapter 10.

There will always be some sound generated by wind turbines when they are operating, but noise can be can minimized through careful design. In general, upwind machines are quieter than downwind machines, and lower tip speed ratio rotors are quieter than those with higher tip speed ratios. Selection of airfoils, fabrication details of the blades, and design of tip brakes (if any) will also affect noise. Gearbox noise can be reduced by including sound proofing in the nacelle or eliminated by using a direct drive generator. Variable speed turbines tend to make less noise at lower wind speeds, since the rotor speed is reduced under those conditions.

In general, it appears that turbines with lower tip speeds and towers with few perching opportunities are the least likely to adversely affect birds.

Visual appearance is very subjective, but there are reports that people prefer the sight of three blades to two, slow rotors to faster ones, and solid towers to lattice ones. A neutral color is often preferred as well.

Electromagnetic interference created by wind turbines has sometimes been the source of considerable concern. Experience has shown, however, that the impact is usually fairly minimal if the blades are not made of metal. Since most horizontal axis wind turbines now have non-metallic blades, the preferred design already reduces the possible adverse effects.

6.4 Materials

Many types of materials are used in wind turbines. Two of the most important of these are steel and composites. The composites are typically comprised of fiberglass or wood together with a matrix of polyester or epoxy. Other common materials include copper and concrete. The following provides an overview of some of the aspects of materials most relevant to wind turbine applications.

6.4.1 Review of basic mechanical properties

In this text it is assumed that the reader has a familiarity with the fundamental concepts of material properties, as well as with the most common materials themselves. The following is a list of some of the essential concepts, (for more details, see a text on mechanical design, such as Spotts, 1985):

- Hooke's Law
- Modulus of elasticity
- Yield strength, breaking strength
- Ductility and brittleness
- Hardness and machinability
- Failure by yielding or fracture

6.4.1.1 Fatigue properties

Most materials can withstand a load of a certain magnitude when applied once, but cannot withstand the same load when it is applied and then removed multiple times. The decreasing ability to survive repeated loads is called fatigue. Fatigue is of great significance to wind turbine design, and was discussed in greater length in Chapter 4. The most important fatigue properties of a material are summarized in the S–N curve, as described previously (Section 4.2.3.2).

6.4.2 Steel

Steel is one of the most widely used materials in wind turbine fabrication. Steel is used for many structural components including the tower, hub, main frame, shafts, gears and gear cases, fasteners as well as the reinforcing in concrete. Information on steel properties can be found in Spotts (1985), Baumeister (1978) and data sheets from steel suppliers.

6.4.3 Composites

Composites are described in more detail in this text than are most other materials, because it is assumed that they may be less familiar to many readers than are more traditional materials. They are also the primary material used in blade construction. Composites are materials comprising at least two dissimilar materials, most commonly fibers held in place by a binder matrix. Judicious choice of the fibers and binder allows tailoring of the composite properties to fit the application. Composites used in wind turbine applications include those based on fiberglass, carbon fiber, and wood. Binders include polyester, epoxy and vinyl ester. The most common composite is fiberglass reinforced plastic, known as GRP. In wind turbines, composites are most prominently used in blade manufacture, but they are also used in other parts of the machine, such as the nacelle cover. The main advantage of composites is that they have a high strength and high stiffness to weight ratio. They are also corrosion resistant, are electrical insulators, and lend themselves to a variety of fabrication methods.

6.4.3.1 Glass fibers

Glass fibers are formed by spinning glass into long threads. The most common glass fiber is known as E-glass. It is a low-cost material, with reasonably good tensile strength.

Fibers are sometimes used directly, but are most commonly first combined into other forms (known as 'preforms'). Fibers may be woven or knitted into cloth, formed into continuous strand or chopped strand mat, or prepared as chopped fibers. Where high strength is required, unidirectional bundles of fibers known as 'tows' are used. Some fiberglass preforms are illustrated in Figure 6.1. More information is presented in Chou et al. (1986).

6.4.3.2 Matrix (binder)

There are three types of resins commonly used in matrices of composites. They are: (1) unsaturated polyesters, (2) epoxies, (3) vinyl esters. These resins all have the general property that they are used in the liquid form during the lay up of the composite, but when they are cured they are solid. As solids, all of the resins tend to be somewhat brittle. The choice of resins affects the overall properties of the composite.

Polyesters have been used most frequently in the wind industry because they have a short cure time and low cost. Cure time is from a few hours to overnight at room temperature, but with the addition of an initiator, curing can be done at elevated temperatures in a few minutes. Shrinkage upon curing is relatively high, however. The present cost is in the range of $2.20/kg ($1/lb.)

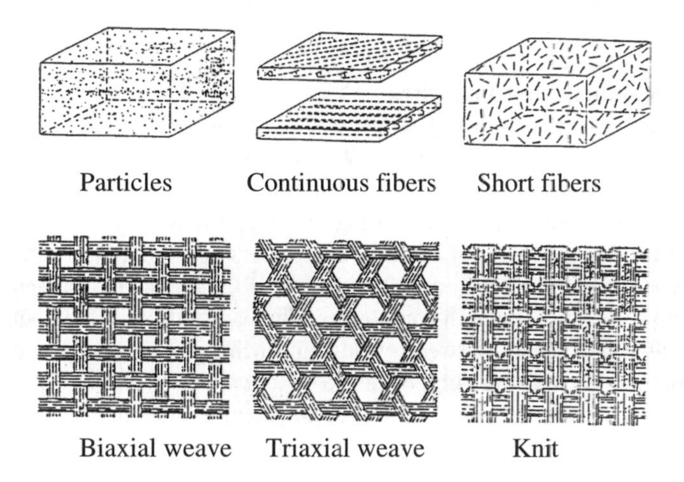

Figure 6.1 Fiberglass performs (National Research Council, 1991)

Epoxies are stronger, have better chemical resistance, good adhesion, and low shrinkage upon curing, but they are also more expensive (almost twice as expensive as polyester) and have a longer cure time than polyesters.

Vinyl esters are epoxy-based resins which have become more widely used over recent years. These resins have similar properties to epoxies, but are somewhat lower in cost and have a shorter cure time. They have good environmental stability and are widely used in marine applications.

6.4.3.3 Carbon fiber reinforcing

Carbon fibers are more expensive than are glass fibers (by approximately a factor of 15), but they are stronger and stiffer. One way to take advantage of carbon fiber's advantages, without paying the full cost, is to use some carbon fibers along with glass in the overall composite.

6.4.3.4 Wood–epoxy laminates

Wood is used instead of synthetic fibers in some composites. In this case the wood is preformed into laminates (sheets) rather than as fibers, or fiber-based cloth. The most common wood for wind turbine laminates is Douglas Fir. Properties of woods vary significantly with respect to the direction of the wood's grain. In general, though, wood has good strength to weight ratio, and is also good in fatigue. One important characteristic of wood is its strong anisotropy in tensile strength. This means that laminates have to be built up with grain going in different directions if the final composite is to be strong enough in all directions. More information on properties of wood may be found in Hoadley (2000).

The use of wood together with an epoxy binder was developed for wind turbine applications based on previous experience from the high-performance boat building industry. A technique known as the wood–epoxy saturation technique (WEST) is used in this process. Wood–epoxy laminates have good fatigue characteristics: according to one source (National Research Council, 1991) no wood–epoxy blade has ever failed in service due to fatigue.

6.4.3.5 Fatigue damage in composites

Fatigue damage occurs in composites as it does in many other materials, but it does not necessarily occur by the same mechanism. The following sequence of events is typical. First, the matrix cracks, then cracks begin to combine and there is debonding between the matrix and the fibers. Then there is debonding and separation (delamination) over a wider area. This is followed by breaking of the individual fibers, and finally by complete fracture.

The same type of analysis techniques that are used for metals (explained in Chapter 4) are also used for predicting fatigue in composites. That is, rainflow cycle counting is used to determine the range and mean of stress cycles, and Miner's Rule is used to calculate the damage from the cycles and the composite's S–N curve. S–N curves for composites are modeled by an equation that has a somewhat different shape than that used in metals, however:

$$\sigma = \sigma_u (1 - B \log N) \tag{6.4.1}$$

where σ is the cyclic stress amplitude, σ_u is the ultimate strength, B is a constant and N is the number of cycles.

The parameter, B, is approximately equal to 0.1 for a wide range of E-glass composites when the reversing stress ratio $R = 0.1$. This is tension-tension fatigue. Life is reduced under fully reversed tension–compression fatigue ($R = -1$) and compression–compression fatigue.

Fatigue strength of glass fibers is only moderate. The ratio of maximum stress to static strength is 0.3 at 10 million cycles. Carbon fibers are much more fatigue resistant than are glass fibers: the ratio of maximum stress to static strength is 0.75 at 10 million cycles, two and half times that of glass. Fatigue life characteristics of E-glass, carbon fiber, and some other common fibers) are illustrated in Figure 6.2.

Owing to the complexity of the failure method of composites and the lack of complete test data on all composites of interest, it is in practice still difficult to predict fatigue life accurately.

6.4.4 Copper

Copper has excellent electrical conductivity and for that reason it is used in nearly all electrical equipment on a wind turbine, including the power conductors. Mechanical properties of copper are in general of much less interest than the conductivity. The weight, however, can be significant. A substantial part of the weight of the electrical generator is due to the copper windings, and the weight of the main power conductors may also be of importance. Information on copper relative to its use in electrical applications can be found in many sources, including Baumeister (1978).

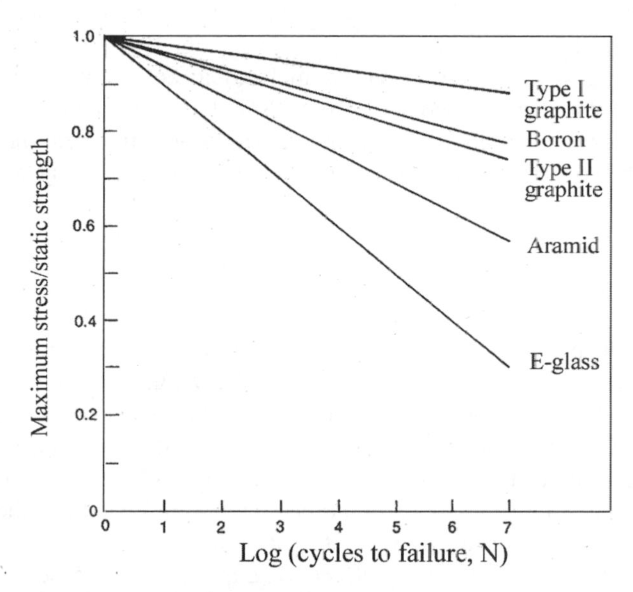

Figure 6.2 Fatigue life of composite fibers (National Research Council, 1991). Reproduced with permission from the National Academy of Science, courtesy of the National Academy Press, Washington, D.C.

6.4.5 Concrete

Reinforced concrete is frequently used for the foundations of wind turbines. It has sometimes been used for the construction of towers as well. Discussion of reinforced concrete, however, is outside the scope of this text.

6.5 Machine Elements

Many of the principal components of a wind turbine are composed, at least partially, of machine elements which have a much wider applicability, and for which there has been a great deal experience. Many of these elements are commercially available and are fabricated according to recognized standards. This section presents a brief overview of some machine elements that are often found in wind turbine applications. For more details, the reader should consult a text on machine design, such as Spotts (1985) or Shigley and Mischke (1989).

6.5.1 Shafts

Shafts are cylindrical elements designed to rotate. Their primary function is normally to transmit torque, and so they carry or are attached to gears, pulleys, or couplings. In wind turbines shafts are typically found in gearboxes, generators, and in linkages.

In addition to being loaded in torsion, shafts are often subjected to bending. The combined loading is often time-varying, so fatigue is an important consideration. Shafts also have resonant natural frequencies at 'critical speeds'. Operation near such speeds is to be avoided, or large vibrations can occur.

Materials used for shafting depend on the application. For the least severe conditions, hot-rolled plain carbon steel is used. For greater strength applications, somewhat higher carbon content steel may be used. After machining, shafts are often heat treated to improve their yield strength and hardness. Under the most severe conditions, alloy steels are used for shafts.

6.5.2 *Couplings*

Couplings are elements used for connecting two shafts together for the purpose of transmitting torque between them. A typical use of couplings in wind turbines is the connection between the generator and the high-speed shaft of the gearbox.

Couplings consist of two major pieces, one of which is attached to each shaft. They are often kept from rotating relative to the shaft by a key. The two pieces are in turn connected to each other by bolts. In a solid coupling the two halves are bolted together directly. In a typical flexible coupling teeth are provided to carry the torque, and rubber bumpers are included between the teeth to minimize effects of impact. Shafts to be connected should ideally be collinear, but flexible couplings are designed to allow some slight misalignment. An example of a solid coupling is shown in Figure 6.3.

Figure 6.3 Typical solid coupling

6.5.3 *Springs*

There are numerous applications for springs in wind turbines. They are particularly useful in passively actuated safety systems. Examples include spring applied brakes, return springs for blade pitch linkages, aerodynamic surfaces or teeter dampers. Rubber bumpers, such as may be used to prevent excessive teeter excursions on two-bladed rotors, are another example.

Springs can be made in a variety of forms, and from a variety of materials. The most common springs are made from spring steel wire, formed into a helical coil shape. Springs may be designed for tension, compression or torsion applications.

6.5.4 *Clutches and brakes*

Clutches are elements intended to transmit torque when applied, but not to do so when they are released. Clutches are used in wind turbines in such applications as pitch linkages and clutch-type brakes. The latter may include drive train shaft brakes, yaw brakes, or erection winch brakes. Clutches are typically applied by spring pressure and released through an active mechanical or electromechanical mechanism.

One common type of clutch is known as a plate clutch. The clutch consists of at least one pressure plate and at least one friction disc. The friction disc is surfaced with a heat-resistant material with a moderately high coefficient of friction, typically in the form of pads. A simple plate clutch is illustrated in Figure 6.4.

Figure 6.4 Simple plate clutch

Two common types of brakes used on wind turbines are disc brakes and clutch brakes. They are both analyzed in a manner similar to that for clutches. The main difference is that heating is a more important consideration for a brake than for a clutch. Disc brakes are used in conjunction with a relatively thin disc. Pressure is applied from brake pads on either side of the disc (to balance the applied load.) The disc is often hollow to help with the cooling.

6.5.5 *Bearings*

Bearings are used to reduce frictional resistance between two surfaces undergoing relative motion. In the most common situations, the motion in question is rotational. There are many bearing applications in wind turbines. They are found in main shaft mountings, gearboxes, generators, yaw systems, blade pitch systems, and teetering mechanisms to name just a few.

Bearings come in variety of forms, and they are made from a variety of materials. For many high-speed applications ball bearings, roller bearings, or tapered roller bearings may be used. These bearings are typically made from steel. In other situations bushings made from plastics or composites may be used.

Ball bearings are widely used in wind turbine components. They consist of four types of parts: an inner ring, an outer ring, the balls, and the cage. The balls run in curvilinear grooves in the rings. The cage holds the balls in place and keeps them from touching each other. Ball bearings are made in a range of types. They may be designed to take radial loads or axial thrust loads. Radial ball bearings can also withstand some axial thrust.

Roller bearings are similar in many respects to ball bearings, except that cylindrical rollers are used instead of balls. They are commonly used in wind turbines in applications such as gearboxes. A typical roller bearing is illustrated in Figure 6.5.

Figure 6.5 Cutaway view of typical roller bearing (Torrington Co., http://howstuffworks.lycos.com/bearing.htm, 2000)

Other types of bearings also have applications in wind turbine design. Two examples are the sleeve bearings and thrust bearings used in the teetering mechanism of some two-bladed wind turbines.

Generally speaking, the most important considerations in the design of a bearing are the load it experiences and the number of revolutions it is expected to survive. Detailed information on all types of bearings may be found in manufacturers' data sheets

6.5.6 Gears

Gears are elements used in transferring torque from one shaft to another. Gears are described in somewhat more detail in this section than other elements because they are widely used in wind turbines. The conditions under which they operate differ in significant ways from many other applications, and it has been necessary to investigate in some detail these conditions and the gears' response so that they perform as desired.

There are numerous applications for gears in wind turbines. The most prominent of these is probably the drive train gearbox. Other examples include yaw drives, pitch linkages, and erection winches. Common types of gears include spur gears, helical gears, worm gears and internal gears. All gears have teeth. Spur gears have teeth whose axes are parallel to the rotational axis of the gear. The teeth in helical gears are inclined at an angle relative to the gear's rotational axis. Worm gears have helical teeth, which facilitate transfer of torque between shafts at right angles to each other. An internal gear is one which has teeth on the inside of an annulus. Some common types of gears are illustrated in Figure 6.6.

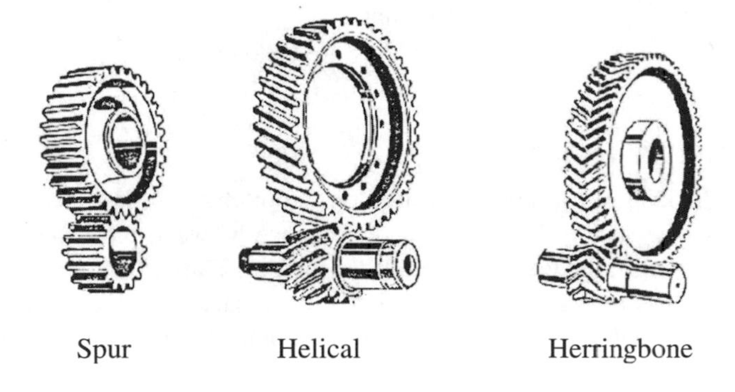

Figure 6.6 Common gear types

Gears may be made from a wide variety of materials, but the most common material in wind turbine gears is steel. High strength and surface hardness in steel gear teeth is often obtained by carburizing or other forms of heat treating.

Gears may be grouped together in gear trains. Typical gear trains used in wind turbine applications are discussed in Section 6.7.

6.5.6.1 Gear terminology

The most basic, and most common gear, is the spur gear. Figure 6.7 illustrates the most important characteristics. The pitch circle is the circumference of a hypothetical smooth gear (or one with infinitesimally small teeth). Two smooth gears would roll around each other with no sliding motion at the point of contact. The diameter of the pitch circle is known as the pitch diameter, d. With teeth of finite size, some of each tooth will extend beyond the pitch circle, some of it below the pitch circle. The face of the tooth is the location that meets the corresponding face of the mating gear tooth. The width of the face, b, is the dimension parallel to the gear's axis of rotation. The circular pitch, p, of the gear is the distance from one face on one tooth to the face on the same side of the next tooth around the pitch circle. Thus $p = \pi d / N$ where N is the number of teeth.

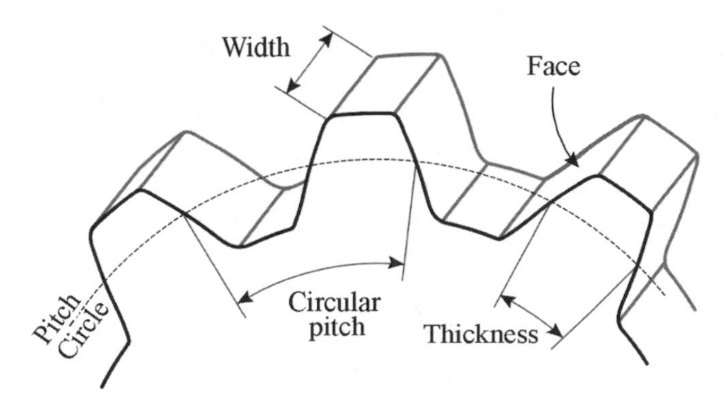

Figure 6.7 Principal parts of a gear

Ideally, the thickness of a tooth measured on the pitch circle is exactly one half of the circular pitch (i.e. the width of the teeth and the space between them are the same on the pitch circle.) In practice, the teeth are cut a little smaller. Thus when the teeth mesh there is some free space between them. This is known as backlash. Excessive backlash can contribute to accelerated wear, so it is kept to minimum. Backlash is illustrated in Figure 6.8.

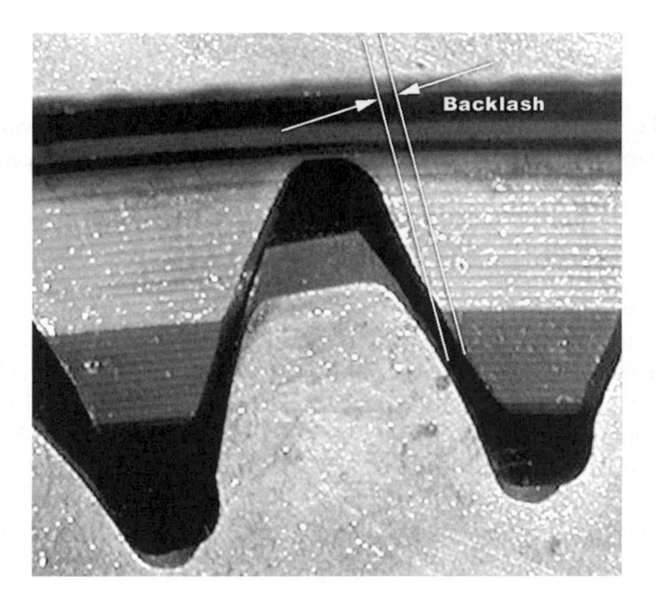

Figure 6.8 Backlash between gears

6.5.6.2 Gear speed relations

When two meshing gears, *1* and *2*, are of different diameter, they will turn at different speeds. The relation between their rotational speeds n_1 and n_2 is inversely proportional to their pitch diameters d_1 and d_2 (or number of teeth). That is:

$$n_1 / n_2 = d_2 / d_1 \tag{6.5.1}$$

6.5.6.3 Gear loading

Loading on a gear tooth is determined by the power being transmitted and the speed of the tooth. In terms of power, *P*, and pitch circle velocity, $V_{pitch} = \pi dn$, the tangential force, F_t, on a tooth is

$$F_t = P / V_{pitch} \tag{6.5.2}$$

As the gear turns, individual teeth will be subjected to loading and unloading. At least one pair of teeth is always in contact, but, at any given time, more than one pair is likely to be in contact. For example, one pair may be unloading while another is taking a greater fraction of the load.

The bending stress, σ_b, on a gear tooth of width *b* and height *h* is calculated by application of the bending equation for a cantilevered beam:

$$\sigma_b = \frac{6M}{bh^2} \tag{6.5.3}$$

The moment, *M*, is based on a load F_b (which is closely related to F_t) applied at a distance *L* to the weakest point on the tooth. The results is:

$$\sigma_b = \frac{F_b}{b} \frac{6L}{h^2} \tag{6.5.4}$$

The factor $h^2 / 6L$ is a property of the size and shape of the gear, and is frequently expressed in terms of the pitch diameter as the form factor (or Lewis factor,) $y = h^2 / 6pL$. In this case, Equation 6.5.4 can be expressed as:

$$\sigma_b = \frac{F_b}{y\,p\,b} \tag{6.5.5}$$

The form factor is available in tables for commonly encountered numbers of teeth and pressure angles. Typical values for spur gears range from 0.056 for 10 teeth/gear to 0.170 for 300 teeth/gear.

6.5.6.4 Gear dynamic loading

Dynamic loading can induce stresses that are also significant to the design of a gear. Dynamic effects occur because of imperfections in the cutting of gears. The mass and spring constant of the contacting teeth, and the loading and unloading of the teeth as the gear rotates, are also contributing factors. Dynamic effects can result in increased bending stresses and can exacerbate deterioration and wear of the tooth surfaces.

The effective spring constant, k_g, of two meshing gear teeth can be important in the dynamic response (natural frequency) of a wind turbine drive train. The following equation gives an approximation to that spring constant. This equation accounts for gears (1 and 2) of different materials. Assuming moduli of elasticity E_1 and E_2, k_g is given by:

$$k_g = \frac{b}{9}\frac{E_1 E_2}{E_1 + E_2} \tag{6.5.6}$$

Dynamic effects and wear are very significant to the design of gears for wind turbine gearboxes. More discussion, however, is beyond the scope of this book. Information on gear tooth wear in general can be found in Spotts (1985) and Shigley and Mischke (1989). Gear tooth wear in wind turbine gearboxes in particular is discussed in McNiff et al. (1990).

6.5.7 Dampers

Wind turbines are subject to dynamic events, with potentially adverse effects. These effects may be decreased by the use of appropriate dampers. There are at least three types of devices that act as dampers and that have been used on wind turbines: (1) fluid couplings, (2) hydraulic pumping circuits, and (3) linear viscous fluid dampers.

Fluid couplings are sometimes used between a gearbox and generator to reduce torque fluctuations. They are used most commonly in conjunction with synchronous generators, which are inherently stiff. Hydraulic pumping circuits consist of a hydraulic pump and a closed hydraulic loop with a controllable orifice. Such circuits may be used for damping yaw motion. Linear viscous fluid dampers are essentially hydraulic cylinders with internal orifices. They may be used as teeter dampers on one- or two-bladed rotors.

Detailed discussion of dampers is beyond the scope of this book. More information on the general topic of hydraulics, on which many damper designs are based, can be found in the Hydraulic Handbook (Hydraulic Pneumatic Power Editors, 1967).

6.5.8 Wire rope

Wire rope is composed of a number of wires combined into a single rope. Depending on the size, some of the wires may first be twisted into strands, and then the strands are twisted together over a central core. Wire rope is used to hold up guyed wind turbine towers or meteorological masts. It is also used with turbine erection systems.

Flexible wire rope, such as used in hoisting, has a relatively large number of small-diameter wires. When wire rope is used for hoisting, it is often used together with sheaves or pulleys for changing direction. Such direction change entails bending, which contributes to fatigue of the rope. Fatigue is accordingly an important consideration in selecting wire rope for such applications. Wire rope for use in guying generally has few wires of larger diameter. It is not intended for use with pulleys.

The primary consideration in selection of wire rope is the tensile stress, σ_t, which is the force in the rope, T, divided by the cross-sectional area, A_c:

$$\sigma_t = T / A_c \tag{6.5.7}$$

Depending on the application, safety factors of from 3 to 8 are used. Thus

$$\gamma \sigma_t < \sigma_{bp} \tag{6.5.8}$$

where σ_{bp} = material breaking stress and γ = a safety factor. Breaking stress for steel wire rope is typically between 1.10 and 1.38 x 10^9 N/m^2

The size of sheaves to be used with wire rope is also an important consideration. The minimum diameter of the sheave will be on the order of 20 to 40 times the rope diameter, depending on the type of rope.

6.5.9 Fastening and joining

Fastening and joining is an important concern in wind turbine design. The most important fasteners are bolts and screws. Their function is to hold parts together, but in a way which can be undone if necessary. Bolts and screws are tightened so as to exert a clamping force on the parts of interest. This is often accomplished by tightening the bolt to a specified torque level. There is a direct relation between the torque on a bolt and its elongation. Thus a tightened bolt acts like a spring as it clamps. Fatigue can be an important factor in specifying bolts. The effects of fatigue can often be reduced by prestressing the bolts.

Bolts and screws on wind turbines are frequently subjected to vibration, and sometimes to shock. These tend to loosen them. To prevent loosening a number of methods are used. These include washers, locknuts, lock wire, and chemical locking agents (such as LockTite®).

There is also a variety of other fasteners, and the use of ancillary items, such as washers and retainers, may be critical in many situations. Joining by means which are not readily disassemblable, such as welding, riveting, soldering, or bonding with adhesives, is also frequently applied in wind turbine design. More details on fastening and joining may be found in Parmley (1997).

6.6 Wind Turbine Loads

6.6.1 Overview

Once the basic layout of the turbine is selected, the next step in the design process is to consider the loads that the turbine must be able to withstand. As is commonly used in mechanics, the loads are the externally applied forces or moments to the entire turbine or to any of the components considered separately.

Wind turbine components are designed for two types of loads: (1) ultimate loads and (2) fatigue loads. Ultimate loads refer to likely maximum loads, multiplied by a safety factor. Fatigue loads refer to the component's ability to withstand an expected number of cycles of possibly varying magnitude. Wind turbine loads can be considered in five categories: (1) steady (here including static loads), (2) cyclic, (3) stochastic, (4) transient (here including impulsive loads), and (5) resonance-induced loads. These loads and their origins are illustrated in Figure 6.9.

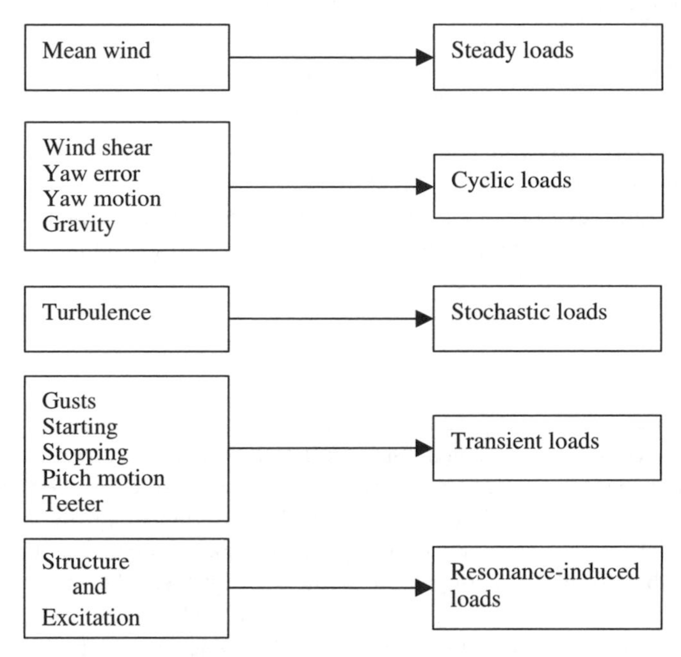

Figure 6.9 Sources of wind turbine loads

Steady loads, which were discussed in detail in Chapter 4, include those due to the mean wind speed, centrifugal forces in the blades due to rotation, weight of the machine on the tower, etc.

Cyclic loads, which were also discussed in Chapter 4, are those which arise due to the rotation of the rotor. The most basic periodic load is that experienced at the blade roots (of a HAWT) due to gravity. Other periodic loads arise from wind shear, cross wind (yaw error), vertical wind, yaw velocity, and tower shadow. Mass imbalances or pitch imbalances can also give rise to periodic loads.

Stochastic loads are due to the turbulence in the wind. Short-term variations in the wind speed, both in time and space across the rotor, cause rapidly varying aerodynamic forces on the blades. The variations appear random, but they can be described in statistical terms. In addition, the nature of the turbulence on the rotor is affected by the rotation itself. This effect is described under the term 'rotational sampling'. Rotational sampling is discussed in detail by Connell (1988).

Transient loads are those which occur only occasionally, and are due to events of limited duration. The most common transient loads are associated with starting and stopping. Other transient loads arise from sudden wind gusts, changes in wind direction, blade pitching motions or teetering.

Resonance-induced loads arise as a result of some part of the structure being excited at one of its natural frequencies. The designer tries to avoid the possibility of that happening, but response to turbulence often inevitably excites some resonant response.

Steady loads and cyclic loads were discussed in detail in Chapter 4. Gusts are discussed in Section 6.6.2. Loads due to starting and stopping can be quite significant, as illustrated in Figure 6.10. A detailed description of transient loads is outside the scope of this text, but some discussion of such loads on the drive train can be found in Manwell et al. (1996). The next section provides some information on resonance-induced loads.

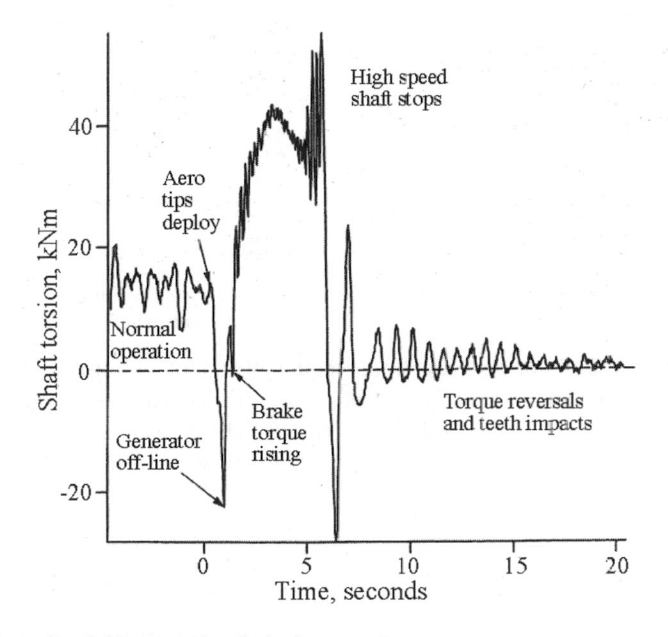

Figure 6.10 Example of drive train loads during stopping

6.6.1.1 Resonance-induced loads

Vibrations and natural frequencies of wind turbine components were discussed in Chapter 4. It was also noted there that operation of the turbine in such a way as to excite those natural frequencies should be avoided. One way to identify points of correspondence between natural frequencies and excitation from the rotor is to use a Campbell diagram. A Campbell diagram illustrates the most important natural frequencies of the turbine as a function of rotor speed. Superimposed on those are lines corresponding to excitation frequency as a function of rotor speed, specifically the rotor rotation frequency ($1P$) and the blade passing frequency (BP), where B is the number of blades and P signifies once per revolution. Points of intersection indicate operating speeds that are to be avoided. A Campbell diagram for a three-bladed turbine is shown in Figure 6.11.

As can be seen in Figure 6.11 there may be a number of different frequencies to consider, and they correspond to a variety of types of motion. For example, the figure includes frequencies of combined rotor and nacelle motion; tower bending, both for and aft and laterally; blade bending, among others.

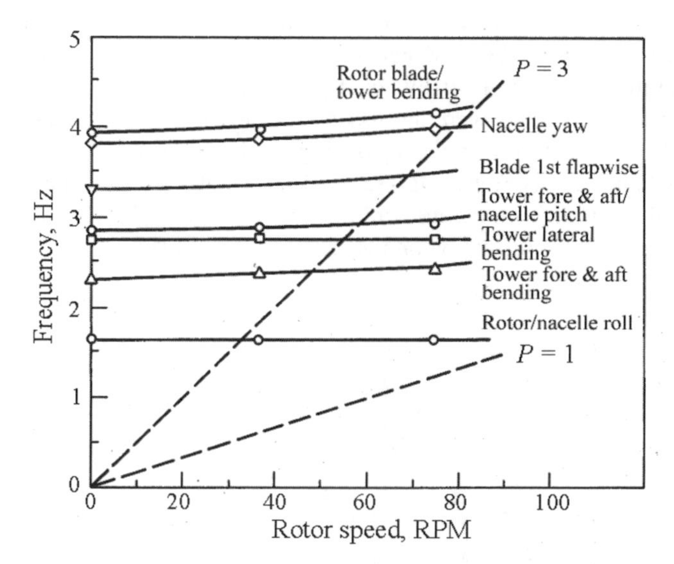

Figure 6.11 Example of Campbell diagram for a wind turbine; *P*, per revolution (Eggleston and Stoddard, 1987). Reproduced by permission of Kluwer Academic Publishers

Sometimes operation at or near a natural frequency cannot be completely avoided. This may occur during start up or shutdown, or at some rotor speeds of a variable speed wind turbine. Effects of operation under such conditions must be considered. Wind turbine design standards developed by Germanischer Lloyd (Germanischer Lloyd, 1993) offer some guidance in this area.

6.6.2 *Wind turbine design loads*

Many manufactured items are designed with reference to a particular 'design point'. This corresponds to an operating condition such that, if the item can meet that condition, it will perform at least adequately at any other realistic set of conditions.

A single design point is not adequate for wind turbine design. Rather, the wind turbine must be designed for a range of conditions. Some of these will correspond to normal operation, where most of the energy will be produced. Others are extreme or unusual conditions which the turbine must be able to withstand with no significant damage. The most important considerations are: (1) expected events during normal operation, (2) extreme events, and (3) fatigue.

The process of incorporating loads into the design process consists of the following:

- Determine a range of design wind conditions
- Specify design load cases of interest, including operating and extreme wind conditions
- Calculate the loads for the load cases
- Verify that the stresses due to the loads are acceptable

Enough experience has been gained with wind turbines over the last 20 years that it has been possible to define a set of design conditions under which a turbine should be able to perform. These have been codified by the International Electrotechnical Commission (IEC)). They are known as the IEC 1400-1 Safety Requirements (Bakker, 1996). The designer should be aware of these standards, since a turbine's ability to meet these conditions must be demonstrated if it is intended for use in any country which enforces those standards.

The following sections provide a summary of the IEC 1400-1 design standards. It should be noted that a complete assessment of a turbine's ability to meet these requirements is not possible until a full design has been completed and analyzed. Knowledge of those standards, however, provides a target for the design, so they should be considered early in the design process.

6.6.2.1 IEC design wind conditions

The IEC defines four classes of conditions, I–IV, ranging from the most windy (10 m/s average) to least windy (6 m/s annual average), under which a wind turbine might reasonably be expected to operate. Within those classes, two ranges of turbulence are defined, 'higher turbulence' and 'lower turbulence'. These classes are summarized in Table 6.1. The important thing to note is that each class is characterized by a reference speed and an annual average speed. Other conditions of interest are referenced to the basic characterisations (i.e. the average or reference wind). To cover special cases a fifth class, S, is also provided where the specific parameters are specified by the manufacturer.

Table 6.1 IEC wind classes

Classes	I	II	III	IV
Reference wind speed, U_{ref} (m/s)	50	42.5	37.5	30
Annual average wind speed U_{ave} (m/s)	10	8.5	7.5	6

Higher and lower turbulence for all classes are characterized by turbulence intensities at 15 m/s of 0.18 and 0.16, respectively. The turbulence classes are further defined by a parameter a which is used with the turbulence intensity to specify the standard deviation of the wind speed. For higher turbulence $a = 2$, and lower turbulence $a = 3$. The use of a is described below, under the heading 'Normal Turbulence Model'.

Normal wind conditions Under normal wind conditions the frequency of occurrences of wind speeds are assumed to be described by the Rayleigh distribution (see Section 2.4.3 in Chapter 2).

Normal wind profile (NWP) The wind profile, $U(z)$, is the variation of wind speed with the height z above ground. For the purposes of the IEC requirements the variation of wind speed with height is assumed to follow a power law model (see Section 2.3.3 in Chapter 2), with an exponent of 0.2. It is then known as the normal wind profile (NWP).

Normal turbulence model (NTM) The standard deviation of the turbulence in the direction of the mean wind, σ_x, is assumed to be given by:

$$\sigma_x = I_{15}(15 + aU_{hub})/(a+1) \tag{6.6.1}$$

where I_{15} = turbulence intensity at 15 m/s, a = turbulence parameter, and U_{hub} = wind speed at hub height.

The power spectral density of the turbulence can be modeled with the von Karman spectrum (see Section 2.3.2 in Chapter 2) or Kaimal spectrum (see Fordham, 1985), among others.

Extreme wind conditions There are five extreme wind conditions to be used in determining extreme loads under the IEC standards: (1) extreme wind speed (EWM), (2) extreme operating gust (EOG), (3) extreme coherent gust (ECG), (4) extreme coherent gust with change in direction (ECD), and (5) extreme wind shear (EWS).

Extreme wind speed (EWM) Extreme wind speeds are very high, sustained winds which will probably occur, but only rarely. Two extreme wind speeds are defined by the frequency with which they are expected to recur, the 50-year extreme wind (U_{e50}) and the 1-year extreme wind (U_{e1}). They are based on the reference wind (See Table 6.1). The 50-year wind is approximately 40% higher than the reference wind, while the 1-year wind is 30% higher than the reference wind.

Extreme operating gust (EOG) An extreme operating gust is a sharp increase and then decrease in wind speed which occurs over a short period of time, while the turbine is operating. The magnitude of the 50-year extreme operating gust (U_{gust50}) is assumed to be 6.4 times the standard deviation. The gust is also assumed to rise and fall over a period of 14 seconds. An illustration of an extreme operating gust is shown in Figure 6.12.

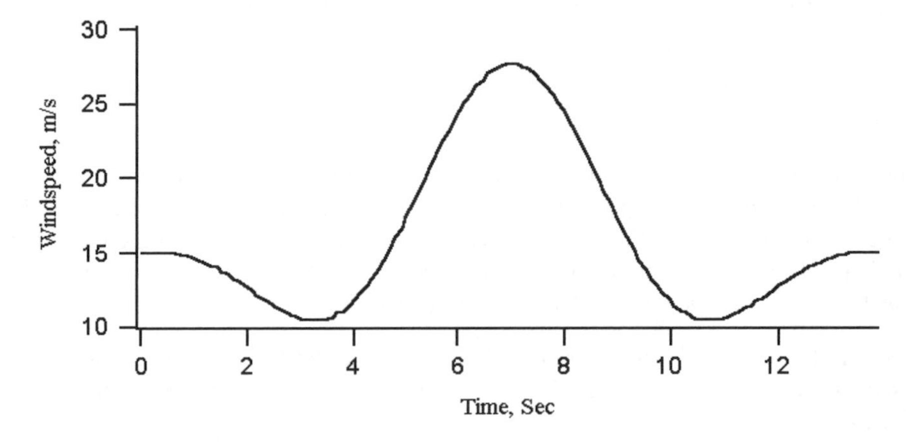

Figure 6.12 Sample extreme operating gust

Extreme direction change (EDC) Extreme direction changes are defined in an analogous manner to extreme gusts. In a typical example, the wind direction may change by 64 degrees over 6 seconds.

Extreme coherent gust (ECG) A coherent gust is a rapid rise in wind speed across the rotor. The IEC extreme coherent gust is assumed to have an amplitude of 15 m/s and to be superimposed on the mean wind. The wind rises sinusoidally to a new level over a period of 10 s.

Extreme coherent gust with change in direction (ECD) In the extreme coherent gust with change in direction a rise in wind speed is assumed to occur simultaneously with a direction change. Details are provided in the IEC standard.

Extreme wind shear (EWS) Two transient wind shear conditions are also defined, one for horizontal shear, and the other for vertical wind shear. Transient wind shears will be much larger than the normal conditions described above.

Rotationally sampled turbulence Rotationally sampled wind speed is normally synthesized by first applying an inverse Fourier transform to the power spectral density, via the Shinozuka technique (Shinozuka and Jan, 1972) to generate a stochastic time series. Then a model of the cross-spectral density is used to estimate the wind that the blade would experience as it rotates through the turbulence. This process is described in Veers (1984). A somewhat simpler approach is given in Stiesdal (1990).

In many situations, however, a simple deterministic model may be used for simulating the rotationally sampled turbulent input. In particular this method can be used where the wind turbines are relatively stiff, such that vibrations are unlikely to be excited by the turbulence. The IEC standard provides the details of this method.

A sample of data using the model is shown in Figure 6.13. For this case a mean of 15 m/s, a turbulence intensity of 0.18, a turbulence length scale of 10 m, and a diameter of 25 m were used. The rotational speed is 0.25 rotations per second.

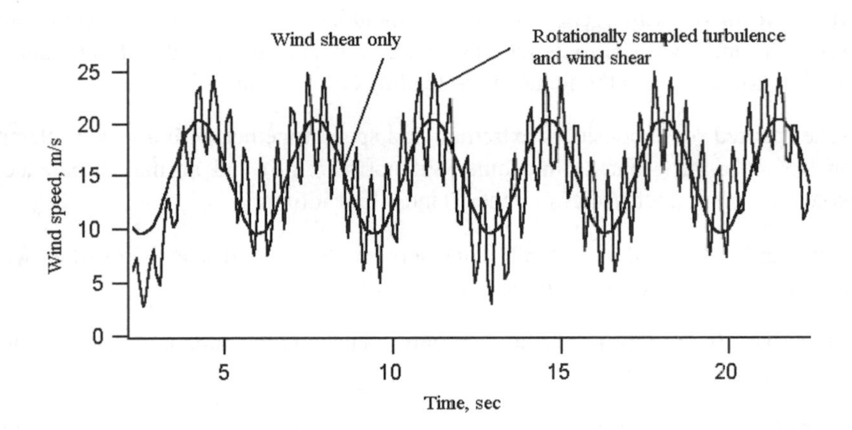

Figure 6.13 Sample deterministic turbulence

6.6.2.2 IEC load cases

The next step is defining the load cases. The load cases are based on the turbine's various operating states, as they are affected by the wind conditions and possible electrical or control system faults. Load cases are defined for eight situations:

1. Power production
2. Power production plus fault
3. Start up
4. Normal shut down
5. Emergency shut down
6. Parked
7. Parked plus fault
8. Transport, assembly, maintenance and repair

Many of the situations have more than one load case. Most of the cases deal primarily with ultimate loads, but also include one fatigue load case.

Power production 'Power production' has nine load cases which cover the full range of design wind conditions, as well as two external electrical faults.

Power production plus fault This has three load cases, which assume normal wind conditions, but include either an internal electrical fault or a control/protection system fault.

Start up 'Start up' load cases include a 1-year extreme operating gust and a 1-year extreme wind direction change for ultimate loading, as well as normal wind conditions (resulting in multiple starts) for fatigue.

Shut down 'Shut down' includes a 1-year extreme operating gust for ultimate loading and normal wind conditions (resulting in multiple stops) for fatigue.

Emergency shut down 'Emergency shut down' includes one case in which normal winds are assumed. Presumably more extreme wind conditions are not considered here since the emergency shut down event is the focus of the loading being evaluated.

Parked The 'parked' case considers extreme wind speed together with a loss of electrical connection (to make sure that the machine will not start up) and normal turbulence for fatigue. Note that 'parked' can refer to either standstill or idling

Parked plus fault 'Parked plus fault' considers extreme wind speed, together with a possible fault (other than loss of electrical connection.)

Transport, assembly, maintenance and repair The eighth category cases are to be specified by the manufacturer.

6.6.2.3 Application of design load cases

The design load cases are used to guide the analysis of critical components to ensure that they are adequate.

For the ultimate loading calculations, four types of analyses are used:

1. Maximum strength
2. Fatigue failure
3. Stability analysis (e.g. buckling)
4. Deflection (e.g. preventing blades from striking tower)

Fundamentally, the analyses first involve calculation of the expected loads for the various operating wind conditions. From the loads and the dimensions of the components, the maximum stresses (or deflections) are then found. Those stresses (or deflections) are then compared with the design stresses (or allowed deflections) of the material from which the component is constructed to make sure that they are low enough.

Calculation of the loads can be a very complex process. The principles that affect the loads are discussed in Chapter 4. Precise predictions of loads involve the use of detailed computer simulations, but such analysis is best applied once a preliminary design has been completed. The simplified methods of Chapter 4 can be used to predict trends, but they are too general to allow accurate load estimates. They can, however, be used in the early stages of design for rough sizing of the components. The estimates from simple methods can also be improved if there are some data available from similar machines with which to 'calibrate' the predictions. When such data are available, scaling methods, discussed below in Section 6.6.3, can help to facilitate the calibrations.

6.6.2.4 Method of partial safety factors

There are usually uncertainties in both the load estimates and the actual characteristics of the material. For this reason, the method of partial safety factors is used in specifying materials and in sizing the various components. The method consists of two parts:

1. Determining design properties for materials by derating their characteristic (or published) properties
2. Selecting safety factors, which in effect increase estimates of the loads

The general requirement for ultimate loading is that the expected 'load function', $S(F_d)$, multiplied by a 'consequence of failure' safety factor, γ_n, must be equal to or less than the 'resistance function', $R(f\ )$. In the most basic case, the load function is the highest value of the expected stress, and the resistance function is maximum allowable design value. The requirement can be expressed as:

$$\gamma_n S(F\ ) \le R(f\ ) \tag{6.6.2}$$

where F_d = design values for loads and f_d = design values for materials

The design values for loads are found from the expected or 'characteristic' values of the loads, F_k, by applying a partial safety factor for loads, γ_f:

$$F_d = \gamma_f F_k \tag{6.6.3}$$

The design values for the materials are found from the characteristic values of the materials, f_k, by applying a partial safety factor for materials, γ_m:

$$f_d = (1/\gamma_m) f_k \tag{6.6.4}$$

Partial safety factors are typically greater than 1.0. Normally, partial safety factors for loads range from 1.0 to 1.5. Partial safety factors for materials are at least 1.1, and partial safety factors for consequence of failure are equal to at least 1.0. More discussion can be found in Bakker (1996). Partial safety factors for materials can also be found in many sources.

6.6.3 Scaling relations

Sometimes design information is available about one turbine, and it is desired to design another turbine which is similar, but of a different size. In this case, one can take advantage of some scaling relations for the rotor in laying out the preliminary design. These scaling relations start with the following assumptions:

- The tip speed ratio remains constant
- The number of blades, airfoil, and blade material are the same
- Geometric similarity is maintained to the extent possible

The scaling relations for a number of important turbine characteristics are described below; first when the radius is doubled, and then for the general case. They are also summarized in Table 6.2.

Table 6.2 Summary of scaling relations

Quantity	Symbol	Relation	Scale dependence
Power, forces and moments			
Power	P	$P_1/P_2 = (R_1/R_2)^2$	$\sim R^2$
Torque	Q	$Q_1/Q_2 = (R_1/R_2)^3$	$\sim R^3$
Thrust	T	$T_1/T_2 = (R_1/R_2)^2$	$\sim R^2$
Rotational speed	Ω	$\Omega_1/\Omega_2 = (R_1/R_2)^{-1}$	$\sim R^{-1}$
Weight	W	$W_1/W_2 = (R_1/R_2)^3$	$\sim R^3$
Aerodynamic Moments	M_A	$M_{A,1}/M_{A,2} = (R_1/R_2)^3$	$\sim R^3$
Centrifugal Forces	F_c	$F_{c,1}/F_{c,2} = (R_1/R_2)^2$	$\sim R^2$
Stresses			
Gravitational	σ_g	$\sigma_{g,1}/\sigma_{g,2} = (R_1/R_2)^1$	$\sim R^1$
Aerodynamic	σ_A	$\sigma_{A,1}/\sigma_{A,2} = (R_1/R_2)^0 = 1$	$\sim R^0$
Centrifugal	σ_c	$\sigma_{c,1}/\sigma_{c,2} = (R_1/R_2)^0 = 1$	$\sim R^0$
Resonances			
Natural frequency	ω	$\omega_{n,1}/\omega_{n,2} = (R_1/R_2)^{-1}$	$\sim R^{-1}$
Excitation	Ω/ω	$(\Omega_1/\omega_{n,1})/(\Omega_2/\omega_{n,2}) = (R_1/R_2)^0 =$	$\sim R^0$

Note: R, radius

6.6.3.1 Power

Power, as discussed previously, is proportional to the swept area of the rotor, so doubling the radius will quadruple the power. In general, power is proportional to the square of the radius.

6.6.3.2 Rotor speed

With the tip speed ratio held constant the rotor speed will be halved when the radius is doubled. In general, rotor speed will be inversely proportional to the radius.

6.6.3.3 Torque

As noted above, when the radius is doubled, the power is quadrupled. Since the rotor speed will drop by half, the torque will be increased by a factor of 8. In general, the rotor torque will be proportional to cube of the radius.

6.6.3.4 Aerodynamic moments

The forces in the blades go up as the square of the radius, and the moments are given by the forces times distance along the blade. When the radius is doubled the aerodynamic moments will increase by a factor of 8. In general, aerodynamic moments will be proportional to cube of the radius.

6.6.3.5 Rotor weight

By the assumption of geometric similarity, as the turbine size gets larger, all dimensions will increase. Therefore, if the radius doubles, the volume of each blade goes up by a factor of 8. Since the material remains the same, the weight must also increase by a factor of 8. In general, rotor weight will be proportional to cube of the radius. Note that the fact that the weight goes up as the cube of the dimension whereas the power output goes up as the square gives rise to the famous 'square–cube law' of wind turbine design. It is this 'law' which may eventually limit the ultimate size that turbines may reach.

6.6.3.6 Maximum stresses

Maximum bending stresses, σ_b, in the blade root due to flapwise moments applied to the blade, M, are related to the thickness of the root, t, and its area moment of inertia, I, by $\sigma_b = M(t/2)/I$, as should be clear from discussions in Chapter 4 (Section 4.2.1.3). For simplicity, consider the blade root to be approximated by a rectangular cross-section of width c (corresponding to the chord) and thickness t. The moment of inertia about the flapping axis is $I = ct^3/12$. If the radius is doubled, then the moment of inertia goes up by a factor 16, and the thickness by a factor of 2. The ratio, $2I/t$, which is given by $2I/t = ct^2/6$, is then increased by a factor of 8, just like the aerodynamic moments. In general, the blade root area moment of inertia scales as R^3.

Maximum stresses due to aerodynamic moments, blade weight and centrifugal force are a function of the area moment of inertia and the applied moments. They are discussed in more detail below.

Stresses due to aerodynamic moments Aerodynamically induced stresses, σ_A, are unchanged with scaling. This is true for both the flapwise and lead–lag directions, as should be apparent from the discussion above. The proof of this for flapwise bending is the subject of one of the problems for this chapter.

Stresses due to blade weight Stresses due to blade weight, unlike most other stresses in the rotor, are not independent of size. In fact, they increase in proportion to the radius. Allowance for that difference must be made during the design process.

Consider a horizontal blade of weight, W, and center of gravity distance, r_{cg}, from the hub. The maximum moment due to gravity, M_g, is:

$$M_g = W r_{cg} \tag{6.6.5}$$

The maximum stress due to gravity, σ_g, in the edgewise direction for a rectangular blade root (here with $I = tc^3/12)$) is therefore:

$$\sigma_g = (W\, r_{cg})(c/2)/I = W\, r_{cg}/(tc^2/6) \tag{6.6.6}$$

Since weight scales as R^3 and the other dimensions scale as R, the stress due to weight also scales as R. The general relation is then:

$$\sigma_{g,1}/\sigma_{g,2} = (R_1/R_2)^1 \tag{6.6.7}$$

Stresses due to centrifugal force Stresses due to centrifugal force are unchanged by scaling. This can be illustrated as follows. The tensile stress, σ_c, due to centrifugal force, F_c, applied across area A_c is given by:

$$\sigma_c = F_c/A_c \tag{6.6.8}$$

Centrifugal force itself is found from:

$$F_c = \frac{W}{g} r_{cg} \Omega^2 \tag{6.6.9}$$

where Ω is the rotor rotational speed. Blade weight scales as R^3, r_{cg} scales as R and Ω scales as R^{-1}. Thus $F_c \sim R^2$. It is also the case that $A_c \sim R^2$, so σ_c is independent of R. In general

$$\sigma_{c,1}/\sigma_{c,2} = (R_1/R_2)^0 = 1 \tag{6.6.10}$$

6.6.3.7 Blade natural frequencies

Blade natural frequencies decrease in proportion to the radius. This can be seen by modeling a blade as a rectangular cantilevered beam of dimension c wide, t thick and R long. As shown in Chapter 4, the natural frequencies of a cantilevered beam are given by:

$$\omega_{n,i} = \frac{(\beta R)_i^2}{R^2}\sqrt{\frac{EI}{\tilde{\rho}}} \tag{6.6.11}$$

where E is the modulus of elasticity, I is the area moment of inertia, $\tilde{\rho}$ is the mass per unit length, and $(\beta R)_i^2$ is series of constants such that $(\beta R)_i^2$ = (3.52, 22.4, 61.7,...).

For the example, $I = ct^3/12$ and $\tilde{\rho} = \rho_b ct$ (where ρ_b = mass density of blade). In this case

$$\omega_n = \frac{(\beta R)_i^2}{R^2}\sqrt{\frac{Ect^3}{12\rho_b ct}} = \frac{(\beta R)_i^2}{R^2}t\sqrt{\frac{E}{12\rho_b}} \tag{6.6.12}$$

Blade thickness is proportional to the radius. Therefore, it is apparent that $\omega_n \sim R^{-1}$. In general, the relation of natural frequencies between two blades (1 and 2) is:

$$\omega_{n,1}/\omega_{n,2} = (R_1/R_2)^{-1} \tag{6.6.13}$$

Since rotor rotational speed also decreases with radius, the propensity of the rotor to excite a particular resonance condition is independent of radius.

It should be emphasized that scaling relations are only useful guidelines, and cannot be used to make exact predictions. Other factors, such as technology development, can also alter the implications. For example, recent developments of larger machines indicate an increase of mass at a rate of somewhat less than the 'square–cube law' (power and mass vs. radius) predicts. More discussion on this topic is provided in Jamieson (1997).

6.7 Wind Turbine Subsystems and Components

The principal component groups in a wind turbine are the rotor, the drive train, the main frame, the yaw system and the tower. The rotor includes the blades, hub and aerodynamic control surfaces. The drive train includes the gearbox (if any), the generator, mechanical brake and shafts and couplings connecting them. The yaw system components depend on whether the turbine uses free yaw or driven yaw. The type of yaw system is usually determined by the orientation of the rotor (upwind or downwind of the tower.) Yaw system components include at least a yaw bearing and may include a yaw drive (gear motor and yaw bull gear), yaw brake, and yaw damper. The main frame provides support for mounting the other components and a means for protecting them from the elements (the nacelle cover). The tower group includes the tower itself, its foundation, and may include the means for self-erection of the machine.

The following sections discuss each of the component groups. Unless specifically noted, it is assumed that the turbine has a horizontal axis.

6.7.1 Rotor

The rotor is unique among the component groups. Other types of machinery have drive trains, brakes, and towers, but only wind turbines have rotors designed for the purpose of extracting significant power from the wind and converting it to rotary motion. As discussed elsewhere, wind turbine rotors are also nearly unique in that they must operate under conditions that include steady as well as periodically and stochastically varying loads. These varying loads occur over a very large number of cycles, so fatigue is a major consideration. The designer must strive to keep the cyclic stresses as low as possible, and to use material that can withstand those stresses as long as possible. The rotor is also a generator of cyclic loadings for the rest of the turbine, in particular the drive train.

The next three sections focus on the topics of primary interest in the rotor: (1) blades, (2) aerodynamic control surfaces, and (3) hub.

6.7.1.1 Blades

The most fundamental components of the rotor are the blades. They are the devices that convert the force of the wind into the torque needed to generate useful power. There are many things to consider in designing blades, but most of them fall into one of two categories: (1) aerodynamics and (2) structure. Underlying all of these, of course, is the need to minimize life cycle cost of energy, which means that the cost of the turbine itself should be kept low, but that the operation and maintenance costs should be kept low as well.

The basic shape and dimensions of the blades are determined primarily by the overall layout of the turbine (as discussed in Section 6.3) and aerodynamic considerations, which were discussed in Chapter 3. Details in the shape, particularly near the root, are also influenced by structural considerations. For example, the planform of most real wind turbines differs significantly from the optimum shape, because the expense of blade manufacture would otherwise be too high. Figure 6.14 illustrates some typical planform options. Material characteristics and available methods of fabrication are also particularly important in deciding upon the exact shape of the blades.

Aerodynamic design The primary aerodynamic factors affecting the blade design are:

- Design rated power and rated wind speed
- Design tip speed ratio
- Solidity
- Airfoil
- Number of blades
- Rotor power control (stall or variable pitch)
- Rotor orientation (upwind or downwind of the tower)

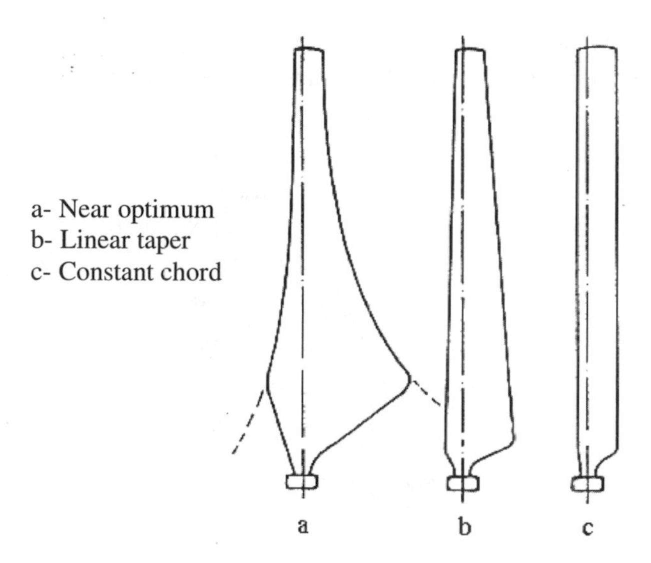

Figure 6.14 Blade planform options (Gasch, 1996). Reproduced by permission of B. G. Teubner GmbH

The overall size of the rotor swept area, and hence the length of the blades, is directly related to design rated power and rated wind speed. Other things being equal it is usually advantageous to have a high design tip speed ratio. A high tip speed ratio results in a low solidity, which in turn results in less total blade area. This in turn should result in lighter, less expensive blades. The accompanying higher rotational speed is also of benefit in the rest of the drive train. On the other hand, high tip speed ratios result in more aerodynamic noise from the turbine. Because the blades are thinner, the flapwise stresses tend to be higher. Thinner blades are also more flexible. This can sometimes be an advantage, but thinner blades may also experience vibration problems, and extreme deflections can result in blade–tower impacts. The tip speed ratio also has a direct effect on the chord and twist distribution of the blade.

As design tip speed ratios increase, selection of the proper airfoil becomes progressively more important. In particular it is necessary to keep the lift-to-drag ratio high if the rotor is to have a high power coefficient. It is also of note that the lift coefficient will have an effect on the rotor solidity and hence the blade's chord: the higher the lift coefficient, the smaller the chord. In addition, the choice of airfoil is to a significant extent affected by the method of aerodynamic control used on the rotor. For example, an airfoil suitable for a pitch regulated rotor may not be appropriate for a stall-controlled turbine. One concern is fouling: certain airfoils, particularly on stall-regulated turbines, are quite susceptible to fouling (due, for example, to a build up of insects on the leading edge). This can result in a substantial decrease in power production. Selection of an airfoil can be done with the help of data bases such as those developed by Selig (1998.)

Wind turbine blades frequently do not have just one airfoil shape along the entire length. See, for example, Figure 6.15. More commonly (but not always), the airfoils are all of the same family, but the relative thickness varies. Thicker airfoils near the root provide greater strength, and can do so without seriously degrading the overall performance of the blade.

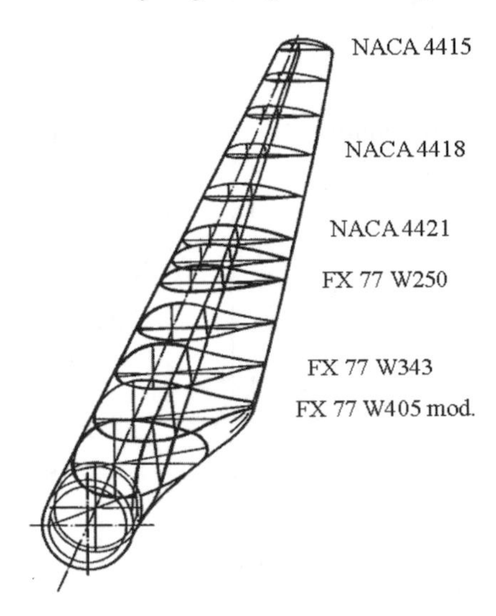

Figure 6.15 Airfoil cross-sections with radius (from Gasch, 1996). Reproduced by permission of B. G. Teubner GmbH

With present manufacturing techniques it is generally advantageous to have as few blades as possible. This is primarily because of the fixed costs in fabricating the blades. In addition, when there are more blades (for a given solidity) they will be less stiff and may have higher stresses at the roots. At the present time all commercial wind turbines have either two or three blades, and that will be assumed to be the case here as well. Two-bladed wind turbines have historically had a lower solidity than three-bladed machines. This keeps the blade cost low, which is one of the presumed advantages of two blades over three blades.

The method of power control (stall or variable pitch) has a significant effect on the design of the blades, particularly in regard to the choice of the airfoil. A stall-controlled turbine depends on the loss of lift which accompanies stall to reduce the power output in high winds. It is highly desirable that the blades have good stall characteristics. They should stall gradually as the wind speed increases, and they should be relatively free of transient effects, such as are caused by dynamic stall. In pitch-controlled turbines, stall characteristics are generally much less important. However, it is important to know that the blades perform acceptably when being pitched in high winds. It is also worth noting that blades can be pitched towards either feather (decreasing angle of attack) or stall (increasing angle of attack).

The rotor orientation with respect to the tower has some effect on the geometry of the blades, but mostly in a secondary manner related to the preconing of the blades. This preconing is a tilting of the blades away from a plane of rotation as defined by the blade roots. As previously noted, most downwind turbines operate with free yaw. The blades must be coned away from the plane of rotation to enable the rotors to track the wind and maintain some yaw stability. Some upwind rotors also have preconed blades. In this case, the purpose is to keep the blades from hitting the tower.

Blade design often involves a number of iterations to properly account for both aerodynamic and structural requirements. In each iteration a tentative design is developed and then analyzed. One approach to expedite this process, known as an inverse design method, has been developed by Selig and Tangler (1995). It involves the use of a computer code (PROPID) to propose designs which will meet certain requirements. For example, as mentioned in Chapter 3, it is possible to specify overall dimensions, an airfoil series, peak power and blade lift coefficient along the span, and then use the code to determine the chord and twist distribution of the blade.

Structural design In addition to the loads which a wind turbine blade must withstand, the primary considerations in its structural design are (1) materials and (2) fabrication options. An additional important concern is the attachment of the blades to the hub.

Historically, wind turbine blades were made from wood, sometimes covered with cloth. Until the middle of this century blades for larger wind turbines were made from steel. Examples include both the Smith–Putnam 1250 kW turbine (1940s) and the Gedser 200 kW turbine (1950s).

Since the 1970s, most blades for horizontal axis wind turbines have been made from composites. The most common composites consist of fiberglass in a polyester resin, but wood–epoxy laminates have also been widely used as well. Typical composites used for wind turbine blades were described in more detail in Section 6.5.

Some wind turbines have used aluminum for blade construction. Aluminum has been a popular choice for vertical axis wind turbines. Their blades normally have a constant chord with no twist, so lend themselves to formation by aluminum pultrusion. Pultrusion is a process whereby material (such as aluminum) is pulled through a forming die to create the desired shape. The shape is uniform with length. A few horizontal axis wind turbines have used aluminum blades, but aluminum is not commonly used for HAWTs at this time.

Blade fabrication details The basic concept in wind turbine blade fabrication is to make a strong, light structure whose exterior shape corresponds to the aerodynamic design. Desired shapes for horizontal axis wind turbine blades are decidedly non-linear. The cross-section at any point has an airfoil shape, so the perimeter includes varying amounts of curvature. In addition, the blade is usually tapered and twisted. In order to make such a shape and have the required strength, the usual method is to make the blade in two types of parts: a skin and a spar. The skin provides the desired airfoil shape and the spar supplies the stiffness. Figure 6.16 illustrates a cross-section of a typical fiberglass blade.

The first step in the fabricating of a blade is normally to build a spar. Spars may take on a variety of forms, but the purpose is to create a lightweight member which can resist the applied moments. The shape of the spar may be that of a web, a box beam or a D. In the case of a box beam or web, its outer dimension in the flapwise direction will be such that it can be bonded to the inside of the skin on both the top and bottom of the blade. With a D spar, the blade skin is bonded to the front of the spar as well. Spars in fiberglass blades are usually made by building up layers of fiberglass and resin around a mandrel, which is then later removed.

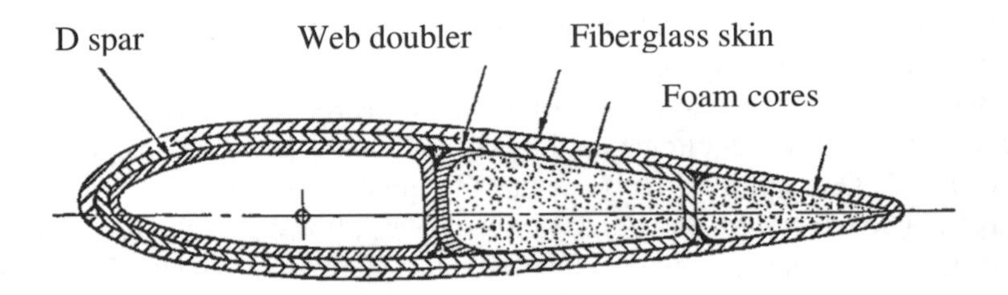

Figure 6.16 Typical fiberglass blade cross-section (Peery and Weingart, 1980). Reproduced by permission of American Institute of Aeronautics and Astronautics

The skin of a GRP blade is made by building up layers of fiberglass cloth and resin inside a mold. In this method there are two parts to mold, one for the upper surface and one for the lower surface. When the two halves of the blade are completed, they are removed from the molds. They are then bonded together, with the spar in between. An example of part of the process is shown in Figure 6.17.

Figure 6.17 Laying fibreglass cloth into blade molds. Reproduced by permission of LM Glasfibre

Fabrication of wood–epoxy blades follows a similar procedure. The main difference is that wood plys are used in the laminate rather than fiberglass cloth. In addition, the thickness of the skin relative to the blade thickness is usually greater than in a GRP blade, and rather than a box beam spar, a plywood strip is used to provide stiffness. Figure 6.18 illustrates the cross-section of a typical wood–epoxy blade.

Note that when using molds of the type described here, any plausible surface can be produced. This includes concave surfaces which are commonly found on some of the newer airfoils, such as the SERI series (which are illustrated in Figure 6.19). The disadvantage of building blades in this way is that the lay-up involves a significant amount of hand labor. This results in high costs, and also makes it difficult to ensure consistency from one blade to another.

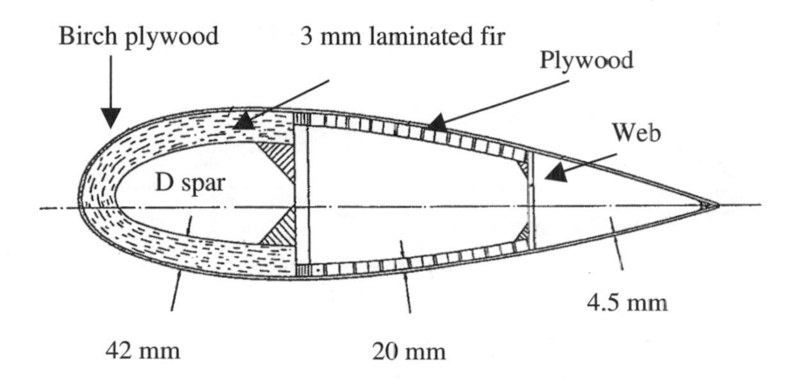

Figure 6.18 Cross-section of wood–epoxy blade (adapted from Hau, 1996). Reproduced by permission of Springer Verlag GmbH

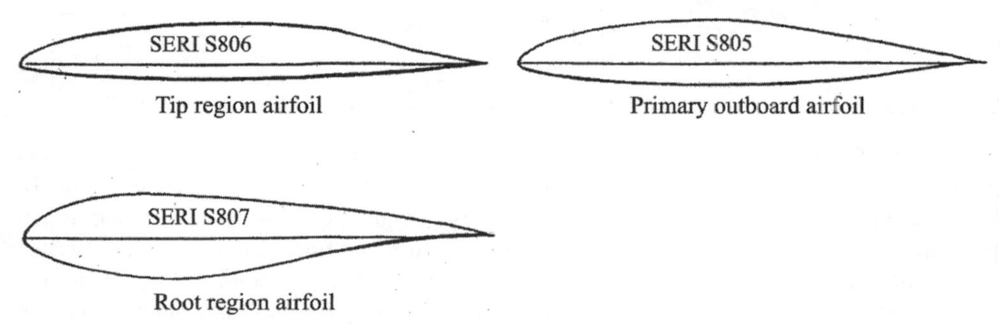

Figure 6.19 Solar Energy Research Institute (SERI) airfoils, thin airfoil family (National Research Council, 1991)

Another method for fabricating blades is known as 'filament winding'. This is a technique for making fiberglass blades, but the process is quite different than that of the mold method described above. In the filament winding method, glass fibers are wound about a mandrel, while resin is applied simultaneously. This method, developed originally in the aerospace industry, can be automated. It is difficult to use with concave shapes, however.

A critical part of the blade is the root, which is the end nearest the hub. The root experiences the highest loads, and is also the location that must provide for the connection to the hub. In order to reduce stresses, the root is generally made as thick as is practical in the flapwise direction. Connection between the root and the hub has often proven to be difficult. This is largely due to dissimilarities in material properties and stiffnesses between the blades, the hub and the fasteners. Highly variable loads also contribute to the problem.

One type of root is known as the Hütter design, named after its inventor, the German wind energy pioneer Ulrich Hütter. In this method long fiberglass strands are bonded into the lower part of the blade. Circular metal flanges are provided at the base of the blade, and attached to these flanges are circular hollow spacers. The fiberglass strands are wrapped around the spacers and brought back into the rest of the blade. Resin keeps all the strands and the flanges in place. The blades are eventually attached to the hub via bolts through the flanges and spacers. As described here this root design is most applicable to fixed-pitch rotors. The method can be modified, however, for variable pitch rotors as well. This root is illustrated in Figure 6.20.

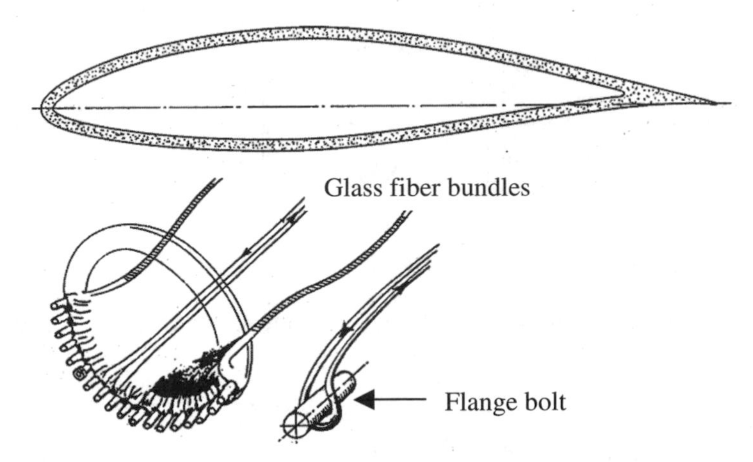

Figure 6.20 Hütter Root (Hau, 1996). Reproduced by permission of Springer Verlag GmbH

Details of a variant of the Hütter root design, which was widely used in the 1970s and 1980s, are shown in Figure 6.21. In that figure, which illustrates part of a cross-section of the root, the lower surface of the base plate is closest to the hub. The base plate and a steel pressure ring form a 'sandwich', inside of which are glass fiber roving bundles (twisted strands of fibers). The roving bundles originate in the fiberglass of the rest of the blade, and wrap around steel bushings. Bolts pass through the pressure plate, bushing, and base plate to complete the connection to the hub.

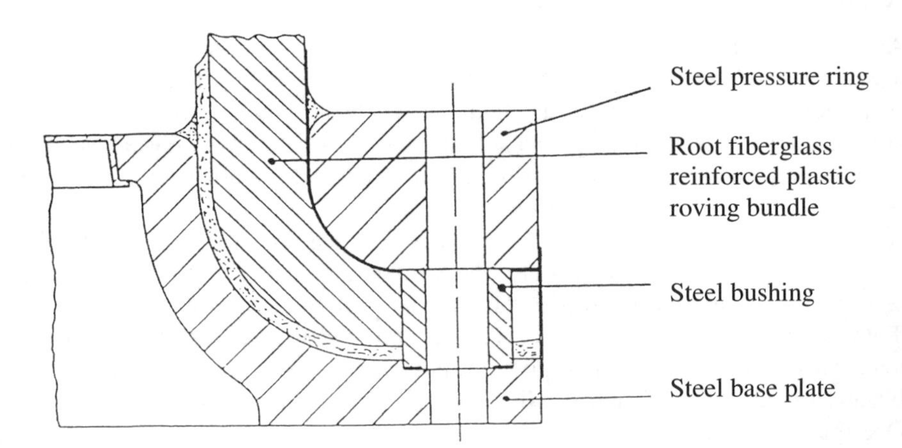

Figure 6.21 Modified Hütter root (National Research Council, 1991). Reproduced with permission from the National Academy of Science, courtesy of the National Academy Press, Washington, D.C.

The modified Hütter root has some limitations. The problem is that it is subject to fatigue. Cyclic stresses during operation have tended to loosen the matrix resin, allowing relative motion of the fiberglass. Movement of the glass then exacerbates the problem.

Voids in the matrix and other manufacturing details appear to be the ultimate source of the problem. Careful quality control reduces the frequency of occurrence.

Another method of attachment is the use of studs or threaded inserts bonded directly into the blades. This method, illustrated in Figure 6.22, was originally developed in conjunction with wood–epoxy blades, but it has proven applicable in GRP blades as well.

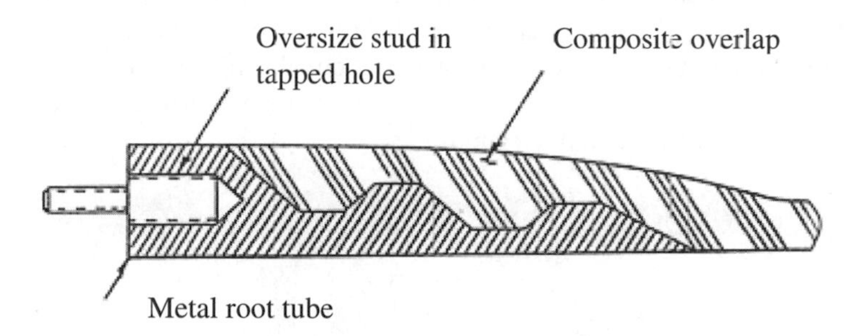

Figure 6.22 Blade root stud in fibreglass reinforced plastic (GRP) blade (National Research Council, 1991). Reproduced with permission from the National Academy of Science, courtesy of the National Academy Press, Washington, D.C.

Fixed-pitch wind turbine blades normally are fastened to the hub with bolts or studs which are aligned radially, and perpendicular to the bottom of the blade root. These fasteners must withstand all the loads arising from the blades.

The construction of a variable pitch blade root is rather different than that of a fixed-pitch blade. In particular, the root–hub connection must incorporate bearings so that the blade can be rotated. These bearings must be able to withstand the bending moments and shear forces imposed by the rest of the blade. In addition, these, or other, bearings must take the centrifugal load resulting from the rotor's rotation.

The blade attachment methods discussed above are most common on medium size or larger turbines. Blades on small turbines normally employ different attachment techniques. In one method the root is thickened, and bolts are placed through the root and a matching part on the hub. The bolts are perpendicular to both the long axis and chord of the blade.

Blade properties Properties of the overall blade, such as total weight, stiffness and mass distributions, and moments of inertia are needed in the structural analysis of the rotor. Important concerns are the blade's strength, its tendency to deflect under load, its natural vibration frequencies, and its resistance to fatigue. These were all discussed in Chapter 4. Some of the blade properties can be difficult to obtain due to the complex geometry of the blade, which varies from root to tip. The normal method used is to divide the blade into sections, in a manner similar to that for aerodynamic analysis. Properties for each section are found, based on the dimensions and material distribution, and then combined to find values for the entire blade.

6.7.1.2 Aerodynamic control surfaces

An aerodynamic control surface is a device which can be moved to change the aerodynamic characteristics of a rotor. There is a variety of types of aerodynamic control surfaces that can be incorporated in wind turbine blades. They must be designed in conjunction with the rest of the rotor, especially the blades. The selection of aerodynamic control surfaces is strongly related to the overall control philosophy. Stall-regulated wind turbines usually incorporate some type of aerodynamic brake. These can be tip brakes, flaps or spoilers. An example of a tip flap is illustrated in Figure 6.23.

Turbines which are not stall-controlled usually have much more extensive aerodynamic control. In conventional pitch-controlled turbines the entire blade can rotate about its long axis. Thus, the entire blade forms a control surface. Some turbine designs use partial span pitch control. In this case the inner part of the blade is fixed relative to the hub. The outer part is mounted on bearings, and can be rotated about the radial axis of the blade. The advantage of partial span pitch control is that the pitching mechanism need not be as massive as it must be for full span pitch control.

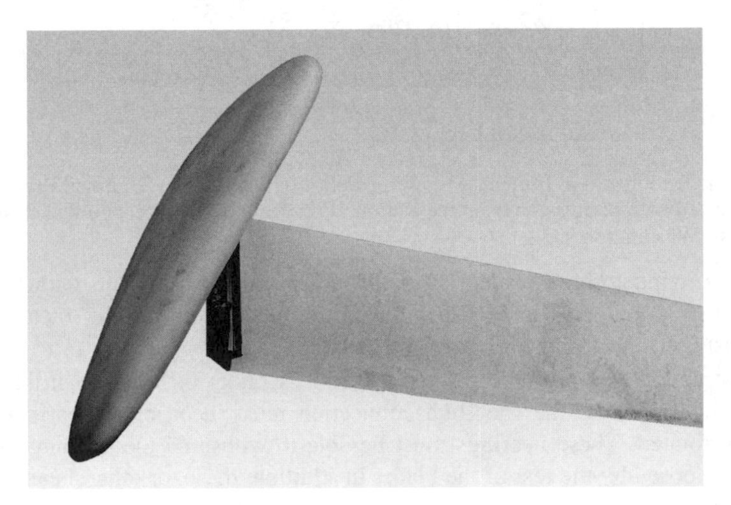

Figure 6.23 Example of a tip flap aerodynamic brake

Another type of aerodynamic control surface is the aileron. This is a movable flap, located at the trailing edge of the blade. The aileron may be approximately 1/3 as long as the entire blade, and extend approximately 1/4 of the way towards the leading edge.

Any control surface is used in conjunction with a mechanism that allows or causes it to move as required. This mechanism may include bearings, hinges, springs and linkages. Aerodynamic brakes often include electromagnets to hold the surface in place during normal operation, but to release the surface when required. Mechanisms for active pitch or aileron control include motors for operating them.

More details on wind turbine control are provided in Chapter 7.

6.7.1.3 Hub

Function The hub of the wind turbine is that component that connects the blades to the main shaft and ultimately to the rest of the drive train. The hub transmits and must withstand all the loads generated by the blades. Hubs are generally made of steel, either welded or cast. Details in hubs differ considerably depending on the overall design philosophy of the turbine.

Types There are three basic types of hub design that have been applied in modern horizontal axis wind turbines: (1) rigid hubs, (2) teetering hubs, and (3) hubs for hinged blades. Rigid hubs, as the name implies, have all major parts fixed relative to the main shaft. They are the most common design, and are nearly universal for machines with three (or more) blades. Teetering hubs allow relative motion between the part that connects to the blades and that which connects to the main shaft. Like a child's teeter-totter (seesaw), when one blade moves one way, the other blade moves the other way. Teetering hubs are commonly used for two- and one-bladed wind turbines. Hubs for hinged blades allow independent flapping motion, relative to the plane of rotation. Such hubs are used on only a few commercial machines but they have been employed on some historically important turbines (Smith–Putnam) and are presently receiving renewed attention. Some of the common types of hubs are illustrated in Figure 6.24.

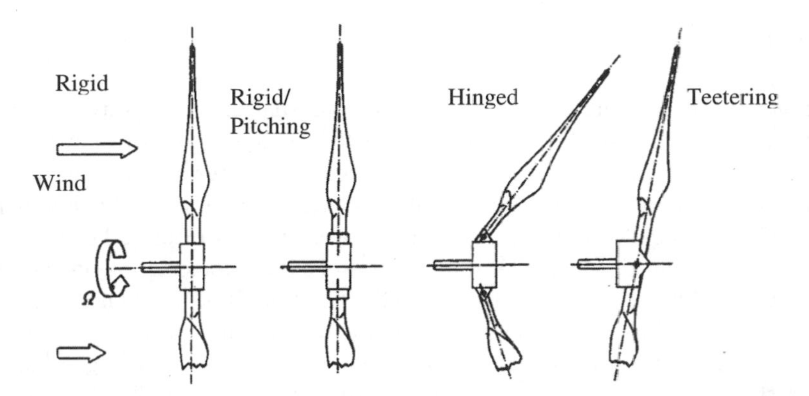

Figure 6.24 Hub options (Gasch, 1996). Reproduced by permission of B. G. Teubner GmbH

Rigid hub As indicated above, a rigid hub is designed to keep all major parts in a fixed position relative to the main shaft. The term rigid hub does, however, include those hubs in which the blade pitch can be varied, but in which no other blade motion is allowed.

The main body of a rigid hub is a casting or weldment to which the blades are attached, and which can be fastened to the main shaft. If the blades are to be preconed relative to the main shaft, provision for that is made in the hub geometry. A rigid hub must be strong enough to withstand all the loads that can arise from any aerodynamic loads on the blades, as well as dynamically induced loads, such as those due to rotation and yawing. These loads are discussed in Chapter 4 as well as in Section 6.6 of this chapter.

A hub on a pitch-controlled turbine must provide for bearings at the blade roots, a means for securing the blades against all motion except pitching, and a pitching mechanism. Pitching mechanisms may use a pitch rod passing through the main shaft, together with a linkage on the hub. This linkage is in turn connected to the roots of the blades. The pitch rod is driven by a motor mounted on the main (non-rotating) part of the turbine. An alternative method is to mount electric gear motors on the hub and have them pitch the blades directly. In this case, power still needs to be provided to the motors. This can be done via slip rings or a rotary transformer. Regardless of the design philosophy of the pitching mechanism, it should be fail-safe. In the event of a power outage, for example, the blades should pitch themselves into a no-power position. An example of a blade pitching mechanism is illustrated in Figure 6.25.

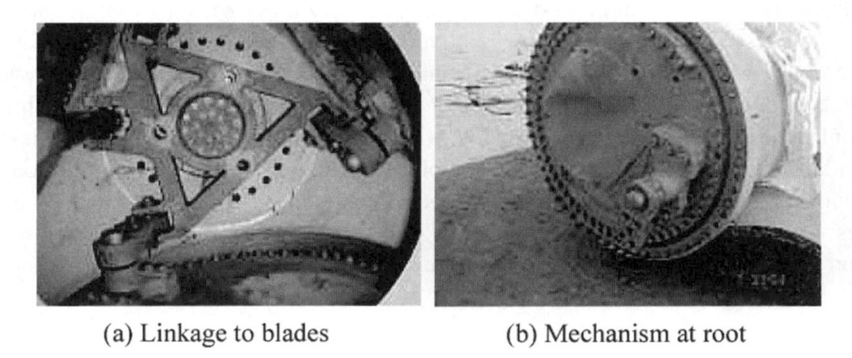

(a) Linkage to blades (b) Mechanism at root

Figure 6.25 Blade pitching mechanism. Reproduced by permission of Vestas Wind Systems A/S)

Hub attachment The hub must be attached to the main shaft in such a way that it will not slip or spin on the shaft. Smaller turbines frequently employ keys, with keyways on the shaft and the hub. The shaft is also threaded and the mating surfaces are machined (and perhaps tapered) for a tight fit. The hub can then be held on with a nut. Such a method of attachment is less desirable on a larger machine, however. First of all, a keyway weakens the shaft. Machining threads on a large shaft can also be inconvenient. One method used to attach hubs to wind turbine shafts is the Ringfeder® Shrink Disc®, which is illustrated in Figure 6.26. In the arrangement shown, a projection on the hub slides over the end of the main shaft. The diameter of the hole in the hub projection is just slightly larger than the end of the main shaft. The Shrink Disc® consists of a ring and two discs. The inner surface of the ring slides over the outside of the hub projection. The outside of the ring is tapered in both axial directions. The two discs are placed in either side of the taper, and then pulled together with bolts. As they approach each other, the ring is compressed and this in turn compresses the hub projection. The compression of the hub projection clamps it to the hub.

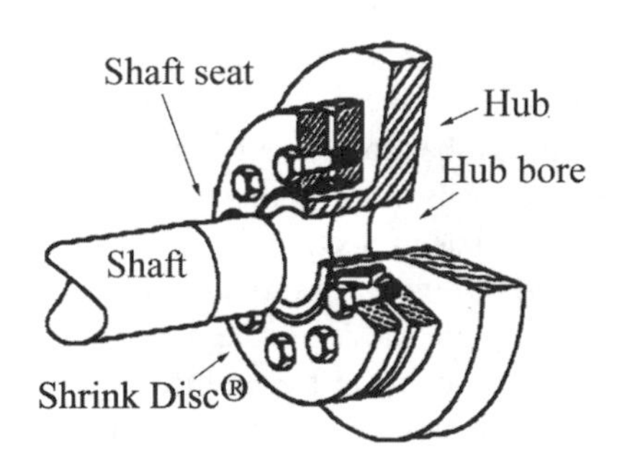

Figure 6.26 Ringfeder® hub attachment. Reproduced by permission of Ringfeder Corp.

Another method of hub attachment involves the use of a permanent flange on the end of the shaft. The flange may be either integral to the shaft or added later. The hub is attached to the flange by bolts.

Teetering Hub Teetering hubs are used on nearly all two-bladed wind turbines. This is because a teetering hub can reduce loads due to aerodynamic imbalances or loads due to dynamic effects from rotation of the rotor or yawing of the turbine. Teetering hubs are considerably more complex than are rigid hubs. They consist of at least two main parts (the main hub body and a trunnion pin), as well as bearings and dampers. A typical teetering hub is illustrated in Figure 6.27. The main hub body is a steel weldment. At either end are the attachment points for the blades. This hub has blades that are preconed downwind from the plane of rotation, so the planes of attachment are not perpendicular to the long axis of the hub. On either side of the hub body are teeter bearings. They are held in place by removable bearing blocks. The arrangement is such that the bearings lie on an axis perpendicular to the main shaft, and equidistant from the blade tips. The teeter bearings carry all of the loads passing between the hub body and the trunnion pin. The trunnion pin is connected rigidly to the main shaft.

In the hub shown in Figure 6.27 a line perpendicular to the axis of the pins is parallel to the long axis of the hub. In general, these lines need not be parallel. The angle between the two is known as the delta-3 angle (δ_3 , a term borrowed from the helicopter industry.) When the lines are parallel ($\delta_3 = 0$) all blade motion is in the flapping direction during teetering. When $\delta_3 \neq 0$ then there is a pitching component as well. There may be some benefit to having a non-zero delta-3 angle, but there is no consensus in the wind energy industry as to if and when it should be employed, and how big the angle should be. A hub with a non-zero delta-3 angle is illustrated in Figure 6.28.

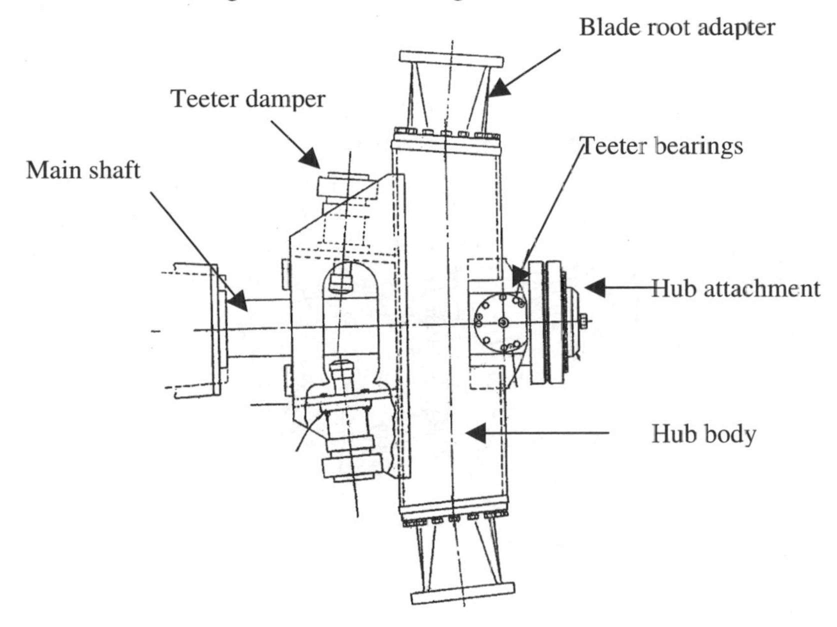

Figure 6.27 Teetering hub

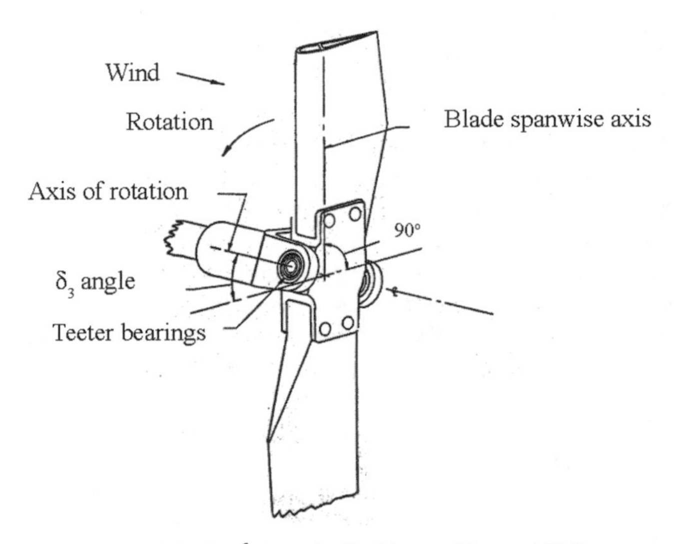

Figure 6.28 Hub with non-zero delta-3 (δ_3) angle (Perkins and Jones, 1981)

Most teetering hubs have been built for fixed-pitch turbines, but they can be used on variable pitch turbines as well. Design of the pitching system is more complex since the pitching mechanism is on the part of the hub which moves relative to the main shaft. A pitching teetering hub is illustrated in Figure 6.29.

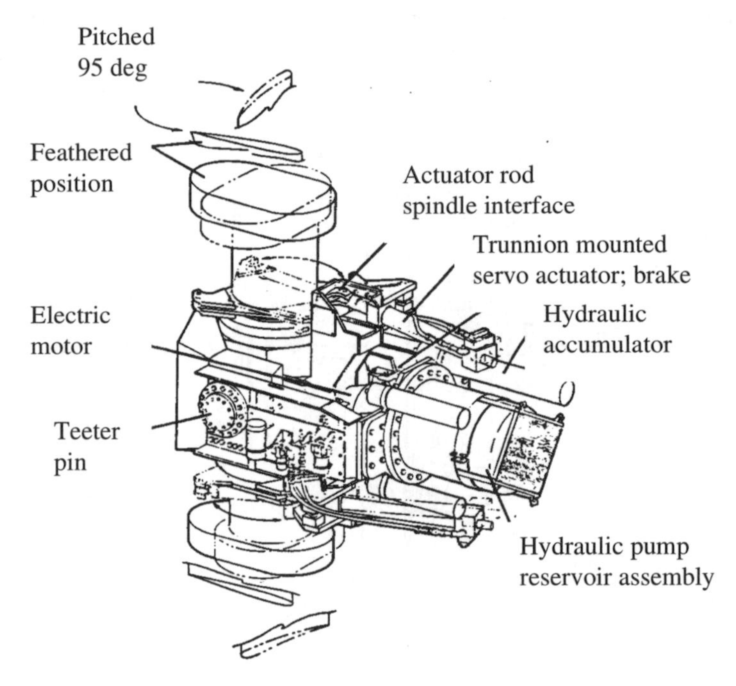

Figure 6.29 Pitching teetering hub (Van Bibber and Kelly, 1985)

Teetering hubs require two types of bearings. One type is a cylindrical, radially loaded bearing; the other is a thrust bearing. There is one bearing of each type on each pin. The cylindrical bearings carry the full load when the pin axis is horizontal. When the pin axis is not horizontal, there is an axial component due primarily to the weight of the rotor. One of the thrust bearings will carry that part of the load. Teeter bearings are typically made of special purpose composites.

During normal operation a teetering hub will move only a few degrees forwards and backwards. During high winds, starts and stops, or high yaw rates, greater teeter excursions can occur. To prevent impact damage under these conditions, teeter dampers and compliant stops are provided. In the hub shown in Figure 6.27 (which has a maximum allowed range of ± 7.0 degrees) the dampers are on the side of the hub opposite the bearings.

The options for attaching a teetering hub to the main shaft are the same as for rigid hubs.

Hinged hub A hinged hub is in some ways a cross between a rigid hub and a teetering hub. It is basically a rigid hub with 'hinges' for the blades. The hinge assembly adds some complexity, however. As with a teetering hub, there must be bearings at the hinges. Teetering hubs have the advantage that the two blades tend to balance each other, so lack of centrifugal stiffening during low rpm operation is not a major problem. There is no such counterbalancing on a hinged blade, however, so some mechanism must be provided to keep the blades from flopping over during low rotational speed. This could include springs. It would almost certainly include dampers as well.

6.7.2 Drive train

A complete wind turbine drive train consists of all the rotating components: rotor, main shaft, couplings, gearbox, brakes, and generator. With the exception of the rotor components which were considered above, all of these are discussed in the following sections. Figure 6.30 illustrates a typical drive train.

6.7.2.1 Main shaft

Every wind turbine has a main shaft, sometimes referred to as the low-speed or rotor shaft. The main shaft is the principal rotating element, providing for the transfer of torque from the rotor to the rest of the drive train. It also supports the weight of the rotor. The main shaft is supported in turn by bearings, which transfer reaction loads to the main frame of the turbine. Depending on the design of the gearbox, the shaft and/or the bearings may be integrated into the gearbox or they may be completely separate from it, connected only by a coupling. The main shaft is sized in accordance with methods described in Section 6.5.1, taking into account the combined loads of torque and bending. Main shafts are normally made of steel. Methods of connecting the main shaft to the rotor were discussed in Section 6.7.1. Figure 6.31 illustrates some options for the main shaft.

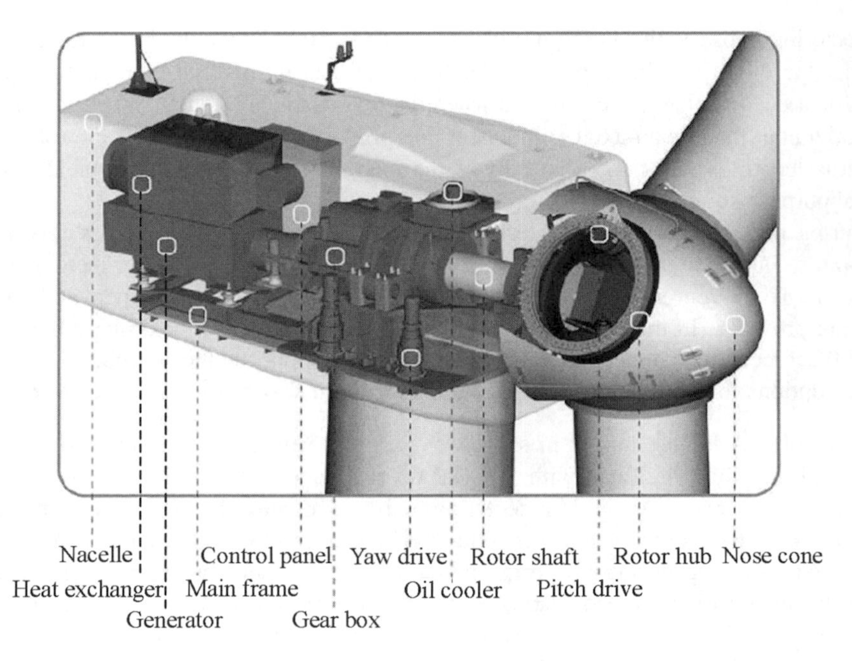

Figure 6.30 Drive train and associated components. Reproduced by permission of Enron Wind

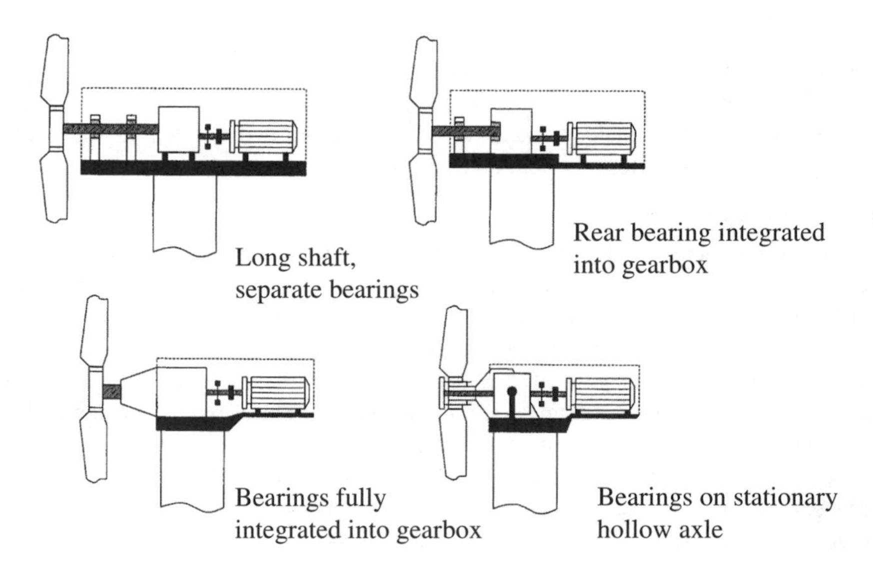

Figure 6.31 Main shaft options (Harrison et al., 2000)

6.7.2.2 Couplings

Function Couplings, as discussed in Section 6.5, are used to connect shafts together. There are two locations in particular where large couplings are likely to be used in wind turbines: (1) between the main shaft and the gearbox, and (2) between the gearbox output shaft and the generator.

The primary function of the coupling is to transmit torque between two shafts, but it may have another function as well. Sometimes it is advantageous to dampen torque fluctuations in the main shaft before the power is converted to electricity. A coupling of appropriate design can serve this role. A fluid coupling (as noted in Section 6.5) may be used for this purpose. Since couplings were described in Section 6.5, more detail will not be provided here.

6.7.2.3 Gearbox

Function Most wind turbine drive trains include a gearbox to increase the speed of the input shaft to the generator. An increase in speed is needed because wind turbine rotors, and hence main shafts, turn at a much lower speed than is required by most electrical generators. Small wind turbine rotors turn at speeds on the order of a few hundred rpm. Larger wind turbines turn more slowly. Most conventional generators turn at 1800 rpm (60 Hz) or 1500 rpm (50 Hz).

Some gearboxes also perform functions other than increasing speed, such as supporting the main shaft bearings. These are secondary to the basic purpose of the gearbox, however.

The gearbox is one of the single heaviest and most expensive components in a wind turbine. Gearboxes are normally designed and supplied by a different manufacturer than the one actually constructing the wind turbine. Since the operating conditions experienced by a wind turbine gearbox are significantly different than those in most other applications, it is imperative that the turbine designer understand gearboxes, and that the gearbox designer understand wind turbines. Experience has shown that underdesigned gearboxes are a major source of wind turbine operational problems.

Types All gearboxes have some similarities: they consist primarily of a case, shafts, gears, bearings and seals. Beyond that there are two basic types of gearboxes used in wind turbine applications: (1) parallel-shaft gearboxes and (2) planetary gearboxes.

In parallel-shaft gearboxes, gears are carried on two or more parallel shafts. These shafts are supported by bearings mounted in the case. In a single-stage gearbox there are two shafts, a low-speed shaft and a high-speed shaft. Both of these shafts, which are parallel, pass out through the case. One of them would be connected to the main shaft or rotor and the other to the generator. There are also two gears, one on each shaft. The two gears are of different size, with the one on the low-speed shaft being the larger of the two. The ratio of the pitch diameter of the gears is inversely proportional to the ratio of the rotational speeds (as described in Section 6.5.)

There is a practical limit to the size ratio of the two gears that can be used in a single-stage parallel-shaft gearbox. For this reason, gearboxes with large speed-up ratios use multiple shafts and gears. These gears then constitute a gear train. A two-stage gearbox, for example, would have three shafts: an input (low-speed) shaft, an output (high-speed) shaft and an intermediate shaft. There would be gears on the intermediate shaft, the smaller driven by the low-speed shaft. The larger of these gears would drive the gear on the high-speed shaft. A typical parallel-shaft gearbox is illustrated in Figure 6.32.

Figure 6.32 Parallel-shaft gearbox (Hau, 1996). Reproduced by permission of Springer Verlag GmbH

Planetary gearboxes have a number of significant differences from parallel-shaft gearboxes. Most notably, the input and output shafts are coaxial. In addition, there are multiple pairs of gear teeth meshing at any time, so the loads on each gear are reduced. This makes planetary gearboxes relatively light and compact. A typical planetary gearbox is illustrated in Figure 6.33.

Figure 6.33 Exploded view of two-stage planetary gearbox

In planetary gearboxes, a low-speed shaft, supported by bearings in the case, is rigidly connected to a planet carrier, which holds three identical small gears, known as planets. These gears are mounted on short shafts and bearings and are free to turn. These planets mesh with a large-diameter internal or ring gear and a small-diameter sun gear. When the low-speed shaft and carrier rotate, meshing of the planets in the ring gear forces the planets to rotate, and to do so at a speed higher than the speed of the carrier. The meshing of the planets with the sun gear causes it to rotate as well. The sun gear then drives the high-speed shaft, to which it is rigidly connected. The high-speed shaft is supported by bearings mounted in the case. Figure 6.34 illustrates the relation between the gears and the angles made during a small angle of rotation. Note that before the rotation the sun and planet gear mesh at point B, while the planet and ring gear mesh at point A. After the rotation the corresponding meshing points are $B1$ and $A1$. The centers of the sun and the planet are at O and OP respectively.

The speed-up ratio for the configuration shown in Figure 6.34 (with the ring gear stationary) is:

$$\frac{n_{HSS}}{n_{LSS}} = 1 + \frac{D_{Ring}}{D_{Sun}} \tag{6.7.1}$$

where n_{HSS} is the speed of high-speed shaft, n_{LSS} is the speed of low-speed shaft, D_{Ring} is the diameter of ring gear, and D_{Sun} is the diameter of sun gear.

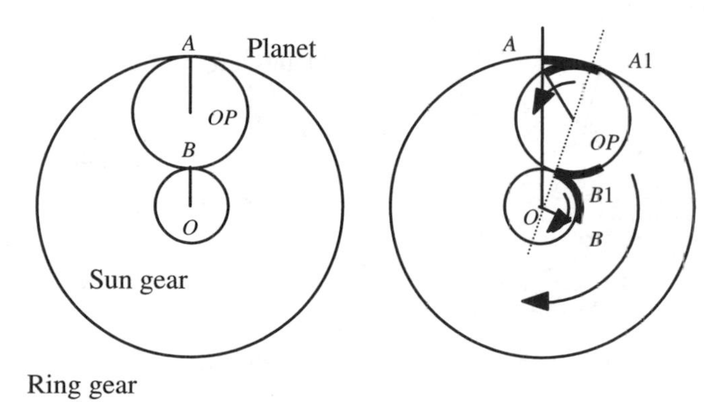

Figure 6.34 Relations between gears in a planetary gearbox

As with the parallel-shaft gearbox there is a limit to the speed-up ratio that can be achieved by a single stage planetary gear set. To achieve a higher speed-up ratio, multiple stages are placed in series. When there are multiple stages in series, the overall speed-up is the product of the speed-up of the individual stages.

Gears in many wind turbine gearboxes are of the spur type, but helical gears are found as well. Bearings are ball bearings, roller bearings, or tapered roller bearings, depending on the loads. Gears and bearings were discussed in more detail in Section 6.5.

Gearbox design considerations There are a great many issues to consider in the design and selection of a gearbox. These include:

- Basic type (parallel-shaft or planetary), as discussed above
- Separate gearbox and main shaft bearings, or an integrated gearbox
- Speed-up ratio
- Number of stages
- Gearbox weight and costs
- Gearbox loads
- Lubrication
- Effects of intermittent operation
- Noise

Wind turbine gearboxes are either separate components, or they are combined with other components. In the latter case they are known as integrated or partially integrated gearboxes. For example, in a number of turbines with a partially integrated gearbox, the main shaft and main shaft bearings are integrated into the rest of the gearbox. A fully integrated gearbox is one in which the gearbox case is really the main frame of the wind turbine. The rotor is attached to its low-speed shaft. The generator is coupled to the high-speed shaft and is also bolted directly to the case. Part of the yaw system is integrated into the bottom of the case. Figure 6.35 illustrates an integrated planetary gearbox.

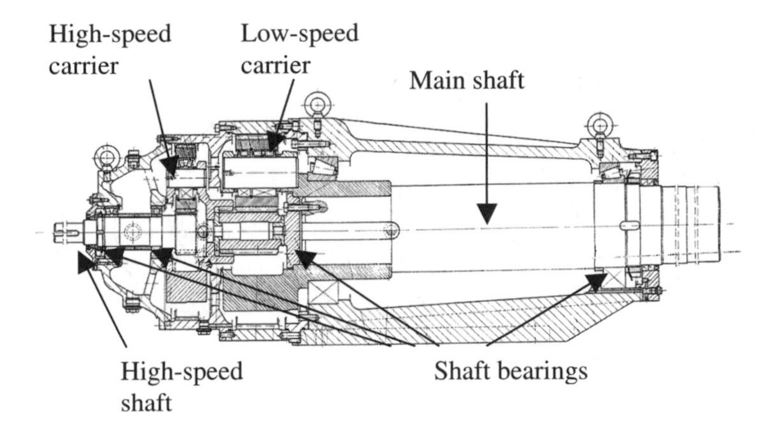

Figure 6.35 Partially integrated, two-stage planetary gearbox

The speed-up ratio of a gearbox is directly related to the desired rotational speed of the rotor and the speed of the generator. As previously indicated the rotor speed is determined primarily by aerodynamic considerations. Generator speed is in most cases 1800 rpm in 60 Hz grids or 1500 rpm in 50 Hz grids, although other speeds are also possible (as is discussed in Chapter 5.) For example, a wind turbine with a rotor designed to operate at 60 rpm and an 1800 rpm generator would need a gearbox with a 30:1 speed-up ratio.

The number of stages in a gearbox is generally of secondary concern to the wind turbine designer. It is important primarily because it affects the complexity, size, weight, and cost of the gearbox. The more stages there are, the more internal components, such as gears, bearings, and shafts, that there are. Generally, any one stage will not provide a speed-up of more than 6:1. The ratios of multiple stages placed in series result in an overall ratio given by the product of the ratios in each stage. For example, one could gain a speed-up of 30:1 by having two stages of 5:1 and 6:1 in series.

The weight of a gearbox increases dramatically with increasing power rating of the turbine. In fact, the gearbox weight will scale approximately with the cube of the radius, as does the weight of the rotor. Since planetary gearboxes are lighter than parallel-shaft boxes, there is a weight advantage to be gained by using them. However, due to their greater complexity they also cost more than would be indicated by their reduced weight.

The loads that the gearbox must withstand are due primarily to those imposed by the rotor. This will include at least the main shaft torque, and may include the weight of the rotor and various dynamic loads, depending on degree of integration of the gearbox with the main shaft and bearings. Loads are also imposed by the generator, both during normal operation and while starting, and by any mechanical brake located on the high-speed side of the gearbox. Over an extended period of time the gearbox, like the rotor, will experience some loads that are relatively steady, other loads that vary periodically or randomly, and still others that are transient. All of these contribute to fatigue damage and wear on the gear teeth, bearings and seals.

Lubrication is a significant issue in gearbox operation, but it will not be dealt with in detail here. Oils must be selected to minimize wear on the gear teeth and bearings, and to function properly under the external environmental conditions in which the turbine will operate. In some cases, it may be necessary to provide filtering or active cooling of the oil. In any event, periodic oil samples should be taken to assess the state of the oil, as well as to check for signs of internal wear.

Intermittent operation, a common situation with wind turbines, can have a significant impact on the life of a gearbox. When the turbine is not running, oil may drain away from the gears and bearings, resulting in insufficient lubrication when the turbine starts.

In cold weather the oil may have too high a viscosity until the gearbox has warmed up. Turbines in such environments may benefit by having gearbox oil heaters. Condensation of moisture may accelerate corrosion. When the rotor is parked (depending on the nature and location of a shaft brake) the gear teeth may move slightly back and forth. The movement is limited by the backlash, but it may be enough to result in some impact damage and tooth wear.

Gearboxes may be a source of noise. The amount of noise is a function of, among other things, the type of gearbox, the materials from which the gears are made and how they are cut. Designing gearboxes for a minimum of noise production is presently an area of significant interest.

More details on wind turbine gearboxes, relating particularly to design, are given in draft design guidelines from the American Gear Manufacturers Association (1997).

6.7.2.4 Generator

The generator converts the mechanical power from the rotor into electrical power. Generator options were described in detail in Chapter 5 and will not be discussed here. One of the important things to recall is that most grid-connected generators turn at constant or nearly constant speed. This is responsible for the fact that most wind turbine rotors also turn at constant or nearly constant speed.

6.7.2.5 Brake

Function Nearly all wind turbines employ a mechanical brake somewhere on the drive train. Such a brake is normally included in addition to any aerodynamic brakes. In fact, some current design standards (Germanischer Lloyd, 1993) require two independent braking systems, one of which is usually aerodynamic and the other of which is on the drive train. In most cases, the mechanical brake is capable of stopping the turbine. In other cases, the mechanical brake is used only for parking. That is, it keeps the rotor from turning when the turbine is not operating. Brakes that can be used only for parking are becoming less common, because of the influence of design standards. Generally, such lightweight brakes would only be used on a turbine which has a fail-safe, pitch-controlled rotor.

Types of brakes There are two types of brakes in common usage on wind turbines: disc brakes and clutch brakes. The disc brake operates in a manner similar to that on an automobile. A steel disc is rigidly affixed to the shaft to be braked. During braking a hydraulically actuated caliper pushes brake pads against the disc. The resulting force creates a torque opposing the motion of the disc, thus slowing the rotor. An example of a disc brake is shown in Figure 6.36.

Figure 6.36 Disc brake. Reproduced by permission of Svendborg Brakes A/S

Clutch type brakes were described in Section 6.5.4. Actuation of clutch brakes is normally via springs, so they are fail-safe by design. These brakes are released by compressed air or hydraulic fluid.

Another, less common type of brake is electrically based and is known as a ‘dynamic brake’. The basic principle is to feed power to a resistor bank after disconnecting the wind turbine’s generator from the electrical grid. This puts a load on the generator, and hence a torque on the rotor, thereby decelerating it. More details on dynamic brakes are presented in Childs et al. (1993).

Location Mechanical brakes can be located at any of a variety of locations on the drive train. For example, they may be on either the low-speed or high-speed side of the gearbox. If on the high-speed side, they may be on either side of the generator.

It is important to note that a brake on the low-speed side of the gearbox must be able to exert a much higher torque than would be the case with one on the high-speed side. It would thus be relatively massive. However, if the brake is on the high-speed side, it will necessarily act through the gearbox, possibly increasing the gearbox wear. Furthermore, in the event of an internal failure in the gearbox, a brake on the high-speed side might be unable to slow the rotor.

Brake activation Brake activation depends on the type of brake used. Disc brakes require hydraulic pressure. This is normally supplied by a hydraulic pump, sometimes in conjunction with an accumulator. There are also designs in which springs apply brake pressure, and the hydraulic system is used to release the brakes.

Clutch-type brakes are normally spring-applied. Either a pneumatic system or hydraulic system is used to release the brake. In the case of pneumatics, an air compressor and storage tank must be provided, as well as appropriate plumbing and controls.

Performance Three important considerations in the selection of a brake include:

- Maximum torque
- Length of time required to apply
- Energy absorption

A brake intended to stop a wind turbine must be able to exert a torque in excess of what could plausibly be expected to originate from the rotor. Recommended standards indicate that a brake design torque should be equal to the maximum design torque of the wind turbine (Germanischer Lloyd, 1993).

A brake intended to stop a turbine should begin to apply almost immediately, and should ramp up to full torque within a few seconds. The ramp-up time selected is a balance between instantaneous (which would apply a very high transient load to the drive train) and so slow that acceleration of the rotor and heating of the brake during deceleration could be concerns. Normally the entire braking event, from initiation until the rotor is stopped, is less than five seconds.

Energy absorption capability of the brake is an important consideration. First of all the brake must absorb all the kinetic energy in the rotor when turning at its maximum possible speed. It must also be able to absorb any additional energy that the rotor could acquire during the stopping period.

6.7.3 Yaw system

6.7.3.1 Function

With very few exceptions, all horizontal axis wind turbines must be able to yaw so as to orient themselves in line with the wind direction. Some turbines also use active yaw as a means of power regulation. In any case, a mechanism must be provided to enable the yawing to take place, and to do so at a slow enough rate that large gyroscopic forces are avoided.

6.7.3.2 Types

There are two basic types of yaw systems: active yaw and free yaw. Turbines with active yaw are normally upwind machines. They employ a motor to actively align the turbine. Turbines with free yaw are normally downwind machines. They rely on the aerodynamics of the rotor to align the turbine.

6.7.3.3 Description

Regardless of the type of yaw system all horizontal axis wind turbines have some type of yaw bearing. This bearing must carry the weight of the main part of the turbine, as well as transmit thrust loads to the tower.

In a turbine with active yaw, the yaw bearing includes gear teeth around its circumference. A pinion gear on the yaw drive engages with those teeth, so that it can be driven in either direction.

The yaw drive normally consists of an electric motor, speed reduction gears, and a pinion gear. The speed must be reduced so that the yaw rate is slow, and so that adequate torque can be supplied from a small motor. Historically, some yaw drives have used small wind rotors mounted at right angles to the main rotor. This has the advantage of not requiring a separate power source or controls. However, it lacks the flexibility of those with motors, and is not now commonly used.

One problem encountered with active yaw has been rapid wear or breaking of the yaw drive due to continuous small yaw movements of the turbine. This is possible because of backlash between the yaw drive pinion and the bull gear. The motion results in many shock load cycles between those gears. In order to reduce these cycles, a yaw brake is frequently used now in active yaw systems. This brake is engaged whenever the turbine is not yawing. It is released just before yawing begins. A typical yaw drive with a brake is illustrated in Figure 6.37.

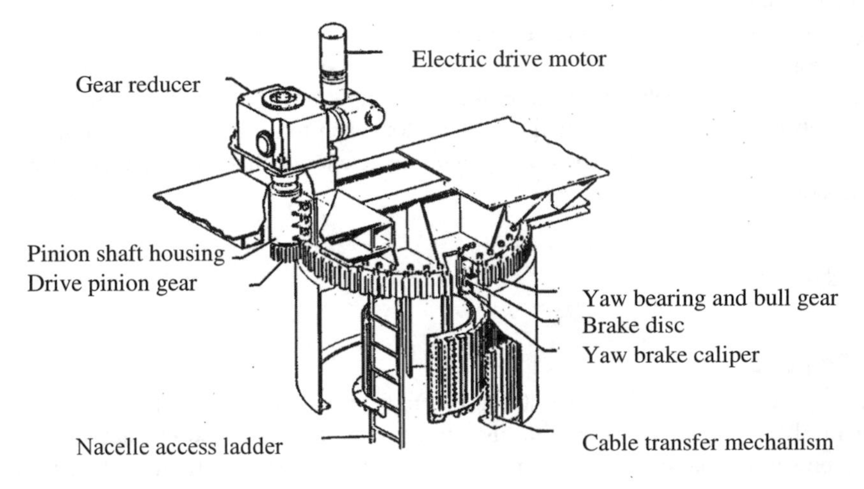

Figure 6.37 Typical yaw drive with brake (Van Bibber and Kelly, 1985)

The yaw motion in an active yaw system is controlled using yaw error as an input. Yaw error is monitored by means of a wind vane mounted on the turbine. When the yaw error is outside the allowed range for some period of time, the drive system is activated, and the turbine is moved in the appropriate direction.

In turbines with free yaw the yaw system is normally much simpler. Often there is nothing more than the yaw bearing. Some turbines, however, include a yaw damper. Yaw dampers are used to slow the yaw rate, helping to reduce gyroscopic loads. They are most useful for machines which have a relatively small polar moment of inertia about the yaw axis.

6.7.4 Main frame and nacelle

The nacelle is the housing for the principal components of the wind turbine (with the exception of the rotor). It includes the main frame and the nacelle cover.

6.7.4.1 Main frame

Function The main frame is the structural piece to which the gearbox, generator and brake are attached. It provides a rigid structure to maintain the proper alignment among those components. It also provides a point of attachment for the yaw bearing, which in turn is bolted to the top of the tower.

Types There are basically two types of main frames. The main frame is either a separate component, or it is part of an integrated gearbox.

Description When the main frame is a separate component, it is normally a rigid steel casting or weldment. Threaded holes or other attachment points are provided in appropriate locations for bolting on the other components. When the main frame is part of an integrated gearbox, the case is made thick enough that it can carry the requisite loads. As with the separate main frame, attachment points are provided for securing the other items.

Main frame loads The main frame must transmit all the loads from the rotor and reaction loads from the generator and brake to the tower. It must also be rigid enough that it allows no relative movement between the rotor support bearings, gearbox, generator and brake.

6.7.4.2 Nacelle cover

The nacelle cover provides weather protection for the wind turbine components which are located in the nacelle. These include especially electrical and mechanical components that could be affected by sunlight, rain, ice or snow.

Nacelle covers are normally made from a lightweight material, such as fiberglass. On larger machines the nacelle cover is of sufficient size that it can be entered by personnel for inspecting or maintaining the internal components. On small and medium-size turbines, a separate nacelle cover is normally attached to the main frame in such a way that it can be readily opened for access to items inside. An example of a nacelle cover is shown in Figure 6.38. A component which some turbines have, and which is closely related to the nacelle cover, is the spinner or nose cone. This is the housing for the hub.

Figure 6.38 Typical nacelle cover. Reproduced by permission of Nordex AG

6.7.5 Tower

Towers are supports to raise the main part of the turbine up in the air. Some of the considerations in selecting a type of tower were discussed in Section 6.3. The height of a tower is normally at least as high as the diameter of the rotor. For smaller turbines the tower may be much higher than that. Generally, tower height should not be less than 24 m because the wind speed is lower and more turbulent so close to the ground.

6.7.5.1 General tower issues

There are three types of towers in common use for horizontal axis wind turbines:

- Free-standing lattice (truss)
- Cantilevered pipe (tubular tower)
- Guyed lattice or pole.

Historically, free-standing lattice towers were used more commonly until the mid-1980s. For example, the Smith–Putnam, US Department of Energy MOD-0, and early US Windpower turbines all used towers of this type. Since that time tubular towers have been used more frequently. With a few notable exceptions (such as the Carter and Wind Eagle turbines) guyed towers have never been very common for machines of medium size or larger. Some tower options are illustrated in Figure 6.39.

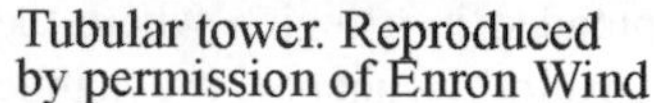

Tubular tower. Reproduced by permission of Enron Wind

Truss tower

Guyed tower. Reproduced by permission of Vergnet SA

Figure 6.39 Tower options

Tubular towers have a number of advantages. Unlike lattice towers, they do not rely on many bolted connections which need to be torqued and checked periodically. They provide a protected area for climbing to access the machine. Aesthetically, they provide a shape which is considered by some to be visually more pleasing than an open truss.

Materials Wind turbine towers are usually made of steel, although sometimes reinforced concrete is used. When the material is steel, it is normally galvanized or painted to protect it from corrosion. Sometimes Cor-Ten® steel, which is inherently corrosion resistant, is used.

Tower loads The tower can experience two major types of load: (1) steady and (2) dynamic. Steady tower loads arise primarily from aerodynamically produced thrust and torque. These were discussed in detail in Chapter 4. The weight of the machine itself is also a significant load. The loading on the tower is evaluated for at least two conditions: (1) operating at rated power and (2) stationary at survival wind speed. In the latter case, IEC standards recommend that the 50-year extreme wind speed be used (Bakker, 1996). The effects of loading must be considered especially on bending and buckling.

Dynamic effects can be a significant source of loads, especially on soft or soft–soft towers. Recall that a stiff tower is one whose fundamental natural frequency is above the blade passing frequency, a soft tower is one whose natural frequency is between the blade passing frequency and the rotor frequency, and a soft–soft tower is one whose natural frequency is below both the rotor frequency and blade passing frequency. For either a soft or soft–soft tower, the tower can be excited during start-up or shutdown of the turbine.

Determination of the tower natural frequency may be done by methods discussed in Chapter 4. For the simple case, when the turbine/tower can be approximated by a uniform cantilever with a point mass on the top, the following equation (Baumeister, 1978) may be used.

$$f_n = \frac{1}{2\pi}\sqrt{\frac{3EI}{(0.23m_{Tower} + m_{Turbine})L^3}} \tag{6.7.2}$$

where f_n is the fundamental natural frequency (Hz), E is the modulus of elasticity, I is the moment of inertia of tower cross-section, m_{Tower} is the mass of tower, $m_{Turbine}$ is the mass of turbine, and L is the height of tower.

For non-uniform or guyed towers, the Rayleigh method may be quite useful. The method is described in general by Thomson (1981) and by Wright et al. (1981) for wind turbines. Comprehensive analysis of towers, including natural frequency estimation, may be done with finite element methods. An example of this is given in El Chazly (1993).

A tower should be designed so that its natural frequency does not coincide with the turbine's excitation frequencies (the rotor frequency or the blade passing frequency). In addition, the excitation frequencies should generally not be within 5% of tower natural frequency during prolonged operation. When operation is intended in a region where the excitation frequencies are between 30% and 140% of tower natural frequency, a dynamic magnification factor, D, should be used to multiply the design loads in evaluating the structure. The magnification factor is determined by the damping properties of the tower and the relation between the excitation frequencies. It is equivalent to the non-dimensional amplitude which was developed in Chapter 4 (Equation 4.2.27):

$$D = \frac{1}{\sqrt{\left[1-(f_e/f_n)^2\right]^2 + \left[2\xi(f_e/f_n)^2\right]}} \tag{6.7.3}$$

where f_e = excitation frequency, f_n = natural frequency, ξ = damping ratio.

The damping ratio is found from the 'logarithmic damping decrement', δ, by the relation:

$$\xi = \frac{\delta}{2\pi} \tag{6.7.4}$$

Damping of tower vibrations is due to both aerodynamic and structural factors. The damping decrement suggested by Germanischer Lloyd (1993) is 0.1 for reinforced concrete and between 0.05 – 0.15 for steel.

A comparative assessment of wind turbine tower options is given in Babcock and Connover (1994).

6.7.5.2 Tower climbing safety

Nearly all wind turbines must be climbed occasionally for doing inspections or maintenance. Provision must be made in the tower design for safe climbing. This typically includes a ladder or climbing pegs and an anti-fall system. Figure 6.40 illustrates tower climbing safety equipment.

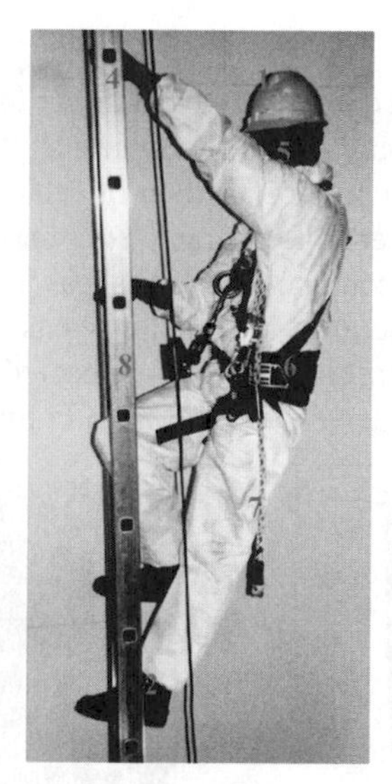

Figure 6.40 Tower climbing safety equipment. Reproduced by permission of Vestas Wind Systems A/S)

6.7.5.3 Tower top

The tower top provides the interface for attaching the main frame of the wind turbine to the tower. The stationary part of the yaw bearing is attached to the tower top. The shape of the tower top depends strongly on the type of tower. It is usually made from cast steel.

6.7.5.4 Tower foundation

The foundation of a wind turbine must be sufficient to keep the turbine upright and stable under the most extreme design conditions. At most sites, the foundation is constructed as a reinforced concrete pad. The weight of the concrete is chosen to provide resistance to overturning under all conditions. Sometimes turbines are installed on rock. In this case the foundation may consist of rods grouted into holes drilled deep into the rock. A concrete pad may be used to provide a level surface, but any tensile loads are taken ultimately by the rods. Some of the possibilities that may be encountered in wind turbine foundations are illustrated in Figure 6.41.

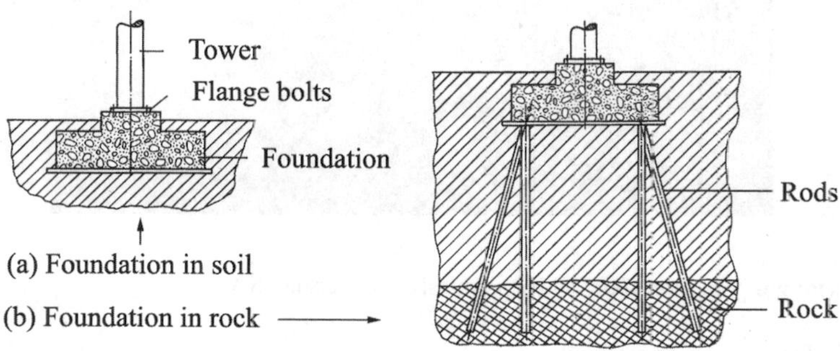

6.41 Wind turbine foundations (adapted from Hau, 1996). Reproduced by permission of Springer Verlag GmbH

6.7.5.5 Tower erection

The intended method of tower erection will have a direct impact on the design of the tower. Larger turbines are most commonly erected with cranes. Small and medium-size turbines are often self-erecting. The most common method of self-erection is to use a gin pole or 'A frame' at a right angle to the tower. The A frame is connected to the top of the tower by a cable. A winch is then used, in conjunction with sheaves to raise the tower. With such a method of erection, the tower base must include hinges as well as a way to secure the tower in place once it is vertical. The turbine itself is connected to the tower before it is raised. Some of the methods for erecting towers are shown in Figure 6.42.

(a) Crane erection of tubular tower. Reproduced by permission of Vestas Wind Systems A/S

(b) Tilt up with gin pole. Reproduced by permission of Vergnet SA

Figure 6.42 Tower erection methods

Regardless of the method of erection, an important consideration in the design of the tower is the loads that it will experience during the installation.

6.7.6 *Interconnection and control*

There are a great number of electrical and control issues associated with the design of wind turbines. These are discussed in Chapters 5 (electrical), 7 (controls), and 8 (systems.)

6.8 Design Evaluation

Once a detailed design for the wind turbine has been developed, its ability to meet basic design requirements, such as those discussed in Section 6.6, must be assessed. This design evaluation should use the appropriate analytical tools. Where possible, validated computer codes should be used. When necessary, models specific to the application may need to be developed.

There are five steps that need to be taken in performing a detailed design evaluation:

- Prepare the wind input
- Model the turbine
- Perform a simulation to obtain loads
- Convert predicted loads to stresses
- Assess damage

Each of these steps is summarized below. An extensive discussion of detailed design evaluations for a number of turbine types is given in Laino (1997).

6.8.1 *Wind input*

Wind input needs to be generated that will correspond to the design input conditions. For extreme winds and discrete gusts, specifying the wind input is relatively straightforward, given the guidelines summarized in Section 6.6. Converting that wind input to time series inputs can also be done fairly simply. Generating rotationally sampled synthetic turbulent wind, however, can be quite complicated. For this purpose, public domain computer codes such as SNLWind or SNLWind3D (Kelley, 1993) can be used.

6.8.2 *Model of turbine*

The next step is developing a detailed model of the wind turbine. This should include both aerodynamics and dynamics. This can be done from basics, using the methods discussed in Chapters 3 and 4, but, when possible, it is preferable to use models that are already available. Some of the presently available models that may be appropriate include YawDyn (Hansen, 1996), FAST_AD (Wilson et al., 1996), and ADAMS/WT (Elliot and Wright, 1994). There are also a number of commercially available codes that could be used.

Once the model has been selected or developed, inputs describing the specific turbine need to be assembled. These generally include weight and stiffness distributions, dimensions, aerodynamic properties, etc.

6.8.3 Simulation

The simulation is the actual running of the computer model to generate predictions. Multiple runs may have to be made to study the full range of design conditions.

6.8.4 Converting simulation outputs to stresses

Outputs from simulation codes are frequently in the form of time series loads, that is, forces, bending moments, and torques. In that case they must be converted to stresses. This can be done with the help of simple programs, which use the loads together with geometric properties of the components of interest. Laino (1997) describes one approach to this task.

6.8.5 Damage assessment

As discussed above, there are two basic aspects of design evaluation: (1) ultimate loads and (2) fatigue loads. If the maximum stresses are low enough during the extreme load design cases, then the turbine passes the ultimate loads test.

The fatigue case is more complicated. For one thing, the total amount of damage that is generated over an extended period of time will depend on the damage arising as a result of particular wind conditions and the fraction of time that those various conditions occur. Thus, the distribution of the wind speed is an important factor which needs to be taken into account. In order to expedite fatigue damage estimates it is advantageous to use such codes as LIFE2 (Sutherland, 1989) to carry out the assessment.

6.9 Power Curve Prediction

Prediction of a wind turbine's power curve is an important step of the design process. It involves consideration of the rotor, gearbox, generator and control system.

The method used in predicting the power curve is to match the power output from the rotor as a function of wind speed and rotational speed to the power produced by the generator, also as a function of rotational speed. The effects of component efficiencies are also considered where appropriate. In this discussion it is assumed that all drive train efficiencies are accounted for by adjusting the rotor power. The process may be done either graphically or in a more automated fashion. The graphical method best illustrates the concept and will be described here.

Rotor power as a function of rotational speed is predicted for a series of wind speeds by applying estimates for the power coefficient, C_p. The power coefficient as a function of tip speed ratio, and hence rpm, may be obtained as described in Chapter 3. The rotor power, P_{rotor}, is then:

$$P_{rotor} = C_p \eta \frac{1}{2} \rho \pi R^2 U^3 \tag{6.9.1}$$

where η = drive train efficiency, ρ = air density, R = rotor radius, and U = wind speed.

The rotor speed, n_{rotor}, in rpm is found from the tip speed ratio, λ :

$$n_{rotor} = \frac{30}{\pi}\lambda\frac{U}{R} \tag{6.9.2}$$

A power vs. rpm relation is found for the generator, and referred to the low-speed side of the gearbox by dividing the generator speed by the gearbox speed-up ratio. This relation is superimposed on a series of plots (for a range of wind speeds) for rotor power vs. rotor rpm. Every point where a generator line crosses a rotor line defines a pair of power and wind speed points on the power curve. These points also define the operating speed of the rotor.

As was explained in Chapter 5, grid-connected generators are usually either of the synchronous or induction type. Synchronous generators turn at a fixed speed, determined by the number of magnetic poles and the grid frequency. Induction generators turn at a nearly fixed speed, determined primarily by the number of poles and grid frequency but also by the power level. For normal operation, power varies directly with 'slip', which was explained in Chapter 5. The relation may also be expressed as:

$$P_{generator} = \frac{g\, n_{rotor} - n_{sync}}{n_{rated} - n_{sync}} P_{rated} \tag{6.9.3}$$

where $P_{generator}$ is the generator power, g is the gearbox ratio, P_{rated} is the rated generator power, n_{sync} is the synchronous speed of the generator, and n_{rated} is the speed of the generator at rated power.

6.9.1 Example

The following example illustrates the process of estimating the power curve for a hypothetical wind turbine. The turbine has a rotor of 20 m diameter with a power coefficient vs. tip speed ratio relation illustrated in Figure 6.43.

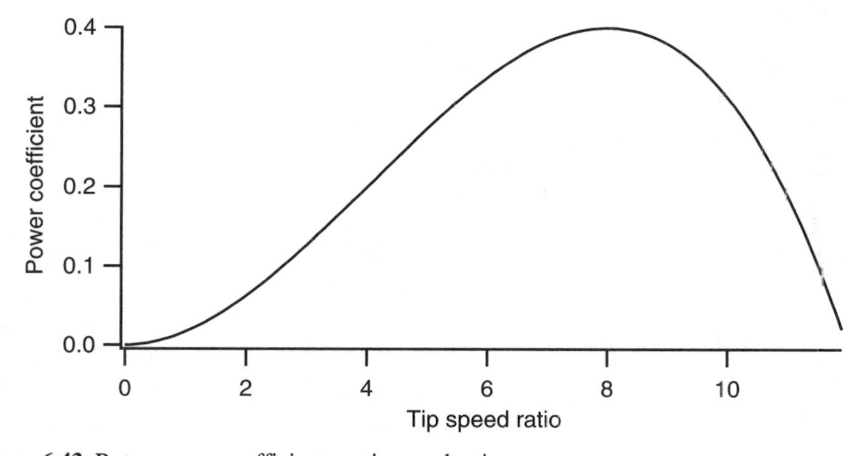

Figure 6.43 Rotor power coefficient vs. tip speed ratio

The overall mechanical and electrical efficiency is assumed to be 0.9. Two possible pairs of gear ratios and generator ratings are considered. Six wind speeds are used, ranging from 6 m/s to 16 m/s. It is assumed that power will be regulated at above rated wind speed (16 m/s), so only the part of the power curve at or below 16 m/s is shown. Gearbox 1 has a speed-up ratio of 36:1, whereas gearbox 2 has a speed-up ratio of 24:1. The rated power of generator 1 is 150 kW and that of generator 2 is 225 kW. Both generators are of the induction type. They have a synchronous speed of 1800 rpm and a speed of 1854 rpm at rated power. Figure 6.44 illustrates the power vs. rotational speed curves for the six wind speeds and two generator/gearbox combinations.

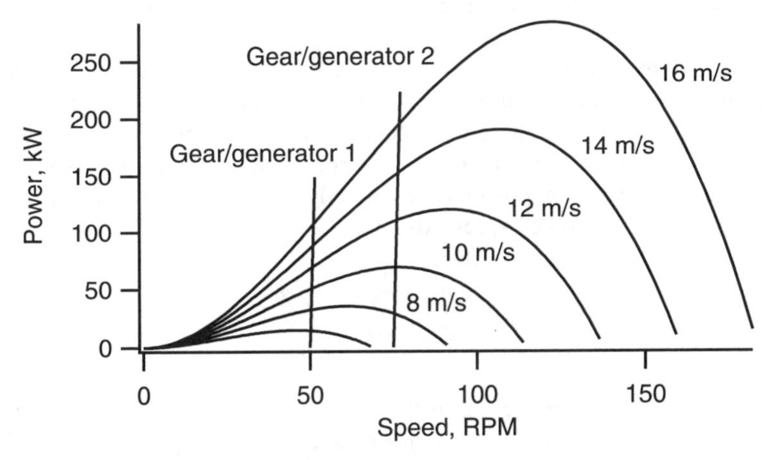

Figure 6.44 Rotor and generator power vs. rotor speed

The power curves that can be derived from Figure 6.44 are shown in Figure 6.45. For comparison an ideal variable speed power curve is shown for the same wind speed range. The ideal curve was obtained by assuming a constant power coefficient of 0.4 over all wind speeds. As can be seen from the figure, gearbox/generator combination 1 would produce more power than combination 2 at winds less than about 8.5 m/s, but less than combination 2 at higher winds.

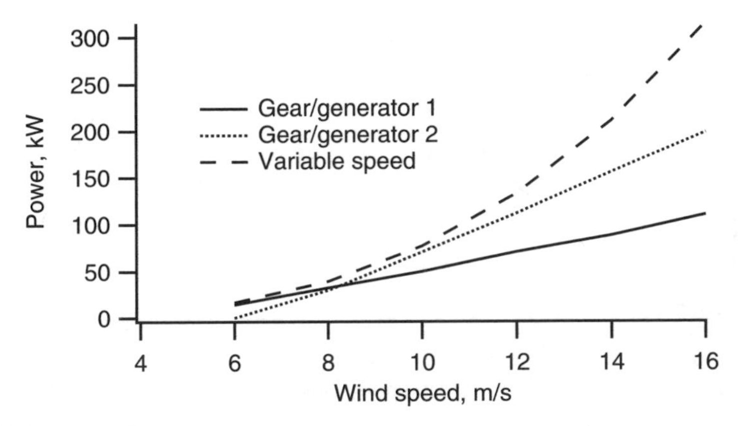

Figure 6.45 Power curves

Curves of the type developed above can be useful in selecting the generator size and gearbox ratio. By combining the power curves with characterizations of prospective wind regimes (as described in Chapter 2), the effect on annual energy production can be estimated. Generally speaking, as illustrated in this example, a smaller generator and slower rotor speed (larger gearbox ratio) will be beneficial when the wind speeds are lower. Conversely, a larger generator and faster rotor speed are more effective in higher winds.

References

American Gear Manufacturers Association (1997) Recommended Practices for Design and Specification of Gearboxes for Wind Turbine Generator Systems. AGMA Information Sheet. AGMA/AWEA 921-A97. American Gear Manufacturers Association, 1500 King St, Suite 201, Alexandria, VA 22314.

Babcock, B. A., Connover, K. E. (1994) Design of Cost Effective Towers for an Advanced Wind Turbine. *Proc. of the 15th ASME Wind Energy Symposium*, American Society of Mechanical Engineers, New York.

Bakker, D. (Secretary). (1996) Wind Turbine Systems, Part 1: Safety Requirements, 88/1400-1, Standards (Draft) Reference Number 88/69/CD. International Electrotechnical Commission.

Baumeister, T. (Ed.) (1978) *Marks' Standard Handbook for Mechanical Engineers*, 8th Edition, McGraw Hill, New York.

Childs, S, Hughes, P., Saeed, A. (1993) Development of a Dynamic Brake Model. *Proc. of the 1993 American Wind Energy Association Annual Conference*, American Wind Energy Association, Washington DC.

Chou, T. W., McCulloch, R. L., Pipes, R. B. (1986) Composites, *Scientific American*, 254, 193.

Connell, J. (1988) A Primer of Turbulence at the Wind Turbine Rotor, *Solar Energy*, 41, (3), 281–293.

Doerner, H. (1988) Philosophy of the Wind Power Plant Designer a Posteriori. Internet: http://129.69.67.105/~doerner/edesignphil.html.

Eggleston, D. M., Stoddard, F. S. (1987) *Wind Turbine Engineering Design*, Van Nostrand Reinhold, New York.

El Chazly, N. (1993) Wind Turbine Tower Structural and Dynamic Analysis Using the Finite Element Method. *Proc. of the 15th British Wind Energy Association Annual Conference*. Mechanical Engineering Publications, London.

Elliot, A. S., Wright, A. D. (1994) ADAMS/WT: An Industry Specific Interactive Modeling Interface for Wind Turbine Analysis. *Proc. of the 15th ASME Wind Energy Symposium*, American Society of Mechanical Engineers, New York.

Fordham, E. J. (1985) Spatial Structure of Turbulence in the Atmosphere, *Wind Engineering*, 9, 95–135.

Gasch, R. (Ed.) (1996) *Windkraftanlagen* (Windpower Plants), B. G. Teubner, Stuttgart.

Geraets, P. H., Haines, R. S., Wastling, M. A. (1997) Light Can Be Tough. *Proc. of the 19th British Wind Energy Association Annual Conference*, Mechanical Engineering Publications, London.

Germanischer Lloyd (1993) Regulation of the Certification of Wind Energy Conversion Systems, Rules and Regulations IV: Non Marine Technology Part 1, Wind Energy, Germanischer Lloyd, Hamburg.

Hansen, C. (1996) User's Guide to the Wind Turbine Dynamics Computer Programs YawDyn and AeroDyn for ADAMS®, Version 9.6, University of Utah, Salt Lake City. Prepared for the National Renewable Energy Laboratory under Subcontract No. XAF-4-14076-02.

Harrison, R., Hau, E., Snel, H. (2000) *Large Wind Turbines: Design and Economics*. John Wiley, Chichester.

Hau, E. (1996) *Windkraftanlagen* (Windpower Plants), Springer, Berlin.

Hoadley, R. B. (2000) *Understanding Wood: A Craftsman's Guide to Wood Technology*. The Taunton Press, Newtown, CT.

Hydraulic Pneumatic Power Editors (1967) *Hydraulic Handbook*. Trade and Technical Press, Ltd., Morden, Surrey, England.

Jamieson, P. (1997) Common Fallacies in Wind Turbine Design. *Proc. of the 19th British Wind Energy Association Annual Conference*. Mechanical Engineering Publications, London.

Kelley, N. (1993) Full Vector (3-D) Inflow Simulation in Natural and Wind Farm Environment Using an Expanded Version of SNLWIND (Veers) Turbulence Code. *Proc. of the 14th ASME Wind Energy Symposium*. American Society of Mechanical Engineers, New York.

Laino, D. J. (1997) *Evaluating Sources of Wind Turbine Fatigue Damage*, PhD Dissertation, University of Utah.

Manwell, J. F., McGowan, J. G., Adulwahid, U., Rogers, A., McNiff, B. P. (1996) A Graphical Interface Based Model for Wind Turbine Drive Train Dynamics. *Proc. of the 1996 American Wind Energy Association Annual Conference*, American Wind Energy Association, Washington DC.

McNiff, B. P., Musial, W. D., Erichello, R. (1990) Variations in Gear Fatigue Life for Different Braking Strategies. *Proc. of the 1990 American Wind Energy Association Annual Conference*, American Wind Energy Association, Washington DC.

National Research Council (1991) Assessment of Research Needs for Wind Turbine Rotor Materials Technology. Committee on Assessment of Research Needs for Wind Turbine Rotor Materials Technology, Energy Engineering Board, National Research Council, National Academy Press, Washington, DC.

Parmley, R. D. (1997) *Standard Handbook of Fastening and Joining*, 3rd Edition, McGraw Hill, New York.

Peery, D. J., Weingart, O. (1980) Low Cost Composite Blades for Large Wind Turbines. *Proc. of the AIAA/SERI Conference*, Boulder CO.

Perkins, F., Jones, R.W. (1981) The Effect of Delta 3 on a Yawing HAWT Blade and on Yaw Dynamics. *Wind Turbine Dynamics*, NASA Conference Publication 2185.

Pytel, A., Singer, F. L. (1987) Strength of Materials, Harper and Row, New York.

Selig, M. (1998) UIUC Airfoil Coordinates Data Base, UIUC Airfoil Data Site, Internet: http://amber.aae.uiucc.edu/~m-selig/ads.html.

Selig, M., Tangler, J. L. (1995) Development of a Multipoint Inverse Design Method for Horizontal Axis Wind Turbines. *Wind Engineering*, 19 (2), 91-105.

Shigley, R. G., Mischke, C. R. (1989) *Mechanical Engineering Design*, 5th Edition, McGraw Hill, New York.

Shinozuka, M., Jan C. M. (1972) Digital Simulation of Random Processes and its Application. *Journal of Sound and Vibration*, 25, 111-128.

Spotts, M. F. (1985) *Design of Machine Elements*, Prentice Hall, Englewood Cliffs, NJ.

Stiesdal, H. (1990) The 'Turbine' Dynamic Load Calculation Program. *Proc. of the 1990 American Wind Energy Association Conference*, American Wind Energy Association, Washington, DC.

Sutherland, H. J., Schluter L. L. (1989) The LIFE2 Computer Code, Numerical Formulation and Input Parameters. *Proc. of Windpower '89, SERI/TP-257-3628*, American Wind Energy Association, Washington, DC.

Thomson, W. T. (1981) *Theory of Vibrations with Applications*, 2nd Edition, Prentice-Hall, Englewood Cliffs, NJ.

Van Bibber, L. E., Kelly, J. L. (1985) Westinghouse 600 kW Wind Turbine Design. *Proc. of Windpower 1985*, American Wind Energy Association, Washington, DC.

Veers, P. (1984) Modeling Stochastic Wind Loads on Vertical Axis Wind Turbines, SAND83-1909, Sandia National Laboratories, Albuquerque, NM.

Wilson, R. E., Freeman, L. N., Walker, S. N., Harman, C. R. (1996) Final Report for the FAST Advanced Dynamics Code, OSU/NREL Report 96-01, Department of Mechanical Engineering, Oregon State University, Corvallis, Oregon.

Wright, A. D., Sexton, J. H., Butterfield, C. P. (1981) SWECS Tower Dynamics Analysis Methods and Results. *Proc. of the Wind Turbine Dynamics Workshop*, Cleveland, OH.

7

Wind Turbine Control

7.1 Introduction

The previous chapters have discussed the numerous components of wind turbines and their operation. To successfully generate power from these various components, wind turbines need a control system that ties the operation of all the subsystems together. For example, a control system might sequence wind speed measurements, check the health of system components, release the parking brake, implement blade pitch settings and close contactors to connect a wind turbine to the grid. Control systems may dynamically adjust blade pitch settings and generator torque to control power in high winds on variable speed wind turbines. Without some form of control system, a wind turbine cannot successfully and safely produce power.

There are three levels of control system operation, the two most important being supervisory and dynamic control. Supervisory control manages and monitors turbine operation and sequences control actions (e.g. brake release and contactor closing). Dynamic control manages those aspects of machine operation in which the machine dynamics affect the outcome of control actions (e.g. changing blade pitch in response to turbulent winds). The operation and design of all these aspects of turbine control systems is reviewed in this chapter.

This information here is intended to provide the reader with an overview of the important aspects of control systems that are specifically relevant to wind turbine control. Section 7.1 starts with a description of the levels of wind turbine control and examples of control systems in commercial wind turbines. The section includes information on the various subsystems, actuators, and measurement sensors that comprise the turbine control system. In Section 7.2 a basic wind turbine model is developed that is used to explain control system components in general and the specifics of control system components in wind turbines. This is followed, in Section 7.3, by the important aspects of common turbine operating strategies that are found in modern turbines, and then, in Section 7.4, by the details of the supervisory control systems that are used to implement these strategies. Finally, Section 7.5 presents an overview of dynamic control system design approaches and dynamic control issues that are specifically important in wind turbines.

Control system design and issues related to control system design for wind turbines are very large topic areas. The material presented here provides an overview of the more relevant issues. Introductory information on the implementation of control systems in wind turbines can be found in Grimble et al. (1990). Discussion of various aspects of wind turbine control systems can also be found in Gasch (1996), Heier (1996), Hau (1996) and Freris (1990).

7.1.1 Types of control systems in wind turbines

The purpose of the control system of a wind turbine is to manage the safe, automatic operation of the turbine. This reduces operating costs, provides consistent dynamic response and improved product quality, and helps to ensure safety. This operation is usually designed to maximize annual energy capture from the wind while minimizing turbine loads.

Wind turbine control systems are typically divided, functionally, if not physically, into three separate parts: (1) a controller that controls numerous wind turbines in a wind farm, (2) a supervisory controller for each individual turbine and, (3) if necessary, separate dynamic controllers for the various turbine subsystems in each turbine. These separate controllers operate hierarchically with interlocking control loops (see Figure 7.1).

The wind farm controller, often called a supervisory control and data acquisition (SCADA) system, can initiate and shut down turbine operation and coordinate the operation of numerous wind turbines. These SCADA systems communicate with the supervisory controllers for each wind turbine. More information on SCADA systems is included in Chapter 8.

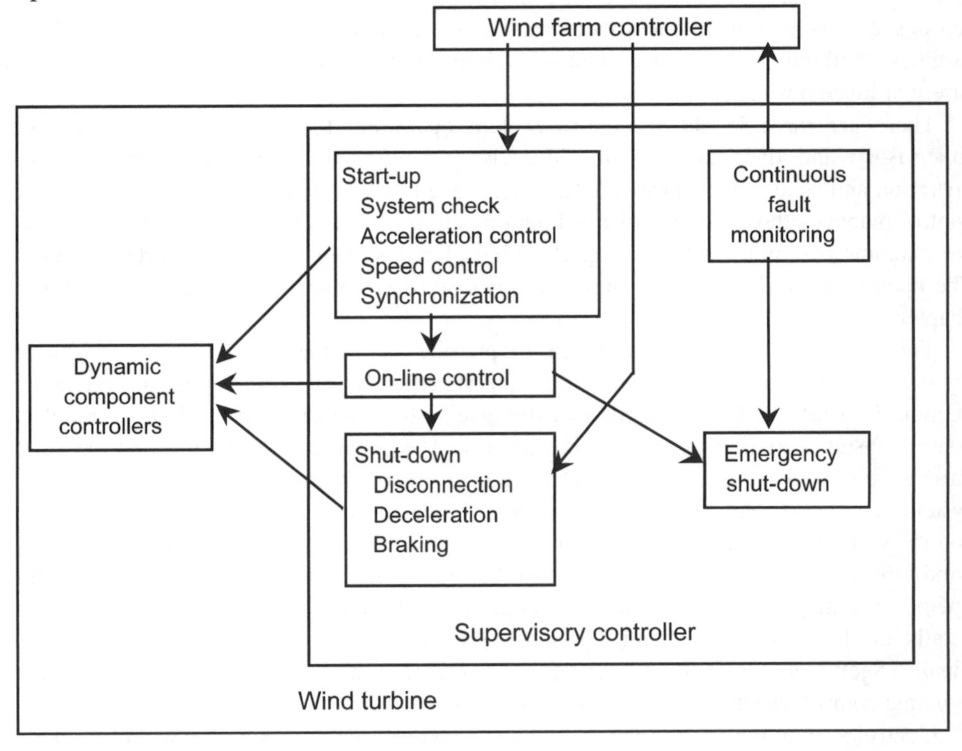

Figure 7.1 Control system components

The functions delegated to the supervisory controllers in the individual turbines are characterized by reactions to medium- and long-term changes in environmental and operating conditions. Thus, there may be a relatively long time between supervisory controller actions. Typically, the supervisory controller switches between turbine operating states (power production, low wind shutdown, etc.), monitors the wind and fault conditions such as high loads and limit conditions, starts and stops the turbine in an orderly fashion, and provides control inputs to the turbine dynamic controllers, for example, the desired tip speed ratio or rpm.

In contrast, dynamic controllers for the various turbine components make continuous high-speed adjustments to turbine actuators and components as they react to high-speed changes in operating conditions. Dynamic control is used for control systems in which the larger system dynamics affect the outcome of control actions. Typically, a dynamic controller will manage only one specific subsystem of the turbine, leaving control of other subsystems to other dynamic controllers and coordination of the various dynamic controllers and other operations to the supervisory control system. Dynamic control systems are used to adjust blade pitch to reduce drive train torques, to control the power flow in a power electronic converter, or to control the position of an actuator. Each of these controllers operates actuators or switches which affect some aspect of a turbine subsystem and thereby the overall operation of the wind turbine. The effect of controller actions is often measured and used as an input to the dynamic control system.

7.1.2 Examples of wind turbine control systems

Wind turbine control systems vary significantly from turbine to turbine. The choice of control system components and configurations depends significantly on the specific wind turbine design. Before examining control systems in general, a few highlights of some aspects of wind turbine control systems illustrate the variety of possibilities. These examples include two turbines with mechanical pitch control, and three different microprocessor controlled turbines.

7.1.2.1 10 kW Bergey Excel

The Bergey Windpower Company's 10 kW Excel wind turbine has a 7 m diameter variable speed rotor with a direct drive permanent magnetic alternator providing variable frequency three-phase power (see Figure 7.2). Depending on the application, that power is used directly for water pumping, retified to regulated DC power for battery charging or converted, through an inverter, to 240 V AC power for grid connection. The turbine has three hardware control systems plus an electronic controller that controls the power for the required application.

The three hardware control systems limit power in high winds, keep the turbine oriented into the wind and protect against rotor overspeed in extreme winds. The first controls aerodynamic torque with a unique system that is part of the blade design. The rotor has a rigid hub with three torsionally flexible blades with pitch weights located near the tips. Aerodynamic and centrifugal forces change the twist of the blades (and therefore the angle of attack) as the wind speed changes. The second control system orients the turbine into the wind with a tail vane. The downwind vane keeps the upwind rotor facing into the wind by means of aerodynamic forces on the vane surface. Finally, the rotor is protected against overspeed in high winds by another hardware-based control system. Above wind speeds of about 15 m/s the rotor is turned partially out of the wind through the action of aerodynamic and gravitational forces, without the use of springs.

An additional electronic controller manages the interface between the turbine generator and the intended application. For battery charging, a dedicated controller monitors battery voltage and controls current to ensure that the battery is not overcharged. Water pumping motors are usually driven directly by the alternator output, with the pump controller turning the pump on when sufficient voltage of enough frequency is being produced to power the pump without damage. For grid connection, a controller in the inverter manages the power flow to the grid and includes diagnostics to ensure safe operation of the inverter.

Figure 7.2 Bergey Excel. Reproduced by permission of Bergey Windpower Co.

7.1.2.2 Lagerwey LW18/80

The 80 kW Lagerwey LW18/80 has an 18 m diameter two-bladed, upwind, variable speed rotor (see Figure 7.3). Power produced by an induction generator is fed through an inverter before the connection to the grid. The turbine includes a unique system to control blade pitch. Both blades are hinged at the connection to the hub to allow the blades to cone downwind as the aerodynamic thrust increases in high winds. A linkage system in the hub converts this coning motion into pitching motion of the blades, controlling the rotor power in high winds. The result is a system that limits power to 80 kW between wind speeds of 12.5 and 25 m/s. Additional controls on the converter control the generator speed and current to the grid. The converter controls allow the variable speed rotor to operate between 50 and 120 rpm. This allows the rotor to operate efficiently in low wind speeds. The result is a very low cut-in wind speed of 3 m/s.

Figure 7.3 Lagerwey LW18/80. Reproduced by permission of Atlantic Wind Test Site Inc.

7.1.2.3 ESI-80

The ESI-80 is a two-bladed, stall-regulated turbine (see Figure 7.4). The turbine has a four-pole induction generator, a 30:1 planetary gear box and a teetered rotor. The teetered rotor is free to teeter up to +/- 3 degrees. Beyond that, the teeter motion is damped, with hard stops at +/1 7 degrees. The turbine is a free yaw design with a downwind rotor. As originally manufactured in the 1980s, the turbine had a hardware relay logic control system.

Power regulation is solely a function of the blade design. The higher the wind speed, the larger the angle of attack along the blade. As the angle of attack increases, the airflow over the blade is increasingly stalled and the rotor power coefficient, C_p, decreases. Thus, as the wind speed increases, the power increases to about 275 kW and then decreases in higher wind speeds. Overspeed control is handled by a pneumatically actuated disk brake and aerodynamic tip brakes. The tip brakes are latched with electromagnets. The tip brakes deploy when the power to the magnets is cut off or when the centrifugal forces on the brakes exceed the holding force of the magnets, as a result of high rotational speeds.

As originally manufactured in the 1980s, the turbine had a hardware relay logic control system. This monitored wind speed, turbine vibration, brake pressure and grid conditions.

Figure 7.4 ESI-80 turbine

7.1.2.4 Vestas V47-660/200 kW

The Vestas V47-660/200 kW turbine includes active pitch control, a small range of variable speed operation, and a control system with supervisory and dynamic control functions distributed between the top and bottom of the tower. It comes in a version with two generators for low-wind sites (see Figure 7.5).

In contrast to the ESI-80, the Vestas V47-660/200 changes the blade pitch in order to control the mean power output from the turbine in high winds. Fluctuations in power about the mean are reduced by allowing the rotor speed to vary a little. In high winds, the rated slip of the generator is changed by electronically changing the rotor resistance. This allows the rotor speed to vary by up to 10% as it absorbs the energy in gusts. The two-generator version operates at two distinct rotor speeds, depending on which generator is being used.

On the Vestas V47-660/200, the computer control is separated into two main processors, one on the tower top and one in the tower base. The tower top controller controls all aspects of the yaw motion and the generator and pitch systems, and monitors the turbine operation. The controller at the tower base controls the generator grid connection and the power factor correction capacitors, logs operating data, and communicates with remote operators.

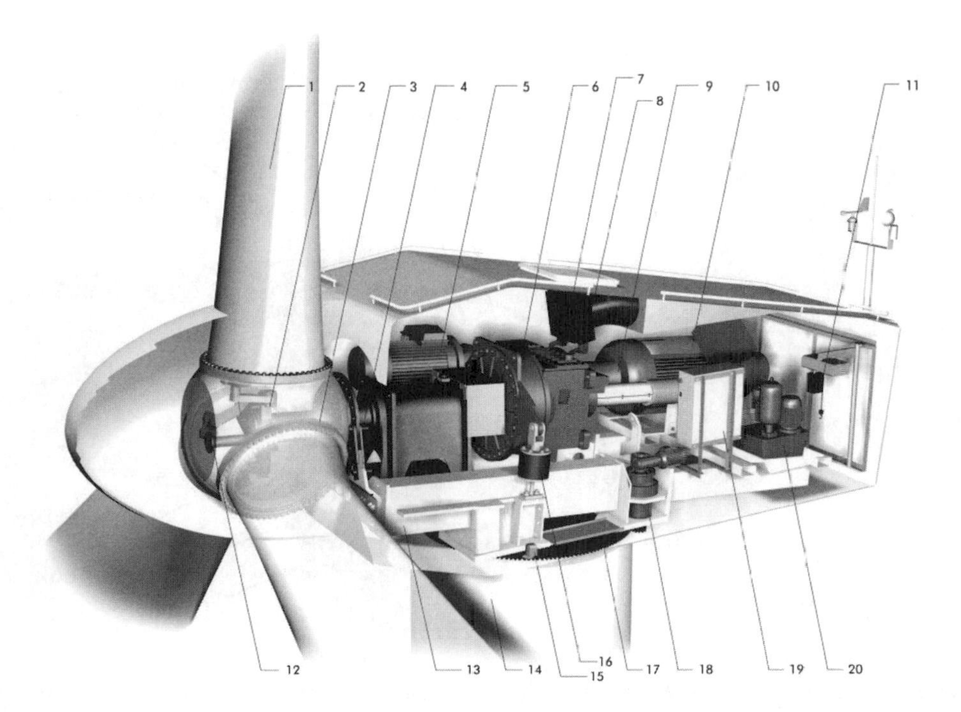

Figure 7.5 Vestas V47-660/200 kW. Legend: 1. Blade, 2. Blade hub, 3. Blade bearing, 4. Main shaft, 5. Secondary generator, 6. Gearbox, 7. Disc brake, 8. Oil cooler, 9. Cardan shaft, 10. Primary generator , 11. Service crane, 12. Pitch cylinder, 13. Machine foundation, 14. Tower, 15. Yaw control, 16. Gear tie rod, 17. Yaw ring, 18. Yaw gears, 19. Control unit, 20. Hydraulic unit. Reproduced by permission of Vestas Wind Systems A/S

7.1.2.5 Enron Wind 750i

The 750 kW Enron Wind 750i turbine is a fully variable speed, variable pitch machine with a doubly fed induction generator (see Figure 7.6). In contrast to the fixed-speed ESI-80 and the Vestas V47 with its discrete speed ranges, the rotor speed of the Enron Wind 750i can be varied continuously up to its peak operating speed. Also, in contrast to the Lagerwey LW18/80, which is also variable speed, the blade pitch is also fully controllable. The blade pitch is held constant below the rated wind speed and the rotor speed is varied for maximum aerodynamic efficiency. Above rated wind speed, the pitch is slowly varied to control the average power input to the rotor. Rotor speed variations absorb the energy in gusts. At the same time, the converter/generator control system controls for constant output power.

In this turbine, multiple distributed microcontrollers control the pitch and speed regulation, the high-speed shaft brake and yaw brake, the yaw motor and the hydraulic pump motor and monitor the turbine operation. The variable speed controller that controls the generator speed and blade pitch is located in the nacelle. It communicates with the master controller through a high-speed serial port over fiber optic lines to eliminate electrical noise and interference. The generator torque is controlled by the controller for the converter that is connected to the generator rotor. The controller is able to adjust the voltage and frequency of the generator current, thus controlling torque and generator efficiency. It is also able to set the power and power factor of the energy supplied to the grid.

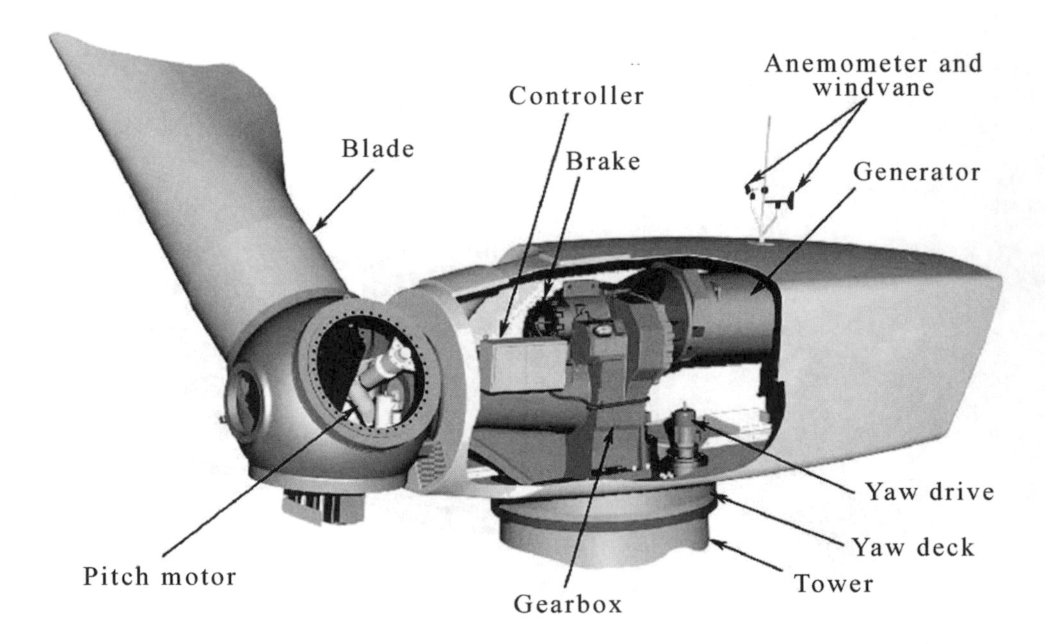

Figure 7.6 Picture of Enron Wind 750i turbine. Reproduced by permission of Enron Wind Corp.

7.2 Overview of Wind Turbine Control Systems

While the details of the control systems on different turbines vary, all the turbines considered here have a common purpose: the conversion of wind energy into electrical energy. This common purpose defines common elements that need to be considered in any control system design. This section starts with a simple turbine model that can be used to illustrate these turbine components and then reviews the basic functional elements common to all control systems and the forms those elements take in wind turbines.

7.2.1 Basic turbine model

A simplified horizontal axis wind turbine model is useful for understanding the integration of control systems into a modern wind turbine. A typical wind turbine can be modeled as a drive shaft with a large rotor inertia at one end and the drive train (including the generator) inertia at the other end (see Figure 7.7). An aerodynamic torque acts on the rotor and an electrical torque acts on the generator. Somewhere on the shaft is a brake.

The aerodynamic torque affects all turbine operations and provides the power that is delivered to the load. As discussed in Chapters 3 and 4, the aerodynamic torque is the net torque from the wind, consisting of contributions related to the rotor tip speed ratio, blade geometry, wind speed, yaw error, and any added rotor drag. Each of these inputs to the aerodynamic torque, except wind speed, may be able to be changed by a control system. Variable speed turbines can operate at different speeds (and different tip speed ratios);

pitch-regulated turbines can change the rotor or blade geometry; turbines with yaw drives or yaw orientation systems can control yaw error; and turbines with auxiliary drag devices can modify rotor drag. Below rated wind speed, control systems might attempt to maximize aerodynamic torque (or power), whereas above rated wind speed a control system would attempt to limit aerodynamic torque.

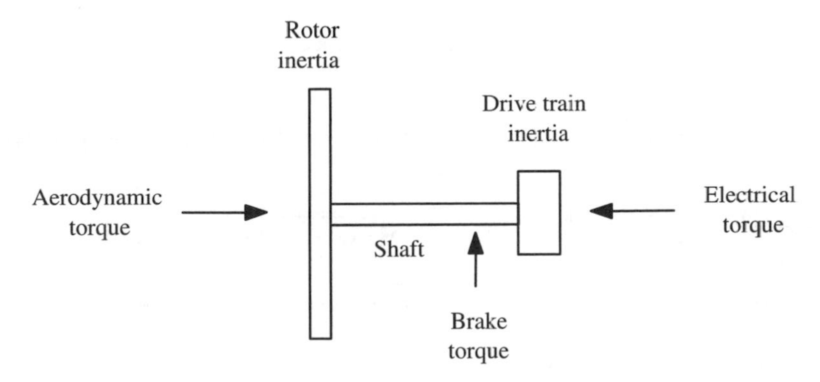

Figure 7.7 Simple wind turbine model

In a turbine designed to operate at nearly constant speed, the generator torque is a function of the fluctuating aerodynamic torque and drive train and generator dynamics. That is:

$$\text{Constant speed generator torque} = \text{f (aerodynamic torque, system dynamics)} \quad (7.2.1)$$

The drive train and generator dynamics are determined by the design of the various components and are not controllable. Thus, the only method for controlling generator torque in a constant-speed wind turbine is by affecting the aerodynamic torque.

In a variable speed turbine the generator torque can be varied independently of the aerodynamic torque and other system variables. That is:

$$\text{Variable speed generator torque} = \text{f (generator torque control system)} \quad (7.2.2)$$

In such a system the aerodynamic and generator torques could be independently controlled. The speed could be altered by changing either the aerodynamic or generator torque, resulting in either an acceleration or deceleration of the rotor.

7.2.2 Control system components

Control of mechanical and electrical processes requires five main functional components (see Figure 7.8):

1. A process that has a point or points that allow the process to be changed or influenced.
2. Sensors or indicators to communicate the state of the process to the control system.

3. A controller, consisting of hardware or software logic, to determine what control actions should be taken. Controllers may consist of computers, electrical circuits, or mechanical systems.
4. Power amplifiers to provide power for the control action. Typically, power amplifiers are controlled by a low-power input that is used to control power from an external high-power source.
5. Actuators or components for intervening in the process to change the operation of the system.

Examples of each of these functional components are provided in the following sections.

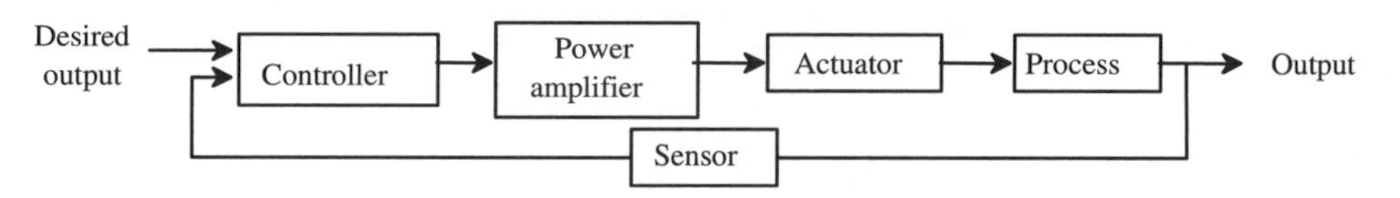

Figure 7.8 Control system components

7.2.2.1 Controllable processes in wind turbines

Controllable wind turbine processes include, but are not limited to:

- The development of aerodynamic torque (see Chapters 3 and 4).
- The development of generator torque (see Chapter 5).
- The conversion of electrical current and fluid flow into motion. Yaw drives and pitch mechanisms often use the control of electrical current or the flow of hydraulic fluid to control valves and the direction and speed of mechanical motion.
- Overall conversion of wind energy into electrical power. The successful conversion of the kinetic energy in the wind into useful electrical energy requires the monitoring and sequencing of a number of sub-processes. These larger aspects of turbine operation are also subject to control system actions. These might include connecting the generator to the grid, turning on compressors and pumps or opening valves.

7.2.2.2 Wind turbine sensors

On a large modern wind turbine many sensors are used to communicate important aspects of turbine operation to the control system. These measured variables might include:

- Speeds (generator speed, rotor speed, wind speed, yaw rate, direction of rotation)
- Temperatures (gearbox oil, hydraulic oil, gearbox bearing, generator bearing, generator winding, ambient air, electronics temperatures)
- Position (blade pitch, teeter angle, aileron position, blade azimuth, yaw position, yaw error, tilt angle, wind direction)
- Electrical characteristics (grid power, current, power factor, voltage, grid frequency, ground faults, converter operation)
- Fluid flow parameters (hydraulic or pneumatic pressures, hydraulic oil level, hydraulic oil flow)

- Motion, stresses, and strain (tower top acceleration, tower strain, shaft torque, gearbox vibration, blade root bending moment)
- Environmental conditions (turbine or sensor icing, humidity, lightning)

Sensors may also be composed of machine elements that act as part of the control system. For example, in the Lagerwey LW18/80 turbine, blade coning motion is used to control the turbine, just as input from an anemometer could be used to control the turbine.

7.2.2.3 Wind turbines controllers

Controllers provide the connection between the measurement of an aspect of turbine operation and actions to affect that turbine operation. Typical controllers in a wind turbine include:

- Mechanical mechanisms. Mechanical mechanisms, including tail rotors, linkages, springs, fly ball governors, etc., can be used to control blade pitch, yaw position and rotor speed.
- Electrical circuits. Electrical circuits may provide a direct link from the output of a sensor to the desired control action. For example, sensor inputs could energize coils in relays or switches. Electrical circuits can also be designed to include a dynamic response to input signals in order to shape the total system dynamic operation.
- Computers. Computers are often used for controllers. Computers can be configured to handle digital and analog inputs and outputs, and can be programmed to perform complicated logic and to provide dynamic responses to inputs. The ease with which control code, and thus control operation, can be changed by reprogramming the computer is a major advantage of computer control systems.

More detail on the different types of controllers is presented in Sections 7.4.4 and 7.5.4.

7.2.2.4 Power amplifiers in wind turbines

When the control signal from the controller does not have enough power to power the actuator, then an amplifier is needed between the controller and the actuator. Typical power amplifiers in a wind turbine include:

- Switches. There is a variety of switches that can be controlled with a small amount of current or a small force but which act as amplifiers in that they can switch high currents or high forces. These include relays, contactors, power electronic switches such as transistors and silicon-controlled rectifiers (SCRs) and hydraulic valves.
- Electrical amplifiers. Electrical amplifiers that directly amplify a control voltage or current to a level that can drive an actuator are often used as power amplifiers in a control system.
- Hydraulic pumps. Hydraulic pumps provide high-pressure fluid that can be controlled with valves that require very little power.

Note that power amplifiers are not always needed. In the Lagerwey LW18/80, for example, the coning motion of the blades, driven by a aerodynamic forces, develops enough power to change the blade pitch without amplification.

7.2.2.5 Wind turbine actuators

Actuators in a wind turbine may include:

- Electromechanical devices. Electromechanical devices include DC motors, stepper motors, AC motors with solid-state controllers, linear actuators and magnets.
- Hydraulic pistons. Hydraulic pistons are often used in positioning systems that need high power and speed.
- Resistance heaters and fans. Resistance heaters and fans are used to control temperature.

Actuator systems may include gears, linkages and other machine elements that modify the actuating force or direction.

7.2.3 Control of turbine processes

Wind turbine processes such as the development of aerodynamic and generator torque and the conversion of current into motion can be affected by controller action. The details of some of the typical approaches to affecting these processes in wind turbines are described in this section.

7.2.3.1 Aerodynamic torque control

As mentioned above, aerodynamic torque consists of contributions related to the rotor tip speed ratio and C_P (determined by blade design, wind speed and rotor speed), rotor geometry (blade pitch and aileron settings), wind speed, yaw error and any added rotor drag. All of these, except wind speed, can be used to control aerodynamic torque.

Tip speed ratio variations can be used to change the rotor efficiency and thereby the rotor torque. In stall-regulated fixed-speed wind turbines, low tip speed ratios (and accompanying low C_P) are used to regulate the aerodynamic torque in high winds. In variable speed wind turbines, the rotor speed can be changed to either maintain a favorable tip speed ratio or to decrease the tip speed ratio and power coefficient as in a stall-regulated turbine.

Changing the rotor geometry changes the lift and drag forces on the blade, affecting the aerodynamic torque. Aerodynamic torque control through rotor geometry adjustments can be accomplished with full span pitch control or by changing the geometry of only a part of the blade, as described below.

Full span pitch control requires rotating the blade about its long axis. Full span pitch control can be used to regulate aerodynamic torque by either pitching the blade to feather (reducing the angle of attack) or toward stall (increasing the angle of attack) to reduce loads. Blades for pitch-regulated wind turbines are usually designed for optimum power production, with no provision for gradually increasing stall as the wind increases. These blades are usually operated at the most efficient point with relatively high angles of attack. From this position, rotating the blade toward stall can often be accomplished faster than rotations to feather. Rotations to feather result in quieter operation and more exact control because each angle of attack is associated with one operating condition. In contrast, inducing stall results in unsteady loads, greater thrust forces on the turbine, less accurate control due to the unsteady nature of stalled flow and more noise.

Ailerons can be used to affect rotor geometry by changing the blade geometry over part of the blade. They are used to reduce the lift coefficient and increase the drag coefficient over the length of the blade with the aileron. Ailerons do not require the use of as powerful an actuator as full span pitch control does, but at least some of the actuating mechanism must be installed in the blade. The necessity of separating the blade into articulated pieces and providing actuation inside the blade significantly affects the blade design.

Auxiliary rotor devices such as tip brakes or spoilers can also be used to modify the rotor torque (see Figure 7.9). Tip flaps and pitchable tips add a negative torque to the rotor and spoilers disrupt the flow around the blade, decreasing lift and increasing drag.

Increasing the yaw error (turning the rotor out of the wind) and tilting the rotor and/or drive train can also be used to decrease or regulate the aerodynamic torque.

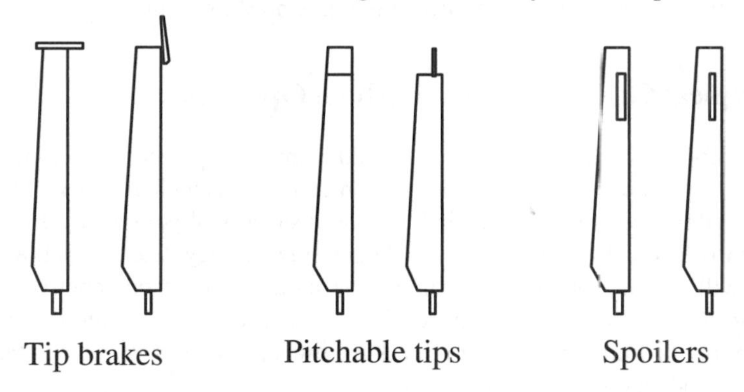

Figure 7.9 Aerodynamic drag devices (Gasch, 1996). Reproduced by permission of B. G. Teubner GmbH

7.2.3.2 Generator torque control

Generator torque may be regulated by the design characteristics of the grid-connected generator or independently controlled with the use of power electronic converters.

Grid-connected generators operate over a very small or no speed range and provide whatever torque is required to maintain operation at or near synchronous speed (see Chapter 5). Grid-connected synchronous generators have no speed variations and, thus, any imposed torque results in an almost instantaneous compensating torque. This can result in high torque and power spikes under some conditions. Grid-connected induction generators change speed by as much as a few percent of the synchronous speed. This results in a softer response and lower torque spikes than with a synchronous generator.

Alternately, the generator can be connected to the grid through a power electronic converter. This allows the generator torque to be very rapidly set to almost any desired value. The converter determines the frequency, phase, and voltage of the current flowing from the generator, thus controlling generator torque.

7.2.3.3 Brake torque control

Parking a wind turbine and stopping a stall-regulated wind turbine are often accomplished with a brake system on either the high-speed or the low-speed shaft (more on brakes can be found in Chapter 6). Brakes are typically pneumatically, hydraulically, or spring applied. Thus, control of brakes usually requires the activation of solenoid valves or, possibly, controllable valves to actively control the braking torque. Additional braking methods include (1) braking the rotor using the generator torque on machines controlled by power converters and (2) dynamic brakes, which are auxiliary electrical components that provide a electrical braking torque to the generator.

7.2.3.4 Yaw orientation control

A number of different designs have been used for controlling wind turbines by changing the direction of the wind entering the rotor. This approach is typically used in small wind turbines and may involve either yawing the rotor out of the wind or rotating the nacelle upwards to limit power output. Gyroscopic loads need to be considered in the design of a yaw power regulation system. If gyroscopic loads are a concern, yaw rate can be limited, but limiting yaw rate may affect the ability to regulate power output.

7.3 Typical Grid-connected Turbine Operation

Each of the processes mentioned above (aerodynamic torque control, generator torque control, brake torque control, and yaw orientation control), as well as others, can be used in a variety of combinations to enable a wind turbine to successfully convert the kinetic energy in the wind into electrical energy. The overall operating strategy determines how the various components will be controlled. For example, as part of an overall control strategy, the control of rotor torque can be used to maximize energy production, minimize shaft or blade fatigue, or simply to limit peak power. Blade pitch changes can be used to start the rotor, control energy production, or to stop the rotor.

In general, the goals of wind turbine control strategies are (1) maximizing energy production, (2) ensuring safe turbine operation and (3) minimizing operation and maintenance costs by reducing loads and increasing fatigue life. The control scheme used for operating a turbine depends on the turbine design. Within the limits of the design, the best overall strategy to meet these goals is chosen. Typical wind turbine operating strategies are explained in the Section 7.3.1.

The exact approach to wind turbine control and the immediate goals for the control strategy depend on the operating regime of the turbine. Below rated wind speed, one generally attempts to maximize energy production. Above rated wind speed, power limitation is the goal. Typical wind turbine control strategies for pitch- and stall-regulated machines, illustrated in Figure 7.10, are a function of wind speed and the options for control input. Fixed-speed stall-regulated wind turbines have no options for control input. Fixed-speed pitch-regulated wind turbines typically use the pitch regulation for start-up and, after start-up, only to control power above rated wind speed. Variable speed wind turbines typically use pitch control, if it is available, only above rated wind speed, but use generator torque control over the whole operating range of the turbine. It should be remembered that there are other approaches to turbine control that are used or are possible, but that are less frequently found in commercial designs, including, for example, yaw control to limit output power.

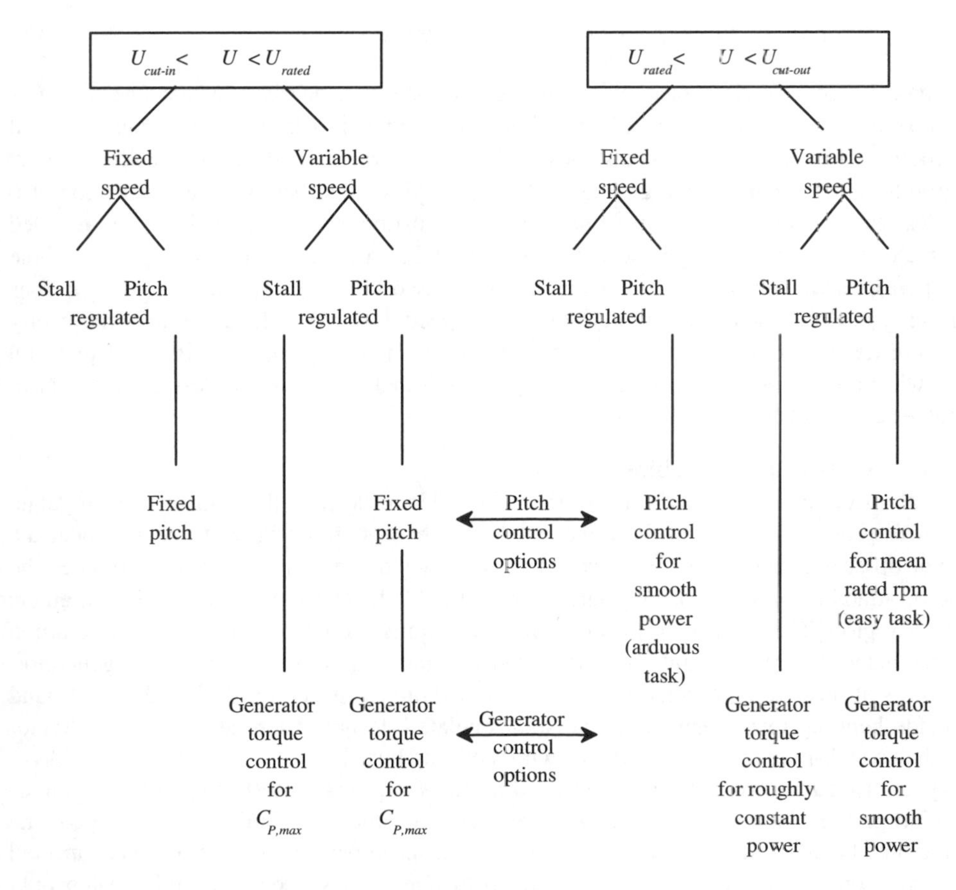

Figure 7.10 Overview of typical control strategies; U , mean wind velocity; U_{cut-in} , $U_{cut-out}$, U_{rated} , cut-in, cut-out, and rated wind speed, respectively

The details of typical turbine control operating strategies are summarized in the next section, with constant speed turbine and variable speed turbine descriptions described separately.

Turbine start-up strategy is also a function of available control options. Fixed-speed pitch-regulated wind turbines usually adjust the blade pitch to accelerate the rotor to operating speed, at which point the generator is connected. Many fixed-speed stall-regulated wind turbines cannot count on aerodynamics to accelerate the rotor. These turbines start by connecting the generator to the grid and motoring to operating speed. Variable speed wind turbines can use the same start-up strategies as fixed-speed wind turbines, but with the generator connected to the grid through a power converter.

7.3.1 Typical constant speed operating schemes

The majority of grid-connected wind turbines operate at a nearly constant speed that is predetermined by the generator design and gearbox gear ratio (discussed in Chapters 5 and 6). Among constant-speed wind turbines, a few different standard designs have emerged, each with their typical operating strategy. These are described below. The description also includes one two-speed design that operates at two distinct speeds. This design is included with the constant-speed designs because it is not a true variable speed design, but a turbine that operates as a constant speed machine at one of two selectable speeds. The descriptions of these typical turbine designs (and the variable speed designs) include the most frequently occurring designs and their control strategies. Again, it should be kept in mind that any given wind turbine design has its unique aspects and that other turbine designs and control strategies exist and are possible.

7.3.1.1 Stall-regulated turbines

Constant-speed stall-regulated wind turbines have blade designs that intrinsically regulate the power produced by the turbine. The fixed-pitch blades are designed to operate near the optimal tip speed ratio at low wind speeds. As the wind speed increases, so, too, does the angle of attack and an increasingly large part of the blade, starting at the blade root, enters the stall region. This reduces rotor efficiency and limits output power. The most common stall-regulated design is a rigid-hub, three-bladed, wind turbine with an induction generator. The rotors of these turbines tend to be heavy welded or cast structures designed to withstand the blade bending loads inherent in the stall-regulated design. There are also some lighter two-bladed stall-regulated wind turbines with teetering hubs.

Typically, control of stall-regulated wind turbines requires only starting and stopping of the wind turbine based on wind and power criteria. Once the brake is disengaged, the turbine may be allowed to freewheel up to operating speed before the generator is connected to the grid or the turbine may be motored up to operating speed. Thus, this design only requires control for the generator or soft-start contactor (see Section 5.6) and brake operation.

7.3.1.2 Two-speed stall-regulated turbines

A variation of the stall-regulated concept involves operating the wind turbine at two distinct, constant operating speeds. In low winds, the slower operating speed is chosen to improve the rotor efficiency and reduce noise. The higher rotor speed is chosen for moderate and high winds. One way to do this is to use a generator with switchable poles and therefore a switchable synchronous operating speed. Another approach is the use of two generators of different sizes. The operating speeds are determined by choosing the number of generator poles and/or the gear ratio for connecting these generators to the rotor. The smaller generator is used in low winds and the larger in high winds. Both generators operate close to maximum efficiency. Two-speed wind turbines require more complicated equipment for transferring power between generators and for avoiding transient power and current spikes when switching generators or generator poles.

7.3.1.3 Active pitch-regulated turbines

Rotors with adjustable pitch are often used in constant-speed machines to provide better control of turbine power. Blade pitch can be changed to provide power smoothing in high winds. Because pitchable blades are often designed for optimum power production, with no provision for stall regulation, the aerodynamic torque can be sensitive to gusts. One solution is to use fairly fast pitch mechanisms. The faster the pitch mechanism responds to gusts, the smoother the power in high winds will be. However, blade rotation velocities are limited by the strength of the pitching mechanism and blade inertia. In practice, power is only controlled in the average and some power fluctuations still exist. Below rated power the blade pitch is usually, though not always, held constant in these machines to limit pitch mechanism wear. This reduces energy capture but may improve the overall system economics and reliability.

7.3.2 Typical variable speed operating schemes

Variable speed grid-connected wind turbine operation has been made possible by improvements in power electronics. The cost of power electronics for grid connection tends to limit variable speed operation to larger wind turbines. However, small variable speed wind turbines with dedicated loads have successfully operated for generations. The descriptions below of typical variable speed turbine operation are limited to grid-connected machines.

7.3.2.1 Stall-regulated turbines

Variable speed operation of stall-regulated wind turbines is currently an active topic of research at a number of locations in Europe and the United States. Variable speed stall-regulated wind turbines are controlled by using power electronics to regulate the generator torque. By using the generator torque to regulate the rotor speed, the turbine can be operated at any desired tip speed ratio within the limits of the generator and rotor design constraints. By decreasing generator torque below the aerodynamic torque, the rotor is allowed to accelerate. The rotor decelerates when the generator torque is set higher than the aerodynamic torque.

Variable speed stall-regulated wind turbines operate in one of three modes (see Figure 7.11). At low wind speeds the turbine operates with variable speed to maintain optimum power coefficient. Once the maximum design rotor speed is reached the turbine is operated in a constant speed mode similar to normal stall-regulated operation. As the wind speed increases, the power will increase and the blade will be increasingly stalled. Above a predetermined power level the turbine is operated in a constant power mode in which the rotor speed is regulated to limit rotor power. This involves reducing the rotor speed in high winds to increase stall and reduce the rotor efficiency.

Control of variable speed stall-regulated turbines includes the same connection and disconnection logic required for constant speed stall-regulated operation. Once it is grid-connected, the power from a constant speed grid-connected machine is regulated by the rotor aerodynamics and the generator design. In a variable speed machine, grid-connected power is managed by a dynamic controller that regulates the generator torque with the goal of either constant tip speed ratio, constant speed, or constant power, depending on the wind speed.

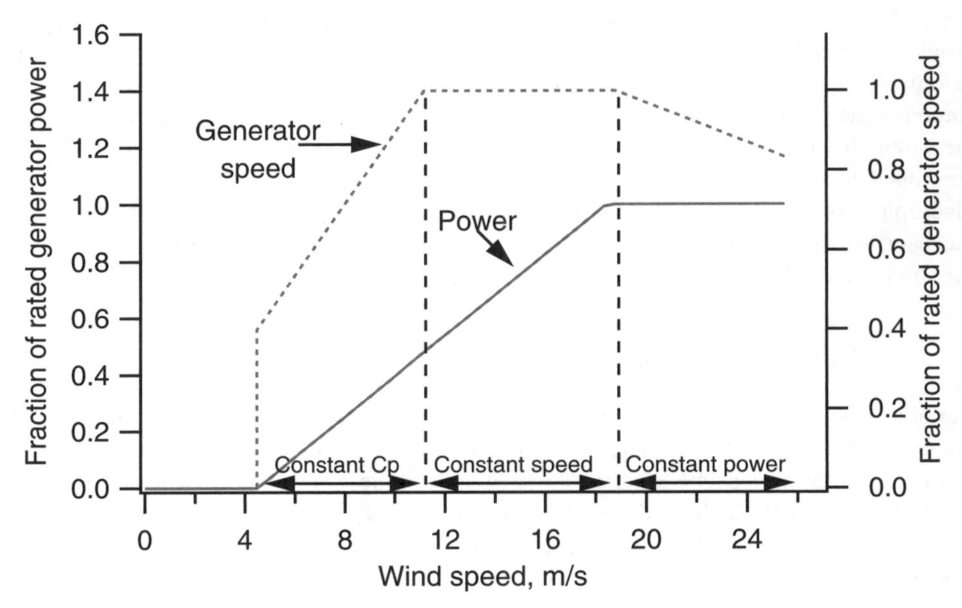

Figure 7.11 Variable speed stall-regulated operation

7.3.2.2 Active pitch-regulated turbines

Variable speed pitch-regulated wind turbines have two methods for affecting turbine operation: speed changes and blade pitch changes. In part load operation these wind turbines operate at fixed pitch with a variable rotor speed to maintain an optimal tip speed ratio. Once rated power is reached, the generator torque is used to control the electrical power output, while the pitch control is used to maintain the rotor speed within acceptable limits. During gusts the generator power can be maintained at a constant level, while the rotor speed increases. The increased energy in the wind is stored in the kinetic energy of the rotor. If the wind speed drops, the reduced aerodynamic torque results in a deceleration of the rotor while the generator power is kept constant. If the wind speed remains high, the blade pitch can be changed to reduce the aerodynamic efficiency (and aerodynamic torque), once again reducing the rotor speed. In this manner the power can be very closely controlled and the pitch mechanism can be much slower than the pitch mechanism used in a constant-speed machine.

7.3.2.3 Small-range variable speed turbines

One approach to approximating some of the advantages of full variable speed without all of the costs is the use of a variable slip induction generator. For example, the Vestas V47 has a wound rotor induction generator with a control system to control the rated slip of the generator (see Section 5.4.4) by changing the rotor resistance. At part load the generator operates as a regular induction generator with 2% rated slip. Once full load is reached, the rotor resistance is changed to increase the slip to allow the rotor to absorb the energy in gusts. (Note that this approach also increases generator losses.) The pitch mechanism is used to modulate the power fluctuations.

7.3.2.4 Passive pitch-regulated turbines

Numerous wind turbine designs, such as the Lagerwey LW30/250, have been produced with one form or another of passive pitch regulation. In these designs the effects of changing rotor speed or wind speed are coupled through linkages to pitch mechanisms. Thus, the wind speed provides the actuator power to adjust the blade pitch to shape the power curve of the wind turbine.

7.4 Supervisory Control Overview and Implementation

This section provides details of the operation of supervisory control systems for wind turbines. The supervisory control system manages the various operating states of the turbine (ready for operation, power production, shutdown, etc.), switching between operating states, and reporting to turbine operators. An overview of supervisory control functions and the details of typically used operating states are described in this section, as well as the various forms for implementing supervisory control systems.

7.4.1 Supervisory control system overview

The supervisory control system manages safe, automatic operation of the turbine, identifying problems and activating safety systems. Automatic turbine operation is typically managed by switching relays to change from one predefined operating state, such as grid-connected power production, to another, perhaps freewheeling in low winds without grid connection. At the same time, operating conditions are continuously checked against preset limits. If any input exceeds a safe value, appropriate action is taken. These tasks may be managed by hardware relay logic, electrical circuits, or most often by an industrial computer. A separate part of a supervisory control system is the fail-safe backup systems needed to safely shut the turbine down if the main supervisory system should fail.

The tasks performed by the supervisory control system may include:

- Monitoring for safe operation. This includes (1) monitoring sensors to insure that no turbine components have failed, (2) monitoring to make sure that operating conditions are within expected limits, (3) monitoring the grid condition, and (4) looking for troublesome environmental conditions
- Information gathering and reporting. This includes gathering information on operation, notification of the necessity for repairs and maintenance, and communication via phone, radio, or satellite, etc. with operators
- Monitoring for operation. This includes monitoring the wind speed and direction and grid condition to determine the appropriate operating condition
- Managing turbine operation. This involves choosing operating states and managing the transition between operating states, sequencing and timing of tasks within operating states, and providing limits and set points for dynamic control subsystems
- Actuating safety and emergency systems. This includes disconnecting the grid, and activating aerodynamic and regular braking systems in emergencies

7.4.2. *Operating states*

Experience has shown that a number of distinct operating states are appropriate for the operation of most wind turbines (see Figure 7.12). These include the system check, ready for operation, start, grid connection, power production, grid disconnection, freewheeling, shutdown, and emergency shutdown operating states. Each of these is described below. Depending on the turbine design, some of these states may be absent, some may be subdivided into multiple separate states, or some may be combined onto one state. The turbine may remain in some operating states for long periods of time, depending on the wind and operating conditions. These states are designated as stationary states. Other operating states may only be transition states that are entered during changes from one stationary operating state to another. The nature of the operating state (whether transitional or stationary) is indicated in Figure 7.12.

7.4.2.1 System check and initialization (transitional)

The system check state is entered when the control system is initialized for operation. This transitional state includes any initial tasks that need to be performed to make sure that the system is ready to operate or to get it ready to operate. When the system is first put into operation, faults may need to be cleared, the rotor and yaw position may need to be determined, variables need to be initialized, and sensor inputs need to be checked to make sure that the turbine systems are operating correctly. In this effort, actuators may be exercised and the results measured to determine that they are ready for operation.

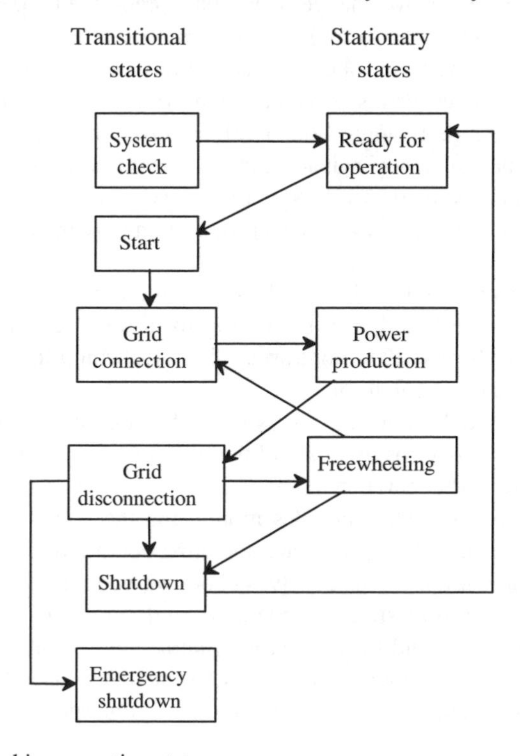

Figure 7.12 Typical turbine operating states

7.4.2.2 Ready for operation (stationary)

The ready for operation state is characterized by a stationary rotor and engaged parking brake. Once put on-line, the supervisory controller must: (1) maintain the readiness of the turbine for operation by monitoring turbine and grid conditions for problems, maintaining pressures in hydraulic and pneumatic reservoirs, and correcting the yaw error, if necessary, and (2) identify appropriate conditions for operation by monitoring the wind speed and direction. When appropriate conditions are identified and system checks find no fault that would preclude operation the start state is entered. Owing to the fluctuating nature of the wind, averages and statistical measures must be used to determine if there is enough wind to start the turbine with the expectation that it will continue to run for a while and not be just shut down in the next lull. Typically, the supervisory controller will determine that the turbine is ready to start when the wind speed, averaged over some time period, is above a predetermined set point and/or if the wind is over some limiting value for a certain time. This is a stationary state in which the turbine may remain for long periods of time.

7.4.2.3 Start and brake release (transitional)

Once conditions are appropriate, the start state is entered and the brake is released. Many turbines, especially pitch-regulated turbines, will accelerate up to operating speed without other intervention, but the pitch mechanism set point may need to be updated, in order to position the blades to accelerate the rotor. The starting procedure for variable speed turbines may require initiating the operation of dynamic controllers and providing speed set points. Meanwhile, system operation and the grid condition are monitored to identify problems that might require a shutdown.

7.4.2.4 Grid connection (transitional)

Some turbines, generally stall-regulated turbines, may need to be motored up to operating speed by connecting the generator to the grid subsequent to taking off the brake. For those turbines not requiring motoring, when the rotor speed approaches operating speed, the generator or converter contactor is closed. Power production commences with the completion of grid connection and the achievement of the correct operating speed. During grid connection, system and grid faults continue to be checked and the turbine continues to be oriented into the wind.

7.4.2.5 Power production (stationary)

During power production current flows into the electric grid. Controller tasks during power production will depend on the turbine design and whether the turbine is operating at part or full load. In constant-speed stall-regulated wind turbines power production may require only monitoring turbine operation and component health. In pitch-regulated turbines, blade pitch may be varied continuously at part load or only at full load. In variable speed turbines different control goals may be required for different load or wind speed ranges. The supervisory controller performs a number of tasks during power production including system fault detection, turbine yaw orientation, and continuous monitoring of power and rotor speed to identify operating problems and to determine set points for speed and power controllers. In high wind gusts, power production may be allowed to exceed the turbine's rated power for short periods of time to limit the duty cycle of the pitch actuator and to allow short period gusts to pass without increasing start–stop cycles.

7.4.2.6 Grid disconnection (transitional)

The tasks of this state may include disconnecting the generator–grid connection, disengaging various control systems, or providing new control goals and set points. This state may be a transition state to either a shutdown or freewheeling mode.

7.4.2.7 Freewheeling (stationary)

In low winds the rotor may be allowed to rotate freely, 'freewheeling,' until the winds drop even further or pick up and power production can occur. During freewheeling the generator is not connected to the grid and the controller monitors conditions for connecting to the grid or shutting down the turbine. System checks are performed. In a controlled-yaw turbine the rotor is oriented into the wind during freewheeling. In a free-yaw turbine the yaw error is monitored during freewheeling. Rotor speed is closely monitored and the blade pitch may be used, where applicable, to maintain the rotor speed within a specific range.

7.4.2.8 Shutdown (transitional)

This state is entered when winds or power are above specified upper levels, when winds drop below specified lower levels, or when system monitoring indicates that the turbine should not be operated. Stopping the turbine may require slowing the rotor with aerodynamic drag devices or by pitching the blades for shutdown, engaging the parking brake, and checking that the rotor has indeed ceased motion. Shutdown may also include parking the blades in a specific orientation and engaging the yaw brake. Upon completely shutting down, unless for a component problem, the turbine is ready for another operating cycle.

7.4.2.9 Emergency shutdown (transitional)

The emergency shutdown state is entered when critical operating or limit conditions are exceeded, when the normal shutdown procedure is deemed too slow to protect the turbine or when the normal shutdown procedure is deemed ineffective because of to a component failure. Typically, emergency shutdown results in a rapid deployment of all brakes, and a shutdown of numerous systems for safety or protection of equipment. Further operation of the turbine is not allowed without operator intervention.

7.4.3 Fault diagnosis

The continuous fault diagnosis capabilities of the supervisory controller must include monitoring for component failures, including sensor failures, operation beyond safe operating limits, grid failure or grid problems, and other undesirable operating conditions.

Component failures may be detected directly or indirectly. For example, the failure of a coupling between the generator and the gearbox could be directly detected if the generator and rotor speeds are known and do not correspond to each other. Such a failure could also be detected by noting that the rotor speed was accelerating or was too high, or by the detection of the deployment of a tip drag device designed to release in an overspeed condition. Generally, monitoring safe operating limits insures that many component failures will be detected, but certain failure modes may need to be monitored specifically to insure that they are detected before operation is adversely affected. Thus, successful failure mode detection requires a full analysis of possible failure modes and the consequences of those failure modes, and an evaluation of the necessary sensors to detect those consequences.

While the most robust and accurate sensors that can be afforded should be chosen for a wind turbine, sensor failures can also occur. Sensor problems can be protected against if the system is designed such that any problem condition would trigger two different sensors. Sensors may need to be selected to withstand cold, wet, or dry weather, vibration, high electrical and magnetic fields, condensation, ice deposits, oil and dirt deposits, high wind, or salt spray.

7.4.4 Supervisory control system implementation

Supervisory controllers can be implemented using hardware logic, electronic circuits, or computers. The choice of which approach to take will typically depend on the size and complexity of the wind turbine. Smaller wind turbines may utilize hardware or electronic controllers, but all large wind turbines are equipped with sophisticated computer controllers.

7.4.4.1 Hardware logic control systems

Hardware logic control systems are most easily implemented for simpler control strategies. These systems often use hardware logic called 'ladder logic.' Ladder logic systems may use a variety of components:

- Industrial relays that may have multiple relay outputs
- Sensors with relay outputs triggered by user-selectable set points
- Industrial timers with relay contacts that close only after a preset time period has elapsed
- A common power system for all actuators and relays

Ladder logic uses a cascading series of relays to control turbine operation. A simple ladder logic design example is illustrated in Figure 7.13. In this example, a small stall-regulated wind turbine is assumed to have the following equipment:

- An induction generator that connects to the grid with a contactor
- A brake that is released by introducing pressurized air from a reservoir tank through a solenoid and a brake pressure switch that indicates if the pressure in the brake is above a preset level
- A compressor that supplies air to the reservoir tank for brake operation. The compressor is switched on when the low tank pressure switch indicates that the air pressure in the tank is low
- A vibration switch that opens and remains open if the turbine vibrates too much
- A wind speed sensor that closes relay contacts when the wind speed is between cut-in and cut-out

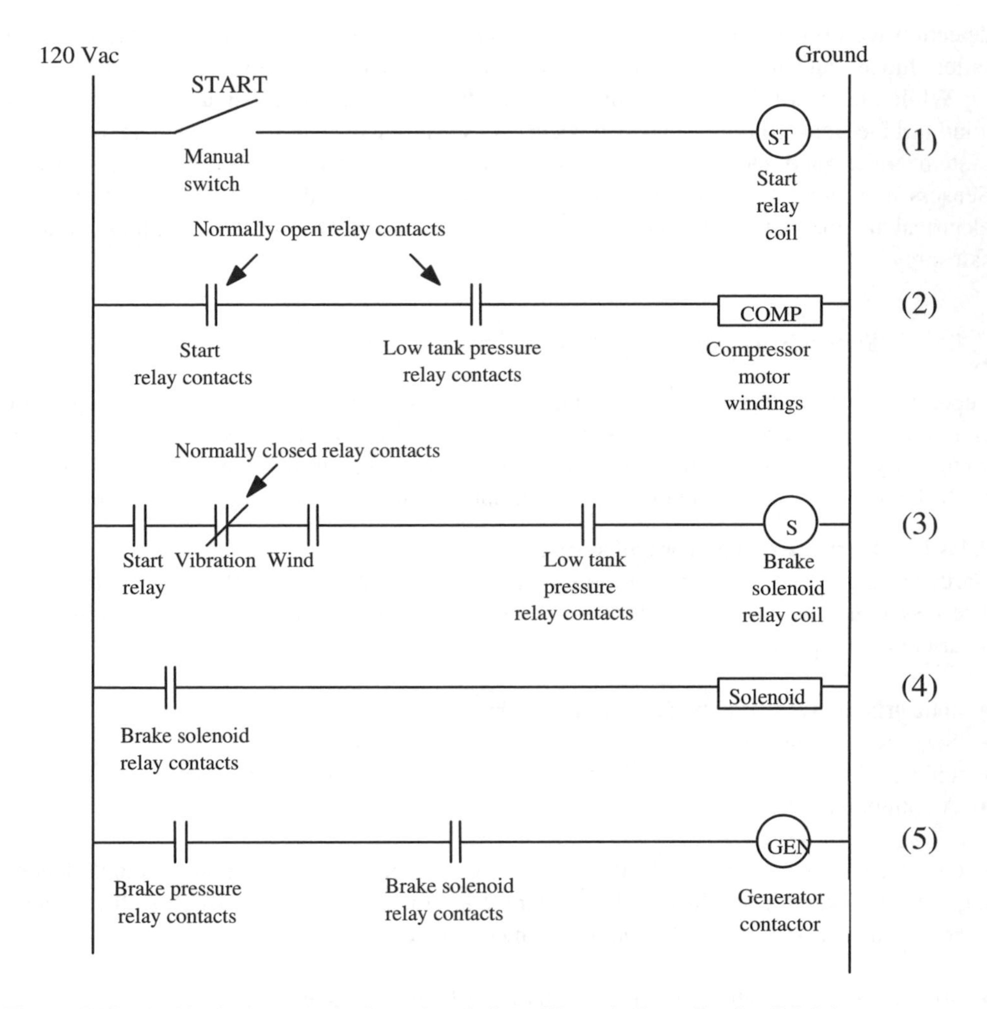

Figure 7.13 Ladder logic example; for a description of Parts 1–5, see Section 7.4.4.1

As shown in Figure 7.13, operator actuation of the start button energizes the coil of the start relay, a relay with multiple sets of contacts (Part 1). Once the start relay is energized, the compressor of the brake will be powered up if the low tank pressure relay on the reservoir tank is closed (Part 2). If there is adequate pressure in the tank the compressor remains off. If at any time the pressure drops, the compressor comes on. If the start relay has been energized *and* the vibration sensor indicates that no excess vibration has occurred *and* the wind speed is within operating range *and* the tank pressure is high enough, then the brake solenoid relay is energized (Part 3). Contacts on this relay open the solenoid to put air into the brake to release the brake (Part 4). Other contacts on this same relay allow the generator contactor to be energized if there is enough air in the brake piston (Part 5). At this point, the turbine accelerates to operating speed and generates power. If the wind drops, the brake solenoid will be closed, releasing air from the brake (Parts 3 and 4). As a consequence of the solenoid closing, the generator contactor will be disconnected from the grid (Part 5).

The brake will stop the wind turbine and the turbine will be ready for operation once the wind is within range and no vibration or low air pressure problems arise.

Some aspects of hardware logic controls can also be used as a backup system to insure that the turbine shuts down, for example, in emergency situations if the main computer has failed.

7.4.4.2 Electrical logic control circuits

Supervisory controllers can also be partially or completely composed of electronic circuits. These might include:

- Switches such as transistors and silicon-controlled rectifiers (SCRs) for switching circuits with high current and power
- Logic chips such as AND and OR gates and 'flip-flops' to implement controller logic
- Circuits to limit currents during start-up (soft-start circuits) or circuits to detect coincident AC waveforms to control contactors when synchronizing a synchronous generator with the power supply

Electronic circuits might range from electronic implementations of ladder logic circuits to much more complicated measuring and timing circuits.

7.4.4.3 Computer control systems

Most wind turbines use computer control systems for supervisory control. These controllers use industrial computers that are designed for dirty environments and with capabilities to interact with other industrial equipment through digital input and output (I/O) ports, analog-to-digital (A/D) and digital-to-analog (D/A) converters, and communication ports. Industrial computers may be similar to desktop computers with slots for user-selected interface boards or they may be designed for dedicated logic control and may come equipped with digital I/O ports and relay contacts ready to switch industrial equipment. Computer control systems may use one central processing unit (CPU) to manage all of the logic and control functions of the controller or may have CPUs distributed around the turbine, with each CPU communicating with a master controller.

7.4.5 Fail-safe backup systems

Wind turbine control systems rely on: (1) power, (2) control logic and (3) sensors and actuators. The control system must include fail-safe backup systems in case any of these elements fail. These fail-safe systems must safely shut down the turbine in cases of grid loss, rotor overspeed, excessive vibration, and other emergency situations.

Fail-safe backup systems should include the following components or functions:

- Orderly shut down on grid loss. If the power grid supplies the power for the actuators on the turbine, the loss of grid power removes the ability of the controller to shut the turbine down with these actuators. If contactors are designed to fail open and brakes are designed to fail closed when power is removed, then a grid loss results in a safe turbine shutdown. Power for shut down may come from springs, hydraulic accumulators or backup power supplies

- Backup controller power. Should the grid fail, backup power for the controller would allow the supervisory controller to set relays in a safe position to ensure that the turbine will not restart when power is restored and to continue to monitor the turbine state and store data for later use
- Independent emergency shutdown. If sensors fail, the supervisory controller may not be aware of a problem situation. Simple fail-safe backup systems could shut down the turbine in case of rotor overspeed and excessive vibration
- Independent hardware shutdown for supervisory controller failure. A software hang-up could leave the turbine in an ambiguous operational state or operating with no supervision. Should the supervisory controller hardware or software fail or should the control computer lose power, the system needs to be designed so that the turbine stops safely

7.5 Dynamic Control Theory and Implementation

This section includes an overview of dynamic control systems, the dynamic control system design process and examples of dynamic control system design issues specific to wind turbine applications. Background information on dynamic control system design can be found in textbooks such as those by Kuo (1987) and Nise (1992).

7.5.1 Purpose of dynamic control

Dynamic control systems are used to control those aspects of machine operation in which machine dynamics have an effect on the outcome of the control action. Typically, control systems can be designed to improve the accuracy and dynamics of machine response, to improve the machine response to unwanted outside disturbances imposed on the system, and to reduce machine sensitivity to variations in machine components or machine operation under different circumstances. To achieve this, control systems use feedback: a measure of the outcome of the control action that is included in the input to the control system. The control system uses this measure of the machine output in determining the next control action to help insure that the machine works properly.

These effects can be illustrated with the example of a pitch control mechanism. A simple pitch-control mechanism might use an electric motor to rotate the blade about its pitch axis. To rotate the blade, a specific amount of current might be applied to the motor for a predetermined amount of time. Whether the blade pitch changes by the desired amount is a function of many factors:

- Changes in machine operation over time, such as changes in motor operation due to temperature, pitch axis bearing friction or wear of components
- Variations in the components installed in different wind turbines (different winding resistance, bearing friction, blade mass and distribution)
- External disturbances such as pitching moments, due to aerodynamic and dynamic forces, or changes in blade inertia due to icing

Thus, it can be seen that such a control system would not always work as desired. As an alternative, a 'closed-loop' control system might be designed which uses a measure of the position of the blade root as an input to the control system. If for any reason the blade had not moved the desired amount, corrections could be made, making machine operation less sensitive to variations and disturbances. Closed-loop control can also be used to improve system dynamics. In the example above the blade would always move about the same distance in about the same amount of time. A properly designed control system might quickly increase the motor current in order to accelerate the blade and then decelerate the blade as it reached its goal, improving the response time of the system.

Wind turbine operation imposes its own unique constraints on control system design. Some of these are related to the turbine dynamics such as the use of long cantilevered blades and tall towers that vibrate, and the use of metal shafts with little inherent damping. Other constraints come from the application itself. A variable speed turbine might attempt to maintain a constant tip speed ratio (constant maximum C_P) over some range of wind speeds. The control system must then try to track variations in the wind speed, changing the generator torque in order to change the rotor speed. In this case the wind speed determines the goal of the control system. Above the rated wind speed, the aim of the control system might be completely different. Above rated wind speed, the generator torque and blade pitch of a pitch-controlled variable speed machine are adjusted to reduce loads and to maintain constant rated power or torque. The purpose of the control system is, then, just the opposite of constant C_P operation: the maintenance of constant power in the face of wind speed changes which are now disturbances to the system operation.

The next section provides an overview of the dynamic control system design process. Section 7.5.3 discusses specific issues in the design of dynamic controllers for wind turbines. Examples illustrate both general dynamic control concepts (open- versus closed-loop control) and some specifics of wind turbine control problems. The final section in the chapter provides examples of the implementation of dynamic control systems.

7.5.2 Dynamic control system design

7.5.2.1 Classic control system design methodology

The classic control system design process, as described in Grimble et al. (1990) and De LaSalle et al. (1990), includes the following steps (see Figure 7.14):

- Problem analysis. The problem analysis must include consideration of the desired machine operation, the available control effort, appropriate sensors and actuators, and any other design constraints. This analysis may indicate the need for an updated machine design with improved control characteristics
- Formulation of specifications. Preliminary design specifications include measures such as system response time, the overshoot in the response of the controlled system to a step input and controlled system stability. There may be hardware tradeoffs to consider. Faster system response requires greater actuator power, increases component loads and decreases component fatigue life

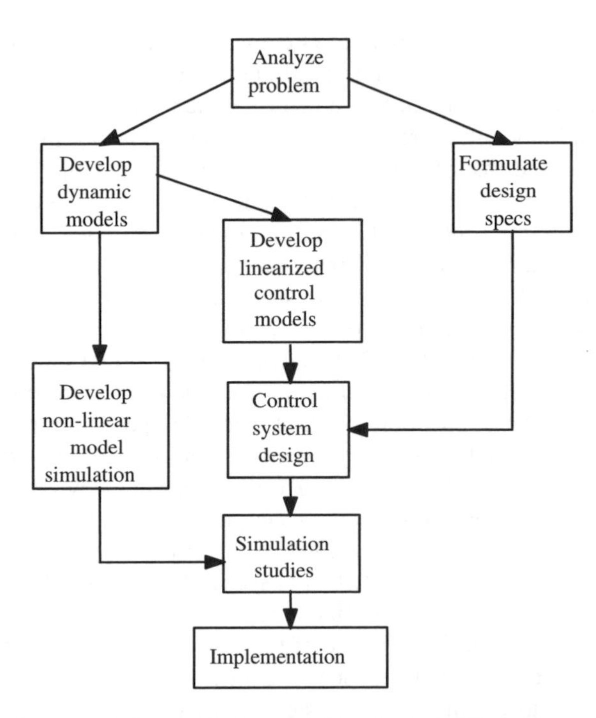

Figure 7.14 Control system design methodology (Grimble et al., 1990). Reproduced by permission of the Institution of Mechanical Engineers

- Model development. Control system design requires an understanding of system dynamics, usually through the use of mathematical models. Depending on the system, these may be linear or non-linear models. Each of the subsystem models needs to correctly reflect the system dynamics in the frequency range of interest
- Model linearization. For initial control system design, these models are often linearized. This allows the use of numerous straightforward approaches to linear system design
- Control design. During the design process, the engineer attempts to design the dynamics of the controller such that the overall controlled machine dynamics meet the design specifications. In the 'classical' design approach, the dynamic response of the controlled system is designed to conform to certain criteria over three different frequency ranges of interest. Low-frequency system behavior is designed to track the desired control commands. At mid-frequency ranges, the system response is designed to ensure stability and adequate system response time. Higher frequency dynamics must ensure that unmodeled dynamics and sensor measurement noise do not affect system behavior. Other design approaches are described below
- Simulation development. In order to test the operation of a controller design, computer codes for non-linear simulations are developed, based on non-linear system models. If possible, the results of these simulations should be validated against real data

- Simulation studies. Once a tentative controller has been designed for the linearized system, the non-linear simulation model is used to investigate behavior of the more realistic non-linear system including the new controller design
- Implementation. Once an adequate design has been achieved, the controller is implemented on a wind turbine and tested in the field

This process is highly iterative and at any point one might return to a previous step to redefine control objectives, refine models, or redesign the controller.

7.5.2.2 Other control design approaches

Other control design approaches that build on or deviate from classical linear system design include adaptive control, optimal control and search algorithms (see De La LaSalle et al., 1990 and Di Steffano et al., 1967). Each of these approaches may have advantages over linear control design methods, especially in the control of non-linear systems. These time domain approaches can include full non-linear dynamic turbine models in computer simulations for the control system design.

Adaptive control The dynamic behavior of a wind turbine is highly dependent on wind speed due to the non-linear relationship between wind speed, turbine torque and pitch angle. System parameter variations can be accommodated by designing a controller for minimum sensitivity to changes in these parameters. These adaptive control schemes are also useful in systems in which system parameters change, especially if they change rapidly or over a wide range. Adaptive control schemes continuously measure the value of system parameters and then change the control system dynamics in order to make sure that the desired performance criteria are always met.

Optimal control Optimal design is a time domain approach in which variances in the system output (for example, loads) are balanced against variances in the input signal (for example, pitch action). Optimal design approaches are inherently multivariable, making them suitable for variable speed wind turbine control design. Optimal control theory formulates the control problem in terms of a performance index. The performance index is often a function of the error between the commanded and actual system response. Sophisticated mathematical techniques are then used to determine values of design parameters to maximize or minimize the value of the performance index. Optimal control algorithms often need a measurement of the various system state variables or a state estimator based on a machine model.

Search algorithms Search algorithms can also be used to control wind turbines. These algorithms might constantly change the rotor speed in an attempt to maximize measured rotor power. If a speed reduction resulted in decreased power then the controller would slightly increase the speed. In this manner the rotor speed could be kept near the maximum power coefficient as the wind speed changed. The controller would not need to use a machine model and would, thus, be immune to changes in operation due to dirty blades, local air flow effects, or mispitched blades.

7.5.2.3 Wind turbine system models

System models are often mathematical models based on the principles of physics. When such models cannot be developed, an experimental approach termed system identification can be used.

Models based on physical principles Dynamic models are used to understand, analyze and characterize system dynamics for control system design. The models consist of one or more differential equations describing system operation. These differential equations are usually written in one of two forms: the transfer function representation or a state space representation. The transfer function representation involves the use of Laplace transforms and characterizes the system in the frequency domain, while the state space representation characterizes the system in the time domain. Each of these system representations is interchangeable and the choice of approach depends on the degree of complexity of the system and the analytical tools available to the system designer.

It should be remembered that the results of control system design are only as good as the model used to describe the system. A control system based on a model that ignores critical dynamics of one part of the machine can result in catastrophic failure. However, an excessively detailed model adds complexity and cost to the analysis and may require machine input parameters that are unknown. Engineering judgment is required in developing the system model. Simple, yet appropriate, models often describe a system as a set of lumped masses or ideal machine elements, ignoring less important machine details. These models are useful if the masses or stiffness of the idealized elements are chosen so that the model truly represents system behavior. Often data from machine operation are used to determine these parameters. In this way the simple model is 'fit' so that it best represents true operation of the machine. As long as the chosen machine model represents the system dynamics of interest, then a successful control system can be developed. When a non-linear system has been linearized for control system design, simulations of system behavior over a wide range of non-linear operation should be used to check operation of the full controlled system.

The important wind turbine subsystems that need to be modeled are:

- Wind structure
- Drive train dynamics
- Aerodynamics
- Generator dynamics
- Actuator dynamics
- Structural dynamics
- Measurement dynamics
- Controller dynamics

The interactions of these subsystems in a constant-speed pitch-controlled wind turbine are illustrated in Figure 7.15. Some sample mathematical models of wind turbine subsystems can be found in Novak et al. (1995).

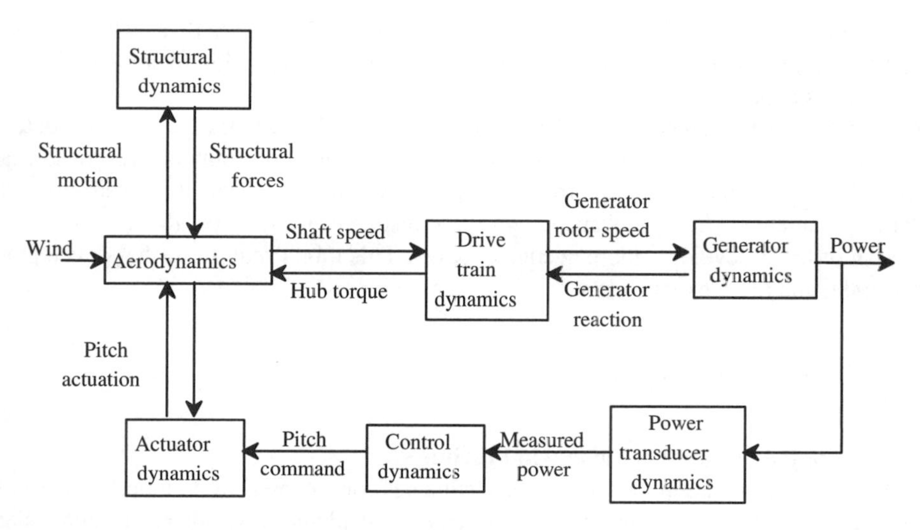

Figure 7.15 Wind turbine dynamics (Grimble et al., 1990). Reproduced by permission of the Institution of Mechanical Engineers

System identification In cases where models for disturbances or for complicated systems cannot easily be determined from physical principals, the experimental approach of system identification can be used (see Ljung, 1999). The system identification approach involves four main steps:

1. Planning the experiments
2. Selecting a model structure
3. Estimating model parameters
4. Validating the model.

System identification involves measuring the system output, given a controlled input signal or measurements of the system inputs. Simpler approaches to system identification require sinusoidal or impulsive inputs. Other approaches can use any input, but may require increased computational capability to determine the system model. In order to correctly identify the dynamics of a system, the input signal should be designed to excite all the modes of the system sufficiently to provide measurable outputs.

The data from experiments are used to fit the parameters for a model of the system. That system model may be based on a priori knowledge of the system and disturbances. If the system is known to behave in a linear manner over a particular operating range, then general representations of linear systems called 'black box models' may be used. In these cases, the parameters as well as the order of the models are to be identified. It may be possible to reduce the number of unknown parameters by using models developed from physical principals.

Parameter estimation is usually formulated as an optimization problem using some criteria for the best fit to the data. There are many different possible approaches to the system identification calculations and the optimization criteria. On-line methods provide parameter estimation as measurements are being obtained. These methods are useful for time-varying systems or for use in adaptive controllers. Off-line methods usually have more reliability and precision. They most often use an optimization criterion in some variation of a least squares approach.

Once the model parameters have been determined, it is important to test and validate the model. The model is often tested by checking the step and impulse responses, and determining model and prediction errors. A sensitivity analysis reveals the sensitivity of the model to parameter variations. Depending on the results of these steps, the process may be iterated in order to develop a more adequate model. This might require further development of the model and more experiments.

7.5.3 Control issues in wind turbine design

7.5.3.1 Control issues specific to wind turbines

Wind turbines present a number of unique challenges for a control system designer (see De LaSalle et al., 1990). Conventional power generation plants have an easily controllable source of energy and are subject to only small disturbances from the grid. In contrast, the energy source for a wind turbine is subject to large rapid fluctuations, resulting in large transient loads in the system. The consequences of these large fluctuations and other aspects of wind turbines introduce unique issues in control system design for wind turbines:

- The penalty for a poorly regulated wind turbine is the requirement that the turbine structure and components (shafts, gearbox, etc.) withstand high loads. This results in heavier components, a heavier structure, and increased turbine costs
- A wind turbine system consists of numerous lightly damped structures (to avoid energy dissipation in the system) that are excited by forcing functions at the frequency of the rotor rotation and its harmonics. In addition, the component dynamics, including the wind-turbulence fluctuations, aerodynamics, rotor, drive train, tower, and control system dynamics, may all have significant responses in the frequency range of the rotor rotation. These potential resonances impose severe constraints on system operation. Resonances between the speed and torque control system, the pitch control system, and the many system natural frequencies (blades, drive train, tower) must be avoided. System natural frequencies at the frequency of the system forcing functions (harmonics of the blade rotation frequency and possibly frequencies of fluctuations in the wind field) must be avoided
- The aerodynamics are highly non-linear. This results in significantly different dynamic descriptions of turbine behavior at different operating conditions. These differences may require the use of non-linear controllers or different control laws for different wind regimes
- Transitions under dynamic operation from one control law or algorithm to another require careful design
- The control goal is not only the reduction of transient loads, but also of fatigue loads, caused by the load fluctuations about the mean load
- Control measurement and actuator hardware cost and weight must be minimized
- Reliable torque measurements for feedback are often difficult to obtain

- Adequate system models are often difficult to determine. System models need to be compared with measurements to confirm that the model parameters represent the true machine operation. The control system designer attempts to make a system model that represents the dynamics of the system without being overly complex. Complexity increases design time and also results in models with numerous parameters that may be difficult to determine. Simple models may represent the system well, but not with the expected parameters. A drive train includes gearbox dynamics (backlash, flexing of gear teeth), shafts with low and difficult to quantify damping ratios, and a non-rigid rotor subject to non-linear aerodynamics. These can result in behavior that deviates from that of a simple model (see Novak et al., 1995)
- Extensive research has been required and is still ongoing into the exact loads that a wind turbine experiences

Some examples of these issues are presented below. First, the advantages of closed-loop over open-loop control systems are illustrated using a simple pitch control system. The system response to outside disturbances is analyzed. Then, issues of natural frequencies and resonance are illustrated using the same closed-loop pitch control system. In the following two subsections, issues related to variable speed turbine operation are considered, including control for constant optimum tip speed ratio in low winds and issues related to transitioning to other operating modes at higher wind speeds. Finally, some of the details of modeling disturbances in a pitch control system are provided.

7.5.3.2 Open-loop and closed-loop response to disturbances

The basic differences between open- and closed-loop control systems are illustrated in Figure 7.16. It can be seen that in an open-loop system, the controller actions are based on the desired system state, with no reference to the actual state of the process. In closed-loop feedback systems, the controller is designed to use the difference between the desired and the actual system output to determine its actions.

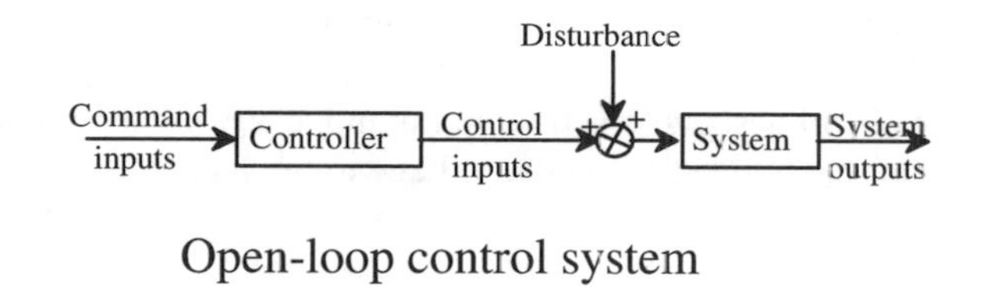

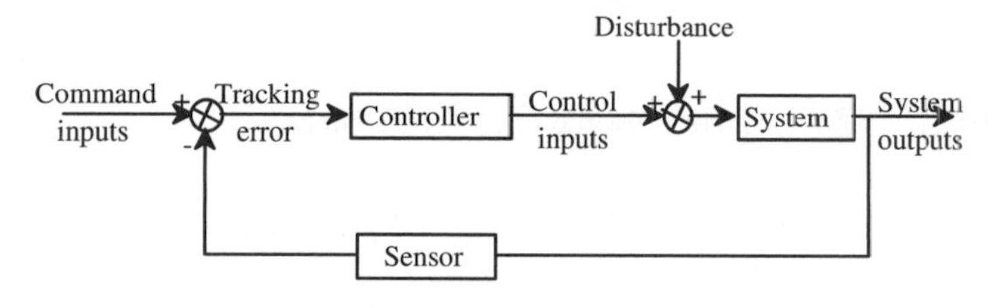

Figure 7.16 Comparison of open-loop and closed-loop control systems

Some aspects of control systems can be illustrated with a simple pitch mechanism driven by an AC servo motor with a spring return and subject to external pitching moments (disturbances to the system). The blade and pitch mechanism model includes moments from the spring and viscous friction. The torque from an AC servo motor can be modeled as a linear combination of terms that are a function of motor speed and of applied voltage (see Kuo, 1987). In the model presented here, the motor torque and speed are referred to the blade side of the pitch mechanism. The differential equation for this system has the system dynamics terms on the left and the external torques from the motor and disturbance on the right:

$$J\ddot{\theta}_p + B\dot{\theta}_p + K\theta_p = kv(t) + m\dot{\theta}_p + Q_p \tag{7.5.1}$$

where θ_p is the angular position of the motor, J is total inertia of the blade and motor, B is the pitch system coefficient of viscous friction, K is the pitch system spring constant, k is the slope of the torque-voltage curve for motor/pitch mechanism combination, $v(t)$ is the voltage applied to the motor terminals, m is the slope of the torque–speed curve for motor/pitch mechanism combination and Q_p is a pitching moment due to dynamic and aerodynamic forces that acts as a disturbance in the system.

If the system is at steady state, then the derivatives of the pitch angle are zero and $v(t)$, the voltage to the motor, is a constant value, v. In this case the differential equation becomes

$$\theta_p = \frac{k}{K}v + \frac{Q_p}{K} \tag{7.5.2}$$

From this it can be seen that the steady state pitch angle will be a function of the voltage applied to the motor, the spring constant, and any pitching moment on the blade. The greater the pitching moment (or the weaker the spring), the greater the error in the pitch angle will be.

Assuming that the airfoil designer has tried to minimize any pitching moment, one could design a control system that applied a specific voltage to the motor for a desired 'reference' pitch angle, $\theta_{p,ref}$:

$$v = \frac{K}{k}\theta_{p,ref} \tag{7.5.3}$$

In this case, the differential equation for the open-loop system is:

$$J\ddot{\theta}_p + (B - m)\dot{\theta}_p + K\theta_p = K\theta_{p,ref} + Q_p \tag{7.5.4}$$

Here the pitch angle is the output and the system has two inputs, the voltage to the pitch motor and a disturbance torque from the aerodynamic and dynamic pitching moments on the blade. If one assumes that all derivatives and the pitching disturbance torque are zero, it can be seen that the steady state response of the system to a desired pitch angle command is, indeed, that desired pitch angle.

The relationships between the various dynamic elements in a controlled system are often illustrated with the use of block diagrams. A block diagram for this system is shown in Figure 7.17.

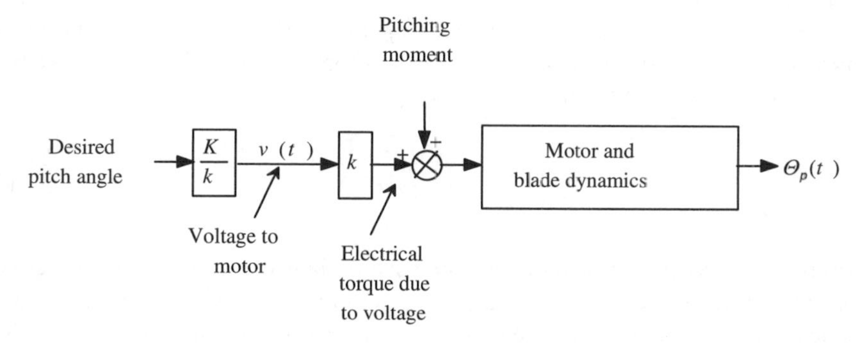

Figure 7.17 Open-loop pitch control mechanism; for notation, see text

The Laplace transform of the system response about steady state operation to an impulse input is referred to as the system transfer function. It is often used in control system design to characterize system dynamics (for more background information, see Kuo, 1987 or Nise, 1992). It can also be used to solve the differential equation for the system. The system transfer function can be found by taking the Laplace transform of the open-loop differential equation, solving for the pitch angle and assuming that initial conditions are all zero. The transfer function of this system would be:

$$\Theta_p(s) = \frac{K\Theta_{p,ref}(s)}{Js^2 + (B-m)s + K} + \frac{Q_p(s)}{Js^2 + (B-m)s + K} \tag{7.5.5}$$

where $\Theta_p(s)$, $\Theta_{p,ref}(s)$ and $Q_p(s)$ are the Laplace transforms of $\theta_p(t)$, $\theta_{p,ref}(t)$ and $Q_p(t)$. The first term in Equation 7.5.5 is the transfer function from the voltage input to the pitch angle output. The second term is the transfer function from the disturbance (the pitching moment) to the pitch angle.

The dynamic response of the open-loop system to a unit step disturbance can be found from the inverse Laplace transform of the transfer function that relates the disturbance to the final pitch angle:

$$\Theta_p(s) = \frac{Q_p(s)}{Js^2 + (B-m)s + K} \tag{7.5.6}$$

where $Q_p(s)$ would be $1/s$ for a step input. If, for example, the motor and blade dynamics are:

$$\Theta_p(s) = \frac{1}{\left(s^2/16\right) + \left(s/4\right) + 1} \tag{7.5.7}$$

where $J = 1/16 = 1/16$, $(B - m) = 1/4$ and K, Q_p, and k are all equal to 1, the response to a step disturbance is shown in Figure 7.18.

In Figure 7.18, the deviation from ideal operation should be zero, but the pitch disturbance results in a steady pitch angle error. From Equation 7.5.2, the steady state positioning error due to the pitching moment is:

$$\theta_p = \frac{Q_p}{K} \tag{7.5.8}$$

In practice open-loop control systems often have failings that significantly affect system operation. Manufacturing variability, wear, changes in operation with temperature and time and outside disturbances can affect open-loop system performance. In this example, changes in the wind speed, rotor speed, blade icing or anything else that might affect the pitching moment would change the pitch. If the effects of these changes in operation become a problem, closed-loop control systems could be used to improve system performance without significantly complicating the control system.

In a closed-loop control system, a measurement of the pitch angle can be incorporated into the input to the pitch control system, and corrections can be made for errors in the position of the blade. Typically, the controller provides a control output that is a function of the 'tracking error', the difference between the desired system output and the measured output. The block diagram of such a closed-loop system is shown in Figure 7.19. The controller would include appropriate dynamics and a power amplifier to adjust the voltage to the motor.

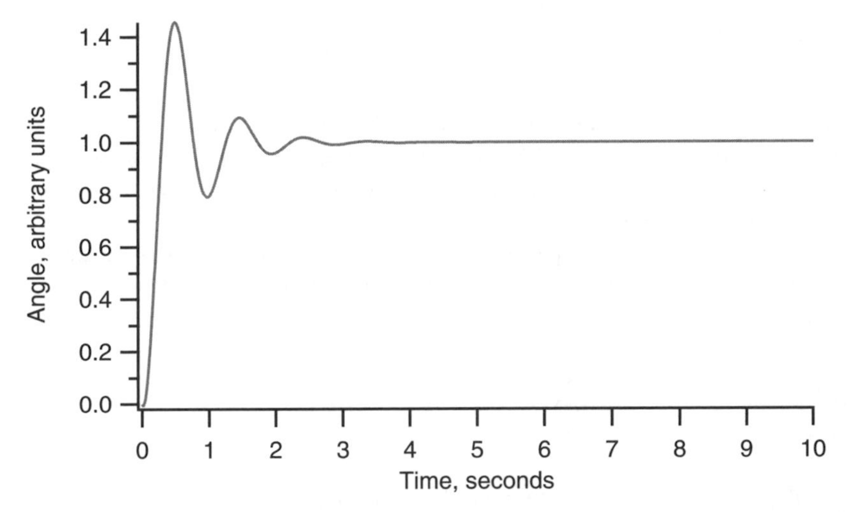

Figure 7.18 Sample pitch system step response

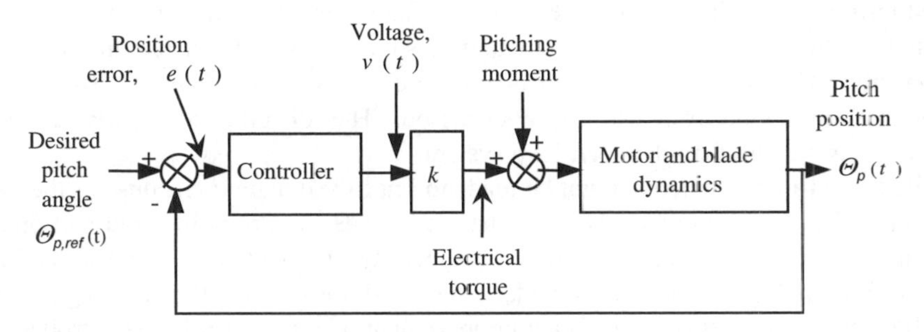

Figure 7.19 Closed-loop pitch mechanism example; for notation, see text

The controller can be designed with any dynamic properties that might help the system achieve the desired operation. There are, however, some standard approaches to control systems that are often implemented and are used as references in considering control system design. Some of these standard approaches include *proportional*, *derivative*, and *integral* control. Often these approaches are combined to yield a proportional–integral (PI) controller or a proportional–integral–derivative (PID) controller.

The differential equation of a PID controller for the pitch system would be:

$$v(t) = K_P e(t) + K_I \int e(t)dt + K_D \dot{e}(t) \tag{7.5.9}$$

where the constants of proportionality are K_P, K_I, and K_D and where $e(t) = \theta_{p,ref}(t) - \theta_p(t)$ is the error, the difference between the desired and the measured pitch angle.

If the closed-loop controller were designed with just the proportional and integral term (a PI controller), then the differential equation for the complete system could be found by substituting the definition of the controller, Equation 7.5.9 without the derivative term, into Equation 7.5.1. One could use the definition of $e(t)$ from above, differentiate the complete equation and rearrange, to get Equation 7.5.10. The resulting controlled system is now a third-order system with two controller constants, K_P and K_I :

$$J\dddot{\theta}_p + (B - m)\ddot{\theta}_p + (K + kK_P)\dot{\theta}_p + kK_I\theta_p = kK_P\dot{\theta}_{p,ref} + kK_I\theta_{p,ref} + \dot{Q}_p \tag{7.5.10}$$

Once again, the dynamic response of the closed-loop system to a unit step disturbance can be found from using inverse Laplace transforms. For example, if it is assumed that the motor and blade dynamics remain the same, and that $K_P = K_I = 2$ then the closed-loop system transfer function is:

$$\Theta_p(s) = \frac{32(s+1)\Theta_{p,ref}(s)}{(s+0.7)(s^2+3.3s+45.7)} + \frac{(16s)Q_p(s)}{(s+0.7)(s^2+3.3s+45.7)} \tag{7.5.11}$$

The first term on the right is the transfer function that relates the commanded position to the final pitch position, and the second term determines the effect of the disturbance on the final pitch position. The response to a step disturbance is shown in Figure 7.20. The figure includes the open-loop response for comparison. The disturbance clearly affects the closed-loop system less than the open-loop system.

While further improvements might be made to improve dynamic response of the system, the PI controller would correctly position the blade under a variety of wind and operating conditions. Thus, without adding too much complexity, the blade can be positioned at any desired position in high or low winds, under icing conditions, and with a sticking bearing on the pitch mechanism. This is a significant improvement over the open-loop controller.

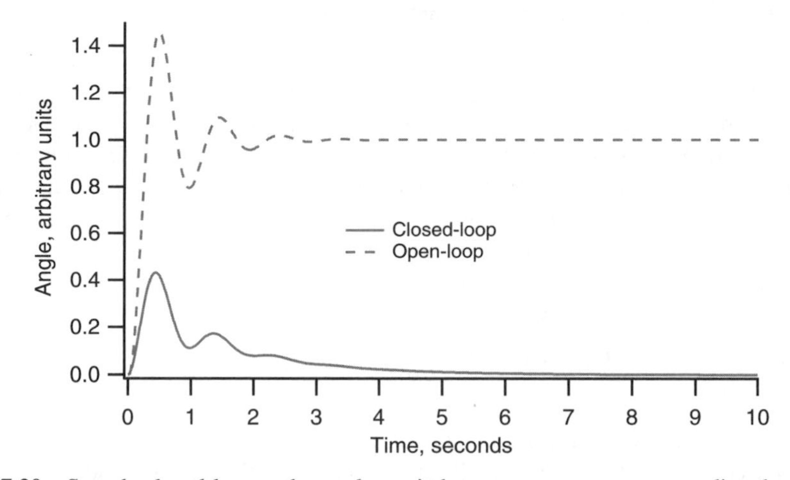

Figure 7.20 Sample closed-loop and open-loop pitch system response to a step disturbance

7.5.3.3 Resonances

Wind turbines consist of long lightly damped members that vibrate and operate in an environment in which gusts, wind shear, tower shadow, and dynamic effects from other turbine components can excite vibrations. Control systems need to be designed to ensure that wind turbines avoid excitation at certain frequencies and must not excite the turbine at those frequencies as a result of its own operation. The control system also needs to be designed to avoid being one of the turbine components that are excited by the numerous possible forcing functions. This can occur in otherwise acceptable control system designs.

The pitch controller of the previous example provides reasonable disturbance rejection of a step input. This indicates that step changes and less severe changes in operating conditions can be compensated for by the pitch control system. Analysis of the differential equation for the closed-loop system (Equation 7.5.10) or of the system transfer function (Equation 7.5.11) shows that the closed-loop system has a natural frequency of 6.76 radians/second (1.08 Hz), and a damping ratio of 0.24 (see Section 4.2.2 for information on natural frequencies and damping ratios). If this pitch control system experienced a sinusoidal disturbance, as it might with wind shear, with a frequency near the natural frequency of the closed-loop system, then the system response might be significantly magnified over that due to other disturbances. For example, the response to a sinusoidal disturbance with a magnitude of 1 and a frequency of 1.04 Hz is shown in Figure 7.21.

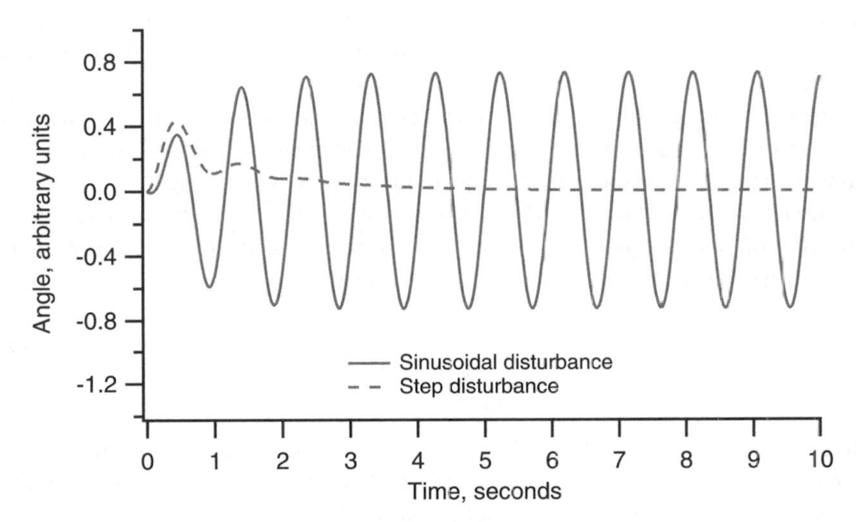

Figure 7.21 Closed–loop pitch system response to unit sinusoidal and step disturbances

The closed-loop system clearly does a much poorer job of managing disturbances near the system response frequency than it does managing step disturbances. The response magnitude depends on the frequency of the disturbance and the damping in the system. Not only could the pitch fluctuations induced by the disturbance wear out the pitch mechanism, but oscillating shaft torque from the aerodynamics caused by the oscillating blade pitch could introduce resonances in other feedback paths. This sample control system was designed specifically to illustrate these issues. A well-designed control system would avoid exciting system natural frequencies and might even be able to provide additional damping at those frequencies.

7.5.3.4 Optimum tip speed ratio control issues

As previously discussed, wind turbines are sometimes operated at variable speed to reduce loads and to maximize energy capture in low speeds. Maximizing energy capture in low speeds requires operating the rotor at its most efficient tip speed ratio (see Section 3.3). To do this the rotor speed must vary with the wind speed. A number of issues arise with such optimum tip speed ratio operation:

- Design tradeoffs. The efficiency of the turbine over time depends on the controller's success in changing rotor speed as the wind speed changes. For maximum efficiency, rotor changes should occur rapidly, but this has the disadvantage of increasing torque fluctuations in the drive train. The flatter the $C_p - \lambda$ curve, the less tight the tracking of the wind needs to be.
- Determining rotor tip speed ratio. Tip speed ratio is the ratio of the rotor tip speed to the wind speed. It is difficult to determine accurately. Turbulent winds vary in time and location over the rotor area. Wind speed measurements on the hub do not measure the undisturbed free stream wind speed and, in any case, they only measure the wind speed at one point on the rotor. Wind speed measurements on a measuring tower sample the wind only at one location in the wind field and at a distance from the turbine. Tip speed

can also be inferred from rotor speed, torque or power measurements, and machine models. Efforts to determine tip speed from rotor operation may be difficult because of noise on torque sensors and inaccurate rotor models.
- Rotor energy. Energy stored in the rotor rotation causes power spikes during rotor decelerations that can confuse measurements of rotor power or torque and that have the potential to cause an overpower situation if the rotor is decelerated too rapidly. The contribution to rotor shaft power, P_r, that is due to changes in stored energy in the rotor is a function of rotor inertia, J_r, and changes in rotor speed, Ω. The more rapid the rotor speed changes, the greater the power fluctuations:

$$P_r = \frac{\mathrm{d}}{\mathrm{d}t}\left(\frac{1}{2} J_r \Omega^2\right) = J_r \dot{\Omega} \tag{7.5.12}$$

- Non-linear aerodynamics. The change in rotor aerodynamic torque or rotor aerodynamic power as a function of speed is highly non-linear, often requiring a non-linear controller for control in both low and high winds.

Tracking the optimum tip speed ratio is usually accomplished with a rotor model, in spite of the possibility that the model may be a poor representation of rotor operation. Another approach that is sometimes considered is a search algorithm that seeks to find the rotor speed for maximum power at each moment. Such an approach, if successful, will maximize rotor energy capture in spite of poorly understood rotor performance, icing, mispitched blades, etc.

One common approach (see Novak et al., 1995) which does not directly attempt to determine the tip speed uses a rotor model to specify the desired rotor torque, Q_{ref}, as a function of rotor speed:

$$Q_{ref} = \frac{\rho \pi R^5 C_{p,\max}}{2(\lambda_{opt})^3} \Omega^2 \tag{7.5.13}$$

where Q_{ref} is the desired rotor torque, ρ is the air density, R is the rotor radius and $C_{p,\max}$ is the rotor power coefficient at the rotor's optimum tip speed ratio, λ_{opt}.

The rotor speed is measured and the generator torque is continually set to the rotor torque that corresponds to the present rotor speed. If the rotor speed is low for the existing wind speed, the torque is set to a low value, allowing the rotor speed to increase. If the rotor speed is high for the wind speed, the rotor torque is set to a high value, slowing the rotor to make it more efficient. The torque command is filtered to avoid high-speed rotor changes.

This approach works fairly well, but the filtering to avoid rapid power or torque fluctuations may need to change for different wind regimes. The change in rotor power with changes in rotor speed varies significantly over the operating range of a turbine and may need to be considered to control power and torque fluctuations. This requires knowing the tip speed ratio.

Linders and Thiringer (1993) have proposed a method of determining the rotor tip speed ratio from a turbine model and measurements of turbine power. Once the operating tip speed ratio is determined, the rate of change of power due to a change in rotor speed can be determined from a rotor model. The difficulty in using power or torque measurements to determine the tip speed ratio is that there are two tip speed ratios that correspond to any C_p. Linders and Thiringer define a monotonic function from which the tip speed ratio can be determined unambiguously:

$$\frac{C_p(\lambda)}{\lambda^3} = \frac{P_r}{\frac{1}{2}\rho A \Omega^3 R^3} = \frac{P_{el}}{\eta \frac{1}{2}\rho A \Omega^3 R^3} \tag{7.5.14}$$

where P_{el} is the electrical power, η is the drive train efficiency, ρ is the air density, A is the rotor area and C_p is the rotor power coefficient at a tip speed ratio of λ.

The derivative of the power with respect to the generator speed is then:

$$\frac{dP_{el}}{d\Omega_{el}} = \left(\frac{dC_p(\lambda)}{d\lambda}\right)\frac{\eta \frac{1}{2}\rho A V^2 R}{np} \tag{7.5.15}$$

where Ω_{el} is the generator speed, p is the number of generator poles and n is the gearbox gear ratio.

The machine control can then be improved by modeling power flows in and out of the rotor as the rotor speed changes and by subtracting these from the measured power.

7.5.3.5 Transitions between variable speed operating modes

Numerous other issues arise in the design of variable speed wind turbines. In general, variable speed wind turbines may have three different control goals, depending on the wind speed. In low to moderate wind speeds, the control goal is maintaining a constant optimum tip speed for maximum aerodynamic efficiency. This is achieved by varying the rotor speed as the wind speed changes. In moderate winds, if the rotor reaches its rated speed before the rated power is reached, the rotor speed must be limited while the power fluctuates. The control goal is then the maintenance of approximately constant rotor speed. In moderate to high winds the control goal is the maintenance of constant rated power output. This is required as the available power in the wind increases over the power conversion capabilities of the wind turbine. Transitions between these operating strategies must be managed smoothly by any successful control system design.

These same general control goals apply to both stall-regulated and pitch-regulated variable speed wind turbines. The advantage that pitch-regulated wind turbines have is access to two control inputs, pitch angle and generator torque, to achieve these control goals. The blade pitch is typically held fixed during constant tip speed ratio and constant speed operation, but during constant power operation the blade pitch can be used to control rotor speed while the generator torque can be used to control output power. During constant speed operation, there may be large fluctuations in generator torque and thus output power, as the wind speed and rotor torque change. During constant power operation, generator power and torque fluctuations are minimized, because the control goal is the maintenance of rated power. The pitch controller does most of the work of maintaining average rated rotor speed.

The turbulence in the wind makes successfully designing a controller to transition between these control goals difficult. In low winds with constant tip speed operation, the change in output power, P, as a function of changes in rotor speed, Ω, $\mathrm{d}P/\mathrm{d}\Omega$, is relatively small. In moderate winds with approximately constant speed operation, $\mathrm{d}P/\mathrm{d}\Omega$ can be quite large. In high winds, $\mathrm{d}P/\mathrm{d}\Omega$ should be close to zero. With two control inputs in moderate and high winds (pitch angle and generator torque) and significantly different controller behavior in different wind speeds, fluctuating wind speeds can result in rapid actuator actions and large power excursions in a poorly designed control system.

An example of such difficulties is presented below. The example uses a turbine with the common controller configuration illustrated in Figure 7.22. In this design, the desired, or 'reference,' generator power is set to be a function of rotor speed, based on a desired power versus generator speed curve.

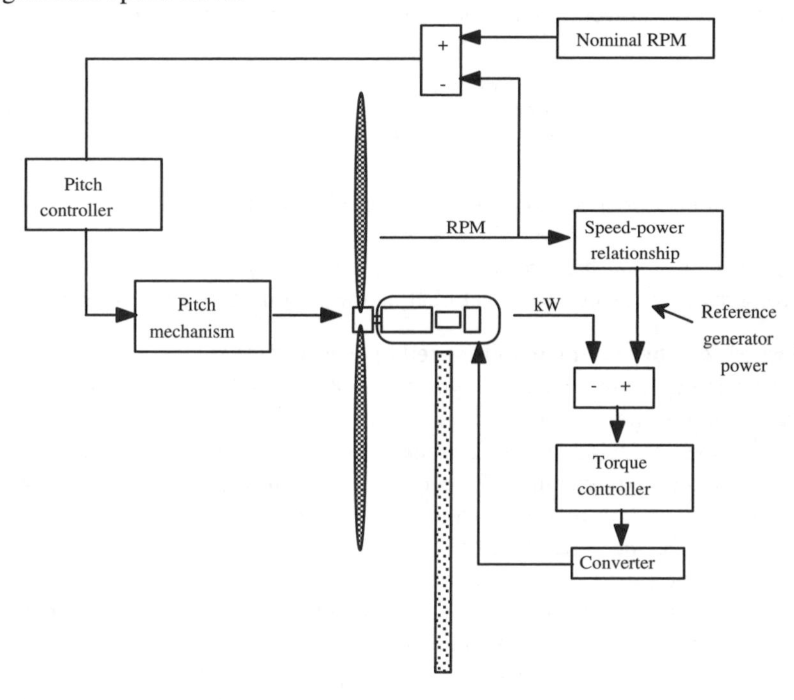

Figure 7.22 Variable speed closed-loop control system

This desired power vs. generator speed relationship for the sample wind turbine is shown in Figure 7.23 (see Hansen et al., 1999). The variable speed operation in the lower part of the curve is based on the aerodynamic performance of the rotor and is intended to maintain a constant tip speed ratio. Once the generator speed approaches the nominal rated speed, 1000 rpm, the slope of the desired power vs. speed curve is very steep. Finally, once the turbine power reaches 225 kW, the control goal is constant power with some small variation of the rotor speed. Only during constant power operation is the pitch control loop trying to control the rotor speed to be the near rated rotor speed.

Difficulties have arisen with this control approach when fluctuating winds cause the generator speed to decrease from over to under nominal rated generator speed. Because of the sharp change in slope ($dP/d\Omega$), even small changes in generator speed cause large changes in the commanded power from the generator. This has caused sudden, sharp reductions in power of as much as 150 kW. Improvements to the control strategy have been made by limiting the allowed rate of change of the desired generator power. This has the effect of significantly limiting power fluctuations near rated power, but it also increases rotor speed fluctuations. The net effects of decreasing generator torque fluctuations and allowing the rotor speed to vary over a wider range are lower power fluctuations that might affect the grid and decreased drive train loads.

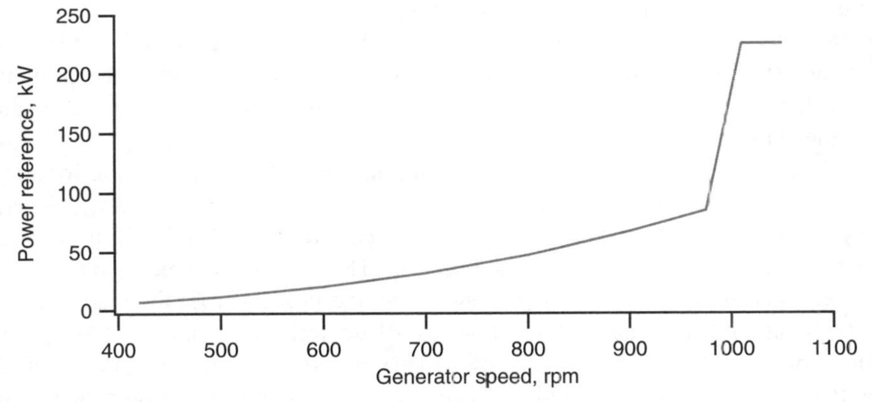

Figure 7.23 Example of grid power–generator speed control relationship (Hansen et al., 1999). Reproduced by permission of James & James (Science Publishers) Ltd. and A. D. Hansen, H. Bindner and A. Rebsdorf

7.5.3.6 Wind turbine loads and disturbances

The design of wind turbine controllers requires information on the magnitude and frequency characteristics of disturbances that affect the components being designed. This information affects both actuator design and the design of the control algorithm itself.

For example, the design of pitch mechanisms and controllers for pitch mechanisms requires knowledge of the magnitude and kinds of loads and disturbances that the mechanisms will experience. Knowledge of which loads are or are not significant in control system design ensures an acceptable control system with as little complexity as possible.

The loads and disturbances on pitch systems include:

- Gravity loads. Gravity loads include the effect of gravity on the distributed mass of the blade that causes moments that act about the pitch axis
- Centrifugal loads. Centrifugal loads include the effect of inertial forces on the distributed mass of the blade that causes moments that act about the pitch axis
- Pitch bearing friction. Pitch bearing friction is a function of the axial loads and edge-wise and flap-wise moments on the pitch bearings as well as the bearing design
- Actuator torque. Actuator torques must be transferred through linkages or gears from hydraulic or electric systems
- Aerodynamic pitching moment. The aerodynamic pitching moment is a function of integrated effects of the aerodynamics over the length of the blade. It includes the effects

of blade design, blade vibration and bending, turbulence, wind shear, off-axis winds, turbulence and blade rotation

- Loads due to other turbine motions. The effects of other turbine motions such as changes in rotor speed, yaw motions and tower vibrations can be transferred to the blades and can cause additional moments about the pitch axis

In normal operation on a given turbine, some of these moments may be of little significance to control system design. Nevertheless, during emergency pitching events or under unusual operating conditions, the contribution of each of these moments needs to be considered to ensure that the control design works in all situations.

The pitch system response to these loads depends on the compliance and backlash in the pitch actuator system (including any linkages), the stiffness of the blades, the compliance and backlash in the pitch bearing, the inertia of the actuator system and the moment of inertia of the blade.

Research has shown (see Bossanyi and Jameison, 1999) that pitch bearing friction and blade bending can significantly affect overall pitching disturbance moments. Generally, roller bearing-type pitch bearings have a very low coefficient of friction, but net friction forces are also a function of the load on the bearing. The high preload required to minimize bearing wear and the large overturning moments on the bearing from the blade can cause unexpectedly high pitching moment disturbances. Blade deflections can result in significant translations of the locus of the aerodynamic forces and of the center of gravity of blade sections from the pitch axis of the blade. Translations of the blade section center of mass can significantly affect the contribution of gravity to the pitching moment. Blade motions also affect the polar moment of inertia of the blade and the aerodynamics at each blade section. The combination of these effects has been shown to result in significantly higher pitching moment disturbances with the use of flexible blades than would be found with the stiffer blades.

7.5.4 Dynamic control system implementation

Dynamic control can be implemented as mechanical systems, as analog electrical circuits, in digital electronic form, or in combinations of these. Mechanical control systems are commonly used only in relatively small wind turbines. Most wind turbines use some combination of analog and digital circuits or only digital circuits. Examples of some of these and related issues are presented below.

7.5.4.1 Mechanical control systems

Hardware dynamic control systems use linkages, springs, and weights to actuate system inputs in response to some output. Two examples of hardware control systems are tail vanes to orient wind turbines into the wind and pitch mechanisms that vary blade pitch on the basis of aerodynamic forces or rotor speed. The pitch systems used on the Bergey Excel and the Lagerwey turbines mentioned at the beginning of this chapter represent examples of mechanical control systems.

7.5.4.2 Analog electrical circuit control systems

Analog electrical circuits have also been used and are still used in control system implementation. They are often used as distributed controllers in a larger control network. Once a control algorithm has been developed and tested it can be hardwired into circuit boards that are robust and easy to manufacture. The controllers may operate independently of the supervisory controller, making the supervisory control scheme simpler. One disadvantage of using analog circuits is that changes in the control algorithm can only be made by changing the hardware.

Electrical circuits with the appropriate power amplifiers and actuators can be used to control all controllable parts of a wind turbine. Linear dynamic controllers can be easily implemented with electronic components. For example, the operational amplifier circuit in Figure 7.24 is the hardware realization of a PID controller with a differential equation of (see Nise, 1992):

$$\begin{aligned} g(t) &= -\left[K_P e(t) + K_D \dot{e}(t) + K_I \int e(t)\mathrm{d}t\right] \\ &= -\left[\left(\frac{R_2}{R_1} + \frac{C_1}{C_2}\right)e(t) + (R_2 C_1)\,\dot{e}(t) + \frac{1}{R_1 C_2}\int e(t)\mathrm{d}t\right] \end{aligned} \tag{7.5.16}$$

where $g(t)$ is the controller output, and R and C are the resistances and capacitances of the respective circuit elements and $e(t)$ is the error signal that is input to the controller.

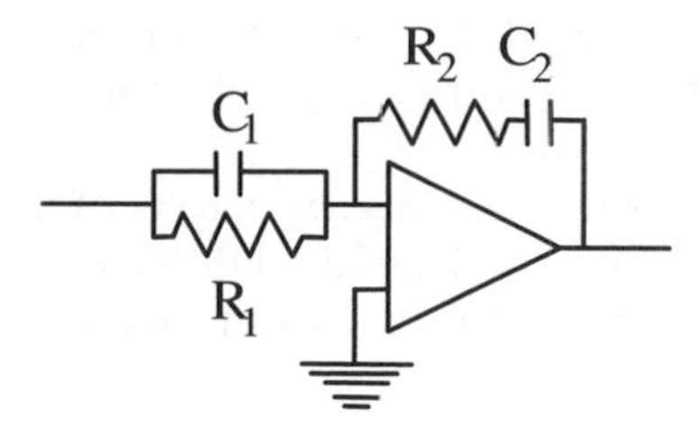

Figure 7.24 Example of proportional–integral–derivative (PID) controller; C, capacitance; R, resistance, 1, 2, circuit elements

7.5.4.3 Digital control systems

The systems described so far have been analog control systems. Analog control systems respond in a continuous manner to continuous inputs such as forces or voltages. Many modern dynamic control systems are implemented in digital controllers. Digital control systems respond periodically to data that are sampled periodically. They are implemented in digital computers. These digital control systems may include controllers that are distributed around the turbine and that communicate with the supervisory controller in a master–slave configuration or there may only be one central controller that handles many supervisory and dynamic control tasks. Digital control algorithms are relatively easy to upgrade and systems with centralized processing result in significantly less control hardware and cost than hardwired systems. Digital control systems also allow for easy implementation of non-linear control approaches. This can result in improved system behavior compared with the same system with a linear controller.

Digital control systems, as illustrated in Figure 7.25, must communicate with analog sensors, actuators and other digital systems (see Astrom and Wittenmark, 1984). Thus the central processing units (CPUs) may need analog-to-digital (A/D) converters to convert analog sensor inputs into digital form or digital-to-analog (D/A) converters to convert digital control commands to analog voltages for amplification for actuators. Depending on the distances that any digital information is being transmitted and the noise immunity desired, different communication standards might be needed.

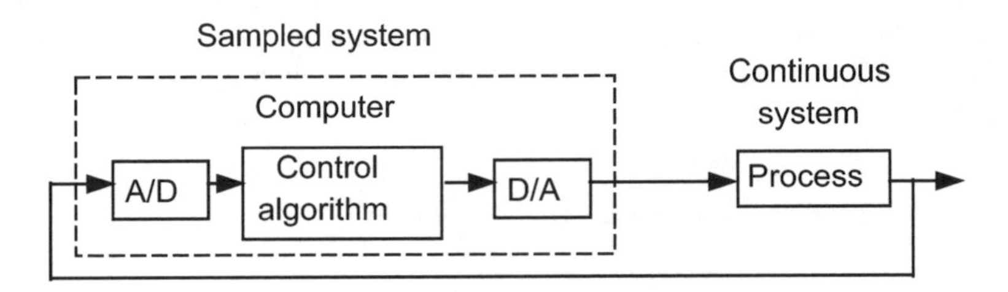

Figure 7.25 Schematic of computer controlled system; A/D, analog–to–digital converter; D/A, digital–to–analog converter

Digital control systems are not continuous, but rather are sampled. Sampling and the dynamics of the A/D converters give rise to a number of issues specific to digital control systems. The sampling rate, controlled by the system clock, affects (1) the frequency content of processed information, (2) the design of control system components and (3) system stability.

The effect that sampling has on the frequency content of the processed information can be illustrated by considering a sinusoidal signal, $\sin(\omega t)$ where ω is the frequency of the sinusoid and t is the time. If this signal is sampled at a frequency of ω, then the ith sample is sampled at time

$$t_i = i\frac{2\pi}{\omega} + t_0 \tag{7.5.17}$$

for some starting time t_0 and for integer, i. The value of each sample, s_i, is then

$$s_i = \sin[\omega(i\frac{2\pi}{\omega} + t_0)] = \sin(2\pi i + \omega t_0) = \sin(\omega t_0) \tag{7.5.18}$$

But $\sin(\omega t_0)$ is a constant. Thus, sampling the signal at the frequency of the signal yields no information at all about fluctuations at that frequency. In fact, in a system that samples information at a frequency ω, there will be no useful information at frequencies above $\omega_n = \omega/2$, referred to as the Nyquist frequency. Furthermore, unless the input signal is filtered with a cut-off frequency below ω_n, then the high-frequency information in the signal will distort the desired lower-frequency information.

Sampling rate also affects the design of the control system. Digital control system dynamics are a function of the sampling rate. Thus, the sampling rate affects the subsequent control system design and operation, including the determination of the values of constants in the controller and the final system damping ratio, system natural frequency, etc. Because of this, changes in sampling rate can also turn a stable system into an unstable one. Stability can be a complicated issue, but, in general, a closed-loop digital control system will become unstable if the sampling rate is slowed down too much.

Digital control may be implemented in small stand-alone single board computers or in larger industrial computers. Single board computers include a central processing unit and often include analog and digital inputs and outputs. They are small enough to be easily housed with other hardware for distributed control systems or as controllers for smaller wind turbines. Larger industrial computers have power supplies and slots for numerous additional communication and processing boards, filters and fans to insure a clean environment, memory and data storage, and a housing to protect the electronics from environmental damage.

References

Astrom, K. J., Wittenmark, B. (1984) *Computer Controlled Systems.* Prentice-Hall, Englewood Cliffs, NJ.

Bossanyi, E. A., Jameison, P. (1999) Blade Pitch System Modeling For Wind Turbines. *Proc. 1999 EWEC*, March 1-5, Nice, France, 893–896.

De LaSalle, S. A., Reardon, D., Leithead, W. E., Grimble, M. J. (1990) Review of wind turbine control. *Int. Journal of Control*, 52 (6) 1295–1310.

Di Steffano, J. J., Stubberud, A. R., Williams, I. J. (1967) *Theory and Problems of Feedback and Control Systems*. McGraw-Hill, New York.

Freris, L. L. (Ed.) (1990) *Wind Energy Conversion Systems*. Prentice Hall International, Hertfordshire, UK.

Gasch, R. (Ed.) (1996) *Windkraftanlagen*, B. G. Teubner, Stuttgart.

Grimble, M. J., De LaSalle, S. A., Reardon, D., Leithead, W. E. (1990) A lay guide to control systems and their application to wind turbines. *Proc. of the 12th BWEA Conference* (Eds. T.D. Davies, J.A. Halliday and J.P. Palutikof), Mechanical Engineering Publications, London, 69–76.

Hansen, A. D., Bindner, H., Rebsdorf, A. (1999) Improving Transition Between Power Optimization and Power Limitation of Variable Speed/Variable Pitch Wind turbines. *Proc. 1999 EWEC*, March 1-5, Nice, France, 889–892.

Hau, E. (1996) *Windkraftanlagen, Grundlagen, Technik, Einsatz, Wirtshaftlichkeit*. Springer, Berlin.

Heier, S. (Ed.) (1996) *Windkraftanlagen im Netzbetrieb*. B. G. Teubner, Stuttgart.

Kuo, B. C. (1987) *Automatic Control Systems, 5th edn*. Prentice-Hall, Englewood Cliffs, NJ.

Linders, J., Thiringer, T. (1993) Control by Variable Rotor Speed of a Fixed-Pitch Wind Turbine Operating in a Wide Speed Range. *IEEE Transactions on Energy Conversion*, 8 (3): September.

Ljung, L. (1999) *System Identification – Theory for the User*. Prentice-Hall, Upper Saddle River, NJ.

Nise N. S. (1992) *Control System Engineering*. Benjamin/Cummings, Redwood City, CA.

Novak, P., Ekelund, T., Jovik, I., Schmidtbauer, B. (1995) Modeling and Control of Variable-speed Wind-turbine Drive-system Dynamics. *IEEE Control Systems*, 15 (4), August, 28–38.

8

Wind Turbine Siting, System Design and Integration

8.1 General Overview

Wind turbines operate as part of larger power producing and consuming systems such as large electrical networks, isolated diesel-powered grid systems or as stand-alone power for a specific load. The process of integrating wind power into such systems includes decisions about where to install the wind turbines, turbine installation and grid connection and turbine operation. Meanwhile, turbine design and operation needs to take into consideration the numerous kinds of interactions between turbines in a wind farm and between wind farms and the connected grid systems. This chapter reviews the issues that need to be considered for the completion of a successful project.

Wind turbines may be installed as single units or in large arrays or 'wind farms.' The installation of individual wind turbines and wind farms requires a significant amount of planning, coordination and design work. Mistakes can be very costly. Before wind turbines can be installed and connected to an electrical system, the exact locations for the future turbines need to be determined. A primary consideration is maximizing energy capture, but numerous constraints may limit where turbines can be situated. The material in Section 8.2 focuses on siting issues for wind turbines and wind farms that are to be connected to large electric grids. Once locations for the turbines are chosen, the installation and integration of wind turbines into large power grids requires obtaining permits, preparing the site, erecting the turbines, and getting them operational. These topics are covered in Section 8.3.

Once installed, significant interactions can occur between individual wind turbines and between wind turbines and the systems to which they are connected. When multiple wind turbines are located together in close proximity, the fatigue life and operation of those wind turbines located downwind of other turbines may be affected. These issues are all considered in Section 8.4.

Many turbines or wind farms are connected to a large electrical grid. Inadequate consideration of the characteristics of the grid at the point of connection can result in unwanted disturbances in the local power system. Some characteristics of electrical grids and possible turbine–grid interactions that can affect other users of the power grid are covered in Section 8.5.

Offshore wind farms have their own unique installation, operational and environmental issues. The most significant differences between land-based and offshore wind farms are the foundation design issues and issues related to the long-distance power transmission. These are covered in section 8.6.

Issues related to operation in severe climates are covered separately in Section 8.7. These include cold weather, high temperatures and lightning.

Wind turbines may also be connected to smaller, isolated electrical grids. In these systems the wind power, other power sources, and any system loads may strongly influence each other. In such cases the power system as a whole needs to be designed and controlled with all of the system components in mind. The characteristics of the components in isolated grid systems and systems with conventional and renewable power sources need to be understood in order to design the complete system. These design issues are covered in Section 8.8.

8.2 Wind Turbine Siting

8.2.1 Overview of wind turbine siting issues

Before wind turbines can be installed, a siting study needs to be undertaken to determine where to locate them. The major objective of a siting study is to locate a wind turbine (or turbines) such that cost of energy is maximized while minimizing such things as noise and visual impacts. The scope of a siting study can have a very wide range, which could include everything from wind prospecting for suitable turbine sites over a wide geographical area to considering the placement of a single wind turbine on a site or of multiple wind turbines in a wind farm (this is generally called micrositing).

The siting of a single turbine or a large-scale wind system for utility interconnection can be broken down into five major stages (after Hiester and Pennell, 1981, and Pennell, 1982):

1. *Identification of geographic areas needing further study* - Areas with high average wind speeds within the region of interest are identified using a wind resource atlas and any other available wind data. The characteristics of turbine types or designs under consideration are used to establish the minimum useful wind speed for each type.
2. *Selection of candidate sites* - Potential windy sites within the region are identified where the installation of one or more wind turbines appears to be practical from engineering and public acceptance standpoints. If the nature of the terrain in the candidate region is such that there is significant variation within it, then a detailed analysis is required to identify the best areas. At this stage, topographical considerations, ecological observations, and computer modelling may be used to evaluate the wind resource. Geologic, social, and cultural issues are also considered.
3. *Preliminary evaluation of candidate sites* - In this phase, each potential candidate site is ranked according to its economic potential, and the most viable sites are examined for any environmental impact, public acceptance, safety, and operational problems that

would adversely affect their suitability as a wind turbine site. Once the best candidate sites are selected, a preliminary measurement program may be required.

4. *Final site evaluation* - For the best remaining candidate sites, a more comprehensive measurement may be required. At this point, the measurements should include wind shear and turbulence in addition to wind speed and prevailing wind directions.
5. *Micrositing* - Once a site is chosen, or possibly as part of the final site evaluation, the exact location of the turbines and their energy production needs to be determined. This may be able to be done with computer programs that can model the wind field and the various aerodynamic interactions between turbines that affect energy capture (see wind farm technical issues below). The more complex the terrain, and the less the available data from nearby sites, the less accurate these models are. A site in complex terrain may require detailed measurements at numerous locations to determine the local wind field for micrositing decisions.

The evaluation of problems that might adversely affect a site's suitability should include:

- *Topographical issues* such as road access and the slope of the terrain at potential turbine sites
- *Legal issues* such as ownership of the land, zoning issues and rights of adjacent land owners (for example, to the wind resource)
- *Permitting issues* such as the number of permits needed, previous rulings of permitting agencies, permitting restrictions and time frames for completion of permitting procedures
- *Geological issues* related to the foundation design, ground resistance for lightning protection and the potential for erosion
- *Environmental issues* such as the presence of environmentally sensitive areas, bird flyways and the presence of endangered species
- *Public acceptance issues* such as visual and noise pollution, distance from residences, the presence of culturally, historically, or archaeologically important areas, competing land uses and interference with microwave links and other communication
- *Safety issues* related to proximity to populated areas or hiking trails
- *Interconnection issues* such as the proximity of power lines and the voltage and current handling capabilities of those power lines

8.2.2 Estimating the wind resource

There are a number of possible approaches to determining the long-term wind resource at candidate sites. Each of these has advantages and disadvantages and, thus, might be used at different stages of the siting process, depending on the information needed. These methods include (1) ecological methods, (2) the use of wind atlas data, (3) computer modelling, (4) statistical methods, and (5) long-term site-specific data collection. These methods build on the material presented in Chapter 2, which described the meteorological effects that create surface level winds and methods for estimating the exploitable wind resource in a region. Some of the methods presented here can also be used for more general estimates of wind resource.

8.2.2.1 Ecological methods

Vegetation deformed by high average winds can be used both to estimate the average annual wind speed and to compare candidate sites, even when no wind data are available. These methods are most useful during initial site selection and in geographic areas with very little available wind data. This technique works best in three regions (Wegley et al., 1980): coastal regions, in river valleys and gorges exhibiting strong channeling of the wind and in mountainous terrain.

Ecological indicators are especially useful in remote mountainous terrain not only because there is usually little wind data there, but also because the winds are highly variable over small areas and are difficult to characterize. Among the many effects of wind on plant growth, the effects of wind on trees are the most useful for the wind prospecting phases of siting (Hiester and Pennell, 1981). Trees have two advantages – height and a long lifetime in which to gather evidence. Research work has produced numerous potential indices relating tree deformation to long-term average wind speeds. Three of the more common ones are the Griggs–Putnam index for conifers, which is explained below, the Barsch index for hardwoods, and the Deformation Ratio, which applies to both hardwoods and conifiers (see Hiester and Pennell, 1981, for more information on these indexes).

The Griggs–Putnam index, for example (Putnam, 1948 and Wade and Hewson, 1980), applies to conifers and defines eight classes of tree deformation (see Figure 8.1) ranging from no effect (class 0) to the predominance of lateral growth in which the tree takes the form of a shrub (class VII). Deformation ratios for different types of trees can be related to mean wind speed at the tree top, as shown in Table 8.1.

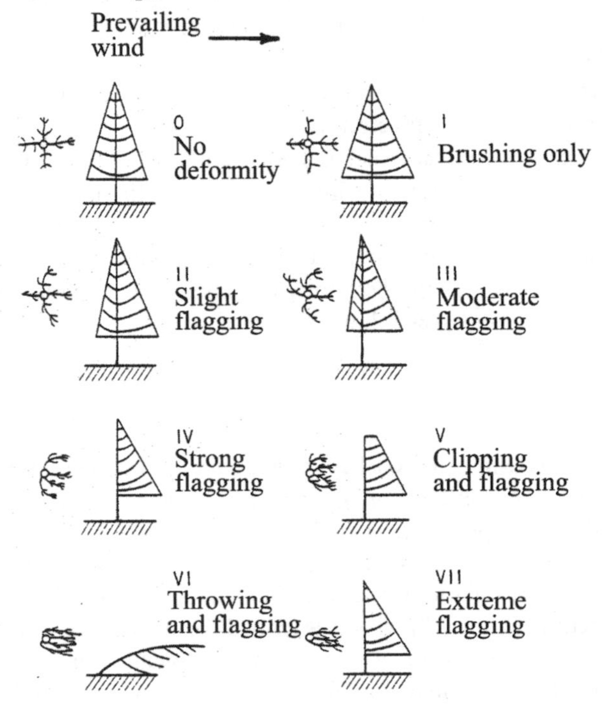

Figure 8.1 The Griggs–Putnam index of wind deformation (Wegley et al., 1980)

Table 8.1 Quantification of the Griggs–Putnam index

Deformation type (Griggs–Putnam index)	Description	Tree height m (ft)	Velocity at tree height m/s (mph)
Brushing (I)	Balsam not flagged	12.2 (40)	6.9 (15.5)
Flagging (II)	Hemlock and white pine show minimal flagging	12.2 (40)	4.7 (10.6)
Flagging (II)	Balsam minimally flagged	12.2 (40)	7.7 (17.3)
Flagging (III)	Balsam moderately flagged	9.1 (30)	8.0 (17.9)
Flagging (IV)	Balsam strongly flagged	9.1 (30)	8.3 (18.6)
Clipping (V)	Balsam, spruce and fir held to 1.3 m	1.2 (4)	9.6 (21.5)
Throwing (VI)	Balsam thrown	7.6 (25)	8.6 (19.2)
Carpeting (VIII)	Balsam, spruce and fir held to 0.3 m	0.3 (1)	12.1 (27.0)

8.2.2.2 Using wind atlas data

As mentioned in Chapter 2, wind atlas data (or other archived data) from nearby sites may be able to be used to determine local long-term wind conditions. For example, the data in the *European Wind Atlas* (Troen and Petersen, 1989 and Petersen and Troen, 1986) has been prepared specifically for this purpose. The atlas includes data from 220 sites spread over Europe, including information on the terrain at each site, wind direction distributions and Weibull parameters for each wind direction. The effects of surface roughness, topography and nearby obstacles have been removed from the data in order to have data that represent the basic surface flow patterns. The atlas also includes a description of a procedure to use these data to estimate the long-term resource at a specific candidate site. The procedure starts with the selection of appropriate comparison sites and includes formulas for corrections to the Weibull parameters to account for surface roughness effects, upwind structures, and changes in elevation. The basic procedure involves determining the estimated Weibull parameters for the wind at the candidate site for each of 12 wind direction sectors. These parameters, in conjunction with the wind direction distribution information, can be used to determine the annual long-term wind speed and wind speed distribution. If a turbine has been selected, then these data can be used to estimate turbine power production.

8.2.2.3 Computer modelling

There are now computer modelling programs that can be used to estimate the local wind field and to optimize turbine layout in a wind farm. Programs to model the local long-term wind field at a site use topographical information and long-term upper-level meteorological data and/or nearby surface level wind data. The more nearby data that are available and the more smooth the terrain is, the more accurate these predictions are. Numerous such programs are commercially available. Only one example will be described here.

In Europe, the database in the *European Wind Atlas* is often used in conjunction with the WA^{S}P computer program to determine the wind resource at a candidate site. WA^{S}P was developed, as part of the international effort that produced the *European Wind Atlas*, to provide a tool to use the data in the atlas (see Petersen et al., 1988). WA^{S}P includes the effects of the atmospheric stability, surface roughness, obstacles and topography at the candidate site in the determination of the site-specific wind conditions. The modelling uses a simple fluid flow model that includes mass and momentum conservation to determine the flow field at the candidate site, based on the nearby reference sites. One advantage of WA^{S}P is that it determines local wind field using a polar coordinate grid centered on the candidate site. This provides a high resolution around the candidate site, allowing the wind field for a complete wind farm to be predicted. Another advantage is that it does not require on-site measurements. It also works at any height and location. Disadvantages are that the method may give inexact results in complex terrain and that it does not include thermally driven effects such as sea breezes.

8.2.2.4 Statistical methods

Statistical wind resource estimation methods use data from nearby sites to predict the wind resource at a candidate site. If long-term data are available for a nearby site and shorter duration, but concurrent data are available for the candidate site, then the long-term wind speed can be estimated for the proposed candidate site. The most common method used is the measure–correlate–predict method (MCP) described by Derrick (1993), Landberg and Mortensen (1993) and Joensen, et al. (1999). The MCP method can not only be used to determine an estimate of the long-term mean wind speed, but also an estimate of confidence intervals about that estimate.

Typically, the MCP method uses hourly mean wind speeds that have been binned by velocity and direction. For each wind direction bin, linear correlations for the predicted wind speed, U_c^*, and predicted direction, θ_c^*, at the candidate site are based on concurrent data of the mean hourly wind speed, U_r, and direction, θ_r, at the reference site and the mean hourly wind speed, U_c, and direction, θ_c, at the candidate site:

$$U_c^* = aU_r + b \tag{8.2.1}$$

$$\theta_c^* - \theta_r = c \tag{8.2.2}$$

where a, b, and c are functions of θ_r at the midpoint of each direction bin. The standard deviation of the estimate of U_c^* can be determined from:

$$\sigma(U_c^*) = \sqrt{U_r^2 \operatorname{var}(a) + \operatorname{var}(b) + 2U_r \operatorname{cov}(a,b)} \tag{8.2.3}$$

The variance and covariance of a and b can be determined from data used to determine the linear fit. Each of these terms is a function of s^2, the estimate of the error sum of squares. It is a function of the sum of the square of the differences of each observation of U_c from the predicted values using the linear fit. If there are n pairs of concurrent wind speed observations at each site, then:

$$s^2 = \frac{\sum_n \left(U_c - U_c^*\right)^2}{n-2} \tag{8.2.4}$$

The variance of the estimated slope, a, is a function of s^2 and the sum of the squares of the differences between each measured wind speed at the reference site and the mean wind speed (indicated with an overbar) of all of the n values at the reference site:

$$\operatorname{var}(a) = \frac{s^2}{\sum_n \left(U_r - \overline{U}_r\right)^2} \tag{8.2.5}$$

The equation for the variance of the estimated constant, b, includes the sum of each of the measured wind speeds at the reference site:

$$\operatorname{var}(b) = \frac{s^2 \sum_n U_r^{\,2}}{n \sum_n \left(U_r - \overline{U}_r\right)^2} \tag{8.2.6}$$

Finally, the covariance term is:

$$\operatorname{cov}(a,b) = \frac{s^2 \overline{U}_r}{\sum_n \left(U_r - \overline{U}_r\right)^2} \tag{8.2.7}$$

The correlations based on the concurrent data are then used with the full set of data from the reference site to predict the mean annual wind speed at the candidate site. The method proceeds in three steps:

1. Determine the probability distributions for winds from each wind direction
2. Calculate the mean predicted wind speed for each wind direction
3. Find the annual mean wind speed from the means for each wind direction

The predicted long-term wind speed probability distribution at the candidate site, for any direction bin, $p(U_c^*)$ can be determined from Equation 8.2.1, using the long-term measured wind data from the reference site. The variance of that estimate can be determined from Equation 8.2.3. The wind direction distribution can be determined from Equation 8.2.2 using the long-term reference data. This should be done for each wind direction bin.

The predicted long-term mean wind speed for each sector, $\overline{U}_{c,k}^*$, where k is the sector index, is a function of the probability distribution of the predicted hourly wind speed, $p_i(U_c^*)$, where i is the index of the wind speed bin:

$$\overline{U}_{c,k}^* = \sum_i p_i\left(U_c^*\right) U_{c,i}^* \tag{8.2.8}$$

The standard deviation of the estimate of the predicted wind speed for each wind direction is:

$$\sigma\left(\overline{U}_{c,k}^{\,*}\right) = \sqrt{\sum_i \left(p_i \sigma_i \left(U_c^{\,*}\right)\right)^2 + \sum_i \sum_j 2 p_i p_j \operatorname{cov}\left(U_{c,i}^{\,*}, U_{c,j}^{\,*}\right)} \tag{8.2.9}$$

The covariance term is required because the values of the predicted wind speed are all derived from the same fit to the data and are not independent:

$$\operatorname{cov}\left(U_{c,i}^{*}, U_{c,j}^{*}\right)=U_{r,i}U_{r,j}\operatorname{var}(a)+\left(U_{r,i}+U_{r,j}\right)\operatorname{cov}(a,b)+\operatorname{var}(b) \quad (8.2.10)$$

The overall predicted long-term mean wind speed for each site, U_c^*, can now be calculated from the wind direction probability distribution, p_k:

$$\overline{U}_c^{*}=\sum_k p_k U_{c,k}^{*} \quad (8.2.11)$$

The overall standard deviation of the predicted wind speed at the candidate site is then:

$$\sigma\left(\overline{U}_c^{*}\right)=\sqrt{\sum_k\left[p_k\sigma_k\left(\overline{U}_c^{*}\right)\right]^2} \quad (8.2.12)$$

where σ_k is the standard deviation of wind speed in each direction bin. The sector predicted means are all independent and thus there is no covariance term in the final equation.

The accuracy of the MCP method depends on a variety of factors. First, it assumes that the long-term reference data and candidate data are accurate. Errors in these data sets will obviously result in erroneous predictions. Also, the longer the data set at the candidate site, the more accurate the results. In any case, the data collection at the candidate site should include representative data from all wind speeds and directions. The results also depend on the correlation coefficient between the time series at the two sites. A low correlation coefficient will result in predictions with a large standard deviation. Correlation coefficients can be affected by diurnal effects, time delays between weather patterns arriving at the two sites, unique weather patterns, the distance between the two sites and stability differences and topographic effects that create unique flow patterns at one site or the other. Finally, the results are only good for the height of the measurements and for the specific mast location at the candidate site. More detailed modelling may still be necessary to define the specific flow patterns at candidate sites if the terrain is such that the predictions cannot be applied to nearby proposed turbine locations.

8.2.2.5 Site-specific data collection

The best way to determine the long-term wind resource at a site is by measuring the winds at the exact locations of interest. These measurements should include wind speed and direction, wind shear, turbulence intensity, and temperature (for determining air density and the potential for icing). Long-term site-specific measurements are the ideal approach to determining the wind resource, but the most costly and time consuming. See Chapter 2 for more information on resource measurement.

8.2.3 Micrositing

Micrositing is the use of resource assessment tools to determine the exact position of one or more wind turbines on a parcel of land to maximize the power production. There are numerous computer codes available for micrositing of wind turbines. Wind farm design and analysis codes used for micrositing use wind data for the potential site, turbine data, and information on the site constraints to determine an optimum layout for the wind turbines on the site. Site constraints include turbine exclusion areas (for example, due to geological or environmental concerns), noise limits at points adjacent to the wind farm, etc. Site information is usually provided using digital contour maps. The outputs of these programs include turbine locations, noise contours and contours of energy capture predictions, energy yield estimates for individual turbines and the overall site, and related economic calculations. Some wind farm optimization codes also determine the visibility of the wind farm from nearby locations and can optimize the wind farm layout to minimize visual impact in addition to maximizing energy capture and minimizing noise.

8.3 Installation and Operation Issues

The installation of a wind power project is a complex process involving a number of steps and legal and technical issues. The process starts with securing legal rights and permit approvals. Once permits are obtained, the site needs to be prepared and the turbine transported to the site and erected. Only after the turbine is connected to the grid and commissioned does regular operation commence. At this point the owner is responsible for oversight of the turbine, safe operation and maintenance.

8.3.1 Predevelopment work and permitting

The first steps in the development process include finding investors, securing legal rights to land and access to power lines, starting to line up power purchase agreements and obtaining permits. Most of these legal and financial aspects of the development process are beyond the scope of this text, but the permitting process involves significant engineering issues and is addressed here.

Once a site is chosen and a contract is negotiated with the landowner(s), the next step is the successful navigation of the permitting process. The permitting process varies greatly from country to country, state to state, and even from town to town. It is intended to ensure that land is used appropriately, and that operation of the installed wind turbine or wind farm will be safe and environmentally benign. Generally, permits that must be obtained may include those related to: building construction, noise emission, land use, grid connection, environmental issues (birds, soil erosion, water quality, waste disposal, and wetland issues), public safety, occupational safety, and/or valuable cultural or archaeological sites. More information on environmental issues is included in Chapter 10. In the permit approval process the technical certification of the turbine may also be important. Wind power professionals in many countries have published information on issues to be considered in the permitting process. For example, more information on the permitting process in the United States can be found in *Permitting of Wind Energy Facilities, A Handbook* (NWCC, 1998).

8.3.2 Site preparation

Once permits have been obtained, the site needs to be prepared for turbine installation and operation. Roads may need to be built; the site needs to be cleared for delivery, assembly and erection of turbines; power lines need to be installed and foundations need to be built. The extent and difficulty of the site preparation will depend on the site location, proximity to power lines, the turbine design and site terrain. Turbine foundation design (see Chapter 6) needs to be site and turbine specific. Expected turbine loading, tower design, and soil properties (sand, bedrock, etc.) will determine the type and size of foundation needed. Road design will also depend heavily on the size and weight of the loads to be transported, the terrain, local weather conditions, soil properties, and any environmental restrictions. Obviously, rugged terrain can make all aspects of site preparation difficult and costly.

8.3.3 Turbine transportation

The next significant hurdle may be transportation of the wind turbines to the site. Smaller turbines can often be packed in containers for easy transport over roads. Larger turbines must be transported in sub-sections and assembled at the site. In remote locations difficult access may limit the feasible turbine size or design or may require expensive transportation methods such as helicopters.

8.3.4 Turbine assembly and erection

Once at the site, the turbine must be assembled and erected. Issues related to assembly and erection need to be considered during the design phase (see Chapter 6) to minimize installation costs. Ease of erection depends on the turbine size and weight, the availability of an appropriately sized crane, the turbine design and site access. Small to medium-sized turbines can usually be assembled on site with a crane. Some can even be assembled on the ground and the whole tower and turbine placed on the foundation with a crane. Where site access is difficult or cranes are not available, such as in developing countries, it may be advantageous to use a turbine with a tilt-up tower. The complete turbine and tower are assembled on the ground and hydraulic pistons or winches are used to raise the tower about a hinge. This erection method can make maintenance easy, as the turbine can be lowered to the ground for easy access. The erection of large wind turbines can present a significant technical challenge. The tower sections, blades, and the nacelle can each be very heavy and large. Very large cranes are often needed. Turbines have also been designed with telescoping towers for easier tower erection, or with lifting devices integral to the tower. The tower then acts not only as the support for the turbine, but also as the crane for placement of the tower top on the tower.

8.3.5 Grid connection

The turbine–grid connection consists of electrical conductors, transformers, and switchgear to enable connection and disconnection. All of this equipment must be thermally rated to handle the expected current and the electrical conductors must be sized large enough to minimize voltage drops between the turbine and the point of connection (POC) to the electric grid (see Figure 8.2). The POC is a commonly used term to denote the grid connection at the wind turbine owner's property boundary. Another frequently used term is the point of common coupling, PCC. The PCC is the closest point in the grid system at which other users are connected to the grid. Details of the turbine–grid connection are presented below in Section 8.5. Once the turbine is installed and grid connected, it is ready for operation.

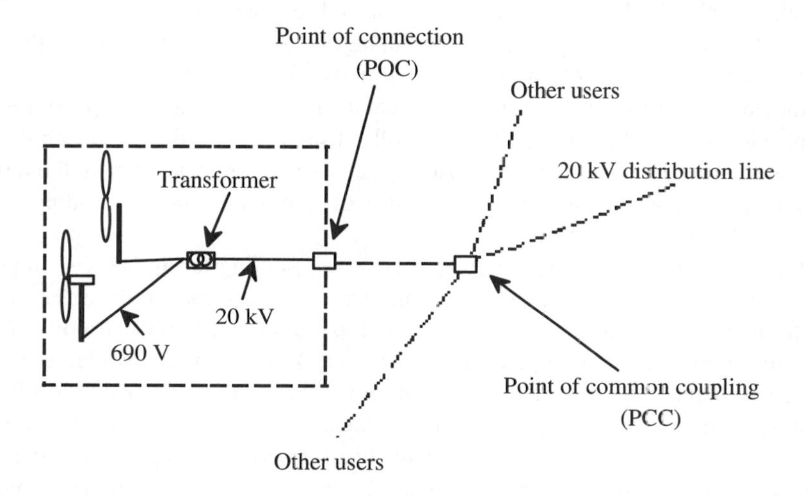

Figure 8.2 Schematic of typical grid connection (with voltages used in Europe)

8.3.6 Commissioning

Wind turbines need to be 'commissioned' before the turbine owner takes control of the turbine operation. Commissioning consists of (1) appropriate tests to ensure correct turbine operation and (2) maintenance and operation training for the turbine owner or operator. The extent of the commissioning process depends on the technical complexity of the turbine and the degree to which the design has been proven in previous installations. For mature turbine designs commissioning consists of tests of the lubrication, electrical, and braking systems, operator training, confirmation of the power curve and tests of turbine operation and control in a variety of wind speeds. Commissioning of (one-of-a-kind) research or prototype turbines includes numerous tests of the various sub-systems while the turbine is standing still (lubrication and electrical systems, pitch mechanisms, yaw drives, brakes, etc.) before any tests with the operating turbine are performed.

8.3.7 Turbine operation

Successful operation of a turbine or wind farm requires (1) information systems to monitor turbine performance, (2) with an understanding of factors that reduce turbine performance and (3) measures to maximize turbine productivity.

Automatic turbine operation requires a system for oversight in order to provide operating information to the turbine owner and maintenance personnel. Many individual turbines and turbines in wind farms have the capability to communicate with remote oversight systems via phone connections. The remote oversight systems (as explained in Chapter 7) receive data from individual turbines and display it on computer screens for system operators. These data can be used to evaluate turbine energy capture and availability (the percent of time that a wind turbine is available for power production).

The availability of wind turbines with mature designs is typically between 97% and 99% (Vachon et al., 1999). Reduced availability is caused by scheduled and unscheduled maintenance and repair periods, power system outages, and control system faults. For example, the inability of control systems to properly follow rapid changes in wind conditions, imbalance due to blade icing, or momentary high component temperatures can cause the controller to stop the turbine. The controller usually clears these fault conditions and operation is resumed. Repeated tripping usually causes the controller to take the turbine off line until a technician can determine the cause of anomalous sensor readings. This results in decreased turbine availability.

Wind turbine manufacturers provide power curves representing turbine power output as a function of wind speed (see Chapter 1). A number of factors may reduce the energy capture of a turbine or wind farm from that expected, based on the published power curve and the wind resource at the site. These include (Baker, 1999) reduced availability, poor aerodynamic performance due to soiled blades and blade ice, lower power due to off-yaw operation, control actions in response to wind conditions, and interactions between turbines in wind farms (see Section 8.4). Soiled blades have been observed to degrade aerodynamic performance by as much as 10–15%. Airfoils that are sensitive to dirt accumulation require either frequent cleaning or replacement with airfoils whose performance is less susceptible to degradation by the accumulation of dirt and insects. Blade ice accumulation can, similarly, degrade aerodynamic performance. Energy capture is also reduced when the wind direction changes. Controllers on some upwind turbine designs might wait until the magnitude of the average yaw error is above a predetermined value before adjusting the turbine orientation, resulting in periods of operation at high non-zero yaw errors. This results in lower energy capture. Turbulent winds can also cause a number of types of trips. For example, in turbulent winds, sudden high yaw errors might cause the system to shut down and restart, also reducing energy capture. In high winds, gusts can cause the turbine to shut down for protection when the mean wind speed is still well within the turbine operating range. These problems may reduce energy capture by as much as 15% from projected values. Operators should not only be prepared to minimize these problems, but should also anticipate them in their financing and planning evaluations.

8.3.8 Maintenance and repair

Wind turbine components require regular maintenance and inspection to make sure that lubrication oil is clean, seals are functioning, and that components subject to normal wear processes are replaced. Problem conditions identified by oversight systems may require that the turbine be taken out of operation for repairs.

8.3.9 Safety issues

Finally, the installed wind turbine needs to provide a safe work environment for operating and maintenance personnel. The turbine also needs to be designed and operated in a manner that it is not a hazard for neighbors. Safety issues include such things as protection against contact with high voltage electricity, protection against lightning damage to personnel or the turbine, protection from the effects of ice buildup on the turbine or the shedding of ice, the provision of safe tower climbing equipment, and lights to warn local night time air traffic of the existence of the wind turbine. Maintenance and repairs may be performed by on-site personnel or turbine maintenance contractors.

8.4 Wind Farms

Wind farms or wind parks, as they are sometimes called, are locally concentrated groups of wind turbines that are electrically and commercially tied together. There are many advantages to this electrical and commercial structure. Profitable wind resources are limited to distinct geographical areas. The introduction of multiple turbines into these areas increases the total wind energy produced. From an economic point of view, the concentration of repair and maintenance equipment and spare parts reduces costs. In wind farms of more than about 10 or 20 turbines, dedicated maintenance personnel can be hired, resulting in reduced labor costs per turbine and financial savings to wind turbine owners.

Wind farms were developed first in the United States in the late 1970s and then in Europe. Recently wind farms have been developed in many other places around the world, most notably in India, but also in China, Japan, and South and Central America.

The oldest existing concentration of wind farms in the United States is in California. The California wind farms, discussed in Chapter 1, originated as a result of a number of economic factors, including tax incentives and the high cost of new conventional generation. These factors spurred a significant boom in wind turbine installation activity in California, starting in about 1980, that levelled off after 1986 when economic forces changed. The result has been the development of three main areas of California: Altamont Pass, east of San Francisco, the Tehachapi mountains and San Gorgonio Pass in southern California (see Figure 8.3). Numerous of these first wind turbines suffered from reliability problems, but in recent years older turbines have been replaced with larger, more reliable, turbines in what is known as 'repowering' of the wind farms. In the 1990s wind farms were also developed in the Midwest and other regions of the United States. As of the end of 2000 the United States had about 2500 MW of installed wind power capacity, almost all of it in wind farms (AWEA, 2000).

Figure 8.3 A wind farm in San Gorgonio Pass, Palm Springs, California (Reproduced by permission of Henry Dupont)

Wind farms in Europe started in the late 1980s in Denmark. In recent years the number of wind turbines installed in Europe has increased tremendously as individual wind turbines and wind farms have been developed, primarily in Denmark, Germany, Spain, the Netherlands and Great Britain. Many of these wind turbines are in coastal areas. As available land for wind power development has become more limited in Europe, smaller installations have been built on inland mountains. As of November 2000 Europe had about 11,000 MW of installed wind power capacity, most of it in wind farms (EWEA, 2000).

8.4.1 Wind farm infrastructure

In addition to the individual wind turbines and their switchgear, wind farms have their own electric distribution system, roads, data collection systems, and support personnel.

8.4.1.1 Electric distribution system

The electrical distribution systems in wind farms typically operate at higher voltages than the turbine generator voltage in order to decrease resistive losses on the way to the substation at the grid connection. They also have switchgear for the whole wind farm at the connection to the grid. The voltage levels of the wind farm distribution system depend on the distances between turbines and transformer and cable costs. Many modern wind turbines come with a transformer installed in the tower base, but groups of lower voltage wind turbines in close proximity could share one transformer for cost reduction. Cost is also an issue when decisions are made as to whether the distribution lines should be overhead or underground. Underground lines, used in Europe and, often, in the United States, are more expensive, especially in rough terrain. Overhead lines are often used in India.

8.4.1.2 Roads

Access roads between wind turbines and maintenance and connecting roads to main highways may represent a significant cost, especially in environmentally sensitive areas with rough terrain. Roads need to be constructed in a manner that disturbs the landscape as little as possible, and that does not result in erosion. Grades and curves should be gentle enough that heavy equipment can reach the turbine sites. The length of the blades or tower sections are important considerations in this regard.

8.4.1.3 Control, monitoring and data collection systems

Modern wind farms include systems for controlling individual turbines and displaying and reporting information on wind farm operation. These systems, as mentioned in Chapter 7, are called SCADA (supervisory control and data acquisition) systems. SCADA systems display operating information on computer screens. Information about the whole wind farm, sets of turbines, or one individual turbine can be displayed. The information typically includes turbine operating states, power level, total energy production, wind speed and direction, and maintenance and repair notes. SCADA systems also display power curves or graphs of other information and allow system operators to shut down and reset turbines. Newer SCADA systems connected to modern turbines may also display oil temperatures, rotor speed, pitch angle, etc. SCADA systems also provide reports on turbine and wind farm operation to system operators, including information on operation and revenue from each turbine based on turbine energy production and utility rate schedules.

8.4.1.4 Support personnel

Once a certain number of turbines are placed in a wind farm, it becomes economical to provide dedicated operating and maintenance staff, sometimes called 'windsmiths.' The staff need to be appropriately trained and provided with suitable facilities.

8.4.2 Wind farm technical issues

Numerous technical issues arise with the close spacing of multiple wind turbines. The most important are related to the question of where to locate and how closely to space the wind turbines (common terms for referring to wind turbine array spacing are illustrated in Figure 8.4). As mentioned in Section 8.2, the wind resource may vary across a wind farm as a result of terrain effects. In addition, the extraction of energy by those wind turbines that are upwind of other turbines results in lower wind speeds at the downwind turbines and increased turbulence. As described in this section, these wake effects can decrease energy production and increase wake-induced fatigue in turbines downwind of other machines. Wind turbine spacing also affects fluctuations in the output power of a wind farm. As described in Section 8.5, the fluctuating power from a wind farm may affect the local electrical grid to which it is attached. This section describes the relationship between wind farm output power fluctuations and the spacing of the turbines in wind farm.

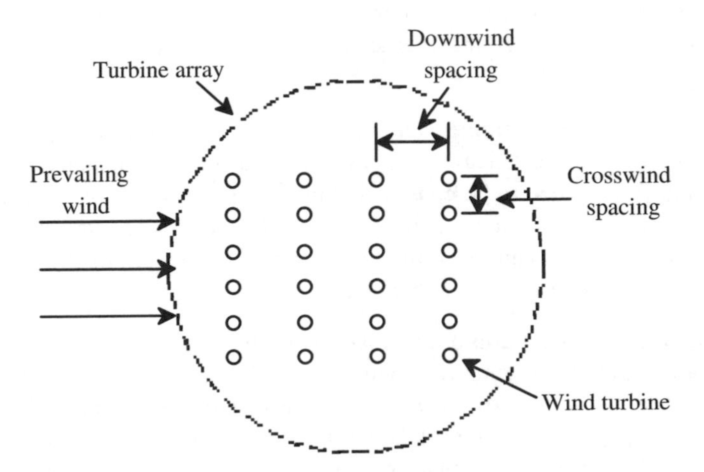

Figure 8.4 Wind farm array schematic

8.4.2.1 Array losses

Wind energy comes from the extraction of kinetic energy from the wind. This results in lower wind speeds behind a wind turbine and less energy capture by the downstream turbines in an array. Thus a wind farm will not produce 100% of the energy that a similar number of isolated turbines would produce in the same prevailing wind. The energy loss is termed 'array loss.' Array losses are mainly a function of:

- Wind turbine spacing (both downwind and crosswind)
- Wind turbine operating characteristics
- The number of turbines and size of the wind farm
- Turbulence intensity
- Frequency distribution of the wind direction (the wind rose)

The extraction of energy from the wind results in an energy and velocity deficit, compared with the prevailing wind, in the wake of a wind turbine. The energy loss in the turbine wake will be replenished over a certain distance by exchange of kinetic energy with the surrounding wind field. The extent of the wake in terms of its length as well as its width depends primarily on the rotor size and power production.

Array losses can be reduced by optimising the geometry of the wind farm. Different distributions of turbine sizes, the overall shape and size of the wind farm turbine distribution, and turbine spacing within the wind farm all affect the degree to which wake effects reduce energy capture.

The momentum and energy exchange between the turbine wake and the prevailing wind is accelerated when there is higher turbulence in the wind field. This reduces the velocity deficits downstream, reducing array losses. Typical turbulence intensities are between 10% and 15%, but may be a low as 5% over water or as high as 50% in rough terrain. Turbulence intensity also increases through the wind farm due to the interaction of the wind with the turning rotors.

Finally, array losses are also a function of the annual wind direction frequency distribution. The crosswind and downwind distances between wind turbines will vary depending on the geometry of the wind turbine locations and the direction of the wind. Thus array losses need to be calculated based on representative annual wind direction data in addition to wind speed and turbulence data.

Field geometry and ambient turbulence intensity have been shown to be the most important parameters affecting array losses. Studies have shown that, for turbines that are spaced 8 to 10 rotor diameters, D, apart in the prevailing downwind direction and 5 rotor diameters apart in the crosswind direction, array losses are typically less than 10% (Lissaman et al., 1982). Figure 8.5 illustrates array losses for a hypothetical 6-6 array of turbines with a downwind spacing of 10 rotor diameters. The graph presents array losses as a function of crosswind spacing and turbulence intensity for wind only from the prevailing wind direction (turbines directly in the wake of other turbines) and for wind that is evenly distributed from all directions.

Array losses may also be expressed as array efficiencies where:

$$\text{Array efficiency} = \frac{\text{Annual energy of whole array}}{(\text{Annual energy of one isolated turbine})(\text{total no. of turbines})} \quad (8.4.1)$$

It can be seen that array efficiency is just 100% minus the array losses in percent.

The design of a wind farm requires careful consideration of these effects in order to maximize energy capture. Closer spacing of wind turbines may allow more wind turbines on the site, but will reduce the average energy capture from each turbine in the wind farm.

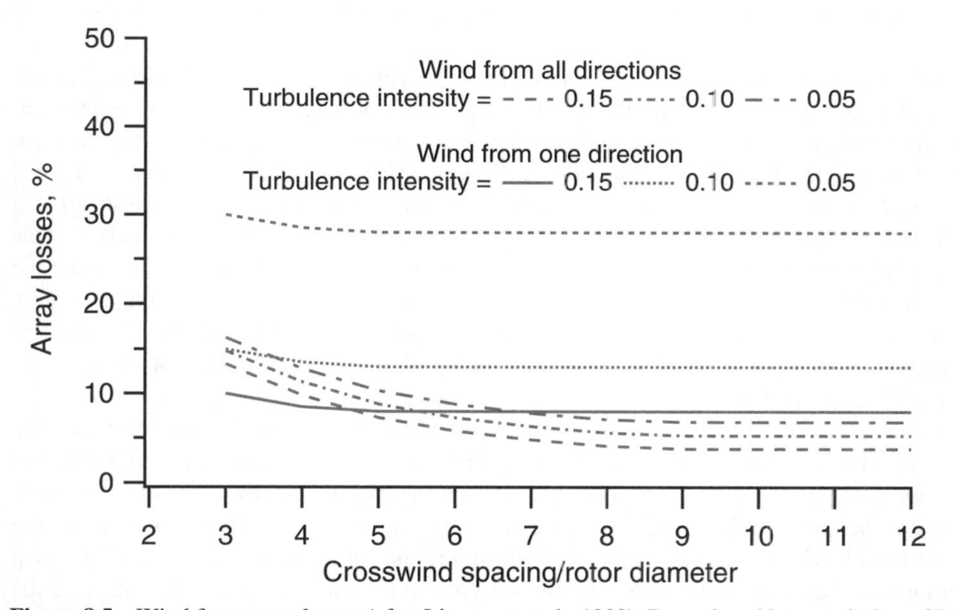

Figure 8.5 Wind farm array losses (after Lissaman et al., 1982). Reproduced by permission of BHR Group Limited

8.4.2.2 The calculation of array losses – wake models

The calculation of array losses requires a knowledge of the location and characteristics of the turbines in the wind farm, a knowledge of the wind regime, and appropriate models of turbine wakes to determine the effect of upstream turbines on downstream ones. A number of turbine wake models have been proposed. These fall into the following categories:

- Surface roughness models
- Semi-empirical models
- Eddy viscosity models
- Full Navier–Stokes solutions

The surface roughness models are based on data from wind tunnel tests. The first models to attempt to characterize array losses were of this type. An excellent review by Bossanyi et al. (1980) describes a number of these models and compares their results. These models assume a logarithmic wind velocity profile upstream of the wind farm. They characterize the effect of the wind farm as a change in surface roughness that results in a modified velocity profile within the wind farm. This modified velocity profile, when used to calculate turbine output, results in appropriately lower power output for the total wind farm. These models are usually based on regular arrays of turbines in flat terrain.

The semi-empirical models provide descriptions of the energy loss in the wake of individual turbines. Examples include models by Lissaman (Lissaman and Bates, 1977), Vermeulen (Vermeulen, 1980) and Katić (Katić et al., 1986). These models are based on simplified assumptions about turbine wakes (based on observations) and on conservation of momentum. They may include empirical constants derived from either wind tunnel model data or from field tests of wind turbines. They are useful for describing the important aspects of the energy loss in turbine wakes, and, therefore, for modelling wind farm array losses.

Eddy viscosity models are based on solutions to simplified Navier–Stokes equations. The Navier–Stokes equations are the defining equations for the conservation of momentum of a fluid with constant viscosity and density. They are a set of differential equations in three dimensions. The use of the Navier–Stokes equations to describe time-averaged turbulent flow results in terms that characterize the turbulent shear stresses. These stresses can be related to flow conditions using the concept of eddy viscosity. Eddy viscosity models use simplifying assumptions such as axial symmetry and analytical models to determine the appropriate eddy viscosity. These models provide fairly accurate descriptions of the velocity profiles in turbine wakes without a significant computational effort and are also used in array loss calculations. Examples include the model of Ainsle (1985 and 1986) and that of Smith and Taylor (1991).

Figures 8.6 and 8.7 illustrate measured wind speed data behind wind turbines. The graphs also include the results of one of these eddy viscosity wake models. Figure 8.6 shows non-dimensionalized vertical velocity profiles at various distances (measured in rotor diameters) behind a wind turbine. The velocity deficit and its dissipation downwind of the turbine are clearly illustrated. Figure 8.7 illustrates the hub height velocity profiles as a function of distance from the rotor axis for the same conditions. The Gaussian shape of the hub height velocity deficit in the far wake can clearly be seen.

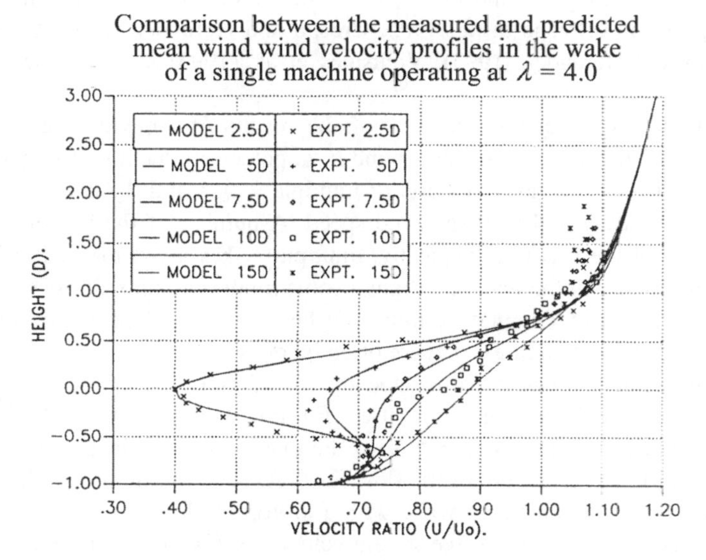

Figure 8.6 Vertical velocity profiles downwind of a wind turbine (Smith and Taylor, 1991); λ, tip speed ratio. Reproduced by permission of Professional Engineering Publishing

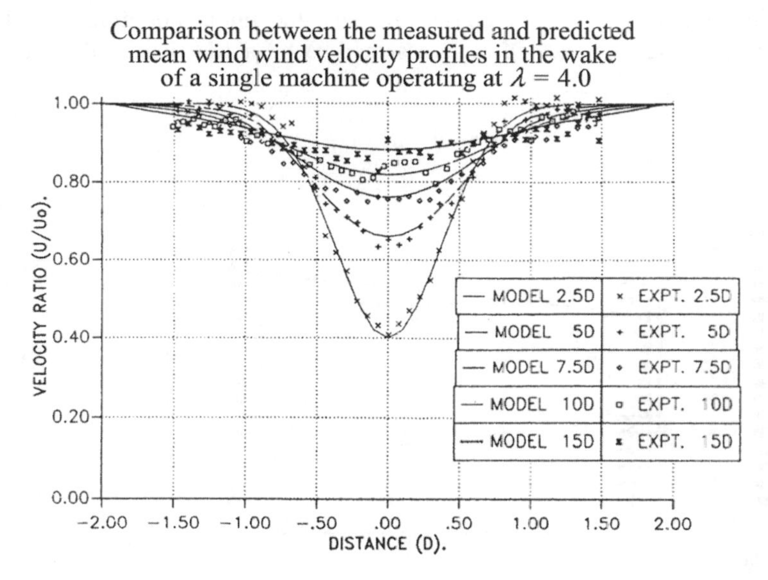

Figure 8.7 Hub height velocity profiles downwind of a wind turbine (Smith and Taylor, 1991); λ, tip speed ratio. Reproduced by permission of Professional Engineering Publishing

Finally, a variety of approaches exist to solving the complete set of Navier–Stokes equations. These models require a significant computational effort and may use additional models to describe the transport and dissipation of turbulent kinetic energy (the k–ε model) to converge to a solution. These models are best suited for research, for detailed descriptions of wake behaviour and to guide the development of simpler models. Examples include models by Crespo et al. (Crespo et al., 1985, Crespo and Herandez, 1990, Crespo et al., 1990, and Crespo and Herandez, 1993), Voutsinas et al. (1993) and Sørensen and Shen (1999).

A number of factors affect the accuracy of the results of applying these models to specific wind farms. When used to calculate wind farm power production, decisions must be made about how to handle the superposition of multiple wakes and the effects of complex terrain on both wake decay and ambient wind speed. A number of the models mentioned above address some of these issues. Typically, multiple wakes are combined based on the combination of the energy in the wakes, although some models assume linear superposition of velocities. The effects of complex terrain may be significant (see Smith and Taylor, 1991) but are more difficult to address and are often ignored.

The use of these models can be illustrated by considering one of the semi-empirical models (Katić et al., 1986) that is often used for micrositing and wind farm output predictions. The model attempts to characterize the energy content in the flow field and ignores the details of the exact nature of the flow field. As seen in Figure 8.8, the flow field is assumed to consist of an expanding wake with a uniform velocity deficit that decreases with distance downstream. The initial free stream velocity is U_0 and the turbine diameter is D. The velocity in the wake at a distance X downstream of the rotor is U_X with a diameter of D_X. The wake decay constant, k, determines the rate at which the wake diameter increases in the downstream direction.

In this and many other semi-empirical models, the initial non-dimensional velocity deficit (the axial induction factor), a, is assumed to be a function of the turbine thrust coefficient:

$$a = \frac{1}{2}\left(1 - \sqrt{1 - C_T}\right) \tag{8.4.2}$$

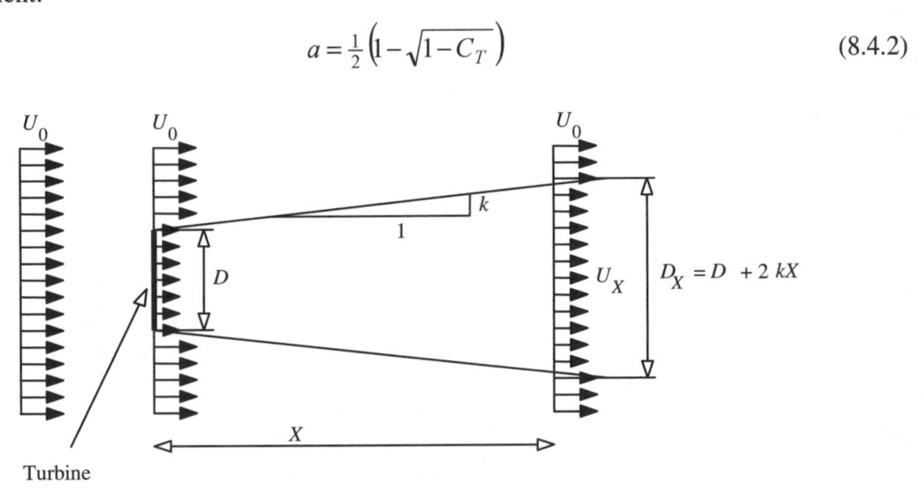

Figure 8.8 Schematic view of wake description (after Katic et al., 1986); U_0, initial free stream velocity, D, turbine diameter, U_X, velocity at a distance X, D_X, wake diameter at a distance X, k, wake decay constant

where C_T is the turbine thrust coefficient. Equation 8.4.2 can be derived from Equations 3.2.16 and 3.2.17 for the ideal Betz model. Assuming conservation of momentum one can derive the following expression for the velocity deficit at a distance X downstream:

$$1-\frac{U_X}{U_0}=\frac{\left(1-\sqrt{1-C_T}\right)}{\left(1+2k\frac{X}{D}\right)^2} \tag{8.4.3}$$

The model assumes that the kinetic energy deficit of interacting wakes is equal to the sum of the energy deficits of the individual wakes (indicated by subscripts 1 and 2). Thus, the velocity deficit at the intersection of two wakes is:

$$\left(1-\frac{U_X}{U_0}\right)^2=\left(1-\frac{U_{X,1}}{U_0}\right)^2+\left(1-\frac{U_{X,2}}{U_0}\right)^2 \tag{8.4.4}$$

The only empirical constant in the model is the wake decay constant, k, which is a function of numerous factors, including the ambient turbulence intensity, turbine-induced turbulence and atmospheric stability. Katić notes that in a case in which one turbine was upstream of another, $k = 0.075$ adequately modelled the upstream turbine, but $k = 0.11$ was needed for the downstream turbine, which was experiencing more turbulence. He notes also that the results for a complete wind farm with wind coming from multiple directions is relatively insensitive to minor changes in the value of k. A small constant gives a large power reduction in a narrow zone, while a large value gives a smaller reduction in a wider zone. The net effect of varying this parameter, when analyzing wind farm performance at many wind speeds from a variety of directions, is small.

The following steps are used to determine the output of a wind farm using the model:

1. The wind turbine radius, hub heights, power and thrust characteristics are determined.
2. The wind turbine locations are determined such that the coordinate system can be rotated for analysis of different wind directions.
3. The site wind data are binned by wind direction with, for example, 45 degree wide bins. Weibull parameters are determined for each bin together with the frequency of the wind occurring from each sector.
4. The annual average wind power is calculated by stepping through all wind speeds and directions. Thrust coefficients are determined from the operating conditions at each turbine.

8.4.2.3 Wake turbulence

The turbines in wind farms that are downwind of other machines experience increased turbulence due to power production upwind. Turbine wakes consist not only of regions of lower average velocity, but of swirling vortices caused by: (1) interaction of the wind over the rotor with the rotor surfaces and (2) differential flow patterns over the upper and lower blade surfaces at the rotor tips. In general, the turbulence intensities in the wake are increased over ambient levels, with an annular area in the far wake of higher relative turbulence (caused by the tip vortices) surrounding the turbulent core of the wake. Figure 8.9 shows measured and predicted turbulence intensities at a variety of rotor distances downstream of a turbine rotor in a wind field with an ambient turbulent intensity of 0.08. The resulting turbulence increases material fatigue, reducing turbine life, in the turbines in a wind farm that are downwind of other machines (for more details, see Hassan et al., 1988). The increased turbulence in the downwind areas of wind farms has also been observed to reduce the energy capture of the turbines in those locations. High turbulence means higher velocity gusts and more extreme wind speed changes over short periods. Control actions to limit loads in gusts cause turbines to shut down more frequently, decreasing overall energy capture (Baker, 1999).

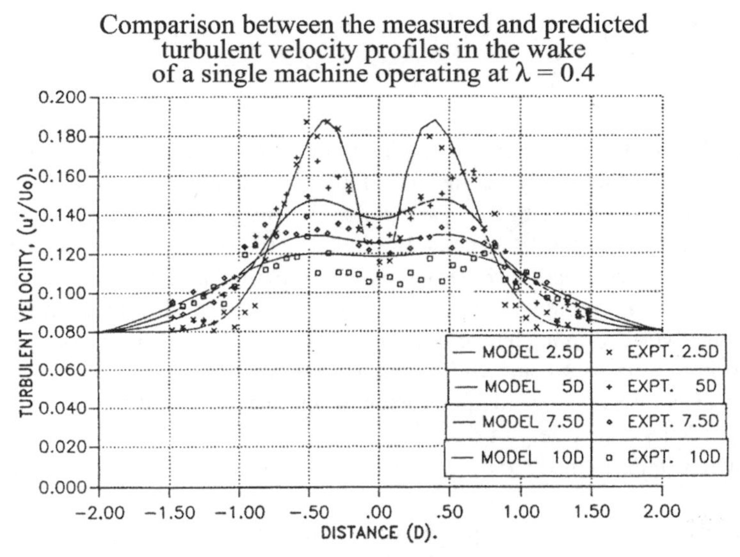

Figure 8.9 Turbulence intensity downwind of a wind turbine (Smith and Taylor, 1991); λ, tip speed ratio. Reproduced by permission of Professional Engineering Publishing

8.4.2.4 Wind farm power curves

The result of array losses and wake turbulence are a modification of the operation of the individual turbines with respect to the overall prevailing wind speed. As the wind approaching an array of wind turbines increases from zero, the first row of turbines will start to produce power. That power production will reduce the wind speed behind the first row and no other turbines will operate. As the wind increases, more and more rows of turbines will produce power until all of the turbines are producing power, with the front row producing the most power per turbine. Once the wind reaches rated wind speed only the first row of turbines will produce rated power. Each turbine will be producing rated power only after the winds are somewhat higher than rated for the turbines in the wind farm. Thus, not only is the total wind farm energy production lower than that of multiple isolated turbines, but the energy production as a function of wind speed has a different shape for the whole wind farm than for an individual turbine (see Figure 8.10).

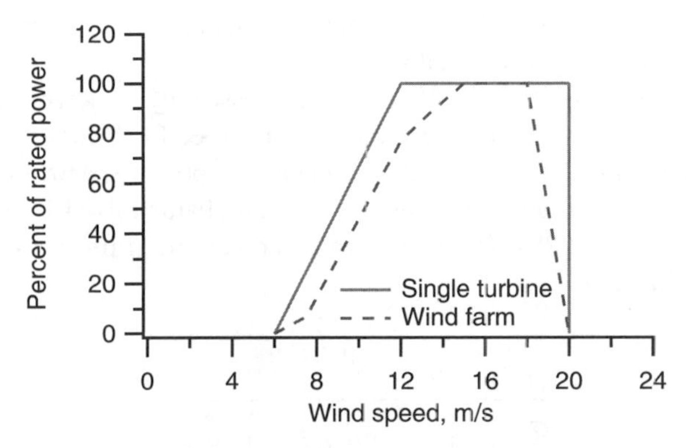

Figure 8.10 Comparison of single turbine and wind farm power curves

8.4.2.5 Power smoothing

The total output power of a wind farm is the sum of the power produced by the individual turbines in a wind farm. Turbulent wind fluctuations result in fluctuating power from each of the wind turbines and, thus, from the wind farm. Turbulent wind conditions result in different winds at widely spaced turbines. This means that the power from one turbine may be rising as the power from another turbine is falling. This results in some reduction in wind farm power fluctuations compared with the power that would be expected from turbines all experiencing the same wind.

For example, assume that one wind turbine produces an average power P_1 over some time interval with a standard deviation of $\sigma_{P,1}$. Then, if N wind turbines in the wind farm experienced the same wind, the total wind farm production would be $P_N = NP_1$ and the standard deviation of the wind farm electrical power output would be $\sigma_{P,N} = N\sigma_{P,1}$. Usually, however, the wind at an individual wind turbine is not well correlated with the wind at any other turbine and, thus, the wind turbines do not all experience the same wind. It can be shown that if N wind turbines experience wind with the same mean wind speed and uncorrelated turbulence with the same statistical description, then the mean power output of the N turbines is still $P_N = NP_1$, but the standard deviation of the of the resulting aggregated power is just:

$$\sigma_{P,N} = \frac{N\sigma_{P,1}}{\sqrt{N}} \tag{8.4.5}$$

Thus, the fluctuation of the total power from the wind farm is less than the fluctuation of the power from individual wind turbines. This effect is termed power smoothing.

In reality, the wind at two different turbine sites in a wind farm is neither perfectly correlated nor perfectly uncorrelated. The degree of correlation depends on the distance between the two locations and on the spatial and temporal character of the wind field. An expression for the variance in wind farm output power as a function of the turbulent length scale and the number and distance between machines is developed here. The analysis assumes that all of the turbines experience the same mean wind speed and that the power from each machine can be assumed to be linear with wind speed about the mean power output. The analysis builds on the material in Chapter 2.

The temporal character of the wind field is expressed by a power spectrum, which measures the variance of the fluctuating winds at one specific location as a function of frequency. As explained in Chapter 2, in the absence of a power spectrum based on actual data collected at a site, atmospheric turbulence is often characterized by the von Karman power spectrum (Fordham, 1985). The von Karman power spectrum $S_1(f)$, repeated here from Chapter 2, is defined by:

$$\frac{S_1(f)}{\sigma_U^2} = \frac{4\frac{L}{U}}{\left[1 + 70.8\left[\frac{Lf}{U}\right]^2\right]^{5/6}} \tag{8.4.6}$$

where $S_1(f)$ is the single point power spectrum, L is the integral length scale of the turbulence, U is the mean wind speed, and f is the frequency (in Hertz).

The spatial character of the wind can be expressed by a coherence function, which is a measure of the magnitude of the power spectrum at one location with respect to the power spectrum at another location. When using the von Karman spectrum, an expression for the coherence in terms of Bessel functions is, strictly speaking, the most appropriate. That expression is significantly more complex than the exponential coherence function approximation for atmospheric turbulence which is often used in micrometeorology (see Beyer et al., 1989), which is given by:

$$\gamma_{ij}^2(f) = e^{-a\frac{x_{ij}}{U}f} \tag{8.4.7}$$

where $\gamma_{ij}^2(f)$ is the coherence between the power spectrums at points i and j, a is the coherence decay constant, taken to be 50, x_{ij} is the spacing between points i and j, U is the mean wind speed, and f is the frequency (in Hz). The coherence decay constant of 50 is most appropriate to the case where the turbines are in a line, perpendicular to the prevailing wind direction (see Kristensen et al., 1981). Equation 8.4.7 illustrates that the similarity in the wind experienced by turbines at different locations decreases with distance between the turbines and the frequency of turbulent fluctuations. At low frequencies (long time periods) and close distances, the winds are similar, but at high frequencies (short time scales) and long distances, they are very different.

The spectra of the total variance in the wind at N turbine locations, $S_N(f)$, is (Beyer et al., 1989):

$$S_N(f) = S_1(f)\frac{1}{N^2}\sum_{i=1}^{N}\sum_{j=1}^{N}\gamma_{ij}(f) \tag{8.4.8}$$

The term $\frac{1}{N^2}\sum_{i=1}^{N}\sum_{j=1}^{N}\gamma_{ij}(f)$ is known as the 'wind farm filter.'

The variance of the fluctuating wind speed is given by the integral of the power spectrum over all frequencies. Note that the total variance is the same value that may be found from time series data in the usual way:

$$\sigma_U^2 = \int_0^\infty S_1(f)\mathrm{d}f \tag{8.4.9}$$

If the standard deviation of the power from each of the wind turbines at the mean power level can be assumed to be k times the standard deviation in the wind speed (that is, if $\sigma_P = k\sigma_U$), then the variance of the fluctuating power from one wind turbine is given by:

$$\sigma_{P,1}^2 = k^2 \int_0^\infty S_1(f)\mathrm{d}f \tag{8.4.10}$$

The variance of the total fluctuating power is given by:

$$\sigma_{P,N}^2 = N^2 k^2 \int_0^\infty S_N(f)\mathrm{d}f = k^2 \int_0^\infty \left\{ S_1(f) \sum_i^N \sum_j^N \gamma_{ij}(f) \right\} \mathrm{d}f \tag{8.4.11}$$

and using the von Karman spectrum:

$$\sigma_{P,N}^2 = k^2 \int_0^\infty \left\{ \frac{4\dfrac{L}{U}}{\left[1 + 70.8\left(\dfrac{Lf}{U}\right)^2\right]^{5/6}} \sum_{i=1}^N \sum_{j=1}^N \exp\left(\frac{-25 x_{ij} f}{U}\right) \right\} \mathrm{d}f \tag{8.4.12}$$

Thus the variability of the total power from a wind farm decreases as the distance between turbines increases. Also, the lack of correlation between the wind at higher frequencies contributes more to the decrease in total variability than do the relatively correlated winds at low frequencies. It is easy to verify, using Equations 8.4.7 and 8.4.11, that, if the wind at the turbines is completely correlated ($x_{ij} = 0$ for all i, j), then turbines all act as one large turbine. Similarly, if the wind at N turbines is completely uncorrelated ($x_{ij} = 0$ when $i = j$ and is otherwise infinite) then Equation 8.4.5 applies.

As an example, Figure 8.11 illustrates the effect of spacing for 2 and 10 wind turbines, assumed to be equally spaced along a line perpendicular to the wind direction. For this example L = 100 m and U = 10 m/s. The figure shows the fractional reduction in power variability as a function of crosswind spacing.

As described above, the wind turbines in a wind farm often do not experience the same mean wind or turbulence of the same statistical description, but the power smoothing effect can nevertheless be seen in wind farm data. Thus, wind farms with a large number of wind turbines can reduce the voltage fluctuations and other problematic effects caused by the power fluctuations of individual wind turbines (see Section 8.5).

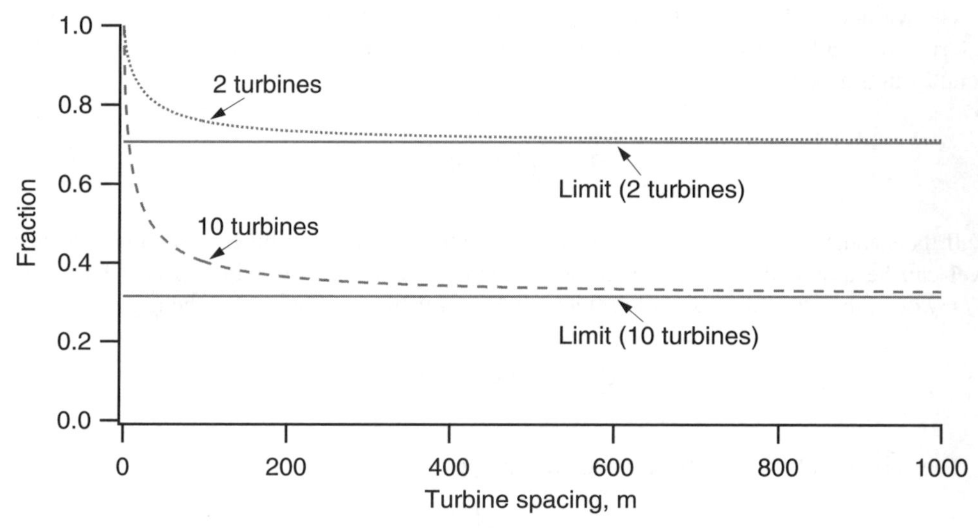

Figure 8.11 Fraction of wind farm power variability as a function of turbine spacing

8.5 Wind Turbines and Wind Farms in Electric Grids

Wind turbines supply power through large electric grids, which are often characterized as "stiff," that is to say totally unaffected by connected loads or generation equipment. In reality grid characteristics can affect and are affected by wind turbines connected to them. To help understand these effects, a brief description of electric grids and grid connection equipment is provided. This is followed by a summary of the electrical behavior of turbines in grids and the types of grid–turbine interactions that can affect wind turbine installations.

8.5.1 Electric grids

Electric grids can be divided into four main parts: generation, transmission, distribution, and supply feeders (see Figure 8.12). The generation function has historically been provided by large synchronous generators powered by fossil or nuclear fuel or hydroelectric turbines. These generators respond to load variations, keep the system frequency stable and adjust the voltage and power factor at the generating station as needed. Generators in large, central power plants produce power at high voltage (up to 25,000 V). These generators feed current into the high-voltage transmission system (110 kV to 765 kV) used to distribute the power over large regions. The transmission systems use high voltage to reduce the losses in the power transmission lines. The local distribution system operates at a lower voltage (10 kV to 69 kV), distributing the power to local neighborhoods. Locally, the voltage is reduced again and the power is distributed through feeders to one or more consumers. Industrial users in the United States typically use 480 V power while commercial and residential users in the United States and much of the rest of the world use 120 or 240 V systems. In Europe industrial users generally use 690 V and residential loads use 230 V.

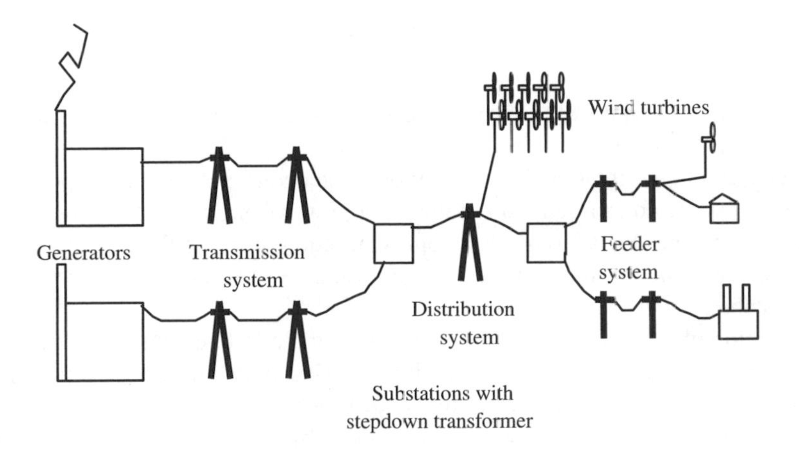

Figure 8.12 Electric grid system schematic

The term electrical load is used to describe a sink for power or a specific device that absorbs power. The total electrical load on the transmission system is the sum of the many fluctuating end loads. Fluctuations in these end loads are mostly uncorrelated, resulting in a fairly steady load on transmission lines that varies according to the time of day and season. The distribution and supply systems are closer to consumer loads, far from the large generators, and, thus, increasingly affected by these changing loads.

The locations of usable wind resources do not usually allow wind turbines to be easily tied into the high-voltage transmission system. Wind turbines are most often connected to the distribution system or, in the case of smaller wind turbines, into the feeder system.

The majority of generators in electrical networks are synchronous generators (see Chapter 5) that are usually driven by prime movers such as steam, hydroelectric, or gas turbines or diesel engines. System operators attempt to control the grid frequency and voltage to be within a narrow range about the nominal system values. Frequency in large electric grids in industrialized countries is maintained at less than +/- 0.1% of the desired value. Depending on the country, voltages at distribution points are allowed to fluctuate from +/-5% to up to +/- 7% of the nominal value, but allowable customer-induced voltage variations are often less (Patel, 1999).

The grid frequency is controlled by the power flows in the system. The torque on the rotor of any given generator consists of the torque of the prime mover, Q_{PM}, the electrical torques due to loads in the system, Q_L, and other generators in the system, Q_O. The equation of motion for the generator, with rotating inertia, J, and speed ω, can be written:

$$J\frac{\mathrm{d}\omega}{\mathrm{d}t} = Q_{PM} + Q_L + Q_O \tag{8.5.1}$$

Each of the generators in the system is synchronized with the other generators in the system. Thus, this same equation of motion can represent the behavior of the system as a whole if each term represents the sum of all of the system inertias or loads.

Recognizing that power is just the product of torque and speed, one can derive the following equation:

$$\frac{\mathrm{d}\omega}{\mathrm{d}t} = \frac{1}{J\omega}\left(P_{PM} + P_L + P_O\right) \tag{8.5.2}$$

where P_{PM} is the prime mover power, P_L is the power from electrical loads and P_O is the power from other generators.

Here changes in the rotational frequency of the generators (which is proportional to grid frequency) is expressed as a function of the input power from the prime movers, the system load, and any power flows from other connected equipment. If the system load changes, then the power from the prime movers is adjusted to compensate and keep the system frequency stable.

The system voltage is controlled by controllers on the field excitation circuits of each of the generators. Changing the field excitation changes both the terminal voltage and the power factor of the power delivered to the load. When the fields are controlled to stabilize system voltage, the power factor is determined by the connected loads. Voltage is controlled, in this manner, at each generating station.

Electrical grids, like other electrical circuits, provide an impedance to current flow that causes voltage changes between the generating station and other connected equipment. This can be illustrated by considering a wind turbine generator connected to a grid system (see Figure 8.13) with a line-to-neutral voltage, V_S, that is assumed to be the same as the voltage at the generating station. The voltage at the wind turbine, V_G, is not necessarily the same as V_S. The difference in voltage is caused by the distribution system impedance, which consists of the distribution system resistance, R, that causes voltage changes primarily because of the real power flowing in the system and distribution system reactance, X, that causes voltage changes because of the reactive power flowing in the system (see Chapter 5 for a definition of electrical terms). The magnitudes of R and X, which are functions of the distribution system, and the magnitudes of the real generated power, P, and reactive power requirements, Q, of the wind turbine or wind farm will determine the distribution system voltage at the wind turbine.

The voltage at the generator can be determined from (Bossanyi et al., 1998):

$$V_G^{\,4} + V_G^{\,2}\left[2(QX - PR) - V_S^{\,2}\right] + (QX - PR)^2 + (PX - QR)^2 = 0 \tag{8.5.3}$$

In lightly loaded distribution circuits, the voltage change can be approximated as:

$$\Delta V = V_G - V_S = \frac{PR - QX}{V_S} \tag{8.5.4}$$

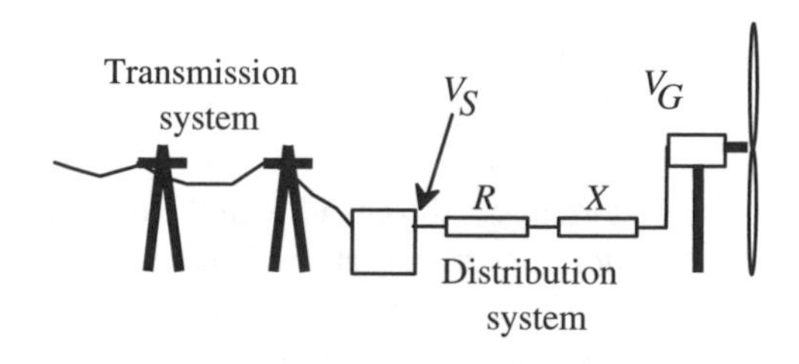

Figure 8.13 Distribution system schematic; R, resistance, X reactance, V_S and V_G, grid and wind turbine voltage respectively.

It can be seen that the voltage increases due to the real power production (PR) in the system. However, the voltage decreases (QX) when reactive power is consumed by equipment on the system.

These voltage changes can be significant. Transformers equipped for automatic voltage control (AVR) are used to provide reasonably steady voltages to end-users. These transformers have multiple taps on the high-voltage side of the transformer. Current flow is switched automatically from one tap to another as needed. The different taps provide different turn ratios and therefore different voltages (for more information, see Rogers and Welch, 1993).

The same cable resistance that causes voltage fluctuations also dissipates energy. The electrical losses in the distribution system, P_{LO}, can be expressed as:

$$P_{LO} = \frac{(P^2 + Q^2)R}{V_S^{\,2}} \tag{8.5.5}$$

Grid 'strength' or stiffness is characterized by the fault level, M, of the distribution system. The fault level at any location in the grid is the product of the system voltage and the current that would flow if there were a short circuit at that location. Using the example above, if there were a short circuit at the wind turbine, the fault current, I_F , would be:

$$I_F = \frac{V_S}{(R^2 + X^2)^{1/2}} \tag{8.5.6}$$

and the fault level, M, would be:

$$M = I_F V_S \tag{8.5.7}$$

This fault level is an indication of the strength of the network, with higher fault levels indicating stronger networks.

8.5.2 Grid connection equipment

The turbine–grid connection consists of equipment to connect and disconnect the turbine or wind farm from the larger grid, equipment to sense problems on the grid or the turbine side of the connection and transformers to transfer power between different voltage levels. This equipment is in addition to the electrical equipment associated with each wind turbine that is described in Chapter 5.

- *Switchgear* - Switchgear to connect and disconnect wind power plants from the grid usually consists of large contactors controlled by electromagnets. Switchgear should be designed for fast automatic operation in case of a turbine problem or grid failure.
- *Protection equipment* - Protection equipment at the point of connection needs to be included to ensure that turbine problems do not affect the grid and visa versa. This equipment must include provision for rapid disconnection in case of a short circuit or overvoltage situation in the wind farm. The wind farm should also be disconnected from the grid in case of a deviation of the grid frequency from the rated frequency due to a

grid failure or a partial or full loss of one of the phases in a three-phase grid (see Chapter 5). The protection equipment consists of sensors to detect problem conditions. Outputs from these sensors control contactor magnets or additional solid state switches such as silicon-controlled rectifiers (SCRs). The ratings and operation of protection equipment should be coordinated with that of other local equipment to ensure that no problems occur. For example, in case of a momentary grid failure, the disconnect at the wind farm should react fast enough to prevent currents from flowing into the grid fault and should remain off long enough to ensure that reconnection only occurs after the other grid faults have been cleared (Rogers and Welch, 1993).

- *Electrical conductors* - Electrical conductors for connecting wind farms to the grid are usually made of aluminium or copper. The electrical conductors used for transformers and grid connections dissipate power because of their electrical resistance. These losses reduce system efficiency and can cause damage if wires and cables get too hot. Cable resistance increases linearly with distance and decreases linearly with the cross-sectional area of the conductor. Economic considerations tend to dictate the resistive losses allowable, given the increased cost of larger cables.
- *Transformers* - Transformers at the substation are used to connect electrical circuits at different voltage levels. The details of transformer operation are covered in Chapter 5. Usually, these transformers have automatic voltage control to help maintain the system voltage.
- *Grounding* – Turbines, wind farms, and substations need grounding systems to protect equipment from lightning damage and short circuits to ground. Providing a conductive path for high currents to the earth can be a significant problem in locations with exposed bedrock and other non-conductive soils. Lightning strikes or faults can result in significant differences in ground potential at different locations. These differences can disrupt grid protection equipment and pose a danger to personnel.

8.5.3 The behavior of wind turbines connected to electric grids

Wind turbine operation results in fluctuating real and reactive power levels and may result in voltage and current transients or voltage and current harmonics. These may contribute to turbine–grid interactions, as explained in the Section 8.5.4. This section builds on the material in Chapter 5 and addresses those aspects of turbine operation that may be significant for turbine–grid interactions.

Wind turbines, especially fixed-speed turbines connected to electrical grids, generally use induction generators. Induction generators provide real power (P) to the system and absorb reactive power (Q) from the system. The relationship of real to reactive power is a function of the generator design and the power being produced. Both real and reactive power are constantly fluctuating during wind turbine operation. Low-frequency real power fluctuations occur as the average wind speed changes. Reactive power needs are approximately constant or increase slowly over the operating range of induction generators. Thus, low-frequency reactive power fluctuations are usually smaller than low frequency real power fluctuations. Higher frequency fluctuations of both real and reactive power occur as a result of turbulence in the wind, tower shadow, and dynamic effects from drive train, tower and blade vibrations.

Wind turbines with synchronous generators operate in a different manner than those with induction generators. When connected to a large electrical network with a constant voltage, the field excitation of the synchronous generators on wind turbines can be used to change the line power factor and to control reactive power, if desired.

Variable speed turbines usually have a power electronic converter between the generator and the grid. These systems can control both the power factor and voltage of the delivered power. Power electronic converters connected to induction generators also need to supply reactive power to the turbine generator. This is done, in effect, by circulating reactive current through the generator coils to support the magnetic field in the generator. The converter components connected to the grid can usually provide current to the grid at any desired power factor. This capability may be used to improve grid operation, if desired.

When generators are connected or disconnected from a power source, voltage fluctuations and transient currents can occur. As explained in Chapter 5, connecting an induction generator to the grid results in a momentary 'in-rush' current as the magnetic field is energized. Also, if the generator is used to speed up the rotor from speeds far from synchronous speed (high slip operation), significant currents can occur. These high currents can be limited, but not eliminated, with the use of a 'soft-start' circuit, which limits the generator current. When induction generators are disconnected from the grid, voltage spikes can occur as the magnetic field decays. Synchronous generators, in contrast, generally have no starting current requirements. Normally they must be accelerated up to operating speed by the turbine rotor before grid connection can occur. Nevertheless, voltage transients may still occur on connection and disconnection as the stator field is energized and de-energized.

8.5.4 Turbine–grid interactions

On the one hand, the introduction of wind turbines into a distribution grid can, at times, lead to problems that limit the magnitude of the wind power that can be connected to the grid. On the other hand, depending on the grid and the turbines, the introduction of turbines may help support and stabilize a local grid. Turbine–grid interactions depend on the electrical behavior of (1) the turbines under consideration and (2) the electric grids to which the turbines are connected. Important aspects of these have been explained above. Interconnection issues include problems with steady state voltage levels, flicker, harmonics and islanding. This section focuses mainly on interactions that affect the local system voltages and currents on medium- to short-term time scales. Larger questions related to overall system control are discussed at the end of the section.

8.5.4.1 Steady state voltages

Changes in the mean power production and reactive power needs of a turbine or a wind farm can cause quasi-steady state voltage changes in the connected grid system. These changes occur over numerous seconds or more and are explained above. The *X/R* ratio of the distribution system and the generator operating characteristics (amount of real and reactive power at typical operating levels) determine the magnitude of the voltage fluctuations. It has been found that an *X/R* ratio of about 2 results in the lowest voltage fluctuations with typical fixed-speed turbines with induction generators. The *X/R* ratio is typically in the range of 0.5 to 10 (Jenkins, 1995).

The weaker the grid, the greater the voltage fluctuations. 'Weak' grids that can cause problematic turbine–grid interactions are those grid systems in which the wind turbine or wind farm rated power is a significant fraction of the system fault level. Studies suggest that problems with voltage fluctuations are unlikely with turbine ratings of 4% of the system fault level (Walker and Jenkins, 1997). Germany limits renewable power generator ratings to 2% of the fault level at the POC and Spain limits them to 5% (Patel, 1999).

Often power factor correction capacitors are installed at the grid connection to reduce the reactive power needs of the turbine and system voltage fluctuations. Power factor correction capacitors need to be chosen carefully to avoid self-excitation of the generator. This occurs when the capacitors are capable of supplying all of the reactive power needs of the generator and the generator becomes disconnected from the grid. In this case the capacitor–inductor circuit, consisting of the power factor correction capacitors and the generator coils, can resonate, providing reactive power to the generator and resulting in possibly very high voltages.

8.5.4.2 Flicker

Flicker is defined as disturbances to the network voltage that occur faster than steady state voltage changes and which are fast enough and of a large enough magnitude that lights noticeably change brightness. These disturbances can be caused by the connection and disconnection of turbines, the changing of generators on two-generator turbines, and by torque fluctuations in fixed-speed turbines as a result of turbulence, wind shear, tower shadow, and pitch changes. The human eye is most sensitive to brightness variations around frequencies of 10 Hz. The blade passing frequency in large wind turbines is usually closer to 1–2 Hz or less, but even at these frequencies the eye will detect voltage variations of +/- 0.5% (Walker and Jenkins, 1997). The magnitude of the flicker due to wind turbulence depends on the slope of the real vs. reactive power characteristics of the generator, the slope of the power vs. wind speed characteristics of the turbine, and the wind speed and turbulence intensity. Flicker is, in general, much less of a problem for fixed-pitch stall-regulated machines than pitch-regulated machines (Gardner et al., 1995). Variable speed system power electronics usually do not impose rapid voltage fluctuations on the network, but still may cause flicker when turbines are connected or disconnected. Flicker does not damage equipment connected to the grid, but in weak grids where the voltage fluctuations are greater it may become an annoyance to other consumers. Numerous countries have standards for quantifying flicker and limits for allowable flicker and step changes in voltage (see, for example, CENELEC, 1993).

8.5.4.3 Harmonics

Power electronics in variable speed wind turbines introduce sinusoidal voltages and currents into the distribution system at frequencies that are multiples of the grid frequency (see Chapter 5). Because of the problems associated with harmonics, utilities have strict limits on the harmonics that can be introduced into the system by power producers such as wind turbines.

The usual measure of the degree of waveform distortion at any point in a system is total harmonic distortion (THD). THD is a function of the magnitude of the fundamental frequency and of the harmonics in the voltage waveform. The instantaneous voltage, v, can be expressed as the sum of the fundamental voltage, v_F (a sinusoidal voltage at the fundamental frequency), and a superimposed harmonic voltage, v_H. The harmonic voltage is the sum of the numerous harmonics, v_n, of order n, ($n > 1$):

$$v_H = \sum_{n=2}^{\infty} v_n \tag{8.5.8}$$

The individual harmonic voltages, v_n, consist of the cosine and sine harmonic components defined in Chapter 5:

$$v_n = a_n \cos\left(\frac{n\pi t}{L}\right) + b_n \sin\left(\frac{n\pi t}{L}\right) \tag{8.5.9}$$

where n is the harmonic number, t is time, L is half the period of the fundamental frequency and a_n and b_n are constants. These harmonics can be expressed as sine functions with an amplitude, c_n, and phase, φ_n:

$$v_n = c_n \sin\left(\frac{n\pi t}{L} + \varphi_n\right) \tag{8.5.10}$$

where:

$$c_n = \sqrt{a_n^2 + b_n^2} \tag{8.5.11}$$

and where the phase is defined by:

$$\sin\varphi_n = \frac{a_n}{c_n}, \qquad \cos\varphi_n = \frac{b_n}{c_n} \tag{8.5.12}$$

The harmonic distortion caused by the nth harmonic of the fundamental frequency, HD_n, is defined as the ratio of the rms value of the harmonic voltage of order n over some time T (an integral number of periods of the fundamental) divided by the rms value of the fundamental voltage, v_F, over the same time T:

$$HD_n = \frac{\sqrt{\frac{1}{T}\int_0^T v_n^2 \mathrm{d}t}}{\sqrt{\frac{1}{T}\int_0^T v_F^2 \mathrm{d}t}} \tag{8.5.13}$$

The THD can be expressed as (see also Stemmler, 1997, and Phipps, et al., 1994):

$$\text{THD} = \frac{\sqrt{\sum_{n=2}^{\infty} \frac{1}{T} \int_0^T v_n^{\,2} \mathrm{d}t}}{\sqrt{\frac{1}{T} \int_0^T v_F^{\,2} \mathrm{d}t}} = \sqrt{\sum_{n=2}^{\infty} (HD_n)^2} \tag{8.5.14}$$

In both the United States and Europe many power companies use the IEEE 519 Standard (ANSI/IEEE, 1992) to determine allowable THD at the point of common connection (PCC). Minimizing problems at this point minimizes problems for other electrical customers. The allowable THD of the voltage wave form according to IEEE 519 are detailed in Table 8.2. Similar restrictions on current harmonics, which depend on the ratio of maximum demand load current to maximum short circuit current at the PCC, can be found in IEEE 519.

Table 8.2 Maximum allowable total harmonic distortion (THD) of voltage at the point of common coupling (PCC)

PCC Voltage	Individual Harmonic, %	THD, %
2.3–69 kV	3.0	5.0
69–138 kV	1.5	2.5
>138 kV	1.0	1.5

8.5.4.4 Islanding

Islanding refers to the isolation of a self-supporting section of an electric grid, subsequent to the action of grid protection equipment in a fault condition. Grid protection equipment at the turbine or wind farm point of connection (POC) should shut down generators in conditions of overload, over- or under-voltage, or over- or under-frequency. Nevertheless, if the connected load and generation are reasonably matched and a source of excitation is available, the islanding may persist for some period of time, undetected by the usual grid protection equipment. Independent excitation of the generators and motors in the islanded system can occur as a result of self-excitation by power factor correction capacitors or by resonances with other equipment in the islanded system. While the risk of islanding is normally low, it can cause current to flow into a grid fault from a disconnected section of the grid, endangering repair personnel and causing synchronization problems upon reconnection of the islanded grid to the main grid. Sensors at the POC need be able to detect the transients that occur in the transition to an islanded condition and shut the generators down (Rogers and Welch, 1993).

8.5.4.5 Grid penetration issues

The inclusion of fluctuating power sources distributed throughout a larger electric grid has large effects on the control of the grid and the delivery of stable power. Utilities need to deliver power, at the nominal system voltage, to all of the system loads. A number of large prime movers are used to provide power, including steam, gas and hydroelectric turbines. Generators attempt to closely follow the fluctuating load in order to minimize voltage and frequency fluctuations. As the load changes during the day, generators are brought on line, but large prime movers may take a while to prepare for generation. Thus, a certain amount of unused generation capacity, called 'spinning reserve,' is kept on line ready to respond to rapid load fluctuations. The greater the anticipated load fluctuations, the greater the required spinning reserve. The long lead times for bringing equipment on line also necessitates load forecasts that can be used to schedule the generators. The introduction of large amounts of wind power into the grid increases the short-term variability of the load as seen by the generators, thus increasing the need for spinning reserve. It also changes the long-term mean load as winds change, disrupting the planning for bringing generation on line.

Wind power grid penetration can be defined as the ratio of the installed wind power to the maximum grid connected load (approximately equal to the total connected generation). The grid penetration of wind is highest in Denmark where over 10% of the total electric generation is from wind. Denmark plans to generate 50% of its electrical needs from wind by 2030. While concern has been raised by utilities about ceilings on the grid penetration of wind power, experience with hybrid power systems (see Section 8.8) suggests that 50% grid penetration is feasible, especially with the inclusion of some additional technology.

The additional technology might include generation capacity that can be brought on line rapidly, energy storage, and new control schemes that take advantage of the capabilities of the available generation equipment. For example, gas turbines can, typically, be brought on line faster than steam generators. The addition of more gas turbines could be used to respond to large-scale load changes without increasing the spinning reserve. Hydropower can be very responsive to load fluctuations and can be brought on line quickly. Pumped storage hydropower is now used by utilities to respond to large and rapid load changes. Pumped storage could also be used to store wind-generated power for use at peak load periods and to smooth out the load fluctuations seen by conventional generators. Control options include the integration of grid control and wind turbine control to enable pitch-regulated turbines to be controlled to pitch their blades to shed power or even to consume power, if need be. Finally, improved 24 and 48 hour wind forecasts are already being used by utilities with high grid penetration to schedule their generation.

8.6 Offshore Wind Farms

Off the coast of many countries lies a significant wind resource. A number of wind turbines have already been installed in offshore European locations in order to gain experience with the problems and potential of offshore wind power. In addition, significant future offshore wind developments are being planned in Europe. Offshore wind farms were first proposed in the 1970s (Heronemus, 1972). The first offshore wind turbine was installed in Sweden in 1991. The first offshore wind farm was installed in 1992 in shallow (2–5 m) water off the coast of Denmark near the town of Vindeby. The Vindeby wind farm consists of eleven 450 kW machines that are at most 3 km from the shore. Since then offshore wind farms have been installed in the Netherlands, Denmark, the United Kingdom and Sweden (see Table 8.3). As of the end of 2001 there were over 80 MW of installed offshore wind capacity. Denmark now has plans to develop 4000 MW of offshore capacity by 2030. This should enable Denmark to provide one half of its electrical needs from wind (Danish Wind Turbine Manufacturers Association, 1999). Other European countries are also exploring expanding the use of offshore wind power. These developments are being driven by reduced visual and noise concerns with offshore turbines and the lack of available land for new turbines in many high wind areas of Europe. These developments have also been made possible by recent reductions in offshore foundation and power transmission costs, the increasing size of available turbines and higher turbine productivity than on land.

Table 8.3 Summary of offshore wind farm experience up to 2001

Location	First operated	MW	Status	Remarks
Nogersud, Baltic (S)	1991	1 x 0.22 = 0.22	Abandoned	Tripod on solid rock
Vindeby (DK)	1991	11 x 0.45 = 4.95	Operating	Box caisson on sandy soil
Medemblik (NL)	1994	4 x 0.5 = 2.0	Operating	Steel tower driven in sandy soil
Tuno Knob (DK)	1995	10 x 0.5 = 5.0	Operating	Box caisson on sandy soil
Dronten (NL)	1996	28 x 0.6 = 16.8	Operating	Turbines near dike in fresh water
Gotland (S)	1997	5 x 0.5 = 2.5	Operating	Pile, seabed drilled and grouted
Blythe Offshore (UK)	2000	2 x 1.9 = 3.8	Operating	Only site in open seas
Utgrunden (S)	2001	7 x 1.5 = 10.5	Operating	Monopile foundation
Middlegrunden (S)	2001	20 x 2.0 = 40	Operating	Gravity foundation

8.6.1 Unique aspects of the offshore wind resource

The relatively smooth surface of the oceans results in low surface roughness and therefore low turbulence intensity and wind shear. This translates into higher low-level winds (allowing greater energy capture and, perhaps, lower tower heights) and lower turbulence intensities, resulting in lower fatigue damage and longer turbine life. These effects increase with distance from land in the downwind direction.

The estimation of vertical wind profiles is very important at offshore sites, as data are often not taken at high elevations and rarely at hub height. The logarithmic wind shear model (described in Chapter 2) that is most often used for offshore winds assumes homogeneous terrain with neutral stability. The mean wind speed, $U(z)$, at a height z is described by:

$$U(z)=\frac{U_*}{\kappa}\ln\left(\frac{z}{z_0}\right) \tag{8.6.1}$$

where U_* is the friction velocity (introduced in Chapter 2), κ is the von Karman constant (κ = 0.4) and z_0 is the roughness length. If the wind speed at a reference height, z_r, is known, then the wind speed at height z can be modelled as:

$$\frac{U(z)}{U(z_r)}=\frac{\ln(z/z_0)}{\ln(z_r/z_0)} \tag{8.6.2}$$

Data from the Vindeby offshore wind farm has been used to show that stable conditions do indeed last much of the year and that a value of z_0 = 0.0002 m (0.2 mm) can often be assumed for offshore sites. Using z_0 = 0.0002 m, the wind speed at a height of 48 m could be predicted from data at a height of 7 m to within +/- 5% (Barthelmie et al., 1996).

Data from the Baltic Sea shows that the effective sea surface roughness length is actually a function of wind speed and fetch (distance from shore). Lange and Hojstrup show that sea surface roughness length increases with wind speed over the range of wind speeds typically used for power production (Lange and Hojstrup, 1999). The Charnock model is often used for modeling the sea surface roughness length as a function of wind speed:

$$z_0 = A_C \frac{U_*^2}{g} \tag{8.6.3}$$

where g is the gravitational constant and A_C is the Charnock constant, often assumed to be to be 0.018 for coastal waters. Another model, the Johnson model (Lange and Hojstrup, 1999), assumes an implicit equation for z_0 as a function of friction velocity and fetch:

$$z_0 = 0.64 \frac{U_*^3}{x^{\frac{1}{2}} g^{\frac{3}{2}} \kappa} \ln\left(\frac{10}{z_0}\right) \tag{8.6.4}$$

where x is the fetch in meters. Comparisons of each of these models with data from locations in the Baltic sea shows that the Johnson model does a better job of modeling z_0 than the Charnock equation for fetches from 10–20 km. At longer distances (>30 m) the Charnock model appears to be better (Lange and Hojstrup, 1999). The friction velocity in these equations can be approximated by:

$$U_* = \sqrt{C_{D,10} U_{10}^2} \tag{8.6.5}$$

where $C_{D,10}$ is the effective surface drag coefficient based on the wind velocity, U_{10}, measured at an elevation of 10 m. Measurements of $C_{D,10}$ indicate that it may vary between 0.001 in low winds and 0.003 in high winds (Garratt, 1994).

The result of the low surface roughness over the ocean is the increase in low-level wind speeds at offshore sites. Data from Vindeby shows that the annual mean wind speed at a 38 m height is about 4% greater at the wind farm, 1400 to 1600 meters from shore, than at the nearby shoreline (Barthelmie et al., 1996).

Offshore turbulence intensity is lower than on land because of the lower surface roughness and lower vertical temperature gradients. Sunlight penetrates several meters into water, whereas on land it only hits the uppermost layer of soil, heating it more. Offshore turbulence intensity also decreases with height. Average turbulence intensity at locations around Denmark with long sea fetches has been measured to be 0.10, 0.09, and 0.08 at 10, 30, and 50 m heights respectively. As wind speed increases, the turbulence intensity at 50 m decreases to about 0.05 and then increases a little with speed. This increase at higher wind speeds reflects a transition from predominantly thermally generated turbulence at lower speeds to predominantly mechanically driven turbulence at higher wind speeds. As mentioned above, lower turbulence intensities require greater spacing between turbines to allow the turbine wakes to be reenergized.

8.6.2 *Turbine and foundation design for offshore*

The offshore environment provides a number of opportunities but imposes a number of constraints on wind turbine designs. Because of the unique environment, a number of turbine manufacturers are designing wind turbines specifically for offshore use. The costs of maintenance, foundations and cable connections and reduced noise and visual concerns provide a different set of design constraints than those found on land. Bad weather and long distances to shore increase maintenance costs and can decrease availability when unexpected repairs are needed. Thus, offshore turbines need to have a proven track record with high reliability and low maintenance requirements. Compared to onshore applications, offshore wind farms may consist of larger arrays of larger machines which are optimized for a more difficult environment, easier and less frequent maintenance and less space for additional equipment. The use of larger turbines lowers per kW infrastructure costs. Less concern about noise would allow higher tip speed ratio machines. Higher tip speed ratios reduce component sizes and costs and improve turbine efficiencies. These considerations may favor two-bladed turbines for offshore applications. It is difficult to increase the tip speed of three-bladed machines above about 75 m/s. The blades become too slender and the stresses too high. With present blade manufacturing technologies, two-bladed rotors are able to withstand tip speeds of up to 90 m/s without incurring dangerously high stresses (Armstrong, 1998). Some cost reductions may also be achieved by reducing tower heights due to the low wind shear far from shore.

A number of different foundation designs have been investigated for offshore installations (see Danish Wind turbine Manufacturers Association, 1999). Some of these designs are illustrated in Figure 8.14. Concrete gravity foundations were used at Vindeby and Tuno Knob. The top of the foundations are conical to break up pack ice and have boulders around the circumference for erosion protection. Concrete foundation costs are approximately proportional to the water depth squared and tend to be prohibitively heavy and expensive at more than 10 m depth. Steel gravity foundations have been considered for depths of 4–10 m. They are made of an upright cylindrical steel tube placed on a flat steel box on the seabed. They have a low transportation weight because they are only filled with the dense olivine ballast on site. There is only a small cost increase per depth as the cost is mostly driven by ice and wave forces. Monopile foundations are steel piles 2.5–4.5 m in diameter driven 10–20 m into the seabed. No seabed preparation is needed, but heavy-duty piling equipment is required and the method will not work in a seabed with large boulders. Costs depend on wave size and pack ice forces. Tripods foundations are anchored to three steel piles driven 10–20 m into seabed. Tripod foundations need minimal site preparation and are suitable for larger water depths with seabed without large boulders. Minimum water depths of 6–7 m are required to allow vessels to approach near the towers.

For deeper seas, floating wind turbines have been proposed using spar buoys to support one or more individual turbines (Heronemus, 1972) or using more complex floating pontoons that support multiple wind turbines (see Halfpenny et al., 1995). Fixed foundations must resist wave action, while floating foundations must be designed to minimize interactions between the periodic excitation by the waves and turbine dynamics. Finally, the turbine tower needs to be high enough to provide clearance between the blades and the highest expected sea, including waves and high tides.

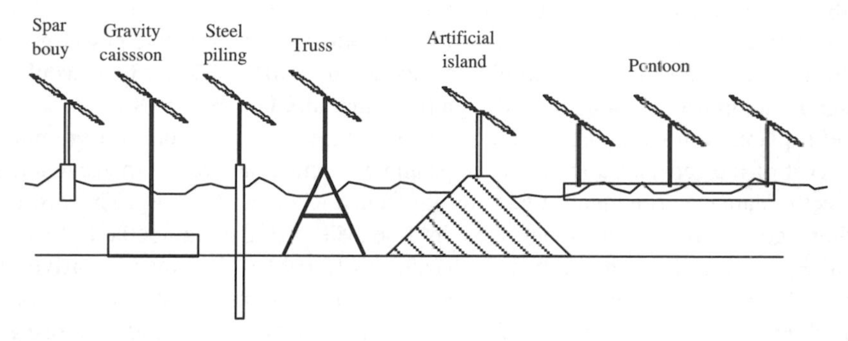

Figure 8.14 Offshore foundation possibilities

8.6.3 Power transmission

Long-distance underwater power transmission requires careful attention to numerous technical and economic issues including:

- Transmission voltage
- Power losses
- Cable electrical characteristics and cost
- Cable burial technologies and costs

Long-distance power transmission can incur significant energy losses. Without efficient long-distance undersea power transfer, the costs of offshore installations can be prohibitive. For short distances, medium-voltage AC connections are suitable. Efficient long-distance power transmission requires large expensive conductors and higher voltages. For especially long distances, high-voltage direct-current (HVDC) transmission has been proposed. Studies have shown that a three-phase AC power transmission system might have losses of 30% over transmission distances of 50 km. The transmission losses for a comparable HVDC transmission system would be only 13% over the same distance (Westinghouse Electric Corp., 1979).

The power system designer needs to consider the cost of switchgear, transformers and cable for different transmission voltages, the most cost-effective type of cable insulation and cable capacitance. Offshore transmission voltages may be limited to 33 kV within the wind farm by switch gear and transformer size and cost, which increase rapidly above that size (Gardner et al., 1998). Cable voltages to shore may be as high as 150 kV, if there is a separate transformer platform. Some possible cable insulation technologies include: cross-linked polyethylene (XLPE), ethylene propylene rubber (EPR) and self-contained fluid-filled (SCFF) insulation. XLPE is in widespread use on land (and is thus cheaper) but needs a moisture barrier under water. EPR does not require a metal sheath and can be designed for underwater use (Grainger and Jenkins, 1998). High-voltage SCFF cable has a copper conductor surrounded by an oil duct and wood pulp paper insulation layer that is surrounded by wire and rubber sheaths for protection (Fermo et al., 1993). Finally, the capacitance of power lines may be significant enough to provide some reactive power or to cause the self-excitation of turbine generators. For example, 33 kV XLPE cables would generate about 100–150 kVAR/km. This would provide only low reactive power generation. In contrast, 132 kV XLPE cables would provide about 1 MVAR/km which would have a more significant impact on wind farm design (Grainger and Jenkins, 1998).

Careful planning for the installation of power cables is very important to minimize the lifetime cost of the cables. Cable laying ships and equipment are very expensive and cables are subject to damage. The greatest hazards are from anchors and fishing. The vast majority of anchors go no more than about 1 m into seabed. All fishing methods that involve dragging equipment along the seabed are hazardous to cables. The most invasive of these methods only engages seabed to a depth of about 30 cm. The most cost-effective solution to these problems may be finding a cable route that avoids fishing and anchoring areas. Another danger, mobile sand waves, can uncover buried cable in a couple of weeks. Cables may need to be buried 2–3 m deep to avoid wave action. Where abrasion on rock is a problem, armored cable or buried cable or a combination may be needed. Cable burial is the most common solution to these problems (Mair, 1999).

A good cable design must balance expensive cable installation and burial costs with the cost of down time and repair. Burial methods include ploughing, air-lifting, water jetting, excavating, and rock-sawing. There are many types of burial machines: towed, free-swimming and tracked remotely operated vehicles (ROVs). There is no ideal method. One may need different methods in different areas. Generally, ploughs are used on long flat routes with clay-based sediments. Jetting tools work best on sands and are cheaper for short runs. Some machines can change tools for changing seabed conditions. Cables may be laid directly on the seabed (not buried), laid and buried simultaneously, or laid and buried later (post-lay burial). The choice of cable laying and burial method will depend on the length of the run, water depth, soil characteristics and equipment available. Cable laying vessels will need a shallow draft, a good mooring and control system, a deck area for cable and equipment and accommodations for personnel. One project may need different vessels for different technologies and areas. A developer may even need to build a vessel for the project. A good design will require an assessment and survey of the route, and contingency plans for bad weather. Finally, the system and routes will need to be designed to allow for economical repairs. Cables must be laid so that damaged section can be identified and located easily and can be jointed in a marine environment. A repair philosophy should be developed during the design phase.

8.6.4 Other offshore design issues

Offshore wind farm siting requires the consideration of a variety of issues including permitting requirements, shipping lanes, fishing areas, fish, mammal and bird breeding and feeding habits, existing underwater communication and power cables, visual concerns (for near-shore wind farms), storm tides and seas, seabed properties, underwater currents, underwater archaeological sites, designated marine sanctuaries, competing uses (recreation, defense), the available infrastructure and staging areas for construction (cable laying vessels, facilities for foundation fabrication, barges for installation) and transportation for maintenance.

8.6.5 Economics of offshore systems

Offshore wind energy costs depend on the wind resource, distance from shore, and water depth. Studies have found that the major cost differences between onshore and offshore wind farms is the cost of foundations and the grid connection (Morthorst and Schleisner, 1997, and Fuglsang and Thomsen, 1998). For example, foundation costs were 23% and grid connection costs were 17% of the total installation costs for Tuno Knob (Morthorst and Schleisner, 1997).

The effect of these greater installation costs may be increased energy costs over the life of the project, in spite of the increased production from the higher resource at sea. For example, one study estimated that the energy costs of the Vindeby wind farm were 56% greater than the average onshore energy costs at the time (EWEA, 1994). Morthorst and Schleisner (1997) estimated costs for a 7.5 MW wind farm and a 200 MW wind farm at two distances offshore. The 7.5 MW wind farm, using 1.5 MW turbines, was expected to generate electricity at about 4.9 US¢/kWh 5 km from the coast. At 30 km from coast, energy costs of the 7.5 MW wind farm rise to 6.9 US¢/kWh. Electricity from a 200 MW wind farm, however, only increases from 4.1 to 4.4 US¢/kWh when moved from 5 km to 30 km offshore. For comparison, energy costs for a typical onshore 600 kW turbine at a 6.9 m/s site were estimated to be about 4.5 US¢/kWh.

Offshore wind energy costs can be reduced by optimising wind turbines for offshore use. The Opti-OWECS project (Kühn et al., 1998) defined a design methodology for the structural and economic optimization of offshore wind turbines. The design methodology considers the correlation between the wind and wave loads to determine design strengths. Fuglsang and Thomsen (1998) studied the possibility of reducing energy costs for offshore wind farms by optimizing the wind turbine design. They concluded that, by optimizing the turbine tower height, diameter, power rating, rotor speed and turbine spacing, the energy costs for an offshore wind farm would only be about 10% greater than those of a comparable onshore stand-alone turbine. The installation costs were higher than those for onshore turbines, but the modeling showed that the offshore wind turbines would produce 28% more energy as a result of the better wind resource.

8.7 Operation in Severe Climates

Operation in severe climate imposes special design considerations on wind turbines. Severe climates may include those with unusually high extreme winds, high moisture and humidity, very high or low temperatures and lightning.

High temperatures and moisture in warm climates cause a number of problems. High temperatures can thin lubricants, degrade the operation of electronics, and may cause excess motion in mechanical systems that expand with the heat. Moisture and humidity can corrode metal and degrade the operation of electronics. Moisture problems may require the use of desiccants, dehumidifiers, and improved sealing systems. All of these problems can be solved with site-specific design details, but these should be anticipated before the system is installed in the field.

Operation in cold temperatures also raises unique design considerations. A number of wind turbines have been installed in cold weather regions of the globe, including Finland, northern Quebec, Alaska, and other cold regions of Europe and North America and in Antarctica. Experience has shown that cold weather locations can impose significant design and operating requirements on wind turbines because of sensor and turbine icing, material properties at low temperatures, and permafrost and snow.

Turbine icing is a significant problem in cold climates. Ice comes in two main forms, glaze ice and rime ice. Glaze ice is the result of rain freezing on cold surfaces and occurs close to 0°C (32°F). Glaze ice is usually transparent and forms sheets of ice over large surfaces. Rime ice results when supercooled moisture droplets in the air contact a cold surface. Rime ice accumulation occurs in temperatures colder than 0°C. Ice accumulation on aerodynamic surfaces degrades turbine performance and, on anemometers and wind vanes, results in either no information from these sensors or misleading information. Ice can also result in rotor imbalance, malfunctioning aerodynamic brakes, downed power lines, and a danger to personnel from falling ice. Attempts to deal with some of these problems have included special blade coatings (Teflon®, black paint) to reduce ice buildup, heating systems, and electrical or pneumatic devices to dislodge accumulated ice.

Cold weather also affects material properties. Cold weather reduces the flexibility of rubber seals, causing leaks, reduces clearances, reduces fracture strength and increases lubricating oil viscosity. Each of these can cause mechanical malfunctions or problems in everything from solenoids to gearboxes. Most turbines designed for cold weather operation include heaters on a number of critical parts to ensure correct operation. Materials also become more brittle in cold climates. Component strengths may need to be de-rated for cold climate operation or special materials may be required for the correct operation of components in cold weather or to insure adequate fatigue life.

Wind turbine installation and operation may be affected by cold weather climate conditions. Wind turbine access may be severely limited by deep snows. This may result in longer down times for turbine problems, or delayed and expensive maintenance. In installation in permafrost, the turbine installation season may be limited to the winter when the permafrost is fully frozen and transportation is easier.

Finally, many regions have frequent thunderstorms. Lightning can damage blades and mechanical and electrical components as it travels to ground. Designing for lightning protection includes providing very low impedance electrical paths to ground that bypass important turbine components, protecting circuitry with voltage surge protectors and designing a low impedance grounding system (IEC, 1999).

8.8 Hybrid Electrical Systems

Many wind turbines are connected not to large electric grids, but to small, independent, diesel powered grids, in which wind generators may be a large fraction of the total generating capacity. Such systems are referred to as wind/diesel power systems (Hunter and Elliot, 1994). Sometimes other renewable generators are added to complement the power from the wind. Power systems that include conventional generation and one or more renewable energy sources are more generally called hybrid power systems. The integration of wind turbines into these hybrid power systems presents unique system design issues. This section provides an overview of relevant design issues.

Numerous communities in isolated locations, islands, and developing countries are connected to small independent electric grids powered by diesel generators. They may range in size from relatively large island grids of many megawatts down to systems with a capacity of a few kilowatts. Isolated and island grids vary significantly. Some isolated grids powered by diesel generators only provide power for a part of the day to conserve fuel. Some have large voltage swings due to the effects of one or two significant loads on the system, such as a saw mill or a fish-processing plant. Large isolated grids provide power at stable voltages and constant frequency. In general, isolated grids are weak grids in which voltage and frequency are susceptible to disruption by interconnected loads and generation.

Wind turbines and other renewable power sources (including wind, solar, biomass or hydropower) can be integrated into these small electric grids. As in larger stable grids, the terms 'wind penetration' or 'renewables penetration' are used to characterize the magnitude of the wind or renewable power in the system compared to the rated load. In typical grid-connected wind turbine application, turbine–grid interactions are limited to part of a distribution system. In contrast, wind turbines in isolated grids may significantly affect the operation of the whole grid. In high wind penetration hybrid systems, the wind turbines might, at times, produce more power than the instantaneous system load. This would require the conventional generator to be shut off completely or cause additional loads to be turned on to absorb the extra power. Because of the significant effects of introducing renewable power into such a grid, these hybrid systems must be designed and analyzed as a complete interacting system.

In this section a number of issues related to hybrid power systems will be considered. The section includes:

- A review of issues of diesel-powered grids
- An overview of hybrid system design issues
- A description of the components of a complete hybrid power system
- Information on computer models for hybrid systems

8.8.1 Independent diesel-powered grids

Independent diesel-powered electric grids include the diesel generators, a power distribution system, electric loads and some form of system supervision.

8.8.1.1 Diesel generators

Generators in independent power systems are normally diesel engines directly coupled to synchronous electrical machines. The frequency of the AC power is maintained by a governor on the engine or on one of the engines in a multi-diesel application. The real and reactive power in a conventional AC system is supplied by the synchronous generator. This is done in conjunction with the voltage regulator on the generator. DC grids typically use a diesel powered AC generator with a dedicated rectifier (see Chapter 5).

Figure 8.15 illustrates fuel consumption (including a linear fit to the data) for a typical small diesel generator set that might be in an existing hybrid grid. It can be seen that the no-load fuel consumption may be a fairly high fraction of the full-load fuel consumption. Large modern diesels have somewhat lower relative no-load fuel consumption than in this example, but still consume a significant amount of fuel at low loads. Obviously substantial fuel savings could be achieved if a lightly loaded diesel generator could be shut down.

Diesel fuel at remote locations is often expensive. These diesels often operate at low load and with poor fuel efficiency. Reducing the load or shutting off diesels reduces fuel costs. This may be the goal of the introduction of renewable power, but it also has negative consequences. Reducing the load on a diesel engine can increase engine maintenance requirements, increase engine wear, and, consequently, decrease engine life. Frequent starts and stops significantly increase engine wear. To improve total system economics, a minimum diesel load is often required while the diesel is running and a minimum diesel run time may be specified. Each of these measures increases fuel consumption compared to operation with frequent starts and stops and operation at no load, but these measures are designed to improve overall system economics by reducing diesel overhaul and replacement intervals.

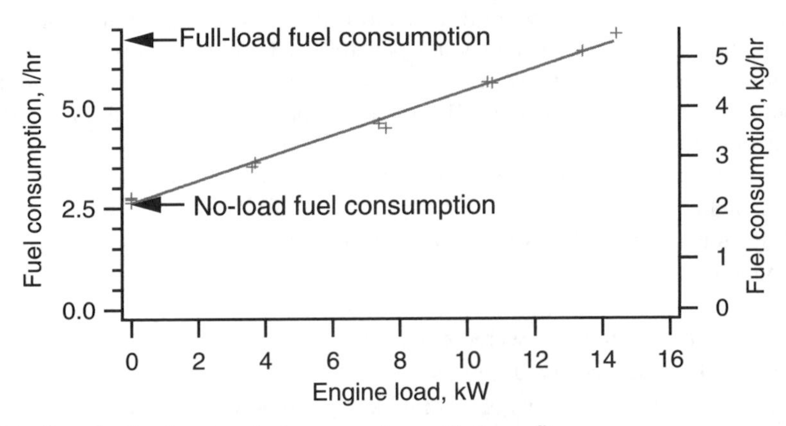

Figure 8.15 Sample diesel engine fuel consumption with linear fit

8.8.1.2 Loads

Electrical loads in independent AC systems are primarily of two types: resistive and inductive. Resistive loads include incandescent light bulbs, space and water heaters, etc. Devices with electric motors are both resistive and inductive. They are a major cause of the need for a source of reactive power in AC systems. DC sources can only supply resistive loads. DC loads may have an inductive component, but this only causes transient voltage and current fluctuations during changes in system operation.

8.8.1.3 System supervision

System supervision in conventional diesel-power systems may be automated, but usually consists of a system operator turning diesel generators on and off as the anticipated load changes, synchronizing them with the other operating diesels and performing engine maintenance when needed.

8.8.2 *Overview of hybrid system design issues*

The design of hybrid power systems depends on the specifics of the existing power system and on the match of the load and the available renewable resource. Given these constraints, there are numerous options including the wind turbine design, the amount of wind penetration, the inclusion of other renewable power systems, the amount of any energy storage, and the nature of any load management. A review of the issues related to hybrid power system design follows.

8.8.2.1 Match of load and resource

The determining factor in hybrid power system design is the interdependence of the system load and the power produced by the hybrid system. Over short time frames the load needs to equal the power produced by the system to ensure system stability. To provide a consistent energy supply over long time periods, a hybrid system may need either longer term energy storage or backup conventional generation.

To maintain system stability power flows need to be balanced over short time frames. As mentioned in Section 8.5, the frequency of an electrical system is a function of the rotating inertia in the system, the fluctuating load and the responsiveness of the prime mover and the control system of the prime mover. The faster the prime mover can respond to changing power flows, the better the frequency regulation. In high-penetration systems, the prime mover may not have the range to respond to the changing power flows. In that case, additional controllable sources or sinks for power need to be used to control the system frequency. These could be loads that can be switched on to balance power flows, systems for the short-term storage and production of power (systems with a few minutes to an hour of storage), or additional generators that can be brought on line. In high-penetration systems, the conventional generators may be turned off when the renewable power can supply all of the load. In that case, the system has fairly low inertia and, without another rapidly controllable power source to control frequency, the system frequency can drift significantly.

Power or energy flows also need to be balanced over longer time frames. If the load peaks in the daytime and the wind blows only at night, then the wind can be used neither to supply energy to the daytime load nor to save expensive fuel. In such a case, the addition of long-term energy storage (systems with a few hours to a day of storage) would allow the wind energy to be stored to be used to supply energy to the daytime load and to save fuel. Long periods of lack of renewable power would deplete any energy storage and require the use of conventional generation, which would need to be capable of supplying the whole load.

Based on the considerations above, it is apparent that a hybrid system might benefit from the addition of energy storage and/or controllable loads. Energy storage could provide power for periods when the wind power is less than the load. When the wind power is greater than the load, energy storage and controllable loads could provide sinks for excess power. With multiple sources and sinks for power, a hybrid power system would also need a system supervisory controller (SSC) to manage power flows to and from system components.

8.8.2.2 System design constraints

A number of factors affect the specifics of the design of a hybrid system, including the nature of the load, the characteristics of the diesel generators and electric distribution system, the renewable resource, fuel costs, availability of maintenance personnel, and environmental factors.

- The load - The magnitude and temporal profile of the local load affect the rated system capacity, energy storage needs, and control system algorithm.
- Diesel generation and electric distribution system – In existing power systems, the fuel consumption and electrical characteristics of any existing power generation system affect the economics of the hybrid system, equipment selection and the control system design. Fuel costs are one important factor in determining system operating cost. In new systems the generation and distribution systems can be designed in conjunction with the design of the hybrid components.
- The renewable resource - The magnitude, variability, and temporal profile of the renewable resource, whether it is wind, solar, hydropower, and/or biomass affects the choice of renewable power system, the control strategy, and storage requirements.
- Maintenance infrastructure - The availability of trained operating and maintenance personnel affects the long-term operability of the system, operating costs and installation costs.
- Site conditions - Site constraints such as the nature of the terrain, local severe weather conditions and the remoteness of the site affect the ability to get equipment to the site, equipment design requirements, and operating system requirements.

In projects with an existing generation and distribution system, the system design objective is usually the minimization of the cost of energy by reducing fuel consumption (often a costly item in remote locations) and the increase of overall system capacity to enable continued local economic development. New systems are often implemented as a cost-effective alternative to other options such as grid extension.

8.8.2.3 Hybrid power system design rules

There are many factors affecting the design of a high-penetration hybrid power system and many choices of components that need to be considered to optimize system cost and efficiency. Design choices include the type, size, and number of wind turbines, solar panels, etc., the instantaneous and long-term energy storage capacity, the size of dump loads, possibilities for other load management strategies, and the control logic needed to decide when and how to use all of the system components. Thus, the problem becomes one of designing a complicated power system with multiple controllable power sources and sinks. The possibilities for controllable loads will depend on the fit of any given load management approach with the daily needs of the local community. Evaluating all of the parameters is most easily done with a computer model intended for hybrid system design (see Section 8.8.4).

Typically, hybrid power systems have been developed by introducing renewable power into existing isolated grids in order to reduce high fuel costs and to provide increased energy. In trying to predict the performance of a hybrid power system, it is worth considering the limiting possibilities. In an ideal diesel grid the diesel fuel consumption would be exactly proportional to the power generated. Thus, the fuel use would be proportional to the load. When renewables are added, the effect is to reduce the load that must be served by the diesel generators. If there were a perfect match between the load and renewable power, the diesel load could be reduced to zero. All the renewable power produced (up to the amount of the load) would be used, but any power produced in excess of that would have to be dumped or otherwise dissipated. If there were a temporal mismatch between the load and the available renewable power, even less of the latter would be used. This gives rise to the following rules:

Rule 1: The maximum renewable energy that can be used is limited by the load.
Rule 2: The use of renewable energy will be further limited by temporal mismatch between the load and the renewables.

The introduction of energy storage increases the use of the renewable resource when there is a temporal mismatch between the load and the renewable resource. Based on Rules 1 and 2, the maximum possible improvement in the use of renewable energy afforded by use of storage is limited by the mismatch between its availability and the load.

In practice the fuel use of diesel generators never varies in exact proportion to the load. Diesel generator efficiency virtually always decreases with decreasing load. However, a diesel generator system which included storage could be optimised to improve its performance. This gives rise to two more rules:

Rule 3: The maximum possible benefit with improved controls or operating strategies is a system approaching the fuel use of the ideal diesel generator - fuel use proportional to the diesel-served load.
Rule 4: The maximum fuel savings arising from the use of renewables in an optimised system is never greater than the fuel savings of an ideal generator supplying the proportional reduction in load resulting from use of renewables.

8.8.3 Hybrid power system components

A hybrid system includes diesel generators, renewable generators, a system supervisory controller and possibly controllable loads and energy storage. Note that storage is actually both a source and a load. To make all of the subsystems work together, it may also include power converters or a coupled diesel system, explained below. A schematic of the possibilities for a hybrid system is illustrated in Figure 8.16. The operation of each of these components and the interactions between components is described below.

Larger systems, usually above 100 kW, typically consist of AC-connected diesel generators, renewable sources, and loads, and occasionally include energy storage. Below 100 kW, combinations of both AC- and DC-connected components are common as is the use of energy storage.

The components in DC systems could include diesel generators, renewable sources, and storage. Small hybrid systems serving only DC loads, typically less than 5 kW, have been used commercially for many years at remote sites for telecommunications repeater stations and other low-power applications.

The operation of each of these components and the interactions between components is described below.

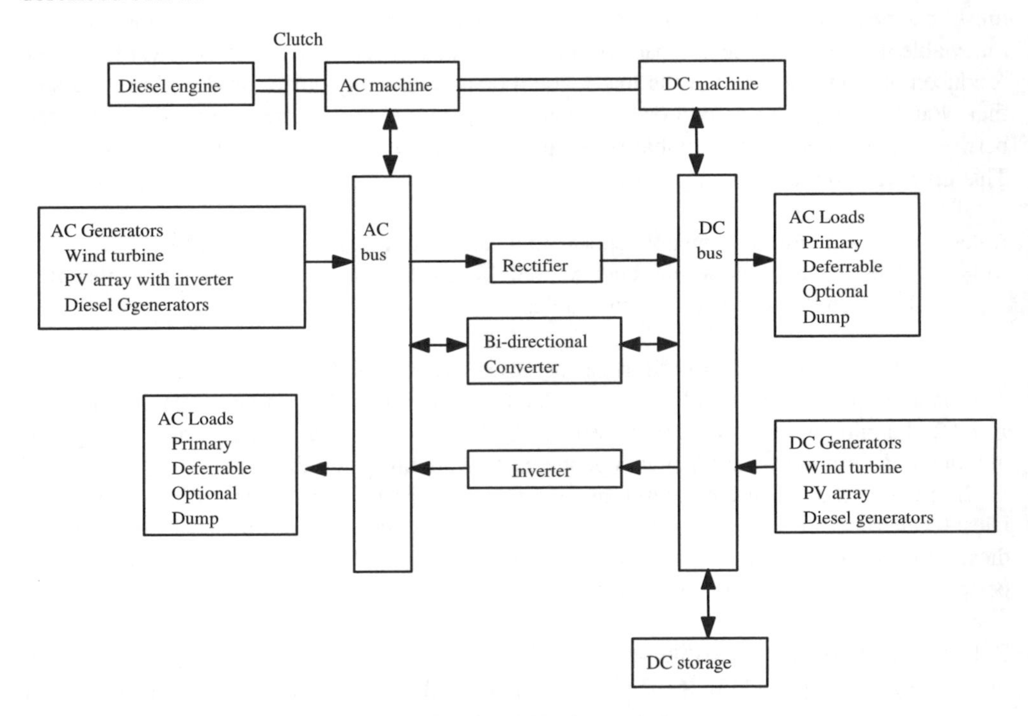

Figure 8.16 Hybrid power system configuration; PV, photovoltaic

8.8.3.1 Diesel generators in hybrid systems

In a conventional AC power system, there must always be at least one diesel generator connected to the network in order to set the grid frequency and to supply reactive power. It is possible to modify the system so that the diesel generator is not always required, but in that case other components must be added. These other components could include an inverter, a rotary converter, a synchronous condenser, or another power producer such as a wind turbine with a synchronous generator. As described in Chapter 5, an inverter is a device (typically a solid state electronic device) that provides AC power from a DC source. A rotary converter is essentially an electromechanical inverter. It requires a separate controller to set the frequency. A synchronous condenser is a synchronous electrical machine connected to the network and allowed to spin at a speed determined by the grid frequency. Operating in conjunction with a voltage regulator, it supplies reactive power to the network, but does not control grid frequency. Reactive power production may also be achieved with a synchronous generator connected to a wind turbine.

8.8.3.2 Wind turbines in hybrid systems

Wind turbines used in larger isolated AC electrical systems typically have a capacity of 10 kW to 500 kW. Most of the wind turbines in this size range are fixed-speed turbines that use induction generators and so require an external source of reactive power. Thus, in hybrid power systems they can operate only when at least one diesel generator is operating or in systems that have separate sources for reactive power. The starting current of turbines with induction generators also needs to be supplied by the system. Some wind turbines use synchronous generators. If they have pitch control, they may be able to be used to control grid frequency and to provide reactive power. In such cases, these turbines could operate without any diesel generators being on line. Other turbines use power electronic converters. Such machines may also be able to run without the presence of a diesel generator.

There are a number of wind turbines that supply DC power as their principal output. These machines are typically in the smaller size range (10 kW or less). With suitable controls or converters they may operate in conjunction with AC or DC loads.

As components in a hybrid power system, control of the wind turbines would need to be integrated into the system supervisory controller.

8.8.3.3 Photovoltaic panels in hybrid systems

Photovoltaic (PV) panels may provide a useful complement to wind turbines in some hybrid power systems. Photovoltaic panels provide electric power directly from incident solar radiation. Photovoltaic panels are inherently a DC power source. As such, they usually operate in conjunction with storage and a separate DC bus. In larger systems they may be coupled with a dedicated inverter and thus act as a de facto AC power source.

Photovoltaic panels provide fluctuating power. Variations of the solar resource on annual and diurnal scales, as well as on the time scales of front-driven weather patterns and the passage of clouds, complicate the design of hybrid systems.

The power generated by photovoltaic panels is determined by the solar radiation level on the panel, the panel characteristics and the voltage of the load to which it is connected. Figure 8.17 shows the current–voltage characteristics of a typical PV panel at a given temperature and solar insolation level. It can be seen that the range of voltages over which a given panel performs effectively is fairly limited. To increase the output voltage multiple panels are connected in series. To increase the current (and power at a given voltage) multiple panels are connected in parallel. The power produced from a photovoltaic panel depends strongly on the load to which it is connected. In particular, the terminal voltage and current of the panel must equal those of the load. In general, photovoltaic panels and loads have different current–voltage relations. At any given operating condition, there is normally only one operating point where both the panel and load have the same voltage and current. This occurs where the panel and load current–voltage curves cross, as illustrated in Figure 8.17.

The power from the panel is equal to the product of the current times the voltage. The maximum power point, where the power is the greatest, occurs at a voltage somewhat less than the open circuit voltage. For the photovoltaic panels to be most effective when used with batteries, the nominal battery voltage should be close to the maximum power point voltage. It is sometimes useful to use power conditioning equipment (maximum power point trackers or MMPT) to match the load with the characteristics of the PV cell. These power electronic converters would adjust the PV array voltage for maximum PV power and convert that resulting PV power so that its voltage is that required by the DC load or the AC grid. The purpose of such a converter is to maximize the power conversion in the electrical parts of the system, whatever the radiation.

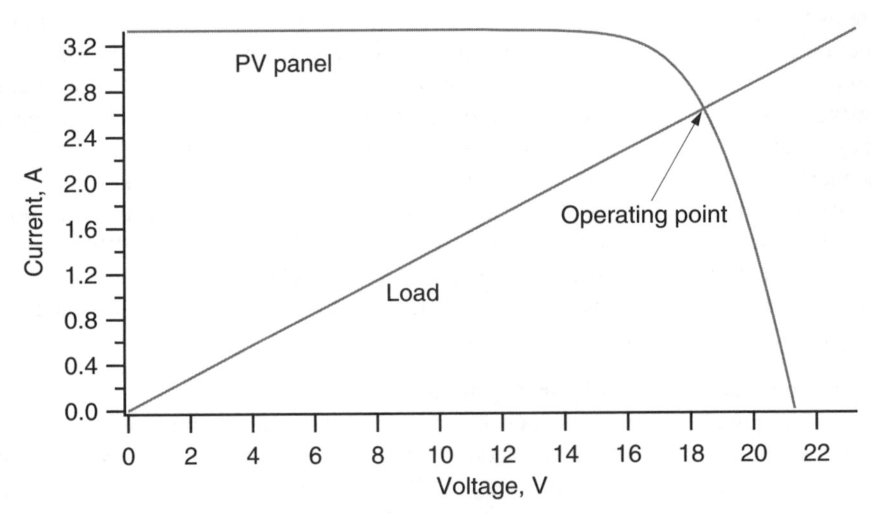

Figure 8.17 Load matching for a photovoltaic PV panel with a given insolation level

8.8.3.4 Controllable loads

One component which may be required in hybrid power systems, but which is uncommon in conventional isolated systems, is a dump load. A dump load is used to protect against an excess of power in the network. Such an excess could arise during times of high renewable contribution and low load. Excess energy could lead to grid instability. The dump load may be based on power electronics or switchable resistors. In some cases dissipation of excess power may be accomplished without the use of a dump load. An example is the dissipation of excess wind power using blade pitch control.

Load management can also be used in hybrid power systems to augment or take the place of storage or to supplant dump loads. For example, optional loads are those that provide a use for surplus power that would otherwise go to waste. An example would be excess energy that was directed toward space heating that would reduce the need for other fuels. Deferrable loads are those which must be supplied at some time, but for which the exact time is flexible. For example, water pumping, for a water storage tank that needs to be refilled at least once a day, is a load that needs to be met, but for which the exact timing of the pumping may not matter. In this case, excess energy could be used for water pumping. If there were no excess energy in a day then energy from battery storage or the diesel generators would need to be used to make sure that the water tank were full at the end of the day.

8.8.3.5 Battery energy storage

Batteries have proven to be the most useful energy storage medium, based primarily on their convenience, and cost. Battery storage systems are modular and multiple batteries can store large amounts of energy. Energy storage is not typically used with larger isolated AC networks, although it certainly can be. Storage is very common with smaller hybrid power systems. Energy storage is most commonly lead acid batteries, although nickel–cadmium batteries are also occasionally used. Batteries are inherently DC devices. Thus, battery energy storage in AC systems requires a power converter.

Typical battery voltage during a discharge–charge cycle is illustrated in Figure 8.18. It can be seen that the terminal voltage drops as the battery is discharged. When charging is initiated, the terminal voltage jumps to a value above the nominal cell voltage. As the cell becomes fully charged, the terminal voltage increases even more before gassing occurs (the production of hydrogen gas in the cells) and the terminal voltage levels off.

A number of aspects of battery behavior affect their use in hybrid power systems (see Manwell and McGowan, 1994):

- Battery capacity - Effective battery capacity is a function of current level. Thus, the amount of storage in a hybrid system is a function of the rate at which the storage is used.
- Terminal voltage - Terminal voltage is a function of state of charge and current level. This affects the operation of the power transfer circuit between the battery storage and the rest of the system.
- Efficiency - Batteries are not 100% efficient. Battery losses can be minimized by intelligent controller operation, but most of the losses are due to differences in voltage during discharging and charging and are inherent to battery operation.
- Battery life - Battery life is a function of the number and depth of charge–discharge cycles, and a function of the battery design.
- Temperature effects - Battery capacity and life are also functions of temperature. Usable battery capacity decreases as the temperature decreases. Typically, battery capacity at 0°C is only half that at room temperature. At temperatures above room temperature, battery capacity increases slightly, but battery life decreases dramatically (Blohm, 1985).

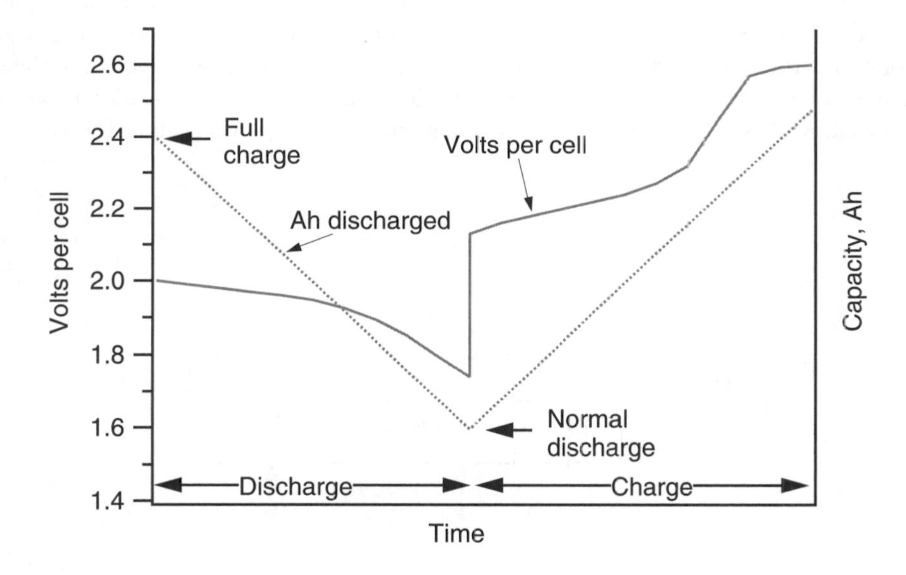

Figure 8.18 Typical battery voltage and capacity curve. Reproduced by permission of McGraw-Hill Companies from Standard Handbook for Electrical Engineers, Fink and Beaty, 1978

8.8.3.6 Power converters

There are two types of power conversion functions of particular significance for hybrid power systems: rectifying and inverting (see Chapter 5). Rectifiers are commonly used to charge batteries from an AC source. Inverters are used to supply AC loads from a DC source.

As mentioned in Chapter 5, most electronic inverters are of one of two types: line commutated or self-commutated. Line-commutated inverters require the presence of an external AC line. Thus they cannot set the grid frequency if, for example, all of the diesel generators in a hybrid power system are turned off. Self-commutated inverters control the frequency of the output power based on internal electronics. They do not normally operate together with another device that also sets the grid frequency. There are inverters that can both operate either in a line-commutated or a self-commutated mode. These are the most versatile, but presently they are also the most expensive.

A rotary converter is essentially an electromechanical rectifier or inverter. It consists of an AC synchronous machine directly coupled to a DC machine. Both electrical machines can act as either a motor or a generator. When one machine is a motor, the other is a generator and vice versa. The advantages of a rotary converter include that it is a well developed piece of equipment and is normally quite rugged. The main disadvantages are that the efficiency is less and the cost is greater than a solid state power converter.

8.8.3.7 Coupled diesel systems

A variant on the diesel generator in a typical hybrid power system is known as the "coupled diesel" system. In this system there are both AC and DC generators and buses. The AC and DC networks are connected together via the AC and DC generators that constitute a rotary converter. A single diesel engine is coupled via a clutch to the rotary converter (See Figure 8.19). The advantage of this concept is that it provides a more efficient means of incorporating a rotary converter than would be the case if the diesel engine were associated only with its own generator. It also offers a convenient way to shut down the diesel when it is not needed, while still providing a means of supplying reactive power to the AC bus.

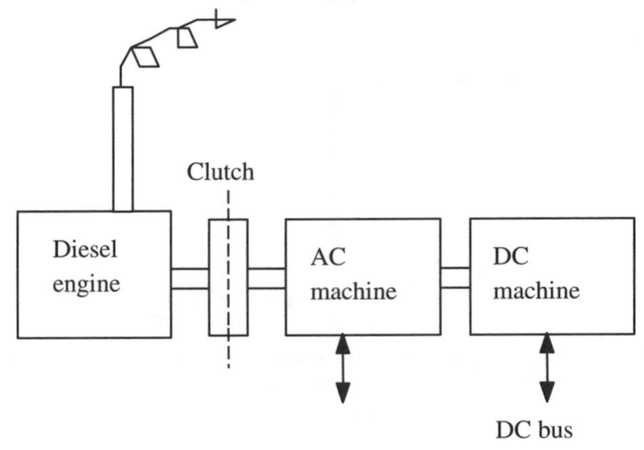

Figure 8.19 Coupled diesel

8.8.3.8 System supervisory control

Most hybrid power systems incorporate numerous forms of control. Some control functions are carried out by dedicated controllers that are integral to the system components. Typical examples include the governor on a diesel, the voltage regulator on a synchronous generator, the supervisory controller on a wind turbine, or the charge controller in a battery bank. Overall system control is accomplished by a separate controller, known as the system supervisory control. The system supervisory control might control some or all of the components indicated in Figure 8.20. This control is usually thought of as automatic, but in reality some of the functions may be carried out by an operator. Specific functions of the system supervisory controller may include turning diesel generators on and off, adjusting their power set points, charging batteries, and allocating power to a controllable loads.

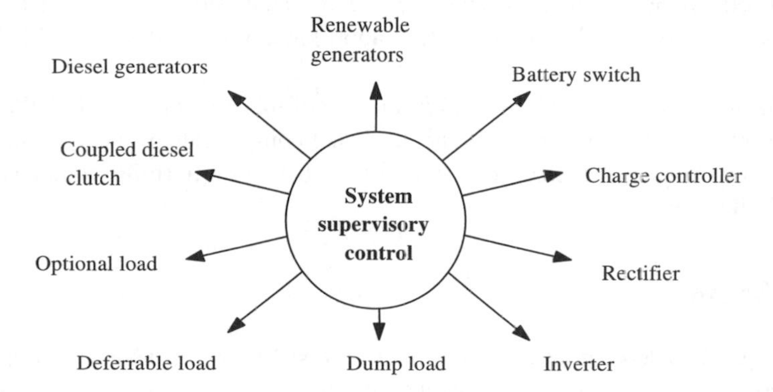

Figure 8.20 Functions of system supervisory control

8.8.4 Hybrid system modeling

Many simulation models have been developed for hybrid power system design. For example, European researchers have developed numerous analytical models of varying sophistication and general use for wind–diesel systems (Infield et al., 1990). Similarly, work at the University of Massachusetts (Manwell et al., 1997) has produced a number of system models for wind–diesel–hybrid power systems. Generally, one can classify these models into two broad categories: logistical and dynamic models.

Logistical models (Infield et al., 1990) are used primarily for long-term performance predictions, component sizing, and for providing input to economic analyses. Generally, they can be divided into the following three categories:

- *Time series (or quasi-steady state)*: this type of model requires the long-term time series input of variables such as wind speed, solar insolation, and load
- *Probabilistic*: models of this type generally require a characterization of long-term load and resource data (e.g. monthly or seasonal) as inputs. The analytical model is based on the use of statistical modeling techniques
- *Time series or probabilistic*: as the name implies, models in this category are based on the use of a combined time series and statistical approach

Dynamic models are used primarily for component design, assessment of system stability, and determination of power quality. They are generally used for hybrid power systems with no storage capability, or systems with minimal storage, such as a flywheel. Depending on the time step size and the number of modeled components they can be divided into the following three categories:

- *Dynamic mechanical model*: this type of model is based on the mechanical equations of motion and power balances. It can be used to get a first approximation of the dynamic behavior of a system and to find such long-term effects as the start-stop behavior of the diesel engine component
- *Dynamic mechanical, steady state electrical model*: this class of model is based on the mechanical equations of motion and the steady state equations of the electrical components of the system. It can give a first approximation of the electrical behavior of the system
- *Dynamic mechanical and electrical model*: models of this type are based on the dynamic equations of motion of the mechanical and electrical components of the system. They are intended to investigate the electrical stability of the system (millisecond scale) and mechanical vibrations

8.9 References

Ainslie, J. F. (1985) Development of an Eddy-Viscosity Model for Wind Turbine Wakes. *Proc. 7th BWEA Conference*, London: Multi-Science Publishing Co.

Ainslie, J. F. (1986) Wake Modelling and the Prediction of Turbulence Properties. *Proc. 8th BWEA Wind Energy Conf*erence, Cambridge, March 1986, 115–120.

ANSI/IEEE (1992) *IEEE Recommended Practices and Requirements for Harmonic Control in Electrical Power Systems*. ANSI/IEEE Std 519-1992.

Armstrong, J. R. C. (1998) Wind Turbine Technology Offshore. *Proc. BWEA Conference*, Cardiff University of Wales, 2–4 September, 301–308.

AWEA (2000) American Wind Energy Association web site, April 2000: *www.awea.org/outlook2000/outlook_2.html*.

Baker, R. W. (1999) Turbine Energy Shortfalls Due to Turbulence and Dirty Blades. *Proc. of AWEA Conference, Windpower '99*. 21–23 June, Burlington, VT.

Barthelmie, R. J., Courtney, M. S., Højstrup, J., Larsen, S. E. (1996) Meteorological Aspects of Offshore Wind Energy: Observations from the Vindeby Wind Farm. *Journal of Wind Engineering and Industrial Aerodynamics*, 62, 191–211.

Beyer, H.G., J. Luther, Steinberger-Willms, R. (1989) Power Fluctuations from Geographically Diverse, Grid Coupled Wind Energy Conversion Systems. *Proc. European Wind Energy Conf.*, 10-13 July 1989, Glasgow, 306–310.

Blohm, R. L. (1985) Selecting and Sizing Stationary Batteries, *Proc. The Power Sources Users Conference*, 15–17 October, 1985.

Bossanyi, E. A., Maclean, C., Whittle, G. E., Dunn, P. D., Lipman, N. H., Musgrove, P. J. (1980) The Efficiency of Wind Turbine Clusters. *Proc. Third International Symposium on Wind Energy Systems*, 26–29 August, 1980, Lyngby, DK, 401–416; BHRA Fluid Engineering, Cranfield, Bedford MK43 OAJ, UK.

Bossanyi, E., Saad-Saoud, Z., Jenkins, N. (1998) Prediction of Flicker Produced by Wind Turbines. *Wind Energy*, vol. 1, John Wiley, September 1998.

CENELEC (1993) Flickermeter – Functional and Design Specifications. European Norm EN 60868: 1993 E: IEC686:1986+A1: 1990, Brussels, Belgium.

Crespo, A., Herandez, J. (1990) Numerical Modelling of Wind Turbine Wakes. *Proc. EWEC*, 1990, 10–14 September 1990, Madrid, Spain.

Crespo, A., Herandez, J. (1993) Analytical Correlations for Turbulence Characteristics in the Wakes of Wind Turbines. *Proc. EWEC*, 1993, 8–12 March, 1993, Lübeck, Germany.

Crespo, A., Manuel, F., Herandez, J. (1990) A Numerical Model of Wind Turbine Wakes and Wind Farms. *Proc. EWEC*, 1990.

Crespo, A., Manuel, F., Moreno, D., Fraga, E., Herandez, J. (1985) Numerical Analysis of Wind Turbine Wakes. *Proc. Workshop on Wind Energy Applications*, Delphi, Greece, 1985.

Danish Wind Turbine Manufacturers Association (1999) www.windpower.dk, July 1999.

Derrick, A. (1993) Development of the Measure–Correlate–Predict Strategy for Site Assessment. *Proc. EWEC*, Lübeck-Travemünde, Germany, 8-12 March 1993, 681–685.

EWEA (1994) Offshore and Onshore Windfarm Costs Compared. *Newsletter of the European Wind Energy Associations*, Volume XIII, No. 3, 1994.

EWEA (2000) *Wind Directions*. European Wind Energy Association, November 2000.

Fermo, R., Guida, U., Poulet, G., Magnani, F., Aleo, S. (1993) 150 kV System for Feeding Ischia Island. *Proc. Third Conference on Power Cables and Accessories, 10 kV–500 kV*, 23–25 November 1993, London, UK, Organized by the Power Division of the Institution of Electrical Engineers.

Fink, D. G., ed. (1978) *Standard Handbook for Electrical Engineers, Eleventh Edition*, McGraw-Hill, New York.

Fordham, E. J. (1985) Spatial Structure of Turbulence in the Atmosphere. *Wind Engineering*, 9, 95–135.

Fuglsang. P., Thomsen, K. (1998) *Cost Optimisation of Wind turbines for Large-scale Off-shore Wind Farms*. Riso National Laboratory, Roskilde, Denmark, February, 1998.

Gardner, P., Craig, L. M., Smith, G. J. (1998) Electrical Systems for Offshore Wind Farms. *Proc. BWEA Wind Energy Conference*, Professional Engineering Publishing Limited, UK.

Gardner, P., Jenkins, N., Allan, R. N., Saad-Saoud, Z., Castro, F., Roman, J., Rodriguez, M. (1995) Network Connection of Large Wind Turbines. *Proc. of the 17th BWEA Conference*, Warwick, UK, 19–21 July 1995.

Garratt, J. R. (1994): *The Atmospheric Boundary Layer*. Cambridge University Press, Cambridge.

Grainger, W., Jenkins, N. (1998) Offshore Wind Farm Electrical Connection Options. *Proc. BWEA Wind Energy Conference*, Professional Engineering Publishing Limited, UK.

Halfpenny, A., Kerr, S., Quinlan, M., Bishop, N. W. M. (1995) A Technical Feasibility Study and Economic Assessment of an Offshore Floating Wind Farm. *Proc. of the 17th BWEA Conference*, 19–21 July 1995, Warwick, UK.

Hassan, U., Taylor, G. J., Garrad, A. D. (1988) The Impact of Wind Turbine Wakes on Machine Loads and Fatigue. *Proc. European Wind Energy Conference*, 6-10 June 1988, Herning, Denmark, 560–565.

Heronemus, W. E. (1972) Pollution-free Energy from Offshore Winds, *Proceedings of 8th Annual Conference and Exposition, Marine Technology Society*, Washington, DC.

Hiester, T. R., Pennell, W. T. (1981) The Meteorological Aspects of Siting Large Wind Turbines. U.S. DOE Report No. PNL-2522.

Hunter, R., Elliot, G. (1994) *Wind–Diesel System*. Cambridge University Press, Cambridge, UK.

IEC (International Electrotechnical Commission) (1999) *IEC 61400-24, Edn. 1: Wind Turbine Generator Systems – Part 24: Lightning Protection for Wind Turbines*. International Electrotechnical Commission, Geneva, Switzerland.

Infield, D. G., Lunsager, P., Pierik, J. T. G, van Dijk, V. A. P., Falchetta, M., Skarstein, O., Lund, P. D. (1990) Wind Diesel System Modelling and Design, *Proc. EWEC 90*, 569–574.

Jenkins, N. (1995) Some Aspects of the Electrical Integrations of Wind Turbines. *Proc. of the 17th BWEA Conference*, Warwick, UK, 19–21 July 1995.

Joensen, A., Landberg, L., Madsen, H. (1999) A New Measure-Correlate-Predict Approach for Resource Assessment. *Proc. EWEC*, 1–5 March 1999, Nice, France, 1157–1160.

Katić, I., Højstrup, J., Jensen, N. O. (1986) A Simple Model for Cluster Efficiency. *Proc. European Wind Energy Conference*, 7–9 October 1986, Rome, Italy.

Kristensen, L., Panofsky, H. A., Smith, S. D. (1981) Lateral Coherence of Longitudinal Wind Components in Strong Winds. *Boundary Layer Meteorology*, 21, 199–205.

Kühn, M., Bierbooms, W. A. A. M., van Bussel, G. J. W., Ferguson, M. C., Göransson, B., Cockerill, T. T., Harrison, R., Harland, L. A., Vugts, J. H., Wiecherink, R. (1998) Opti-OWECS Final Report, Vol. 0-5, Institute for Wind Energy, Delft University of Technology, Report No. IW-98139R.

Landberg, L., Mortensen, N. G. (1993) A Comparison of Physical and Statistical Methods for Estimating the Wind Resource at a Site. *Proc. 15th BWEA Conference*, York, 6–8 October 1993, 119–125.

Lange, B., Hojstrup, J. (1999) The Influence of Waves on the Offshore Wind Resource. *Proc. EWEC*, Nice, 1999.

Lissaman, P. B. S., Bates, E. R. (1977) Energy Effectiveness of Arrays of Wind Energy Conversion Systems. AeroVironment Report AV FR 7050, Pasadena, CA, USA.

Lissaman, P. B. S., Zaday, A., Gyatt, G. W. (1982) Critical Issues in the Design and Assessment of Wind Turbine Arrays. *Proc. of the 4th International Symposium on Wind Energy Systems*, Stockholm, Sweden, September 1982.

Mair, K. (1999) The Installation of Submarine Cables for Offshore Wind Farms. *Proc. 1999 BWEA Conference*.

Manwell, J. F., McGowan, J. G. (1994) A Combined Probabilistic/Time Series Model for Wind Diesel Systems Simulation. *Solar Energy* 53(6), 481–490.

Manwell, J. F., Rogers, A., Hayman, G., Avelar, C. T., McGowan, J. G. (1997) *HYBRID2–A Hybrid System Simulation Model, Theory Summary.* National Renewable Energy Laboratory, Subcontract No. XL-1-11126-1-1, December 1997.

Morthorst. P. E., Schleisner, L. (1997) Offshore Wind Turbines – Wishful Thinking or Economic Reality? *Proc. 1997 EWEC*, October 1997, Dublin, Ireland, 201–205.

NWCC (1998) *Permitting of Wind Energy Facilities, A Handbook*, Prepared by the NWCC Siting Cubcommittee, National Wind Coordinating Committee, Washington, DC, March 1998.

Patel, M. R. (1999) Electrical System Considerations for Large Grid-connected Wind Farms. *Proc. of AWEA Conference, Windpower '99*, 21–23 June 1999, Burlington, VT.

Pennell, W. R. (1982) Siting Guidelines for Utility Application of Wind Turbines. EPRI Report: AP-2795.

Petersen, E. L., Troen, I. (1986) The European Wind Atlas. *Proc. EWEC*, 7–9 October 1986, Rome, Italy.

Petersen, E. L., Troen, I., Mortensen, N. G. (1988) The European Wind Energy Resoursces. *Proc. EWEC*, 6–10 June 1988, Herning, Denmark, 103–110.

Phipps, J. K., Nelson, J. P., Pankaj, K. S. (1994) Power Quality and Harmonic Distortion on Distribution Systems. *IEEE Transactions on Industry Applications*, 30, (2), March/April.

Putnam, P. C. (1948) *Power from the Wind.* Van Nostrand Reinhold, New York.

Rogers, W. J. S., Welch, J. (1993) Experience with Wind Generators in Public Electricity Networks. *Proc. of BWEA/RAL Workshop on Wind Energy Penetration into Weak Electricity Networks*, 10–12 June 1993, Rutherford Appleton Laboratory, Abington, UK, June 1993.

Smith, D., Taylor, G. J. (1991) Further Analysis of Turbine Wake Development and Interaction Data. *Proc. 13th BWEA*, Swansea, 10–12 April 1991, Garrad Hassan & Partners Ltd.

Sørensen, J. N., Shen, W. Z. (1999) Computation of Wind Turbine Wakes using Combined Navier–Stokes Actuator-Line Methodology. *Proc. 1999 EWEC*, 156–159, 1–5 March 1999, Nice, France.

Stemmler, H. (1997) High Power Industrial Drives. In Bose, B. K., ed., *Power Electronics and Variable Frequency Drives, Technology and Applications*, IEEE Press, New York.

Troen, I., Petersen, E. L. (1989) *European Wind Atlas*, Risø National Laboratory, Roskilde, Denmark.

Vachon, W. A., Vachon, R. W., Wade, J. E. (1999) Major Sources of Lost Operating Time in Mature Wind Power Plants. *Proc. of AWEA Conference*, Windpower '99, 21–23 June 1999, Burlington, VT.

Vermeulen, P. E. J. (1980) An Experimental Analysis of Wind Turbine Wakes. *Proc. Third International Symposium on Wind energy Systems*, August 26–29, 1980, Lyngby, Denmark, 431–450. BHRA Fluid Engineering, Cranfield, Bedford MK43 OAJ, UK.

Voutsinas, S. G., Rados, K. G., Zervos, A. (1993) Wake Effects in Wind Parks. A New Modelling Approach. *Proc. EWEC*, 444–447, 8–12 March 1993, Lübeck, Germany.

Wade, J. E., Hewson, E. W. (1980) A Guide to Biological Wind Prospecting. U.S. DOE Report: ET-20316, NTIS.

Walker, J. F., Jenkins, N. (1997) *Wind Energy Technology*, John Wiley, Chichester, UK.

Wegley, H. L., Ramsdell, J. V., Orgill, M. M., Drake, R. L. (1980) *A Siting Handbook for Small Wind Energy Conversion Systems*. Battelle Pacific Northwest Lab., PNL-2521, Rev. 1, NTIS.

Westinghouse Electric Corp. (1979) *Design Study and Economic Assessment of Multi-unit Offshore Wind Energy Conversion Systems Application*, DOE WASH-2830-78/4.

9

Wind Energy System Economics

9.1 Introduction

In the previous chapters, the main emphasis has been on the technical and performance aspects of wind turbines and their associated systems. As noted in Chapter 4, in order for a wind turbine to be a viable contender for producing energy, it must: (1) produce energy, (2) survive, and (3) be cost-effective.

Assuming that one has designed a wind energy system that can reliably produce energy, one should be able to predict its annual energy production. With this result and the determination of the manufacturing, installation, operation and maintenance, and financing costs, the cost-effectiveness can be addressed. As shown in Figure 9.1, in discussing the economic aspects of wind energy, it is also important to treat the costs of generating wind energy and the market value of the energy produced (its monetary worth) as separate subjects. The economic viability of wind energy depends on the match of these two variables: i.e., the market value must exceed the cost before the purchase of a wind energy system can be economically justified.

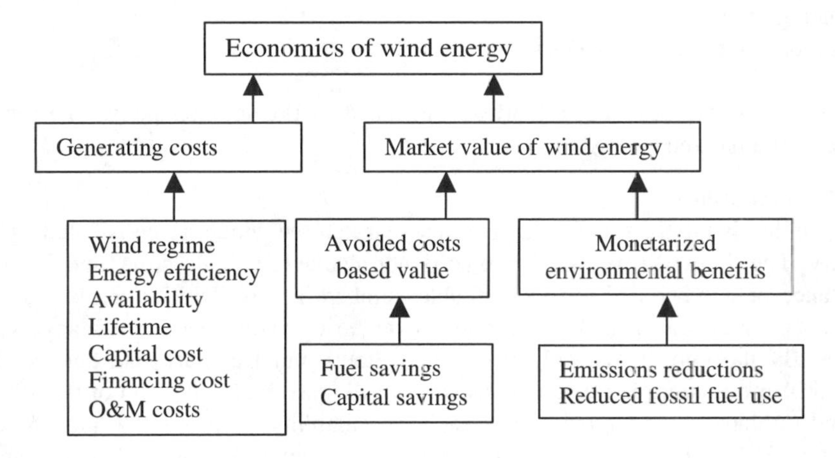

Figure 9.1 Components of wind system economics

The economic aspects of wind energy vary, depending on the application. Grid-connected wind turbines will probably make a larger contribution to the world energy supply than turbines on isolated networks (WEC, 1993). Thus, the primary emphasis in this chapter will be on wind systems supplying electricity to consumers on the main grid. Much of the material here, however, can be used for small to medium sized wind energy installations that are isolated from an electrical grid.

This chapter concentrates on the economics of larger wind energy systems, first introducing the subjects shown in Figure 9.1. Next, details of the capital, operation, and maintenance cost of wind energy systems are summarized in Sections 9.3 and 9.4. Sections 9.5 and 9.6 discuss the value of wind energy and the variety of economic analysis methods that can be applied to determine the economic viability of wind energy systems. These methods range from simplified procedures to detailed life-cycle costing models. The chapter concludes with a section reviewing wind energy system market considerations.

9.2 Overview of Economic Assessment of Wind Energy Systems

This section discusses the overall economics of wind energy systems, covering the topics shown in Figure 9.1.

9.2.1 Generating costs of grid connected wind turbines: overview

The total generating costs for an electricity-producing wind turbine system are determined by the following factors:

- Wind regime
- Energy capture efficiency of the wind machine(s)
- Availability of the system
- Lifetime of the system
- Capital costs
- Financing costs
- Operation and maintenance costs

The first two factors have been addressed in detail in previous chapters. The remaining factors are summarized below.

9.2.1.1 Availability

The availability is the fraction of the time in a year that the wind turbine is able to generate electricity. The times when a wind turbine is not available include downtime for periodic maintenance or unscheduled repairs. Reliable numbers for availability can be determined only if data for a large number of turbines over an operation period of many years are available. By the mid 1990s, only the United States and Denmark had enough data to provide this information. For example, as shown in Figure 9.2, at the end of the 1980s, the best wind turbines in the United States reached availability levels of 95% after 5 years of operation (WEC, 1993). Recent data, as discussed in Chapter 8, indicate that availability is now on the order of 98%.

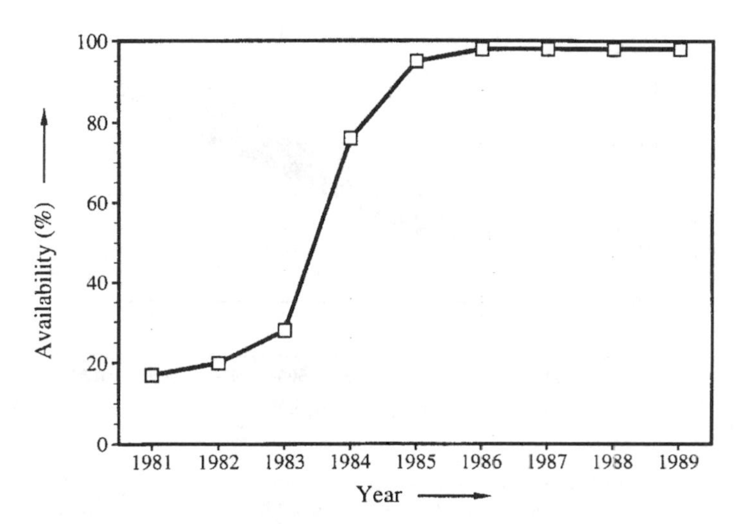

Figure 9.2 Availability of the best wind turbines in California (WEC, 1993). Reproduced by permission of World Energy Council, London

9.2.1.2 Lifetime of the system

It is common practice to equate the design lifetime with the economic lifetime of a wind energy system. In Europe, an economic lifetime of 20 years is often used for the economic assessment of wind energy systems (WEC, 1993). This follows the recommendations of the Danish Wind Turbine Manufacturers Association (1998), that state that a 20-year design lifetime is a useful economic compromise that is used to guide engineers who develop components for wind turbines.

Following recent improvements in wind turbine design, an operating life of 30 years has been used for recent U.S. economic studies (DOE/EPRI, 1997). This assumption requires that adequate annual maintenance be performed on the wind turbines and that 10-year major maintenance overhauls be performed to replace key parts.

9.2.1.3 Capital costs

Details on the determination of the capital cost of wind energy systems are given in Section 9.3. The determination of the capital (or total investment) costs generally involves the cost of the wind turbine(s), and the cost of the remaining installation. Wind turbine costs can vary significantly. For example, Figure 9.3 gives the cost range (not including installation) of production Danish wind machines (Danish Wind Turbine Manufacturers Association, 1998). As shown, costs vary significantly for each rated generator size. This may be due to differing tower height and/or rotor diameter.

In generalized economic studies, wind turbine installed costs are often normalized to cost per unit of rotor area or cost per rated kW. Examples of both these types of normalized wind turbine costs are shown in Figures 9.4 and 9.5. Figure 9.4 gives the specific cost per unit of rotor area for commercial and early experimental utility scale machines in the United States and Europe (Harrison et al., 2000). It should be noted that all of the prototype units have manufacturing costs that are significantly higher than those of commercial or mass-produced units.

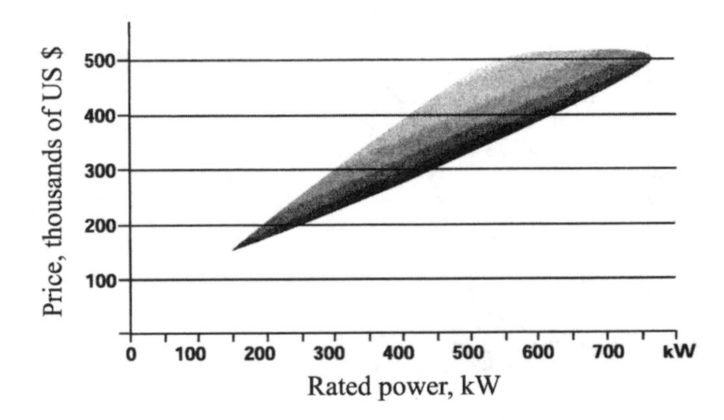

Figure 9.3 Cost (1997) of Danish wind turbines as a function of size: the 'price banana' (Danish Wind Turbine Manufacturers Association, 1998). Reproduced by permission of Danish Wind Turbine Manufacturers Association

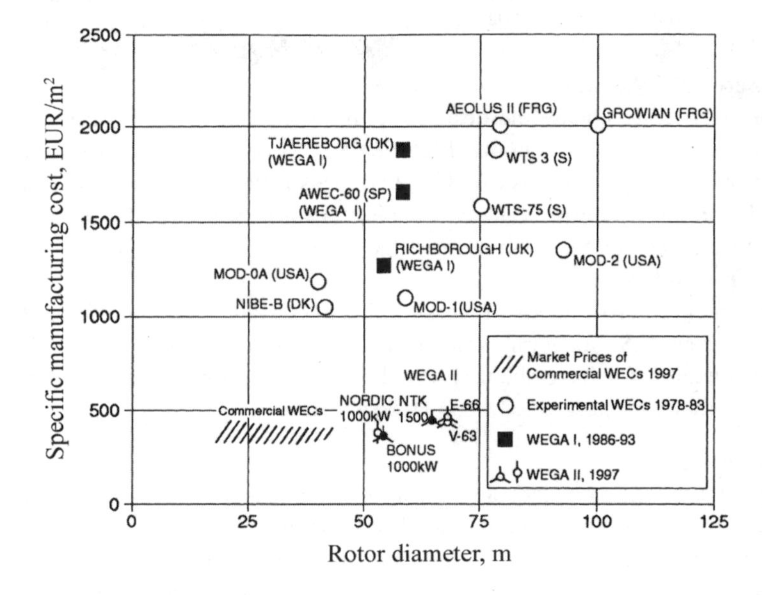

Figure 9.4 Large wind turbine costs per unit rotor area (Harrison et al., 2000) Reproduced with permission of John Wiley & Sons Ltd

The normalization of wind turbine costs to installed cost per unit of rated power (e.g., \$/kW) output follows conventional large-scale electrical generation practice and is often used. For commercial turbines used in wind farms there has been a steady decrease in cost per unit power over the last 25 years. This point is illustrated in Figure 9.5, where installed wind farm costs (\$/kW) are given for the years 1980–95. In this figure, data have been summarized for the California wind farms (CEC PRS), a Danish utility (ELSAM) and Great Britain.

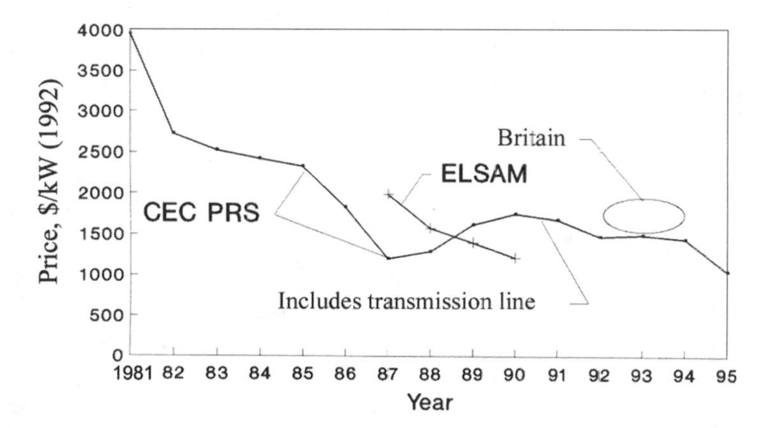

Figure 9.5 Wind farm installed costs (Gipe, 1995) Reproduced with permission of John Wiley & Sons Ltd

Other investigators have projected capital cost data for large-scale wind farms from 1997 to 2030. For example, the U.S. Department of Energy (DOE) and the Electric Power Research Institute (EPRI) projected costs in 1997 dollars (DOE/EPRI, 1997) to decrease from $1000/kW in 1997 to $635/kW in 2030.

As is discussed in Section 9.3.4, the costs of a wind farm installation include costs beyond those of the wind turbines themselves. For example, in developed countries of the world, the wind turbine represents only approximately 65% to 75% of the total investment costs. Wind farm costs also include costs for infrastructure and installation, as well as electrical grid connection costs. It should also be noted that the costs per kW of single wind turbine installations are generally much higher, which is the reason for wind farm development in the first place.

9.2.1.4 Financing costs

Wind energy projects are capital intensive, and the majority of the costs must be borne at the beginning. For that reason, the purchase and installation costs are largely financed. The purchaser or developer will pay a limited down payment (perhaps 10% to 20%), and finance (borrow) the rest. The source of capital may be a bank or investors. In either case, the lenders will expect a return on the loan. The return in the case of a bank is referred to as the interest. Over the life of the project, the cumulative interest can add up to a significant amount of the total costs. Other methods of financing are also used, but they are beyond the scope of this text.

9.2.1.5 Operation and maintenance costs

The Danish Wind Turbine Manufacturers Association (1998) states that annual operation and maintenance (O&M) costs for wind turbines generally range from 1.5% to 3% of the original turbine cost. They also point out that regular service to the turbine constitutes most of the maintenance cost.

Some authors prefer to use a set cost per kWh of output for their operation and maintenance cost estimates. Figure 9.6 gives results from a recent U.S. study (Chapman et al., 1998) where unscheduled and preventive maintenance (U&PM) costs are given as a function of time. Note that these costs are predicted to decline from 0.55 to 0.31 cents per kWh in the 1997 to 2006 time frame. Section 9.4 gives more details on the range of values for O&M costs.

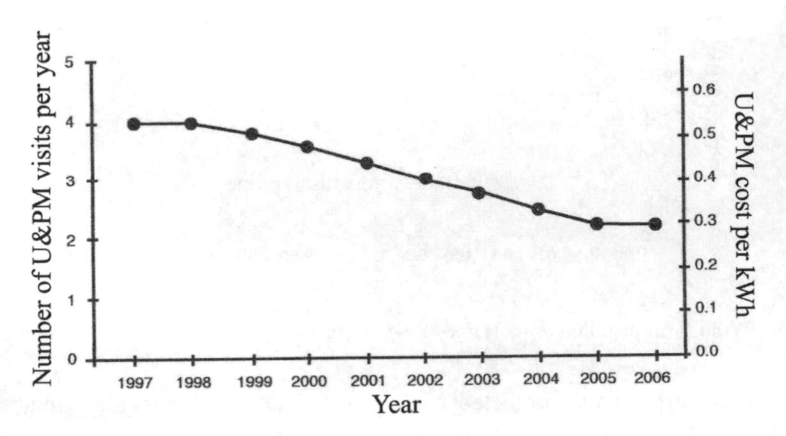

Figure 9.6 Predicted operation and maintenance costs (Chapman et al., 1998); U&PM, unscheduled and preventive maintenance

9.2.2 *Value of grid-generated wind energy: overview*

The value of wind energy depends on the application and on the costs of alternatives to produce the same output. To determine value, one may try to figure 'what the market will bear'. For example, a manufacturer asks a price and adjusts it until he or she finds a buyer (negotiated price). Alternatively, a buyer, such as a distribution utility, may ask for competitive bids to supply power and select those with the lowest prices that look qualified.

For a utility, the value of wind energy is a function primarily of the cost of the fuel that would not be needed or the amount of new system generating capacity that could be deferred. For society as whole, however, the environmental benefits can be quite significant. When these benefits are monetarized, they can contribute significantly to the market value of the electricity. More details on the determination of wind energy's value (including monetarized environmental benefits) are given in Section 9.5.

9.2.3 *Economic analysis methods: overview*

A wind energy system can be considered to be an investment that produces revenue. The purpose of an economic analysis is to evaluate the profitability of a wind energy project and to compare it with alternative investments. The alternative investments might include other renewable electricity producing power systems (photovoltaic power, for example) or conventional fossil fueled systems. Estimates of the capital costs and O&M costs of a wind energy system, together with other parameters which are discussed below, are used in such analyses.

In Section 9.6, the following three different types of overall economic analysis methods are described:

- Simplified models
- Detailed life-cycle cost models
- Electric utility economic models

Each of these economic analysis techniques requires its own definitions of key economic parameters and each has its particular advantages and disadvantages.

9.2.4 Market considerations: overview

At the present time, the market for utility-scale wind energy systems is rapidly expanding. The world market demand has grown from about 200 MW/year in 1990 to a value of 1300 MW/year in 1995, and, more recently, to over 4000 MW/year in 2000. In Section 9.7, the market considerations for wind systems are discussed including the following issues:

- The potential market for wind systems
- Barriers to expansion of markets
- Incentives for market development

9.3 Capital Costs of Wind Energy Systems

9.3.1 General considerations

The determination of the capital costs of a wind energy system remains one of the more challenging subjects in wind engineering. The problem is complicated because wind turbine manufacturers are not particularly anxious to share their own cost figures with the world, or with their competitors. Cost comparisons in the framework of wind turbine research and development projects are particularly difficult. That is, the development costs cannot be compared consistently.

In determining the cost of the wind turbine itself, one must distinguish between the following types of capital cost estimates:

- Cost of wind turbine(s) today. For this type of cost estimate, a developer or engineer can contact the manufacturer of the machine of interest and obtain a formal price quote
- Cost of wind turbine(s) in the future. For this type of cost estimate (assuming present wind turbine state of the art), one has a number of tools that can be applied, namely: (1) historical trends, (2) learning curves (see next section), and (3) detailed examination of present designs (including the total machine and components) to determine where costs may be lowered
- Cost of a new (previously unbuilt) wind turbine design. This type of capital cost estimate is much more complex since it must first include a preliminary design of the new machine. Price estimates and quotes for the various components must be obtained. The

total capital cost estimate must also include other costs such as design, fabrication, testing, etc
- Future cost of a large number of wind turbines of a new design. This type of cost estimate will involve a mixture of the second and third types of cost estimates

It is assumed that one can readily obtain the first type of capital cost estimate. Accordingly, the following discussion concentrates on information that can be used for the last three types of cost estimates.

Recent worldwide experience with wind turbine fabrication, installation, and operation has yielded some data and analytical tools that can be used for capital cost estimates both of wind turbines and/or of the supporting components that are required for a wind farm installation. For example, there have been numerous capital cost estimates for wind turbines based on various simplified scaling techniques. These usually feature a combination of actual cost data for a given machine and empirical equations for the cost of the key components based on a characteristic dimension (such as rotor diameter) of the wind turbine. The same type of generalized scaling analysis has been used to predict wind farm costs.

After a brief discussion of the use of learning curves to predict capital costs of wind turbines, a summary is presented showing how estimates can be made for the capital costs of wind turbines alone and wind turbines in combination with the other components that constitute a wind farm system.

9.3.2 *Use of learning curves to predict capital costs*

One unknown in capital cost estimates is the potential reduction in costs of a component or system when it is produced in large quantities. One can use the concept of learning (or experience) curves to predict the cost of components when they are produced in large quantities. The learning curve concept is based on over 40 years of studies of manufacturing cost reductions in major industries (Johnson, 1985; Cody and Tiedje, 1996).

The learning curve gives an empirical relationship between the cost of an object $C(V)$ as a function of the cumulative volume, V, of the object produced. Functionally, this is expressed as:

$$\frac{C(V)}{C(V_0)}=\left(\frac{V}{V_0}\right)^b \tag{9.3.1}$$

where the exponent b, the learning parameter, is negative and $C(V_0)$ and V_0 correspond to the cost and cumulative volume at an arbitrary initial time. From Equation 9.3.1, an increase in the cumulative production by a factor of 2 leads to a reduction in the object's cost by a progress ratio, s, where $s = 2^b$. The progress ratio, s, when expressed in percent, is a measure of the technological progress that drives the cost reduction.

A graphical example of this relationship is presented in Figure 9.7, which gives a plot of normalized cost $C(V_0) = 1$ for progress ratios ranging from 70% to 95%. As shown, given the initial cost and an estimate of the progress ratio, one can estimate the cost of the tenth, hundredth, or whatever unit of production of the object.

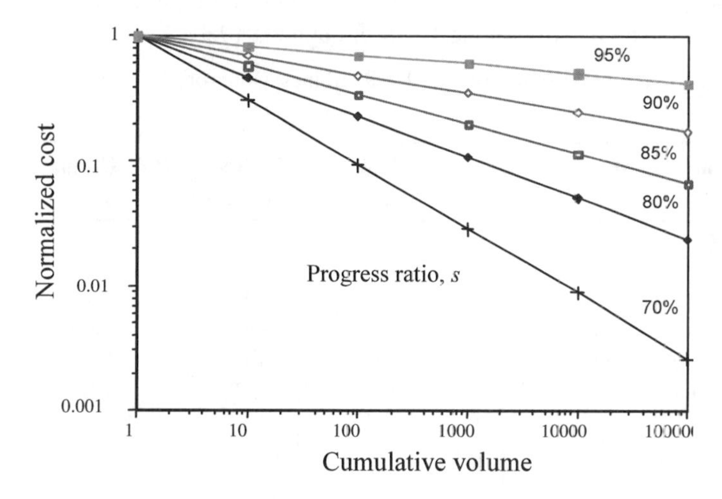

Figure 9.7 Normalized cost vs. cumulative volume for varying progress ratios

Experience has shown that s can range from 70% to 95%. For example, Johnson (1985) gives values of s of 95% for electric power generation, 80% for aircraft assembly, and 74% for hand-held calculators. Various parts of a wind turbine would be expected to have different progress ratios. That is, components such as blades and hub that are essentially unique to wind turbines would have smaller values of s. Electrical generators, however, could represent a mature technology, with higher progress ratio values. For wind turbines, Neij (1999) gives a summary of progress ratios for various utility scale machines.

9.3.3 Wind turbine machine costs

9.3.3.1 General overview

One way to estimate the capital costs of a wind turbine is to use cost data for smaller existing machines normalized to a machine size parameter. Here, the usual parameters that are used are unit cost per kW of rated power or unit cost per area of rotor diameter. Although this methodology can be used for energy generation system or planning studies, it yields little information as to the details of a particular design and potential cost reduction for new designs of wind turbines.

A more fundamental way to determine the capital cost of a wind turbine is to divide the machine into its various components and to determine the cost of each component. This method represents a major engineering task, however, and there are relatively few such studies documented in the open literature. Historically interesting examples here include U.S. work on a conceptual 200 kW machine (NASA Lewis Research Center, 1979) and work on the MOD2 (Boeing, 1979) and MOD5A (General Electric Company, 1984) machines.

The results of such a detailed study depend strongly on the turbine, design assumptions, and method of economic analysis. As summarized in Table 9.1, some interesting historical comparisons between large wind turbine designs can be made if the component costs are expressed as percentages. Note that all these studies were based on horizontal axis wind machines over the time period of the 1950s through the 1980s.

Table 9.1 Projected construction costs for large turbines (Johnson, 1985)

Specification or Cost	British design	Smith–Putnam	P.T. Thomas	NASA MOD-X
Specification				
Rotor diameter (m)	68.6	53.3	61	38.1
Rated power (kW)	3670	1500	7500	200
Rated wind speed (m/s)	15.6	13.4	15.2	9.4
Production quantity	40	20	10	100
Component Cost (%)				
Blades	7.4	11.2	3.9	19.6
Hub, bearings, main shaft, nacelle	19.5	41.5	5.9	15.2
Gearbox	16.7	9.5	2.3	16.5
Electric generator and installation	12.5	3.5	33.6	7.6
Controls	4.4	6.5	8.3	4.4
Tower	8.1	7.7	11.2	20.6
Foundation and site work	31.4	16.6	20.3	16.1
Engineering		3.6	14.5	

The fourth numerical column summarizes the result of a 1980 NASA study of a 200 kW horizontal axis wind turbine (MOD-X) based on their experience with the MOD-0 and MOD-0A machines. This conceptual design featured two pitch-controlled rotor blades, a teetered hub and a passive yaw system. The rotor was connected to the low-speed shaft of a three-stage parallel shaft gearbox, and a synchronous generator rated at 200 kW with a speed of 1800 rpm was used. A significant difference between the three earlier studies and the 1979 NASA study occurred in the area of improved technology in component hardware and analysis techniques.

To further illustrate the differences in component cost fractions for more recent wind turbine designs, one can refer to the text of Hau et al. (1993). Based on this work, Figure 9.8 gives a capital cost breakdown for the main components of three large-scale European wind turbines. A wide variation in component cost fractions can be seen.

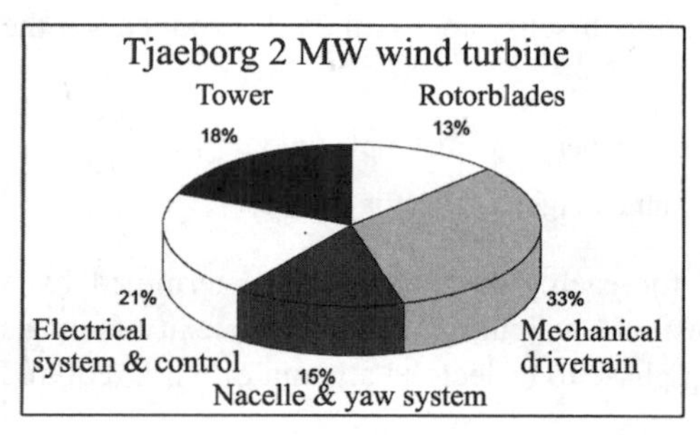

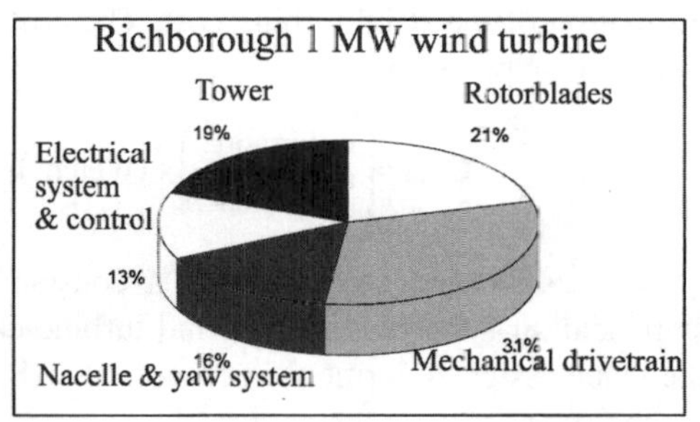

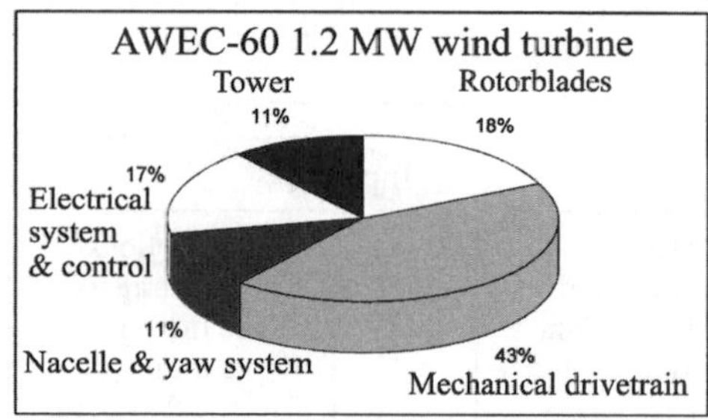

Figure 9.8 Manufacturing cost breakdown of 1990s European wind turbines (Hau et al., 1993). Reproduced by permission of Springer-Verlag GmbH and Co.

9.3.3.2 Detailed capital cost model

In an effort to quantify the capital costs for horizontal axis wind turbines, a detailed cost model for horizontal axis wind turbines has been developed at the University of Sunderland (Hau et al., 1996; Harrison and Jenkins, 1993; and Harrison et al., 2000). Starting with an outline specification of a proposed (two- or three-bladed) wind turbine, and using some basic design options, the computer code can provide a capital cost estimate of a specific wind turbine design. An overview of the key features of the (weight-based) model's operation is shown in Figure 9.9.

Based on the input data, the model uses first principles to develop estimates of the most important influences on cost of the proposed machine. As shown in Figure 9.9, these are called 'design drivers' and include such variables as the blade loadings, the horizontal thrust on the machine, gearbox requirements, and generator specifications. Next, the model estimates the loads on the major subsystems of the wind turbine, allowing the approximate sizes for the major components to be determined. For example, analytical expressions are used to estimate the low-speed shaft diameter such that the shaft will be able to carry both the torsional and axial loading exerted on it by the rotor. For more complex components, where analytical expressions are not sufficient, look-up tables relating component size to loadings are used. In certain cases, components are assigned a complexity rating in order to reflect how much work is involved in their manufacture.

Using the calculated size of each component, the weights of the components (assuming a known material density) are then determined. It should be noted that some of the weights are used to calculate loadings on other components, but their primary purpose is to allow the estimation of the wind turbine cost. The cost of each subsystem is then determined via the following expression:

$$\text{Cost} = \begin{pmatrix}\text{Calibration}\\ \text{coefficient}\end{pmatrix} \text{x} \left(\text{Weight}\right) \text{x} \begin{pmatrix}\text{Cost per}\\ \text{unit weight}\end{pmatrix} \text{x} \begin{pmatrix}\text{Complexity}\\ \text{factor}\end{pmatrix} \tag{9.3.2}$$

The calibration coefficient is a constant for each subsystem. It is determined by a statistical analysis of existing wind turbine cost and weight data. The complexity factor is the value assigned during the component sizing phase to reflect the amount of work required for the subsystem's construction.

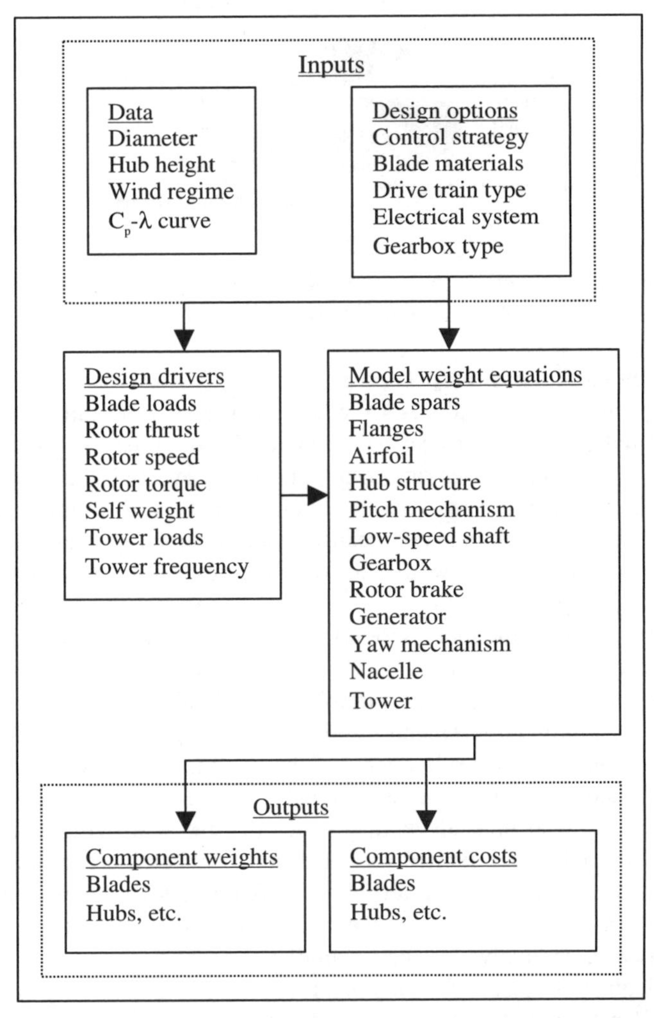

Figure 9.9 Flowchart of Sunderland cost model; C_p, power coefficient; λ, tip speed ratio

The total capital cost of the wind turbine is calculated from a sum of the cost estimates for each subsystem. To validate the model's cost prediction, the code has been used to predict the weight and capital cost of a wide range of actual wind turbines (Harrison et al., 2000). Figure 9.10 gives a comparison between the model predictions of the relationship between blade weight and rotor diameter for three types of blade material (solid or dotted lines) and data from a number of operating wind turbines. Figure 9.11 presents a comparison of total capital cost predictions (for three-bladed machines) and actual factory costs of early small machines and current large machines. Both graphs show good agreement between the model and actual weight and cost data.

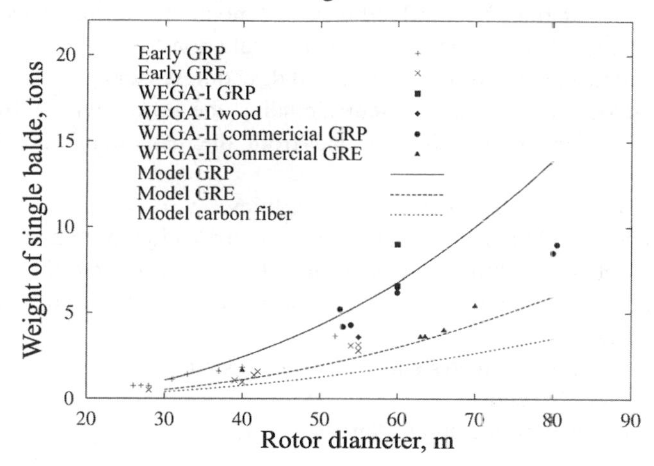

Figure 9.10 Validation of Sunderland cost model weight predictions (Harrison et al., 2000) Reproduced with permission of John Wiley & Sons Ltd

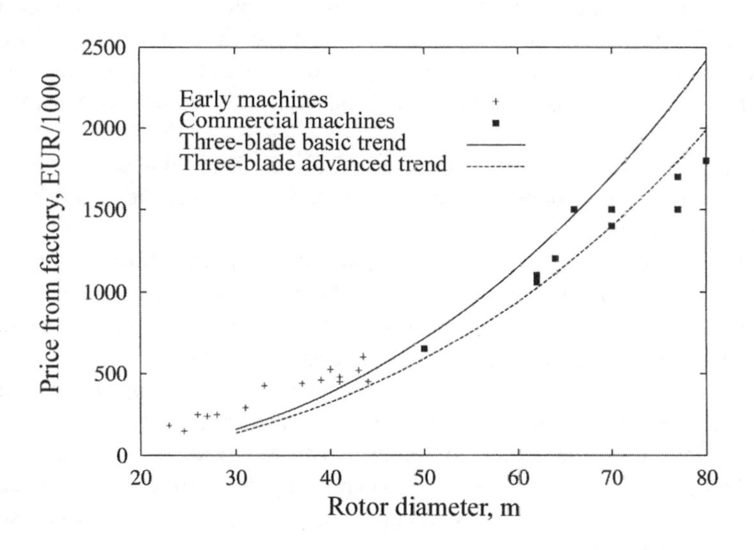

Figure 9.11 Validation of Sunderland wind turbine cost predictions (Harrison et al., 2000) Reproduced with permission of John Wiley & Sons Ltd

9.3.4 *Wind farm costs*

The determination of total system capital costs, or installed capital cost, involves more than just the cost for the turbines themselves. For example, the installed capital cost of a wind farm could include the following (National Wind Coordinating Committee, 1997):

- Wind resource assessment and analysis
- Permitting, surveying, and financing
- Construction of service roads
- Construction of foundations for wind turbines, pad mount transformers and substation
- Wind turbine and tower delivery to the site and installation
- Construction and installation of wind speed and direction sensors
- Construction of power collection system including the power wiring from each wind turbine to the pad mount transformer and from the pad mount transformer to the substation
- Construction of operations and maintenance facilities
- Construction and installation of a wind farm communication system supporting control commands and data flow from each wind turbine to a central operations facility
- Provision of power measurement and wind turbine computer control, display and data archiving facilities
- Integration and checkout of all systems for correct operation
- Commissioning and shakedown
- Final turnover to owner or operating agency

A detailed discussion of the determination of each of these cost components is beyond the scope of this text. A few examples are presented below, however, that give an idea of how these costs are distributed for modern wind farms. A typical breakdown of costs (for a 5 MW wind farm in the U.K.) is given in Table 9.2 (British Wind Energy Association, 1997). In addition to the wind turbine, the major costs include civil works (tower and foundation installation and roads) and the cost associated with the connection to the main electrical grid.

Another example (DOE/EPRI, 1997) is based on larger sized wind farms (on the order of 50 machines). The results for a 1997 base case design, normalized to $/kW and as a percent of total cost, are given in Table 9.3.

The breakdown of costs for the various components of a wind farm can vary among countries. For example, Table 9.4 gives a cost breakdown for wind farms in the U.S. and the Netherlands (WEC, 1993). As shown, in addition to the wind turbines themselves, the costs are divided into those for infrastructure, installation, and engineering.

From these examples of wind farm costs, and other data on the subject, it is reasonable to assume that the cost of wind turbines represents about 65% to 75% of the total capital costs for wind farms in the industrialized countries of the world. In developing countries such as India, however, the relative cost breakdown is slightly different. That is, according to recent data, about 30% to 50% of the total capital costs are for infrastructure, installation, and grid connection (WEC, 1993).

Table 9.2 Wind farm capital cost breakdown (British Wind Energy Association, 1997)

Component	Percent of total cost
Wind turbines	64
Civil works	13
Electrical infrastructure	8
Grid connection	6
Project management	1
Installation	1
Insurance	1
Legal/development costs	3
Bank fees	1
Interest during construction	2

Table 9.3 50-turbine wind farm capital cost breakdown (DOE/EPRI, 1997)

Component	Cost ($/kW)	Percent of Total
Rotor assembly	185	18.5
Tower	145	14.5
Generator	50	5.0
Electrical items, electronics, controls and instrumentation	155	15.5
Transmission, drive train and nacelle	215	21.5
Balance of station	250	25.0
Total installed cost	1000	100.0

Table 9.4 Comparison of wind farm costs (percentage of total cost) in United States and the Netherlands (WEC, 1993)

Component	United States	Netherlands
Wind turbine	74.4	76.8
Engineering and installlation	6.7	3.4
Tower and foundation	11.1	6.6
Grid connection	6.9	11.8
Roads	0.9	1.4

9.4 Operation and maintenance costs

With the installation of operational wind turbines at a given site, it is important to know (and reduce, if possible) the annual operation and maintenance (O&M) costs. It is especially important to project O&M costs if the owner of a wind farm seeks to refinance or sell the project. The operation costs can include a cost for insurance on the wind turbine, taxes, and land rental costs, while the maintenance costs can include the following typical components:

- Routine checks
- Periodic maintenance

- Periodic testing
- Blade cleaning
- Electrical equipment maintenance
- Unscheduled maintenance costs

Until recently, the prediction of the operation and maintenance costs of installed wind turbines was regarded as somewhat speculative (Freris, 1990). Based on experience from the wind farms of California (Lynnette, 1986), and information from the Danish Wind Turbine Manufacturer's Association (1998) and from recent U.S. studies (DOE/EPRI, 1997), this is less so today. However, as will be discussed here, the current data indicate that there still exists a wide range of O&M costs for wind turbines. This difference may depend on the size of the installed wind farm.

Operation and maintenance costs can be divided into 2 categories: fixed and variable. Fixed O&M costs are yearly charges unrelated to the level of plant operation. They must be paid regardless of how much energy is generated (generally expressed in terms of $/kW installed or percentage of turbine capital cost). Variable O&M costs are yearly costs directly related to the amount of plant operation (generally expressed in $/kWh). Probably the best estimates of O&M costs are a combination of these two categories.

One example of data on O&M costs is contained in a 1987 EPRI (Spera, 1994) study of California wind farm costs. This study estimated variable O&M costs, ranging from 0.008 to 0.012 $/kWh (in 1987 dollars). At the time of this study it was found that the variable O&M costs for small-scale machines (up to 50 kW) were higher than medium-scale machines (up to 200 kW). The work also concluded that the O&M expenditures were approximately distributed as shown in Table 9.5.

Table 9.5 Breakdown of O&M Costs (Spera, 1994)

Cost component	Cost percentage
Labor	44
Parts	35
Operations	12
Equipment	5
Facilities	4

The Danish Wind Turbine Manufacturers Association (1998) compiled data on over 4,400 Danish machines installed in Denmark. They found that the newer generation of wind turbines have relatively lower repair and maintenance costs than those of the older generation of wind turbines. Specifically, the older wind turbines (sized from 25 to 150 kW) have annual maintenance costs averaging about 3% of the original capital cost of the turbine(s). For the newer machines, the estimated range of annual O&M costs range from 1.5% to 2% of the original capital cost, or approximately $0.01/kWh. Germanischer Lloyd (Nath, 1998) reports the same trend with machine age in O&M costs. Their variation of O&M costs, however, is much larger, ranging from 2% to 16% of the price of the wind turbine. Other European investigators have stated that, for large turbines, maintenance costs of less than $0.006/kWh are possible (Klein et al., 1994).

Methods have been developed for the prediction of the increase of wind turbine maintenance costs with time (Vachon, 1996). Vachon used a statistical approach for modeling component failures (and maintenance costs) as a function of time. Based on actual field data, Table 9.6 gives results from a Danish Energy Association study (Lemming et al., 1999) that show that wind energy systems O&M costs (expressed as a percent of total wind farm installation cost) vary with turbine size and age. It is important to note the predicted rise in costs with turbine age.

Table 9.6 Comparison of total O&M costs as a function of size and age of turbine (Lemming et al., 1999)

Turbine size	**Years from installation**				
	1–2	3–5	6–10	11–15	16–20
150 kW	1.2	2.8	3.3	6.1	7.0
300 kW	1.0	2.2	2.6	4.0	6.0
600 kW	1.0	1.9	2.2	3.5	4.5

Note: O&M costs are expressed as a percent of total wind farm installation costs

9.5 Value of Wind Energy

9.5.1 Overview

The traditional way to assess the value of wind energy is to equate it to the direct savings that would result due the use of the wind rather than the most likely alternative. These savings are often referred to as ‘avoided costs’. Avoided costs result primarily from the reduction of fuel that would be consumed by a conventional generating plant. They may also result from a decrease in total conventional generating capacity that a utility requires. This topic is discussed in more detail in Section 9.5.3.

Given the cost reductions in wind turbines themselves over the last several decades, there are already some locations where wind could be the most economic option for new generation. Basing wind energy’s value exclusively on the avoided costs, however, would result in many potential applications being uneconomic.

Equating wind energy’s value exclusively to the avoided costs also misses the substantial environmental benefit that results from its use. The environmental benefits are those which arise because wind generation does not result in any significant amount of air emissions, particularly oxides of nitrogen and sulfur and carbon dioxide. Reduction of emissions translates into a variety of health benefits. It also decreases the concentration of atmospheric chemicals that cause acid rain and global warming.

Converting wind energy’s environmental benefits into monetary form can be difficult. Nonetheless, the results of doing so can be quite significant, because when that is done, many more projects can be economic.

The incorporation of environmental benefits into the market for wind energy is done through two steps: (1) quantifying the benefits and (2) ‘monetarization’ of some of those benefits. Quantifying the benefits involves identifying the net positive effects to society as a result of wind’s use. Monetarization involves assigning a financial value to the benefits. It allows a financial return to be captured by the prospective owner or developer of the project. Monetarization is usually accomplished by government regulation. The process frequently considers the cost of alternative measures to reduce emissions (such as scrubbers on coal plants) to assign a monetary value to the avoided emissions. Section 9.5.3 provides more detail on the environmental benefits of wind energy, and the result of monetarizing some of those benefits.

The effect of this method of incorporating environmental benefits is to create two categories of potential revenue for a wind project: (1) revenue based on avoided costs and (2) revenue based on the of monetarized environmental benefits. Figure 9.12 illustrates what the relative magnitude of these two revenue sources can be. In the cases shown there, the effect of including the monetarized environmental benefits ('avoided emissions') is to increase the total value by approximately 50% above that based on avoided fuel and capacity alone.

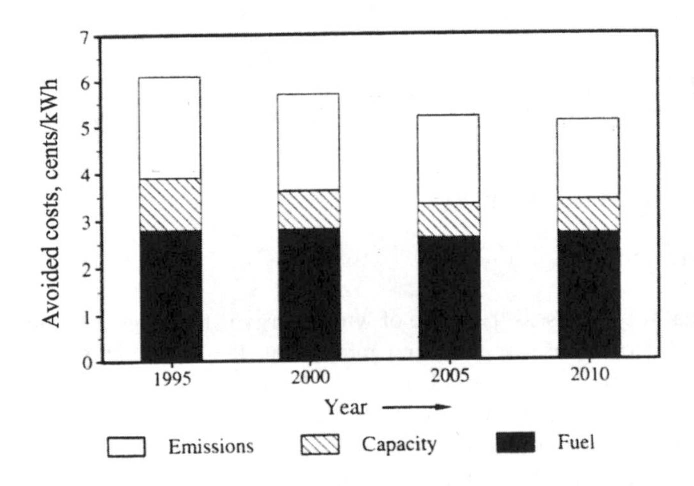

Figure 9.12 Avoided costs of wind energy for the Netherlands for 1995–2010 (WEC, 1993) Reproduced by permission of World Energy Council, London

Because of the very real environmental benefits of wind energy, and because of enormous impact that including these benefits into economic assessments could have, many countries have developed laws and regulations to facilitate the process. These are discussed in Section 9.7.4.

How revenues based on avoided costs and monetarized environmental benefits may factor into an economic assessment of a particular project depends on what may be referred to as the 'market application' and the relation of the project or developer to that market application. Sections 9.5.4.1 and 9.5.4.2 describe various market applications and common types of owners or developers. Section 9.5.4.3 summarizes the most common sources of revenue that may be available to a wind energy project. After that some hypothetical examples are presented to illustrate the difference in value of that may result.

9.5.2 *Avoided cost based value of wind energy*

As stated above, the traditional way of assigning value to wind energy, particularly to a utility, has been to base the value on calculations of avoided fuel and capacity costs. These are discussed in more detail below.

9.5.2.1 Fuel savings

The inclusion of wind turbines in an electricity producing system can reduce the demand for other generating plants that require a fossil fuel input. It might appear that the calculation of fuel saving would be quite straightforward, but this is not generally the case. The type and amount of fuel savings depends on such factors as the mix of different fossil fuel and nuclear plants in the electrical generating system, requirements for 'spinning reserves', and operating characteristics of the fossil fuel components (such as efficiency or heat rate as a function of component load). Also, the fuel savings can depend on the total fraction of wind energy delivered to the system.

The electrical system as a whole must be modeled in order to estimate the avoided consumption of fuel. For example, Figure 9.13 summarizes the results of a study for the Netherlands through the year 2010 (WEC, 1993). Here, the specific amount of avoided fossil fuel is expected to decrease because of an assumed increase in wind generating capacity and because of the higher generation efficiency of new fossil-fired power plants.

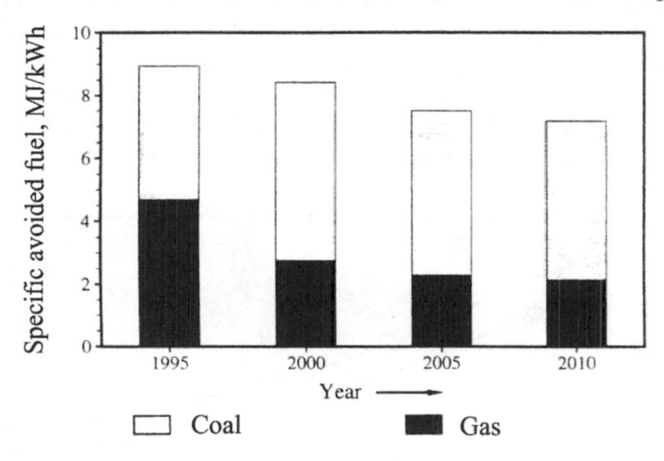

Figure 9.13 Specific avoided fuel consumption for the Netherlands (WEC, 1993) Reproduced by permission of World Energy Council, London

9.5.2.2 Capacity value

The capacity value of a wind system is simply defined as: 'the amount of conventional capacity which must be installed to maintain the ability of the power system to meet the consumers' demand if the wind power installation is deleted' (Tande and Hansen, 1991).

At one extreme, if the utility were absolutely certain that a wind energy plant would produce its full rated power during peak demand hours, its capacity value would equal its rated power. At the other extreme, if the utility were certain that the wind energy plant would never be in the operating range during peak hours then its capacity value would be zero. For a utility, the higher the capacity value of wind power plants, the less new generation capacity is required. In practice, the capacity value of a wind generating plant is somewhere between zero and its rated capacity, depending on the match of the wind energy source to the utility's demand cycles. There are two main methods for the calculation of capacity value (Walker and Jenkins, 1997):

1. The contribution of the wind system during peak demand on a utility is assessed over a period of years and the average power at these times is defined as the capacity value.
2. The loss of load probability (LOLP) or loss of load expectation (LOLE) is calculated, initially with no wind generators in the system (Billinton and Allan, 1984). It is then recalculated with wind generation on the system and then conventional plant capacity is subtracted until the initial level of LOLP is obtained. The capacity of the subtracted power is the capacity value of the wind.

Both methods give similar results. Also, for small penetration of wind systems, the capacity value is generally close to the average output of the wind system.

Most often, capacity value is calculated by the second method (generally using statistical techniques). For example, as shown in Figure 9.14, the capacity value of wind energy in the Netherlands has been determined by two separate studies (WEC, 1993), and can be found as a function of the total size of the wind contribution to the total power grid. For a 1000 MW rated wind capacity, the relative capacity credit is predicted to vary between 16% to 18% (or, equivalent to about 160 to 180 MW of conventional capacity).

Some studies have produced capacity values ranging from a few percent to values approaching 80%. In the middle are other researchers, who propose that renewables with somewhat predictable schedules deserve partial capacity payments (Perez et al., 1997).

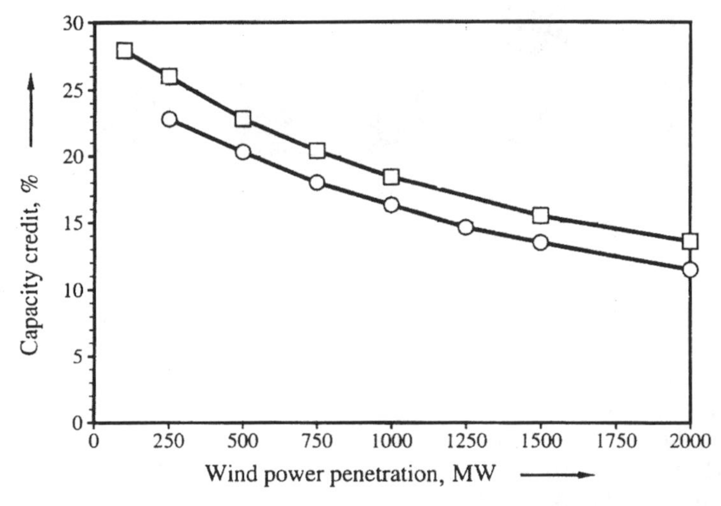

Figure 9.14 Capacity value (%) for the Netherlands vs. wind penetration (WEC, 1993) Reproduced by permission of World Energy Council, London

9.5.3 Environmental value of wind energy

The primary environmental value of electricity generated from wind energy systems is that the wind offsets emissions that would be generated by conventional fossil fueled power plants. These emissions include sulfur dioxide (SO_2), oxides of nitrogen (NO_x), carbon dioxide (CO_2), particulates, slag and ash. The amount of emissions saved via the use of wind energy depends on the types of power plant that are replaced by the wind system, and the particular emissions control systems currently installed on the various fossil-fired plants.

A number of studies have been carried out to determine the emissions savings value of wind power systems. For example, Table 9.7 gives a comparison of results for three European countries (European Commission, 1998). Note that there are major differences in the amount of emissions that can be saved per unit of electricity generated. This depends mainly on the mix of conventional generating units and the emissions control status of the individual units.

At the present time, there is no standard method for computing the emissions savings value of wind energy systems. Most methods, as noted above in Section 9.5.1, make some reference to the costs of alternative measures that could be applied to reduce emissions by the same amount.

Table 9.7 Emissions from fossil-fueled generating plants (g/kWh) (European Commission, 1998)

Pollutant	Netherlands	U.K.	Denmark
CO_2	872	936–1079	850
SO_2	0.38	14.0–16.4	2.9
NO_x	0.89	2.5–5.3	2.6
Slag and ash	not available	not available	55
Dust	not available	not available	0.1

There are also benefits associated with wind energy production other than the direct environmental ones. These include such indirect benefits as improved public health and benefits that would accrue by reducing oil imports. All the benefits (and the corresponding costs) from all sources of energy production are often considered under the general topic of the 'social costs of energy production'. Some studies in this area are discussed by Gipe (1995), WEC (1993) and Hohmeyer (1990). In general, the results show higher social or 'external' costs for conventional electrical power generation plants. As with the value of direct emissions reduction, there is as yet no agreement as how to quantify the exact monetary value. In some countries, however, estimates of the external costs are considered in developing incentives for wind energy, such as renewable portfolio standards, tax credits, or guaranteed prices. These are discussed briefly in Section 9.7.4. The monetary results of those incentives are discussed below.

9.5.4 Market value of wind energy

The market value of wind energy is the total amount of revenue one will receive by selling wind energy, or will avoid paying through its generation and use. The value that can be 'captured' depends strongly on three considerations: (1) the 'market application', (2) the project owner or developer, and (3) the types of revenues available.

The market application affects the type of project that might be built and the magnitude of the avoided costs. The relation of the owner to the market application will determine whether the avoided costs will be accounted for directly or indirectly in the sale of the electricity. The types of revenues are basically in two categories. The first, which includes the sale of electricity and reduced purchases, relate to market value of the energy itself. The second category includes those revenues which are based on monetarized environmental benefits and derive from governmental policies. All of these topics are discussed briefly below.

9.5.4.1 Market applications

The market application refers to the system in which the wind project is located. The most common market applications include: (1) traditional utilities, (2) restructured utilities, (3) customer-owned generation, (4) remote power systems and (5) grid independent applications. The relation between the market application and the person or entity that may own or develop the project has the important effect of determining whether the revenues appear in the form of savings in the costs of production or as sales of energy.

Traditional utilities Traditional electric utilities are entities that generate, transmit, distribute, and sell electricity. They may be privately or publicly owned. When privately owned, they are considered a 'natural monopoly' and are normally regulated by a state or national government. Although most utilities generate at least some of the electricity they sell, many also purchase electricity from other utilities. Two publicly owned types of utilities are (1) municipal electric companies and (2) rural electric co-operatives. These utilities are owned and operated by cities or towns or by their customers (in the case of the co-ops). Frequently they buy most or all of their power from large publicly owned generating entities, such as the Tennessee Valley Authority in the southern United States.

Restructured utilities In recent years, many electric utilities have been 'restructured' or 'deregulated'. They have typically been broken into three parts: (1) a generating company, (2) a 'wires' company, and (3) a marketing company. The wires company only owns and maintains the transmission and distribution wires in a given service area. It does not own generation (at least for that service area). Generation is owned independently. The marketer of energy, not necessarily the utility itself, will purchase energy from a generator, 'rent' the wires for moving the electricity, and will sell the electricity to its customers. Customers are in principal free to choose their supplier (marketer). The marketer sometimes offers electricity 'products' with particular attributes that may be of interest to the customer. These might have to do with reliability, interruptability, or fuel source. The proper functioning of the system is maintained by an 'independent system operator' (ISO), which oversees the process of energy purchase and sale. The wires companies are normally still regulated.

Customer-owned generation Customers of utilities, whether traditional or restructured, may wish to generate some of their own electricity. In this case the customer could purchase a wind turbine and connect it to the local distribution lines, normally on his or her side of the meter. For that reason customer-owned generation is also known as 'behind-the-meter' generation. To what extent this is possible or economically worthwhile depends greatly on the specific laws and regulations that may be in place. In projects of this type, most of the energy is used to reduce purchases of electricity, but in some cases any excess generation may be sold into the electric grid.

Remote power systems Remote power systems are similar in many ways to larger utilities. The main differences are: (1) the type of generators that they use (normally diesel) and (2) their inability to exchange power with any other utilities.

Grid-independent systems Grid-independent wind energy systems are normally relatively small and self-contained, as described in Chapter 8. They may be installed as an alternative to interconnecting to a utility or in lieu of a diesel or gasoline generator.

9.5.4.2 Wind project owners or developers

The nature of the owner or developer, and his or her relation to the market application will have a significant impact on the type of revenues available. Types of owners or developers may include: (1) traditional utilities themselves, (2) independent power producers, (3) electricity consumers wishing to reduce their purchasers, (4) entities responsible for isolated power systems, or (5) individuals needing energy in a grid-independent application.

Traditional utility A traditional utility could develop and own its own wind generation capacity. The wind generation would save some fuel or reduce purchases of electricity from other utilities. The wind project might also provide some capacity value as described in Section 9.5.2.2.

Independent power producers Independent power producers (IPPs) own and operate their own generating facilities and sell the electricity to a utility. They may operate in either traditional or restructured utilities. Wind project IPPs are similar to other generators, except that they may suffer because wind generation is not dispatchable (that is, it can not be turned on and off at will). However, they may benefit by being able to take advantage of incentives not offered to generators that use fossil fuels. They normally require a multi-year contract ('power purchase agreement' or PPA) to secure financing. The details of the PPA are often very important in determining whether or not a project is economically viable.

Customer-generators The customer is normally (but not always) the owner and developer in a behind-the-meter wind energy project. Typical examples include residential customers, farmers, and small businesses.

9.5.4.3 Types of revenue

As described above there are basically two types of revenues that may available, those based on avoided costs and those based on monetarized environmental benefits. The first category includes: (1) reduction in purchases of electricity, fuel, capital equipment or other expenses (avoided costs themselves), (2) sale of electricity (whose price will reflect the avoided costs). The second category includes: (1) sale of renewable energy certificates, (2) tax benefits, (3) guaranteed, above market rates, and (4) net metering.

Reduced purchases When the owner of the turbine is the same as the traditional generator, the revenue from the wind project is actually a savings. In the case of a traditional utility or the manager of remote power system, the revenue would appear as reduced fuel or electricity purchases, reduced operation and maintenance costs on the other generators, and possibly reduced costs for replacements or upgrades to the other generators. The value of the savings (the 'avoided costs) would typically range from about $0.03/kWh to $.04/kWh, depending on the fuel source and type of conventional generators used. For a remote power system, the avoided cost would be in the vicinity of $0.10/kWh. For the owner of a 'behind-the-meter' project, the value of the reduced purchases would be the retail price of electricity. This is much higher than the wholesale price (in the vicinity of $0.10/kWh in a conventional utility to $0.30/kWh or more in a remote power system).

Electricity sale An independent power producer would normally derive revenue by selling electricity into the grid. Ideally the rate would be set by a long-term contract. This rate is close to the expected 'avoided cost' of energy, discussed in Section 9.5.2, for the utility into which the electricity would be sold. The contract price is generally highest for generators which are dispatchable. A wind project operator may also gain a better price if production can be forecast to some extent. Electricity can also be sold on a day-to-day basis (the 'spot market'), but this generally brings the lowest price. In some cases an IPP might be able to sell the energy as 'green power' to a customer willing to pay more to ensure that the electricity comes from an environmentally preferred source.

Renewable energy certificates Over the last few years, new incentives have been developed to facilitate the introduction of more renewable energy into restructured utilities. One of these incentives is the renewable energy or 'green' certificate. In the system on which this incentive is based, the desired attribute of 'greenness' is assumed to be separated from the energy itself and is used to assign value to a certificate. The certificates then have value, typically in the range of a few cents US per kWh. Such certificates may be used to ensure compliance, for example, with mandates known as 'renewable portfolio standards' (RPS). The wind project will receive a quantity of certificates in proportion to the energy generated. The certificates can then be sold, and the proceeds will augment the revenue from the sale of the electricity itself. More details can be found in Section 9.7.4.

Tax benefits Over the years a number of types of tax credits have been used to foster the development of wind energy. At one time, the investment tax credit was widely used. This provides a source of revenue to the project developer based on the cost of the project, rather than its production. A more common incentive nowadays is the production tax credit. It is based on the actual energy produced, and typically amounts to approximately $0.01/kWh.

Guaranteed, above market rates Some countries, such as Germany and Spain, provide for guaranteed, above market rates for generators of wind energy. This arrangement functions similarly to the energy sale method described above. The difference is that the price is set by the government at a value higher than one could expect to receive by conventional sales into the electricity market. The rate for energy in these situations has typically been close to the retail price.

Net metering The situation faced by a small behind-the-meter generator is actually a little more complicated than it first appears. Due to the variability of the wind and the load, it may frequently be the case that the instantaneous generation from the turbine is more than the load, even if the average generation is well below the average load. Depending on the utility rules and type of metering it is possible for an operator to receive little or no revenue from the energy that is produced in excess of the load. This could seriously affect the effective average value for the generation.

In an effort to address this situation, and to provide additional incentives to the small generators, a number of states in the US have developed 'net metering' rules. Under these rules, the key consideration is the net energy that is produced, in comparison to the consumption, over some extended period of time (typically months or more). As long as the net generation is less than the consumption, all the generation will be valued at retail. Any generation in excess of the consumption would be subject to the same rules as would apply to an independent power producer. Net metering rules normally apply to relatively small generators (50 kW or less), although California presently has a much larger limit of 1 MW (see http://www.energy.ca.gov/greengrid/net_metering.html).

9.5.4.4 Examples

It should be apparent that there may a wide range of revenue or value that might apply to a unit of wind generated electricity, depending on the details of the application. The following are some hypothetical examples of this:

1. A private utility using coal-fired generators has a fuel cost of \$0.02/kWh and needs no new capacity. There are no monetarized environmental benefits in the state in which it operates. The value of wind energy would thus be the avoided fuel costs, or \$0.02/kWh.
2. A wind energy project developer in a state with restructured utility can sell his energy into the grid for \$0.02/kWh. The state provides for renewable energy certificates, which he can sell for \$0.015/kWh. He can also take advantage of a federal production tax credit, valued at \$0.01/kWh. The total value of his wind-generated electricity would be \$0.045/kWh.
3. A municipal electric utility buys all its electricity at a fixed rate of \$0.05/kWh. The value of wind energy would thus be the avoided energy purchase costs, or \$0.05/kWh.
4. A farmer is considering whether to install a 50 kW wind turbine to supply part of the farm's electricity. She presently pays \$0.11/kWh. She can take advantage of net metering and a production tax of credit of \$0.01/kWh. The value of her wind-generated electricity would be \$0.12/kWh.

9.6 Economic Analysis Methods

As stated previously, a wind energy system can be considered to be an investment that produces revenue. An economic analysis is used to evaluate the profitability of a wind energy project and to compare it with alternative investments. Economic analysis methods can be applied for wind energy systems, assuming that one has a reliable estimate for the capital costs and O&M costs. The general purpose of such methods is not only to determine the economic performance of a given design of wind energy system, but also to compare it with conventional and other renewable energy based systems. Three different types of overall economic analysis methods will be described in this section. They include:

- Simplified models
- Life-cycle cost models
- Electric utility economic models

Each of these economic analysis techniques has its own definitions of key economic parameters and each has its particular advantages and disadvantages.

It is important to clarify who the owner or developer is, and what the market value can be expected for the energy, as discussed in Section 9.5.4. Depending on the application, one or more economic evaluation methods discussed below may be appropriate.

9.6.1 Simplified economic analysis methods

For a preliminary estimate of a wind energy system's feasibility, it is desirable to have a method for a quick determination of its relative economic benefits. Such a method should be easy to understand, be free of detailed economic variables, and be easy to calculate. Two methods that are often used are: (1) simple payback, and (2) cost of energy.

9.6.1.1 Simple payback period analysis

A payback calculation compares revenue with costs and determines the length of time required to recoup an initial investment. The payback period (in years) is equal to the total capital cost of the wind system divided by the average annual return from the produced power. In it simplest form (simple payback period), it is expressed in equation form as:

$$SP = C_c/AAR \tag{9.6.1}$$

where SP is the simple payback period, C_c is the installed capital cost, and AAR is the average annual return. The latter can be expressed by:

$$AAR = E_a P_e \tag{9.6.2}$$

where E_a is the annual energy production (kWh/year), and P_e is the price obtained for electricity ($/kWh). Thus, the simple payback period is given by:

$$SP = C_c/(E_a P_e) \tag{9.6.3}$$

Consider the following example: C_c = $50,000, E_a = 100,000 kWh/yr, P_e = $0.10 $/kWh. Then SP = 50,000/(100,000 x 0.10) = 5 years.

It should be pointed out that the calculation of simple payback period omits many factors that may have a significant effect on the system's economic cost effectiveness. These include escalating fuel (in a hybrid power system) and loan costs, depreciation on capital costs, O&M, and variations in the value of delivered electricity. Some authors attempt to include some of these variables in their calculations for a simple payback period. For example, one author (Riggs, 1982) divides the capital costs by the net annual savings to obtain simple payback.

9.6.1.2 Cost of energy analysis

The cost of energy, *COE*, is defined as the unit cost to produce energy (in \$/kWh) from the wind energy system. That is:

$$\text{Cost of energy} = (\text{Operating costs})/(\text{Energy produced}) \quad (9.6.4)$$

Based on the previous nomenclature, the simplest calculation of *COE* is given by:

$$COE = [(C_c \times FCR) + C_{O\&M}]/E_a \quad (9.6.5)$$

where $C_{O\&M}$ is the average annual operation and maintenance cost and *FCR* is the fixed charge rate. The fixed charge rate will generally reflect the interest one pays or the value of interest received if money were displaced from savings. For utilities, *FCR* is an average annual charge used to account for debt, equity costs, taxes, etc. (see Section 9.6.3).

Using the values from the previous numerical example, and assuming that $FCR = 10\%$ and $C_{O\&M} = 2\% \times C_c = \$1000/\text{yr}$:

$$COE = [(50{,}000 \times 0.10) + 1000] / 100{,}000 = \$0.06/\text{ kWh}$$

This simplified calculation is based on a number of key assumptions, most of which neglect the time value of capital. They will be addressed in the next section.

9.6.2 Life-cycle costing methods

Life-cycle costing (LCC) is a commonly used method for the economic evaluation of energy producing systems based on the principles of the 'time value' of money. The LCC method summarizes expenditures and revenues occurring over time into a single parameter (or number) so that an economically based choice can be made.

In analyzing future cash flows one needs to consider the time value of money. An amount of money can increase in quantity by earning interest from some investment. Also, money can have a reduced value over time as inflation forces prices upward, making each unit of currency have lower purchasing power. As long as the rate of inflation is equal to the return on investment for a fixed sum of money, purchasing power is not diminished. As is usually the case, however, if these two values are not equal then the sum of money can increase in value (if investment return is greater than inflation) or decrease in value (if the inflation rate is greater than investment return).

The concept of LCC analysis is based on accounting principles used by organizations to analyze investment opportunities. The organization seeks to maximize its return on investment (*ROI*) by making an informed judgment on the costs and benefits to be gained by the use of its capital resources. One way to accomplish this is via a LCC based calculation of the *ROI* of the various investment opportunities available to the organization.

To determine the value of an investment in a wind power system, the principles of LCC costing can be applied to its costs and benefits, that is to say, its expected cash flows. The costs include the expenses associated with the purchase, installation, and operation of the wind system (see Sections 9.3 and 9.4). The economic benefits of a wind system include the use or sale of the generated electricity as well as tax savings or other financial incentives (see Section 9.5). Both costs and benefits may also vary over time. The principles of life-cycle costing can take into account time-varying cash flows and refer them to a common point in time. The result will be that the wind system can be compared to other energy producing systems in an internally consistent manner.

LCC methodology, as described in this section, takes the parameters of inflation and interest applied to money and uses a model based on the 'time value of money' to project a 'present value' for an investment at any time in the future. The important variables and definitions used in life-cycle costing analysis follow.

9.6.2.1 Overview and definition of life cycle costing concepts and parameters

In LCC analysis, some key concepts and parameters include:

- The time value of money and present worth factor
- Levelizing
- Capital recovery factor
- Net present value

A summary description of each follows.

Time value of money and present worth factor A unit of currency that is be to paid (or spent) in the future will not have the same value as one available today. This is true even if there is no inflation, since a unit of currency can be invested and bear interest. Thus its value is increased by the interest. For example, suppose an amount with a present value *PV* (sometimes called present worth) is invested at an interest (or discount) rate r (expressed as a fraction) with annual compounding of interest. (Note that in economic analysis the discount rate is defined as the opportunity cost of money. This is the next best rate of return which one could expect to obtain.) At the end of the first year the value has increased to $PV(1+r)$, after the second year to $PV(1+r)^2$, etc. Thus, the future value, *FV*, after N years is:

$$FV = PV(1+r)^N \tag{9.6.6}$$

The ratio *PV*/*FV* is defined as the present worth factor *PWF*, and it is given by:

$$PWF = PV / FV = (1+r)^{-N} \tag{9.6.7}$$

For illustration purposes, numerical values of *PWF* are given in Table 9.8.

Levelizing Levelizing is a method for expressing costs or revenues that occur once or in irregular intervals as equivalent equal payments at regular intervals. A good way to illustrate this variable is by the following example (Rabl, 1985). Suppose one wants loan payments to be arranged as a series of equal monthly or yearly installments. That is, a loan of present value *PV* is to be repaid in equal annual payments A over N years. To determine the equation for A, first consider a loan of value PV_N that is to be repaid with a single payment F_N at the end of N years. With N years of interest on the amount PV_N, the payment is: $F_N = PV_N(1+r)^N$. In other words, the loan amount PV_N equals the present value of the future payment F_N; $PV_N = F_N(1+r)^{-N}$

Table 9.8 Present worth factor for discount rate r and number of years N

Discount rate, r	Number of years, N					
	5	10	15	20	25	30
0.01	0.9515	0.9053	0.8613	0.8195	0.7798	0.7419
0.02	0.9057	0.8203	0.7430	0.6730	0.6095	0.5521
0.03	0.8626	0.7441	0.6419	0.5537	0.4776	0.4120
0.04	0.8219	0.6756	0.5553	0.4564	0.3751	0.3083
0.05	0.7835	0.6139	0.4810	0.3769	0.2953	0.2314
0.06	0.7473	0.5584	0.4173	0.3118	0.2330	0.1741
0.07	0.7130	0.5083	0.3624	0.2584	0.1842	0.1314
0.08	0.6806	0.4632	0.3152	0.2145	0.1460	0.0994
0.09	0.6499	0.4224	0.2745	0.1784	0.1160	0.0754
0.10	0.6209	0.3855	0.2394	0.1486	0.0923	0.0573
0.11	0.5935	0.3522	0.2090	0.1240	0.0736	0.0437
0.12	0.5674	0.3220	0.1827	0.1037	0.0588	0.0334
0.13	0.5428	0.2946	0.1599	0.0868	0.0471	0.0256
0.14	0.5194	0.2697	0.1401	0.0728	0.0378	0.0196
0.15	0.4972	0.2472	0.1229	0.0611	0.0304	0.0151
0.16	0.4761	0.2267	0.1079	0.0514	0.0245	0.0116
0.17	0.4561	0.2080	0.0949	0.0433	0.0197	0.0090
0.18	0.4371	0.1911	0.0835	0.0365	0.0160	0.0070
0.19	0.4190	0.1756	0.0736	0.0308	0.0129	0.0054
0.20	0.4019	0.1615	0.0649	0.0261	0.0105	0.0042

A loan that is to be repaid in N equal installments can be considered as the sum of N loans, one for each year, the jth loan being repaid in a single installment A at the end of the jth year. Thus, the value, PV, of the loan equals the sum of the present values of all loan payments:

$$PV = \frac{A}{1+r} + \frac{A}{(1+r)^2} + ... + \frac{A}{(1+r)^N} = A\sum_{j=1}^{N}\frac{1}{(1+r)^j} \tag{9.6.8}$$

Or, using an equation for a geometric series:

$$PV = A[1-(1+r)^{-N}]/r \tag{9.6.9}$$

It should be noted that this equation is perfectly general and relates any single present value, PV, to a series of equal annual payments A, given the interest or discount rate, r, and the number of payments (or years), N. Also note that when $r = 0$, $PV = A/N$.

Capital recovery factor The capital recovery factor, CRF, is used to determine the amount of each future payment required to accumulate a given present value when the discount rate and the number of payments are known. The capital recovery factor is defined as the ratio of A to PV and, using Equation 9.6.9, is given by:

$$CRF = \begin{cases} r/[1-(1+r)^{-N}], \text{ if } r \neq 0 \\ 1/N, \text{ if } r = 0 \end{cases} \tag{9.6.10}$$

The inverse of the capital recovery factor is sometimes defined as the series present worth factor, *SPW*.

Net present value The net present value (*NPV*) is defined as the sum of all relevant present values. From Equation 9.6.6, the present value of a future cost, C, evaluated at year j is:

$$PV = C/(1+r)^j \tag{9.6.11}$$

Thus, the *NPV* of a cost C to be paid each year for N years is:

$$NPV = \sum_{j=1}^{N} PV_i = \sum_{j=1}^{N} \frac{C}{(1+r)^j} \tag{9.6.12}$$

If the cost C is inflated at an annual rate i, the cost C_j in year j becomes:

$$C_j = C(1+i)^j \tag{9.6.13}$$

Thus, the net present value, *NPV*, becomes:

$$NPV = \sum_{j=1}^{N} \left(\frac{1+i}{1+r}\right)^j C \tag{9.6.14}$$

As will be discussed in the next section, *NPV* can be used as a measure of economic value when comparing investment options.

9.6.2.2 Life-cycle costing analysis evaluation criteria

In the beginning of this chapter, two economic parameters, simple payback and cost of energy, were introduced. These can be used for a preliminary economic analysis of a wind energy system. With LCC analysis, one can apply a number of other economic figures of merit or parameters to evaluate the feasibility of a wind energy system. Among others, these include net present value of cost or savings and the levelized cost of energy.

Net present value of cost or savings The net present value of a particular parameter is generally used as a measure of economic value when comparing different investment options in a life-cycle cost analysis. Note that it is important to specifically define the *NPV* clearly since numerous previous investigators have used the term 'net present value' to define a wide variety of life-cycle cost analysis parameters. First, one can define the savings version of net present value, NPV_S, as follows:

$$NPV_S = \sum_{j=1}^{N} \left(\frac{1+i}{1+r}\right)^j (S - C) \tag{9.6.15}$$

where S and C represent the yearly gross savings and costs during a project's lifetime.

In evaluating various systems using this criterion, one would look for the systems with the largest value of NPV_S. In practice, a spreadsheet format can be used to evaluate the yearly savings (S) and cost (C), and to calculate the sum of the annual levelized values. As will be shown next, however, it is also possible to use closed form analytical expressions for the direct calculation of net present value. These expressions are valuable tools for use in generalized parametric studies of wind energy systems.

If only cost factors are considered, then a cost version of net present value, NPV_C, may be used. It is the sum of the levelized costs of the energy system. For this version of this parameter, when comparing a number of different systems, the design with the lowest NPV_C is desired.

As an example of costs for wind energy systems, the net present value of costs can be found by calculating the total costs of a system for each year of lifetime. The annual costs should then be levelized to the initial year, and then summed up. If one assumes that both the general and energy inflation rates are constant over the system life, and that the system loan is repaid in equal installments, NPV_C may be found from the following equation:

$$NPV_C = P_d + P_a Y\left(\frac{1}{1+r},\ N\right) + C_c\, f_{OM}\, Y\left(\frac{1+i}{1+r},\ L\right) \tag{9.6.16}$$

where:

P_d = Down payment on system costs
P_a = Annual payment on system costs $= (C_c - P_d)CRF$
CRF = Capital recovery factor, based on the loan interest rate, b, rather than r
b = Loan interest rate
r = Discount rate
i = General inflation rate
N = Period of loan
L = Lifetime of system
C_c = Capital cost of system
f_{OM} = Annual operation and maintenance cost fraction (of system capital cost)

The variable $Y(k, \ell)$ is a function used to obtain the present value of a series of payments. It is determined from:

$$Y(k,\ell) = \sum_{j=1}^{\ell} k^j = \begin{cases} \dfrac{k - k^{\ell+1}}{1-k}, & \text{if } k \neq 1 \\ \ell, & \text{if } k = 1 \end{cases} \tag{9.6.17}$$

Levelized cost of energy In its most basic form, the levelized cost of energy, COE_L is given by the sum of annual levelized costs for a wind energy system divided by the annual energy production. Thus:

$$COE_L = \frac{\sum(\text{Levelized annual costs})}{\text{Annual energy production}} \tag{9.6.18}$$

This type of definition is generally used in a utility-based calculation for cost of energy. Sometimes, the levelized cost of energy is defined as the value of energy (units of \$/kWh) that, if held constant over the lifetime of the system, would result in a cost based net present value (such as the value calculated via Equation 9.6.16). Using this basis, the COE_L is given by:

$$COE_L = \frac{(NPV_C)(CRF)}{\text{Annual energy production}} \tag{9.6.19}$$

Note that the capital recovery factor, *CRF*, is here based on the lifetime of the system, *L*, and the discount rate, *r*. On this basis, the levelized cost of energy times the annual power production would equal the annual loan payment needed to amortize the net present value of the cost of the energy system.

Other life-cycle analysis economic parameters There are a number of other economic performance factors that can be use for evaluating the life-cycle based performance of an energy system. Two of the more common parameters are internal rate of return (*IRR*) and the benefit-cost ratio (*B/C*). They are defined by:

$$IRR = \text{Value of discount rate for } NPV_s \text{ to equal zero} \tag{9.6.20}$$

$$B/C = \frac{\text{Present value of all benefits (income)}}{\text{Present value of all costs}} \tag{9.6.21}$$

The *IRR* is often used by utilities or businesses in assessing investments and is a measure of profitability. The higher the *IRR*, the better the economic performance of the wind energy system in question. Two common forms of *IRR* exist (see Hunter and Elliot, 1994): (1) the financial internal rate of return, *FIRR*, and (2) the economic internal rate of return.

Generally systems with a benefit–cost ratio greater than one are acceptable, and higher values of the B/C value are desired.

9.6.3 Electric utility based economic analysis for wind farms

From a power generating business or utility perspective, the previous definition of cost of energy can be used for a first estimate of utility generation costs for a wind farm system. In this application, *COE* can be calculated from:

$$COE = \frac{[(C_c)(FCR) + C_{O\&M}]}{\text{Annual energy production}} \tag{9.6.22}$$

where C_c is the capital cost of the system, *FCR* is the fixed charge rate (a present value factor that includes utility debt and equity costs, state and federal taxes, insurance, and property taxes), and $C_{O\&M}$ is the annual operation and maintenance cost.

In the United States, electric utilities and the wind industry commonly use either of two methods to estimate the *COE* from a utility-sized wind energy system (Karas, 1992): (1) the Electric Power Research Institute, Technical Analysis Group (EPRI TAG) method (EPRI, 1979, 1989) or, (2) a cash flow method (CFM). Summary details of each follow.

9.6.3.1 EPRI TAG Method

This method produces a levelized cost of energy (*COE*), and in its simplest form for wind energy systems, *COE* (\$/kWh) is calculated from:

$$COE = FCR\left(\frac{\overline{C}_c}{8{,}760 \times CF}\right) + \overline{C}_{O\&M} \tag{9.6.23}$$

where:

$\overline{C}_c$ = Total cost of constructing the facility normalized by rated power (\$/kW)
CF = Capacity factor
$\overline{C}_{O\&M}$ = Cost of operation and maintenance normalized per unit of energy (\$/kWh)

Since this method produces a levelized energy cost it can be applied to a number of technologies, including conventional power plants (with the addition of fuel costs) for a useful comparison index. Some limitations of the EPRI TAG method include:

- It assumes a debt term life equal to the life of the power plant
- It does not readily allow for variable equity return, variable debt repayment, or variable costs.

The second limitation generally excludes the use of this method for independent power producers (IPPs), under which many wind farm projects fall.

9.6.3.2 Cash flow method

The cash flow method is based on the use of an accounting type spreadsheet that requires an annual input of estimated income and expenses over the lifetime of the project. The cost of energy is calculated via the following operations:

- Each cost component of the plant and operation is identified (e.g., plant construction, O&M, insurance, taxes, land, power transmission, administrative costs, etc.)
- A cost projection of each of the above components for each year of the plant's service life
- An annual estimate of depreciation, debt services, equity returns, and taxes
- Discounting of the resulting cash flows to present values using the utility's cost of capital (as established by appropriate regulatory agency)
- Levelizing the cost by determination of the equal payment stream (annuity) that has a present value equal to the results of the previous calculations

The cash flow method allows for the real variations that can be expected in cost, operational, and economic data, such as price increases, inflation, and changing interest rates.

Karas (1992) notes that the *COE* estimates can be determined in a wide variety of ways. His examples show that wind energy system *COE*s are extremely sensitive to actual plant cost and operating assumptions, especially capacity factor and plant installation costs.

Regardless of the method used to calculate a levelized or other type of *COE*, it should be emphasized that one plant can have a number of equally valid *COE*s depending on the assumptions used. Thus, it is important that in the determination of any cost of energy for a wind system, the terminology, methods of parameter determination, and assumptions be made explicit.

9.6.4 Economic performance sensitivity analysis

The previous sections have described a number of methods for the determination of economic performance parameters (such as cost of energy) that can be used to evaluate various wind systems, or to compare their performance with other types of power system. In practice, however, one soon realizes that a large number of machine performance and economic input variables exist that can influence the desired parameter's value. In addition, the potential variation or range of uncertainty of many input parameters may be large.

A complete economic study of a wind energy system should include a series of sensitivity studies for calculation of the output variable of interest. Specifically, each parameter of interest is varied around some central or 'best estimate'. The desired figure of merit (such as cost of energy) is the calculated. The calculations are repeated for a range of assumptions. For example, the following economic and performance variables could be varied:

- Plant capacity factor (or average wind speed at the site)
- Machine availability
- Operation and maintenance costs
- Capital or installed cost
- Lifetime of system and components
- Length of loan repayment
- Interest or discount rate

With calculations of this type, the results are often summarized graphically. This can be accomplished via the use of a 'spider' or 'star' diagram, illustrated in Figure 9.15. It is intended to show how a percentage change in a given parameter changes the cost of energy, expressed in percentage. This type of diagram readily shows the more important variables by lines with the steeper slopes. Note, however, that it may not be reasonable to vary all parameters by the same percentage amount, since some may be known quite accurately (Walker, 1993).

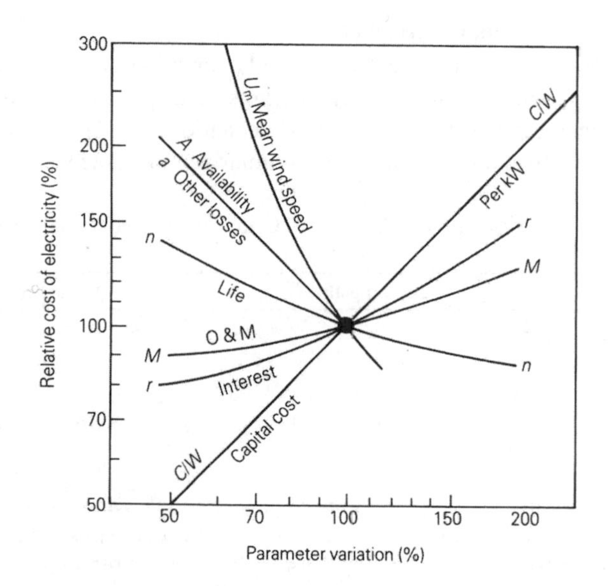

Figure 9.15 Sensitivity of cost of electricity to various parameters (Freris, 1990). Reproduced by permission of Pearson Education Limited

9.7 Wind Energy Market Considerations

9.7.1 Overview

There exists a large potential world wide market for wind energy systems. In this section a review of market considerations for wind energy systems is presented. There are three major issues to consider:

1. Potential market for wind systems
2. Barriers to expansion of markets
3. Incentives for market development

There are numerous technical publications on these subjects in the open literature. This is especially true for utility scale or wind farm applications (e.g., OECD/IEA, 1997, DOE/IEA, 1997, European Commission, 1998). The potential market for small or dispersed wind hybrid systems has also been addressed in recent studies (e.g., see WEC, 1993). A summary of some recent work on the major marketing issues (for utility scale systems) follows.

9.7.2 The market for utility scale wind systems

An overview of utility scale market projections from an international and US perspective is presented in this section.

9.7.2.1 International wind energy market

The World Energy Council carried out a market study for wind energy (WEC, 1993). In their work, a range of future wind energy markets and potential development was established via two scenarios: (1) 'current policies' scenario, and (2) 'ecologically driven' scenario. The assumptions for these scenarios are summarized in Table 9.9.

Table 9.9 Different assumptions in World Energy Council market scenarios (WEC, 1993)

Assumptions	Current policies	Ecologically driven
Growth rate of world electricity consumption	2.8% per year	2.0% per year
Oil price	20–25 U.S. $ per barrel in 2000 30–35 U.S. $ per barrel in 2020	20–25 U.S. $ per barrel in 2000 30–35 U.S. $ per barrel in 2020
Carbon tax	No	Yes, starts in 2000 up to $10 US per barrel in 2005 and thereafter
Fossil fuel plants increase in efficiency	15% in the period 1990–2020	15% in the period 1990–2020
Wind power generation costs	15% reduction in 1990–2010	30% reduction in 1990–2010 10% reduction in 2020–2030
Wind power investment costs	7.5% reduction in 1990–2010	15% reduction in 1990–2010 5% reduction in 2020–2030
Financial constraint	Total annual investment per country or region not more than 1% of GNP	Total annual investment per country or region not more than 3% of GNP
Wind energy penetration into the grid	Up to 2020 not more than 10% of electricity consumption	Up to 2020 not more than 10% of electricity consumption
Growth rate of manufacturing capacity world wide	Less than or equal to 15% per year	Less than or equal to 30% per year
Growth rate of installed wind turbine capacity	Less than or equal to 15% per year	Less than or equal to 30% per year

As an example of this work, Table 9.10 presents the 'ecologically driven' scenario results for electricity consumption and wind energy capacity for the eight International World Energy Council regions for 1987 and projected to 2020.

In this study, for each WEC region, the amount of wind turbine capacity and production was estimated using the previously shown assumptions and constraints for the two economic growth scenarios. For the 'ecologically driven' scenario, the total amount of wind capacity in operation projected for the year 2020 was projected to be about 470 GWe, which was assumed to produce about 970 TWh/yr. This represents about 5% of the world's estimated production of electrical energy in the year 2020. It should be noted that all these projections are speculative and subject to large potential errors. For example, at the time of writing of this text, the WEC's projections of installed wind turbine capacity for Western Europe have already been significantly exceeded.

Table 9.10 Electricity consumption and wind energy production by region in the 'ecologically driven' scenario (WEC, 1993)

WEC region	Electricity use 1987 (TWh)	Annual growth (%/yr)	Electricity use 2020 (TWh)	Potential wind capacity (GWe)	Projected wind capacity 2020 (GWe)	(TWh)
North America	2,990	1.5	4,887	2360	212	423
Latin America and Caribbean	525	3	1,530	1,005	11	21
Western Europe	2.101	1.5	3,435	635	78	165
Eastern Europe and CIS	2,046	2	3,932	2,295	54	120
Middle East and North Africa	271	4.8	1,302	930	12	25
Sub-Saharan Africa	210	2	404	2,170	9	19
Pacific	1972	2.2	4,067	975	80	160
Central and South Asia	271	3	718	80	18	34

Note: CIS, Commomwealth of Independent States; WEC, World Energy Council

9.7.2.2 United States wind energy market

Typical projections of market growth, along with past market history, are presented in the Annual Renewable Energy Reports published by the Energy Information Office of the U.S. Department of Energy (for example, DOE/IEA, 1997).

An example of a regionalized prediction of the wind energy market in the United States is given in the work of Schweizer et al. (1990). In this work a methodology and a computer model that projects the growth of the U.S. electric utility market for wind turbines on a regional basis are described. Their model is also based on the consideration of numerous barriers to wind energy deployment, an issue that will be discussed next. A summary of their market projection through the year 2030 is shown in Figure 9.16. As shown, they estimated the total U.S. wind turbine market to the year 2030 at approximately 21,500 MW. Table 9.11 summarizes their market penetration predictions on a regional basis.

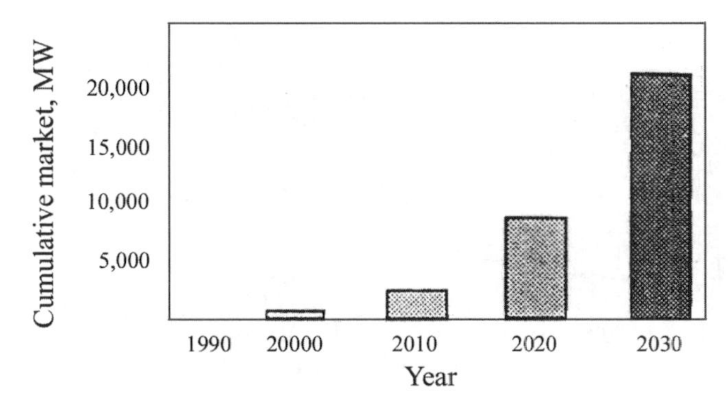

Figure 9.16 Predicted U.S. wind turbine market capture to 2030 (Schweizer et al., 1990); market growth only, current installed capacity not included

Table 9.11 Summary of regional market predictions through 2030 (Schweizer et al., 1990)

Region	Cumulative new market penetration (Rated MW of wind turbines)					Percentage of total regional baseload capacity in 2030	Year constrained by 10% of baseload	Year in which penetration constrained by 25% of available resource		
	1990	2000	2010	2020	2030			7.3 m/s	6.4 m/s	5.9 m/s
New England	0	15	210	970	2330	7.4				
Middle Atlantic	0	25	210	780	2140	2.9		2000	2020	
South Atlantic	0	15	300	990	2910	1.8		2020	2030	
East North Central	0	0	120	1190	3420	2.6		2000	2030	2030
West North Central	0	0	5	50	570	0.9		2010		
East South Central	0	0	55	170	350	0.4		2020	2020	2030
West South Central	0	400	1500	2500	3800	3.4		2010	2020	
Mountain	0	0	35	430	2250	3.3				
Pacific Contiguous	0	60	600	1760	3610	4.2*				
Pacific Non-contiguous	0	10	55	130	150	8.8*	2020			
U.S. Total	0	525	3090	8970	21530					

Note: Includes current installations, 5.8% for Pacific Contiguous region and 10.0% for Pacific Non-continuous region

On a US national basis, there have been numerous predictions of future market penetration for utility-scale wind energy systems. For example, market predictions for US wind energy have been presented by Hoffman and Garges (1996). They project the wind energy market to about 2050 and conclude that, as shown in Figure 9.17, wind energy may eventually contribute from 3% to 6% of the total electrical generation in the United States.

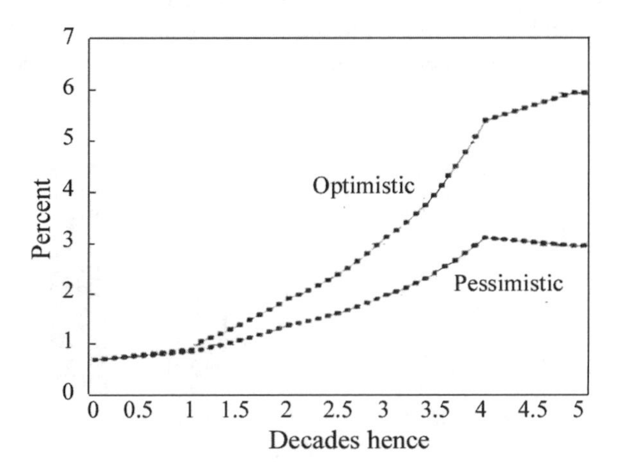

Figure 9.17 Predicted U.S. wind turbine market share (Hoffman and Garges, 1996). Reproduced by permission of Institution of Electrical and Electronic Engineers

9.7.3 Barriers to wind energy deployment

In spite of the potential for wind energy development, there is nonetheless a number of barriers that will slow the process. The barriers to the deployment of utility–scale wind energy can be divided into the following areas:

- Direct costs. A general disadvantage of wind turbines is that their total cost is highly concentrated at the point of initial construction as a result of high capital costs
- System integration. The integration of wind power into electric power systems is readily accomplished with existing engineering tools, but there may be major costs involved
- Regulatory and administrative processes. The responsibilities for approving the siting and other aspects of wind project implementation are usually assigned to existing bodies (i.e., local zoning boards) that have had little experience with wind power - with corresponding regulatory and administrative bottlenecks
- Environmental impacts. All wind energy projects have effects on the environment (as will be discussed in Chapter 10) and resulting barriers that must be overcome
- Electricity system planning. One barrier to the introduction of large-scale wind generation into some utility systems is their very novelty. Those utilities in particular may lack appropriate planning experience. Other utilities may already have wind generation in their service area, but have not yet addressed the issue of how much more wind generation they can absorb before they need to consider additional measures, such as forecasting or short-term energy storage

More details on these potential barriers to wind energy development can be found in an OECD/IEA (1997) publication. The author of this study conclude that barriers in the way of deployment of wind energy are comparatively low relative to market barriers facing some other renewable energy technologies. They also note that the key issue here remains the cost of wind energy relative to other energy sources.

9.7.4 Incentives to wind energy development

The largest amount of wind energy generation has been installed in countries that have offered various incentives to its market deployment. These have either been in the form of financial incentives (e.g., capital tax credits, production tax credits, or premium energy prices) or regulatory options (e.g., renewable portfolio standards and net metering). These are mechanisms that monetarize the environmental benefits of wind energy, as discussed in Section 9.5. Countries where such incentives have made the largest impact have included some in Europe (Denmark, Germany, Spain, the Netherlands, and the United Kingdom) and the United States. A detailed discussion of the various incentives for the first three countries is included in an OECD/IEA (1997) publication.

A number of national support programs have been implemented in Europe. Many different incentives have been used and the level of economic support has differed considerably from country to country and with time. Some of the most common incentives for market stimulation follow (European Commission, 1998):

- Public funds for R&D programs
- Public funds for demonstration projects
- Direct support of investment cost (% of total cost or based on kW installed)
- Support through the use of premium price of electricity from wind systems
- Financial incentives – special loans, favorable interest rates, etc.
- Tax incentives (e.g., favorable depreciation)
- The use of renewable energy or 'green' certificates

Regulatory and institutional measures carried out recently in Germany are of particular note. The German system has provided a framework which has resulted in a rapid increase in the amount of installed wind generation. Among other things, the German system mandates a guaranteed, above market rate for wind-generated electricity. See Gutermuth (2000) for more details on these measures.

In the US, most wind energy development occurred in the California wind farms in response to federal and state legislation that provided a market and favorable tax incentives that attracted private capital. Incentives for wind farms were accomplished via three separate, yet coincidental federal government actions and one act passed by the state of California. They were the Public Utilities Regulatory Policies Act (PURPA) of 1978, the Crude Oil Windfall Profits Act of 1980, the Economic Recovery Tax Act of 1981, and California state tax credits. These incentives were either in the form of investment or production tax credits.

At the present time, the major incentives for wind energy development in the United States are the production tax credits under the National Energy Policy Act and the planned implementation of renewable portfolio standards (RPSs). An RPS is a policy mechanism to insure a minimum level of renewable energy generation in the portfolios of power suppliers in an implementing jurisdiction (e.g., a state or nation). Rules for the implementation of renewable portfolio standards have been adopted by a number of states in the United States, several European countries and Australia (see Berry and Jaccard, 2001). One feature which is emerging as a way to track and account for renewable electrical generation is the renewable energy or 'green' certificate. These will be issued to generators in an amount proportional to production and are expected to be tradable on a commodities exchange. One state that has adopted a certificate based RPS is Massachusetts. Details can be found at http://www.state.ma.us/doer/rps/.

References

Berry, T., Jaccard, M. (2001) The Renewable Portfolio Standard: Design Considerations and an Implementation Survey, *Energy Policy*, 29, 263–277.

Billinton, R., Allen, R. (1984) *Reliability Evaluation of Power Systems*, Plenum Press, New York.

Boeing (1979) MOD-2 Wind Turbine System Concept and Preliminary Design Report, *Report No. DOE/NASA 0002-80/2,* Boeing Engineering and Construction Co.

British Wind Energy Association (1997) The Economics of Wind Energy, *Information Fact Sheet.*

Chapman, J., Wiese, S., DeMeo, E., Serchuk, A. (1998) Expanding Windpower: Can Americans Afford It? *Renewable Energy Policy Project, Report No. 6.*

Cody, G., Tiedje, T. (1996) A Learning Curve Approach to Projecting Cost and Performance in Thin Film Photovoltaics. *Proceedings of 25th Photovoltaic Specialists Conference*, IEEE, 1521–1524.

Danish Wind Turbine Manufacturers Association (1998) Guided Tour on Wind Energy, Internet: http://www.windpower.dk.

DOE/EPRI (US Department of Energy/Electric Power Research Institute) (1997) Renewable Energy Technology Characterizations. *EPRI Report: TR-109496*, EPRI.

DOE/IEA (1997) Renewable Energy Annual 1996, *DOE/IEA Report: 0603(96)*, April.

EPRI (Electric Power Research Institute) (1979) Technical Assessment Guide, Vols 1–3, *EPRI Report: EPRI-PS-1201-SR*, EPRI.

EPRI (Electric Power Research Institute) (1989) Technical Assessment Guide, Vol 1, Rev. 6, *EPRI Report: EPRI P-6587-L*, EPRI.

European Commission (1998), *Wind Energy – The Facts*.

Freris, L. L. (1990) *Wind Energy Conversion Systems*, Prentice Hall, New York.

General Electric Company (1984) MOD-5A Wind Turbine Generator Program Design Report- Vol II, Conceptual and Preliminary Design. *Report No. DOE/NASA/0153-2*, NTIS.

Gipe, P. (1995) *Wind Energy Comes of Age*, Wiley, New York.

Gutermuth, P-G. (2000) Regulatory and Institutional Measures by the State to Enhance the Deployment of Renewable Energies: German Experiences, *Solar Energy*, 69, No. 3, 205–213.

Harrison, R., Jenkins, G. (1993) Cost Modelling of Horizontal Axis Wind Turbines. *UK DTI Report ETSU W/34/00170/REP*.

Harrison, R., Hau, E., Snel, H. (2000) *Large Wind Turbines: Design and Economics*, Wiley, Chichester, UK.

Hau, E., Harrison, R., Snel, H., Cockerill, T. T. (1996) Conceptual Design and Costs of Large Wind Turbines, *Proc. EWEC 96*, 128–131.

Hau, E., Langenbrinck, J., Paltz, W. (1993) *WEGA Large Turbines*. Springer-Verlag, Berlin.

Hoffman, G., Garges, K. (1996) Forecasting the U.S. Market for Windpower, *Proc. 1996 IECEC*, 1767–1772.

Hohmeyer, O. H. (1990) Latest Results of the International Discussion on the Social Costs of Energy – How does Wind Compare Today? *Proc. 1990 EWEC*, 718–724.

Hunter, R., Elliot, G. (1994) *Wind–Diesel Systems*, Cambridge University Press, Cambridge.

Johnson, G. L. (1985) *Wind Energy Systems*, Prentice-Hall, Englewood Cliffs, N.J.

Karas, K. C. (1992) Wind Energy: What Does It Really Cost? *Proc. Windpower '92*, AWEA, 157–166.

Klein, H., Herbstritt, M., Honka, M. (1994) Milestones of European Wind Energy Development – EUROWIN in Western, Central, and Eastern Europe. *Proc. EWEC*, 1275–1279.

Lemming, J., Morthorst, P. E., Hansen, L. H., Andersen, P., Jensen, P. H. (1999) O&M Costs and Economical Life-time of Wind Turbines. *Proc. European Wind Energy Conf.*, 387–390.

Lynnette, R. (1986) Wind Power Stations: 1985 Performance and Reliability, *EPRI Report AP-4639*, EPRI, June.

NASA (1979) 200 kW Wind Turbine Generator Conceptual Design Study. *Report No. DOE/NASA/1028-79/1*, January, Lewis Research Center.

Nath, C. (1998) Maintenance Cost of Wind Energy Conversion Systems. Germanischer Lloyd, Internet: http//www.germanlloyd.de/Activities/Wind/public/mainten.html.

National Wind Coordinating Committee (1997) Wind Energy Costs. *Wind Energy Series Report No. 11*.

Neij, L. (1999) Cost Dynamics of Wind Power. *Energy*, 24, 375–389.

OECD/IEA (1997) *Enhancing the Market Deployment of Energy Technology: A Survey of Eight Technologies*, OECD, Paris.

Perez, R., Seals, R., Wenger, H., Hoff, T., Herig, C. (1997) Photovoltaics as a Long-term Solution to Power Outages Case Study: The Great 1996 WSCC Power Outage. *Proceedings of the 1997 Annual Conference*, American Solar Energy Society, Washington, DC, 309–314.

Rabl, A. (1985) *Active Solar Collectors and Their Applications*. Oxford University Press, Oxford.

Riggs, J. L. (1982) *Engineering Economics*, McGraw-Hill, New York.

Schweizer, T. C., Johnson, B. L., Cohen, J. M. (1990) Regional Perspectives on Wind Turbine Market Penetration in the U.S. *Proc. Windpower '90*, AWEA, 59–64.

Spera, D. (ed.) (1994) *Wind Turbine Technology*, ASME, New York.

Tande, J.O.G., Hansen, J. C. (1991) Determination of the Wind Power Capacity Value. *Proc. EWEC '91,* 643–648.

Vachon, W. A. (1996) Modeling the Reliability and Maintenance Costs of Wind Turbines Using Weibull Analysis. *Proc. Wind Power 96*, AWEA, 173–182.

Walker, J. F., Jenkins, N. (1997) *Wind Energy Technology*, Wiley, Chichester, UK.

Walker, S. (1993) Cost and Resource Estimating, *Open University Renewable Energy Course Notes - T521 Renewable Energy*.

WEC (World Energy Council) (1993) *Renewable Energy Resources: Opportunities and Constraints 1990–2020*, World Energy Council, London.

10

Wind Energy Systems: Environmental Aspects and Impacts

10.1 Introduction

This chapter will review the environmental aspects associated with the deployment of a single wind turbine or a wind farm. Wind energy development has both positive and negative environmental impacts. On the positive side, wind energy is generally regarded as environmentally friendly, especially when the environmental effects of emissions from large-scale conventional electrical generation power plants are considered. For example, estimates of the stack emissions (sulfur and nitrogen oxides, particulates, and carbon dioxide) of conventional coal and gas-fired power plants as compared to those of wind systems (zero in all cases) are shown in Table 10.1.

Table 10.1 Stack emissions of coal, gas, and wind power plants (kg/MWh)

Pollutant	Conventional coal w/ controls	Conventional gas w/ controls	Wind
Sulfur oxides	1.2	0.004	0
Nitrogen oxides	2.3	0.002	0
Particulates	0.8	0.0	0
Carbon dioxide	865	650	0

The stack (or direct) emissions of wind systems are essentially zero, although there are indirect emissions associated with the actual production of wind turbines and the erection and construction of wind turbine systems and wind farms. Estimates of indirect emissions from wind systems (in Germany) are presented in the review paper of Ackerman and Soder (2000), where the emissions values are shown to be generally small (one or two orders of magnitude less than those of conventional power plants).

The determination of the overall cost to society of the various emission pollutants is difficult to measure and open to much debate. In general, the environmental benefits of wind power are calculated by the avoided emissions from other sources. These benefits generally come from the displacement of generation (MWh) as opposed to capacity (MW). The displacement of capacity does have its benefits, however, especially if extension of the fuel and water supply infrastructure can be avoided (Connors, 1996).

As more wind turbines and wind farm systems have been planned or installed in the United States and Europe, the importance of their environmental effects has increased. This is well documented in many publications on the positive environmental aspects of wind power. However, some negative environmental aspects have been shown to be especially important in populated or scenic areas. Here, a number of potential wind energy projects have been delayed or denied permits because of strong environmental opposition based on some potential negative environmental effects of these systems.

Keeping the positive aspects in mind, this chapter will review the potential negative environmental aspects associated with the deployment of a single wind turbine or wind farm. It should be noted that it ought to be one of the objectives of wind energy system engineers to maximize the positive impacts while, at the same time, to minimize the negative aspects. In addition to their importance in the siting process for wind systems, the potentially adverse environmental effects of wind turbines tie in directly with much of the wind turbine design material discussed in previous chapters. With appropriate design of the wind turbine and the wind energy system, these adverse environmental factors can be minimized.

The potential negative impacts of wind energy can be divided into the following categories:

- Avian interaction with wind turbines
- Visual impact of wind turbines
- Wind turbine noise
- Electromagnetic interference effects of wind turbines
- Land-use impact of wind power systems
- Other impact considerations

At the present time, the first three topics include most of the major environmental issues affecting wind system deployment, but the other topics are also important and should be addressed. In the next sections, these potential impacts will be reviewed, including the following general topic areas of each environmental impact:

- Problem definition
- Source of the problem
- Quantification or measurement of the problem
- Environmental assessment examples
- Reference resources or tools that address the problem

Many of these problems have to be addressed in the permitting phases of a wind project. Furthermore, depending on the particular site's environmental assessment regulations, one or more of these areas may require detailed investigation and/or an environmental impact statement. The scope of this chapter does not permit a full summary of the various regulations and laws on this subject, but, where possible, the reader will be referred to the appropriate literature on the subject.

10.2 Avian Interaction with Wind Turbines

10.2.1 Overview of the problem

The environmental problems associated with avian interaction and wind systems surfaced in the United States in the late 1980s. For example, it was found that birds, especially federally protected golden eagles and red-tailed hawks, were being killed by wind turbines and high voltage transmission lines in wind farms in California's Altamont pass. This information caused opposition to the Altamont Pass project among many environmental activists and aroused the concern of the US Fish and Wildlife Service, which is responsible for enforcing federal species protection laws.

There are two primary concerns related to this environmental issue: (1) effects on bird populations from the deaths caused by wind turbines, and (2) violations of the Migratory Bird Treaty Act, and/or the Endangered Species Act. These concerns, however, are not confined to the United States alone. Bird kill problems have surfaced at other locations in the world. For example, in Europe, major bird kills have been reported in Tarifa, Spain (a major point for bird migration across the Mediterranean Sea), and at some wind plants in Northern Europe. Wind energy development can adversely affect birds in the following adverse manners (Colson, 1995):

- Bird electrocution and collision mortality
- Change to bird foraging habits
- Alteration of migration habits
- Reduction of available habitat
- Disturbance of breeding, nesting, and foraging

It should also be pointed out that the same author states that wind energy development has the following beneficial effects on birds:

- Protection of land from more dramatic habitat loss
- Provision of perch sites for roosting and hunting
- Provision and protection of nest sites on towers and ancillary facilities
- Protection or expansion of prey base
- Protection of birds from indiscriminate harassment

At the present time, the long-term implications of the bird issue on the wind industry are not clear. Problems may arise in areas where large numbers of birds congregate or migrate, as in Tarifa, or where endangered species are affected, as in Altamont Pass (NWCC, 1998). This could include a large number of locations, however, since some of the traits that characterize a good wind site also happen to be attractive to birds. For example, mountain passes are frequently windy because they provide a wind channel through a mountain range. For the same reason, many times they are the preferred routes for migratory birds.

The key factors of this important environmental impact area will be reviewed next. This includes the characterization of the problem, mitigation concepts, and available resources that can be used to assess the problem.

10.2.2 Characterization of the problem

There is a close correlation between a locality and its bird fauna. Many bird species are very habitat-specific and often particularly sensitive to habitat changes. Furthermore, there is a correspondingly close correlation between the site and the location of the wind turbines, which is especially dependent on the wind conditions. In European (Clausager and Nohr, 1996) and US studies (Orloff and Flannery, 1992) the impact on birds is generally divided into the following two categories: (1) direct impacts, including risk of collision, and (2) indirect impacts, including other disturbances from wind turbines (e.g., noise). Such effects may result in a total or partial displacement of the birds from their habitats and deterioration or destruction of the habitats.

The risk of collision is the most obvious direct effect, and numerous studies have focused on it. These studies include the estimation of bird numbers that collide with the rotor or associated structures, and they have concentrated on the development of methods for the analysis of the extent of the collisions. Indirect effects include:

- Disturbance of breeding birds
- Disturbance of staging and foraging birds
- Disturbing impact on migrating and flying birds

10.2.3 Current mitigation concepts and case studies

10.2.3.1 Mitigation concepts

A number of recent studies have specifically addressed the subject of measures for the minimization of the impact of wind turbine systems on birds. The most detailed work has taken place in California and has been supported by the California Energy Commission and the American Wind Energy Association. Typical mitigation measures from some of these studies (Colson, 1995; and Wolf, 1995) include the following:

- *Avoid migration corridors*. Major bird migration corridors and areas of high bird concentrations should generally be avoided when siting wind facilities unless there is evidence that because of local flight patterns, low bird usage of the specific area, or other factors, the risk of bird mortality is low
- *Fewer, larger turbines*. For a desired energy capacity, fewer large turbines may be preferred over many smaller turbines to reduce the number of structures in the wind farm
- *Avoid micro habitats*. Microhabitats or fly zones should be avoided when siting individual wind turbines.
- *Alternate tower designs*. Where feasible, tower designs that offer few or no perch sites should be employed in preference to lattice towers with horizontal members that are

suitable for perching. Unguyed meteorological and electrical distribution poles should be used over guyed structures. Also, existing lattice towers should be modified to reduce perching opportunities

- *Remove nests.* With agency approvals, raptor nests found on structures should be moved to suitable habitat away from wind energy facilities
- *Prey base management.* Where appropriate, prey base management should be investigated as an option. A humane program, such as live trappings should be established to remove unwanted prey from existing wind farms
- *Bury electrical lines.* Electrical utility lines should be underground when feasible. New overhead electrical distribution systems should be designed to prevent bird electrocutions. Existing facilities should employ techniques that significantly reduce bird electrocutions
- *Site specific mitigation studies.* The cause and effect of bird interactions with wind energy facilities should be examined to determine the manner in which collisions occur and appropriate mitigation identified and employed
- *Conservation of alternative habitats.* Habitat conditions that will benefit affected species should be maintained to protect them from other more damaging land-uses

10.2.3.2 Case studies

Numerous technical reports and papers exist that have focused on the potential or actual environmental effects of wind turbine systems on birds at a specific site. Although it is beyond the scope of this text to review them in detail, a few of the more appropriate ones (and their particular applications) are listed below:

- *California wind farm studies.* A review of this subject, as applied to the wind farms in Altamont Pass and Solano County, is contained in a technical report by the California Energy Commission (Orloff and Flannery, 1992).
- *Recent US studies.* Sinclair (1999) and Sinclair and Morrison (1997) give an overview of the recent US avian research program through 1999. Also, Goodman (1997) presents a summary of the avian studies (and other environmental issues) carried out for a 6 MW wind farm in Vermont.
- *European studies.* Gipe (1995) presents an overview of bird-related environmental impact studies in Europe up to 1995. In addition, representative examples of European environment impact studies of birds (especially for Denmark and the U.K.) are given in the technical papers of Clausager and Nohr (1996) and Still et al. (1996).

10.2.4 Resources for environmental assessment studies of avian impacts

A review of the previously referenced case studies regarding the environmental impacts of wind systems on birds reveals that such studies can require specialized expertise, be very detailed, and can add to the cost and deployment time of a potential wind site. As mentioned previously, a representative example of this process is given in the technical report of the California Energy Commission (Orloff and Flannery, 1992). In general, this type of work can be divided into two parts: (1) a complete definition of the study area, and (2) an assessment of bird risk.

The first part primarily consists of a detailed definition of the site topography and machine siting layout of the wind farm. In addition, when different types of wind turbines and towers are used, it is important to group turbines into different categories. For example, Figure 10.1 illustrates the eight categories of wind turbines that were installed at the Altamont Pass in California in the 1980s. Another part of this phase of the work involves the selection of sample sites where detailed bird – wind turbine interaction data are to be collected.

The second, and much more detailed part of this type of study, the assessment of bird risk, generally involves a comprehensive methodology for accomplishing this task. In the United States, a number of sponsors including the National Renewable Energy Laboratory (NREL) and the National Wind Coordinating Committee (NWCC) Avian subcommittee have recently worked to develop this type of methodology. This work has focused on the development of a standardized method for determination of the factors responsible for avian deaths from wind turbine facilities, and scientific methods that can be used to reduce fatalities. The following summary presents some of the important methods, measurements, and relationships (metrics) that have been used for this type of work (Anderson et al., 1997; NWCC, 1999):

Bird utilization counts. Under this part of a study, an observer notes the location, behavior, and number of birds using an area. This is done in repeatable ways, using standard methods, so that results can be compared with bird utilization counts from other studies. Bird behaviors to be noted include: flying, perching, soaring, hunting, foraging, and actions close (50 meters or less) to wind farm structures.

Bird utilization rate. The bird utilization rate is defined as the number of birds using the area during a given time. It is based on bird utilization count. Thus:

$$\text{Bird utilization rate} = \frac{\text{No. of birds observed}}{\text{Time}} \tag{10.2.1a}$$

or

$$\text{Bird utilization rate} = \frac{\text{No. of birds observed}}{\text{Time x area}} \tag{10.2.1b}$$

Bird mortality. The bird mortality is defined as the number of observed deaths, per unit area (again, this is based on the bird utilization count). Therefore:

$$\text{Bird mortality} = \frac{\text{No. of dead birds}}{\text{Defined search area}} \tag{10.2.2}$$

Bird risk. Bird risk is a measure of the likelihood that a bird using the area in question will be killed. It is defined as follows:

$$\text{Bird risk} = \frac{\text{Bird mortality}}{\text{Bird utilization rate}} = \frac{\text{No. of dead birds/defined area}}{\text{No. of birds observed/time}} \tag{10.2.3}$$

Bird Risk can be used to compare risk differences for many different variables: i.e., distances from wind facilities; species, type, and all birds observed; seasons; and turbine structure types. It can be used to compare risks between wind resource areas and with other types of facilities such as highways, power lines, and TV and radio transmission towers.

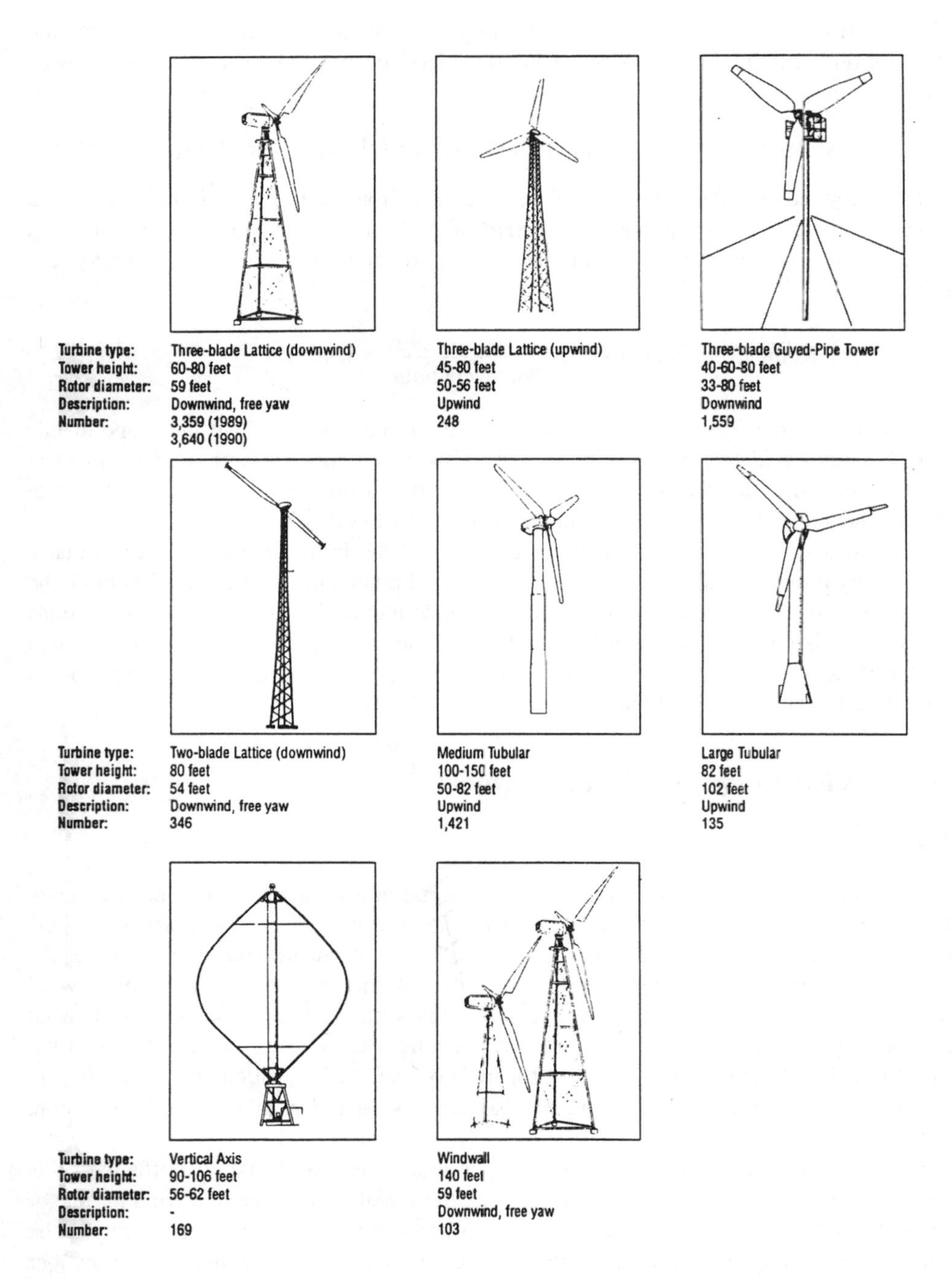

Figure 10.1 Eight categories of wind turbines in the Altamont pass (Orloff and Flannery, 1992)

Rotor swept area metrics. Under this category two measurements that consider the effects of different wind turbine sizes and designs have been defined. The first, rotor swept hours is defined as:

$$\text{Rotor swept hours} = [\text{Rotor swept area } (m^2)] \times (\text{Hours of operation}) \quad (10.2.4)$$

This parameter combines the size of the area of the rotor with the time it operates. The second parameter, rotor swept hour risk (RSHR) allows for a comparison of risks associated with different rotor swept areas, or turbine sizes, in relation to the time they operate. It is defined by:

$$\text{Rotor swept hour risk} = \frac{1}{\text{Rotor swept hours}} \times \text{Bird risk} \quad (10.2.5)$$

A detailed discussion of the example use of these metrics is beyond the scope of this work. It is expected that many of them will be used in future research studies aimed at developing methods to measure the risk to birds not only from wind systems, but also from other human created hazards such as buildings and highways (NWCC, 1998).

In summary, it should be pointed out that even if the initial research indicates that a wind energy project is unlikely to seriously affect bird populations, further studies might be needed to verify this conclusion. These could include monitoring baseline bird populations and behavior before the project begins, then simultaneously observing both a control area and the wind site during construction and initial operation. In certain cases, operational monitoring might have to continue for years.

10.3 Visual Impact of Wind Turbines

10.3.1 Overview

One of wind power's perceived adverse environmental impact factors, and a major concern of the public, is its visibility (Gipe, 1995). Compared to the other environmental impacts associated with wind power, the visual impact is the least quantifiable. For example, the public's perceptions may change with knowledge of the technology, location of wind turbines, and many other factors. Although the assessment of a landscape is somewhat subjective, professionals working in this area are trained to make judgments on visual impact based on their knowledge of the properties of visual composition and by identifying elements such as visual clarity, harmony, balance, focus, order, and hierarchy (Stanton, 1995).

Wind turbines need to be sited in well exposed sites in order to be cost effective. It is also important for a wind engineer to realize that the visual appearance of a wind turbine or a wind farm must be considered in the design process at an early stage. For example, the degree of visual impact is influenced by such factors as the type of landscape, the number and design of turbines, the pattern of their arrangement, their color, and the number of blades.

10.3.2 Characterization of the problem

Visual or aesthetic resources refer to the natural and cultural features of an environmental setting that are of visual interest to the public. An assessment of a wind project's visual compatibility with the character of the project setting can be based on a comparison of the setting and surrounding features with simulated views of the proposed project. In this light, the following parameters and questions can be considered (NWCC, 1998):

- Viewshed alteration: will the project substantially alter the existing project setting (generally referred to as the 'viewshed'), including any changes in the natural terrain or landscape?
- Viewshed consistency: will the project deviate substantially from the form, line, color, and texture of existing elements of the viewshed that contribute to its visual quality?
- Viewshed degradation: will the project substantially degrade the visual quality of the viewshed, affect the use or visual experience of the area, or intrude upon or block views of valuable visual resources?
- Conflict with public preference: will the project be in conflict with directly identified public preferences regarding visual and environmental resources?
- Guideline compatibility: will the project comply with local goals, policies, designations, or guidelines related to visual quality?

The scope of a visual impact assessment depends on whether a single or a number of wind turbines are to be sited in a particular location. Visual impact is not directly proportional to the number of turbines in a wind farm development. It will, however, vary a great deal between a single turbine and a wind farm. That is, a single wind turbine has only a visual relationship between itself and the landscape, but a wind farm has a visual relationship between each turbine as well as with the landscape (Stanton, 1995).

An important task here is the characterization of the location where the proposed wind system is to be sited. As one example of this process, the work of Stanton (1994) as applied to wind farms in the U.K uses the design principles shown in Figure 10.2. It should be noted that although this visual impact study is based on widely known and accepted principles (such as those summarized in Figure 10.2), the judgements made are subjective in nature.

10.3.3 Design of wind systems to minimize visual impact

In the United States and Europe, there are numerous publications that suggest potential designs for wind systems that minimize visual impact (e.g., see Ratto and Solari, 1998). In many cases, the subjective nature of this topic appears and there are major differences of opinion between researchers. An example of this problem includes the choice of turbine colors (Aubrey, 2000). Two examples of previous design strategies (one from the United States and the other European) for the reduction of the visual impact of wind energy systems follow.

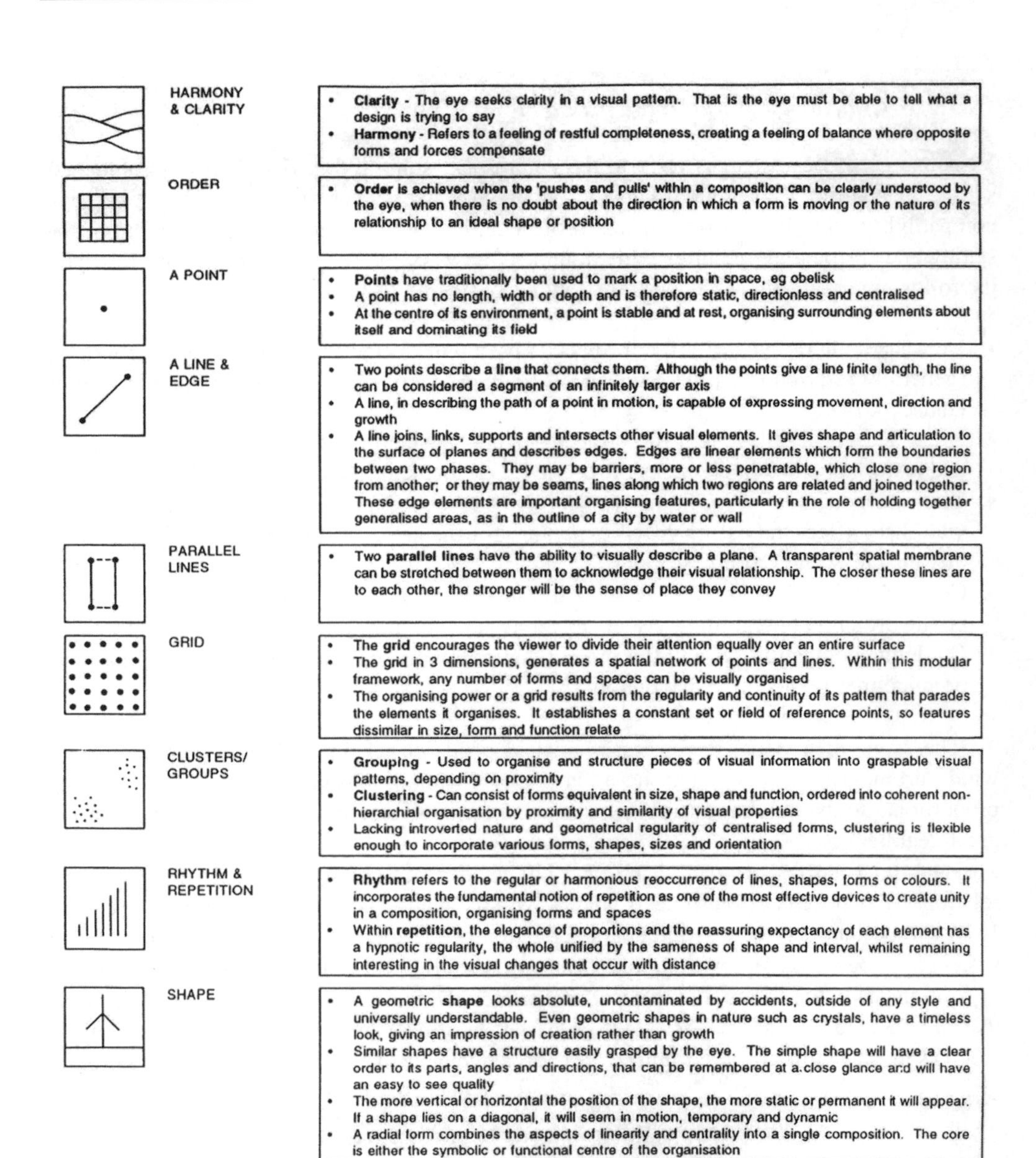

Figure 10.2 Summary of fundamental design principles significant to wind farm development (Stanton, 1994). Reproduced by permission of Mechanical Engineering Publications Ltd.

10.3.3.1 Visual impact mitigation design strategies for US sites

From work carried out in the US (NWCC, 1998), a number of design strategies for reducing the environmental impact of wind systems have been proposed. They include the following:

- Using the local land form to minimize visibility of access and service roads, and to protect land from erosion
- Use of low profile and unobtrusive building designs to minimize the urbanized appearance or industrial character of projects located in rural or remote regions
- Use of uniform color, structure types, and surface finishes to minimize project visibility in sensitive areas with high open spaces. Note, however, that the use of non-obtrusive designs and colors may conflict with efforts to reduce avian collisions and may be in direct conflict with aircraft safety requirements for distinctive markings
- Selecting the route and type of support structures for above-ground electrical facilities as well as the method, mode, and type of installation (below- vs. above-ground). Where multiple generation units are to be sited close together, consolidation of electrical lines and roads into a single right of way, trench, or corridor will cause fewer impacts than providing separate access to each unit
- Controlling the placement and limiting the size, color, and number of label markings placed on individual turbines or advertising signs on fences and facilities
- Prohibiting lighting, except where required for aircraft safety, prevents light pollution in otherwise dark settings. This may incidentally minimize collision by nocturnal feeders which prey on insects attracted to lights
- Controlling the relative location of different turbine types, densities, and layout geometry to minimize visual impacts and conflicts. Different turbine types and those with opposing rotation can be segregated by buffer zones. Mixing of types should be avoided or minimized

10.3.3.2 European wind farm visual impact design characteristics

The assessment of the visual impact of a proposed wind farm development is an important design step in recent European work. One example of this type of work is summarized by the following list of wind farm design characteristics quoted from Stanton's (1994, 1995) U.K. studies:

- Wind turbine form
- Blade number
- Turbine nacelle and tower
- Turbine size
- Wind farm size
- Spacing and layout of wind turbines
- Color

Note that this work complements Stanton's previously described landscape character types. Again, the subjective nature of this example should be emphasized.

10.3.4 Resources for visual impact studies

In both the US and Europe there are numerous references and handbooks where one can find design guidelines for minimizing the visual environmental impact of wind turbines and wind farms (e.g., see Ratto and Solari, 1998). A summary of some of the most representative examples follows.

10.3.4.1 US visual impact resources

A good resource on the subject is contained in a handbook for the permitting of wind energy facilities (NWCC, 1998). Some of the guidelines developed in this handbook have been previously discussed. In addition, this reference contains much additional information of potential use in this and many other phases of environmental impact assessment. For example, the authors point out that a valuable process tool for the assessment of potential project impacts to sensitive visual resources is the preparation and use of visual simulations.

Goodman (1997) gives another good source of information on visual resources. This paper documents the environmental impacts of a wind farm installation in rural Vermont, and summarizes the methods used to minimize the environmental impact of a 6 MW wind farm.

10.3.4.2 European visual impact resources

As discussed in numerous publications on the subject, the visual impact of wind turbines and wind farms have been under significant study in many European countries. A great deal of this work has been carried out in the U.K. For example, the visual impact assessment work of Stanton (1994, 1995) has already been mentioned.

In Europe, a number of investigators have developed some very sophisticated and useful techniques that can be used to illustrate the visual intrusion of a potential wind farm installation. Furthermore, many of these have been successfully used in actual wind farm development projects (see Ratto and Solari, 1998). For example, one uses digital computer techniques based on topographical information and wind turbine design characteristics. With this type of technique it is possible to plot 'zones of visual impact' on a map to illustrate the locations where the development might be seen. One disadvantage to this technique is that it takes no account of local screening, e.g., by buildings or trees, so it is a worst case scenario. An example of the use of this type of technique, using geographical information systems (GIS), is given by Kidner (1996).

Another method (see Taylor and Durie, 1995) is based on the use of photomontages. Here, use is made of panoramic photographs taken from certain key locations of varying distances from the proposed site, and superimposing suitably scaled wind turbines in the photographs. A limitation to this technique is that the visibility of a wind system will change with time of day, season, and with certain weather conditions – especially when the turbine blades are rotating. A third method attempts to overcome some of these disadvantages by creating a video montage of a proposed site and then super-imposing rotating wind turbines on the scene (Robotham, 1992). While this creates a most effective visual presentation, its major disadvantage is its high cost compared to the other two methods. Today, a number of commercial software packages now include all or some of these capabilities.

10.4 Wind Turbine Noise

10.4.1 Overview of problem

The problems associated with wind turbine noise have certainly been one of the more studied environmental impact areas in wind energy engineering. Noise levels can be measured, but, as with other environmental concerns, the public's perception of the noise impact of wind turbines is a partly subjective determination.

Noise is defined as any unwanted sound. Concerns about noise depend on the level of intensity, frequency, frequency distribution and patterns of the noise source; background noise levels; terrain between emitter and receptor; and the nature of the noise receptor. The effects of noise on people is classified into three general categories (NWCC, 1998):

- Subjective effects including annoyance, nuisance, dissatisfaction
- Interference with activities such as speech, sleep, and learning
- Physiological effects such as anxiety, tinnitus (ringing in ear), or hearing loss

In almost all cases, the sound levels associated with environmental noise produce effects only in the first two categories. Workers in industrial plants and those who work around aircraft can experience noise effects in the third category. Whether a noise is objectionable will depend on the type of noise (see the next section) and the circumstances and sensitivity of the person (or receptor) who hears it. Mainly because of the wide variation in the levels of individual tolerance for noise, there is no completely satisfactory way to measure the subjective effects of noise, or of the corresponding reactions of annoyance and dissatisfaction.

Operating noise produced from wind turbines is considerably different in level and nature than most large-scale power plants, which can be classified as industrial sources. Wind turbines are often sited in rural or remote areas that have a corresponding ambient noise character. Furthermore, while noise may be a concern to the public living near wind turbines, much of the noise emitted from the turbines is masked by ambient or the background noise of the wind itself.

The noise produced by wind turbines has diminished as the technology has improved. For example, with improvements in blade airfoils and turbine operating strategy, more of the wind energy is converted into rotational energy, and less into acoustic noise. Even a well designed turbine, however, can generate some noise from the gearbox, brake, hydraulic components or even electronic devices.

The significant factors relevant to the potential environmental impact of wind turbine noise are shown in Figure 10.3 (Hubbard and Shepherd, 1990). Note that this technology is based on the following primary elements: noise sources, propagation paths, and receivers. In the following sections, after a short summary of the basic principles of sound and its measurement, a review of noise generation from wind turbines, its prediction, and propagation, as well as noise reduction methods are given.

10.4.2 Noise and sound fundamentals

10.4.2.1 Characteristics of sound and noise

Sound is generated by numerous mechanisms and is always associated with rapid small-scale pressure fluctuations (which produce sensations in the human ear). Sound waves are characterized in terms of their wavelength, λ, frequency, f, and velocity u, where u is found from:

$$u = f\lambda \tag{10.4.1}$$

The velocity of sound is a function of the medium through which it travels, and it generally travels faster in denser media. The velocity of sound is about 340 m/s in atmospheric air. Sound frequency determines the 'note' or pitch that one hears, which, in many cases, corresponds to notes on the musical scale (Middle C is 262 Hz).

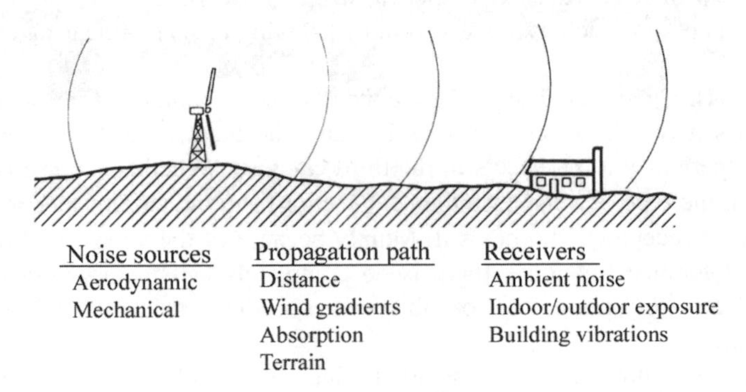

Figure 10.3 Wind turbine noise assessment factors (Hubbard and Shepherd, 1990)

An octave denotes the frequency range between sound with one frequency and one with twice that frequency. The human hearing frequency range is quite wide, generally ranging from about 20 Hz to 20 kHz. Sounds experienced in daily life are usually not a single frequency, but are formed from a mixture of numerous frequencies from numerous sources.

Sound turns into noise when it is unwanted. Whether sound is perceived as a noise depends on the response to subjective factors such as the level and duration of the sound. There are numerous physical quantities that have been defined which enables sounds to be compared and classified, and which also give indications for the human perception of sound. They are discussed in numerous texts on the subject (e.g., for wind turbine noise see Wagner et al., 1996) and are reviewed in the following sections.

10.4.2.2 Sound power and pressure measurement scales

It is important to distinguish between *sound power level* and *sound pressure level*. Sound power level is a property of the source of the sound and it gives the total acoustic power emitted by the source. Sound pressure level is a property of sound at a given observer location and can be measured there by a single microphone. In practice, the magnitude of an acoustical quantity is given in logarithmic form, expressed as a level in decibels (dB) above or below a zero reference level. For example, using conventional notation, a 0 dB sound power level will yield a 0 dB sound pressure level at a distance of 1 m.

Because of the wide range of sound pressures to which the ear responds (a ratio of 10^5 or more for a normal person), sound pressure is an inconvenient quantity to use in graphs and tables. In addition, the human ear does not respond linearly to the amplitude of sound pressure, and, to approximate it, the scale used to characterize the sound power or pressure amplitude of sound is logarithmic (see Beranek and Ver, 1992).

The sound power level of a source, L_W, in units of decibels (dB), is given by:

$$L_w = 10 \log_{10}(W/W_0) \tag{10.4.2}$$

where W is the source sound power and W_0 is a reference sound power (usually 10^{-12} W).

The sound pressure level of a noise, L_p , in units of decibels (dB), is given by:

$$L_P = 20 \log_{10}(p/p_0) \tag{10.4.3}$$

where p is the instantaneous sound pressure and p_0 is a reference sound pressure (usually 2×10^{-5} Pa).

Figure 10.4 gives some examples for various sound pressure levels on the decibel scale. The threshold of pain for the human ear is about 200 Pa, which corresponds to a sound pressure level of 140 dB.

10.4.2.3 Measurement of sound or noise

Sound pressure levels are measured via the use of sound level meters. These devices make use of a microphone, which converts pressure variations to a voltage signal, which is then recorded on a meter (calibrated in decibels). The decibel scale is logarithmic and has the following characteristics (NWCC, 1998):

- Except under laboratory conditions, a change in sound level of 1 dB cannot be perceived
- Outside of the laboratory, a 3 dB change in sound level is considered a barely discernible difference
- A change in sound level of 5 dB will typically result in a noticeable community response.
- A 10 dB increase is subjectively heard as an approximate doubling in loudness, and almost always causes an adverse community response

A sound level measurement that combines all frequencies into a single weighted reading is defined as a broadband sound level (see Section 10.4.3). For the determination of the human ear's response to changes in noise, sound level meters are generally equipped with filters that give less weight to the lower frequencies. As shown in Figure 10.5, there are a number of filters (referenced to A, B, and C) that accomplish this. The most common scale used for environmental noise assessment is the A scale. Measurements made using this filter are expressed in units of dB(A). Beranek and Ver (1992) discuss details of these scales.

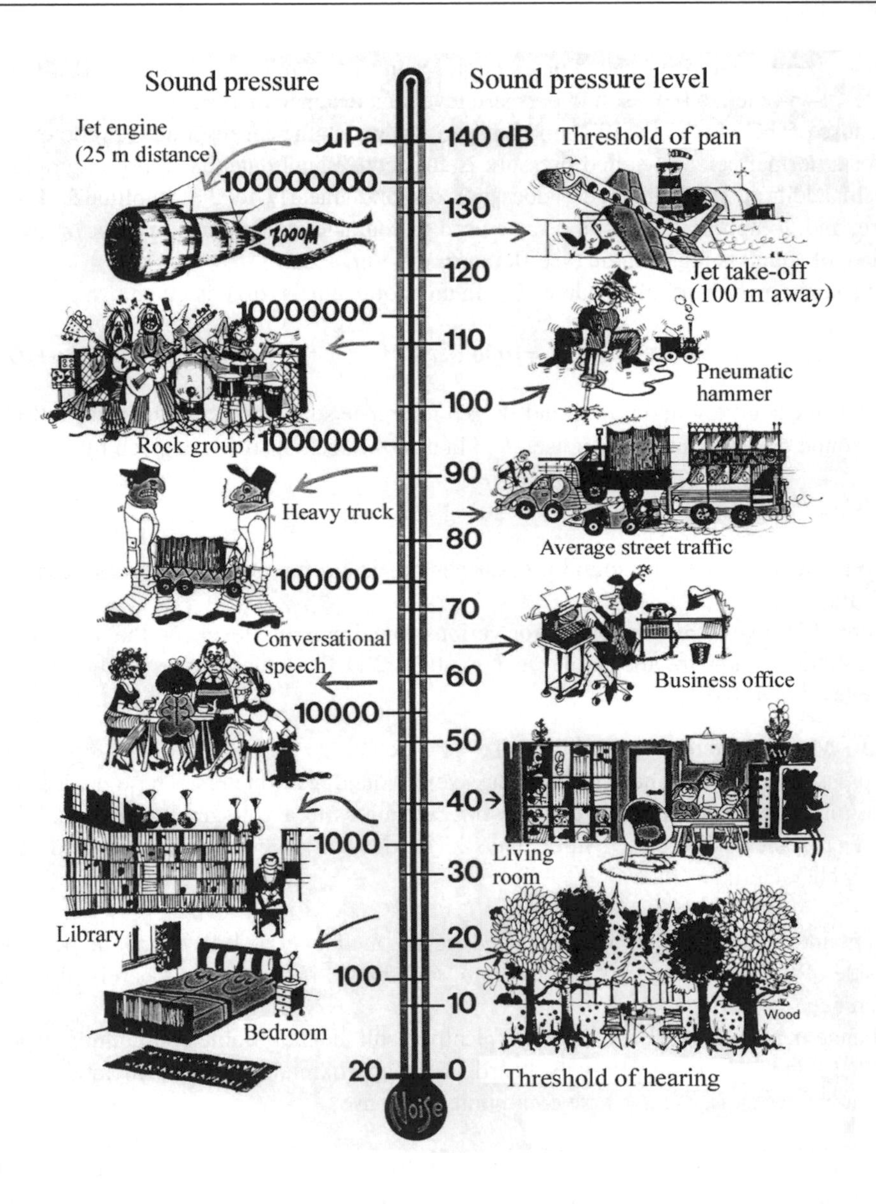

Figure 10.4 Sound pressure level (SPL) examples. Reproduced by permission of Bruel and Kjaer Instruments

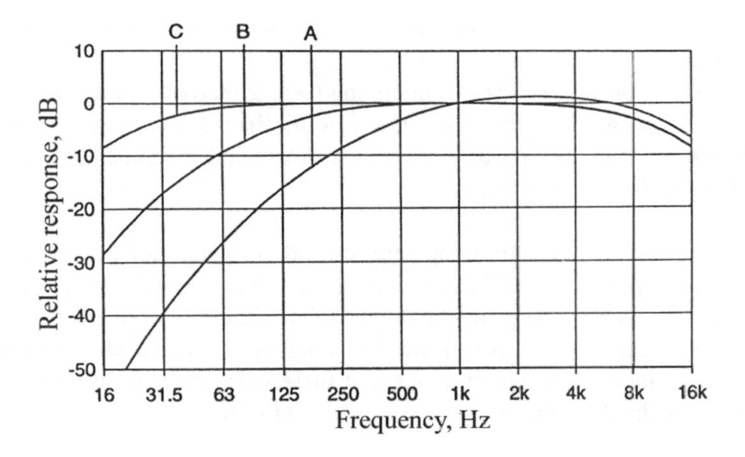

Figure 10.5 Definition of A, B, and C frequency weighting scales (Beranek and Ver, 1992)

Once the A weighted sound pressure is measured over a period of time, it is possible to determine a number of statistical descriptions of time-varying sound and to account for the greater sensitivity to nighttime noise levels. Common descriptors include:

- L_{10}, L_{50}, L_{90}. The A-weighted noise levels that are exceeded 10%, 50%, and 90% of the time, respectively. During the measurement period L_{90} is generally taken as the background noise level.
- L_{eq} (equivalent noise level). The average A-weighted sound pressure level which gives the same total energy as the varying sound level during the measurement period of time.
- L_{dn} (day–night noise level). The average A-weighted noise level during a 24 hour day, obtained after addition of 10 dB to levels measured in the night between 10 p.m. and 7 a.m.

10.4.3 Noise mechanisms of wind turbines

There are four types of noise that can be generated by wind turbine operation: tonal, broadband, low-frequency, and impulsive. They are described below:

- *Tonal.* Tonal noise is defined as noise at discrete frequencies. It is caused by wind turbine components such as meshing gears, non-linear boundary layer instabilities interacting with a rotor blade surface, by vortex shedding from a blunt trailing edge, or unstable flows over holes or slits
- *Broadband.* This is noise characterized by a continuous distribution of sound pressure with frequencies greater than 100 Hz. It is often caused by the interaction of wind turbine blades with atmospheric turbulence, and is also described as a characteristic 'swishing' or 'whooshing' sound
- *Low-frequency.* This describes noise with frequencies in the range of 20 to 100 Hz mostly associated with downwind turbines. It is caused when the turbine blade

encounters localized flow deficiencies due to the flow around a tower, wakes shed from other blades, etc
- *Impulsive*. Short acoustic impulses or thumping sounds that vary in amplitude with time characterize this noise. They may be caused by the interaction of wind turbine blades with disturbed air flow around the tower of a downwind machine, and/or the sudden deployment of tip breaks or actuators

The cause(s) of noise emitted from operating wind turbines can be divided into two categories: (1) aerodynamic and (2) mechanical. Aerodynamic noise is produced by the flow of air over the blades. The primary sources of mechanical noise are the gearbox and the generator. Mechanical noise is transmitted along the structure of the turbine and is radiated from its surfaces. A summary of each of these noise mechanisms follows. A more detailed review is included in the text of Wagner et al. (1996).

10.4.3.1 Aerodynamic noise

Aerodynamic noise originates from the flow of air around the blades. As shown in Figure 10.6 (Wagner et al., 1996), a large number of complex flow phenomena generate this type of noise. This type of noise generally increases with tip speed or tip speed ratio. It is broadband in character and is typically the largest source of wind turbine noise. When the wind is turbulent, the blades can emit low-frequency noise as they are buffeted by changing winds. If the wind is disturbed by flow around or through a tower before hitting the blades (on a downwind turbine design), the blade will create an impulsive noise every time it passes through the 'wind shadow' of the tower.

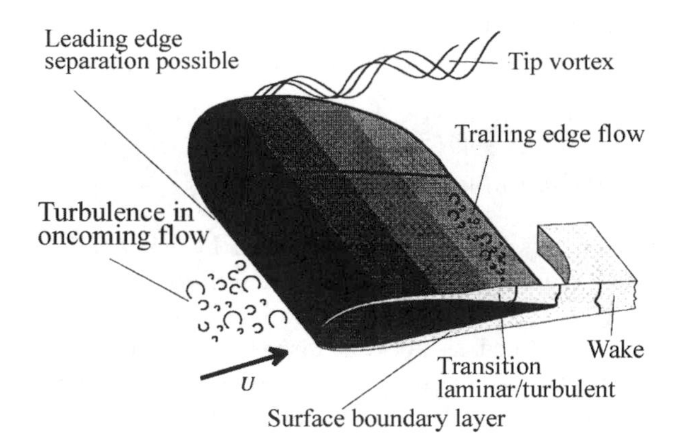

Figure 10.6 Schematic of flow around a rotor blade (Wagner et al., 1996); *U*, wind speed. Reproduced by permission of Springer-Verlag GmbH and Co.

The various aerodynamic noise mechanisms are shown in Table 10.2 (Wagner et al., 1996). They are divided into three groups: (1) low-frequency noise, (2) inflow turbulence noise, and (3) airfoil self noise. A detailed discussion of the aerodynamic noise generation characteristics of a wind turbine is beyond the scope of this work.

Table 10.2 Wind turbine aerodynamic noise mechanisms (Wagner et al., 1996). Reproduced by permission of Springer-Verlag GmbH and Co.

Type or indication	Mechanism	Main characteristics and importance
Low-frequency noise		
Steady thickness noise; steady loading noise	Rotation of blades or rotation of lifting surfaces	Frequency is related to blade passing frequency, not important at current rotational speeds
Unsteady loading noise	Passage of blades through tower velocity deficit or wakes	Frequency is related to blade passing frequency, small in cases of upwind turbines/ possibly contributing in case of wind farms
Inflow turbulence noise	Interaction of blades with atmospheric turbulence	Contributing to broadband noise; not yet fully quantified
Airfoil self-noise		
Trailing-edge noise	Interaction of boundary layer turbulence with blade trailing edge	Broadband; main source of High–frequency noise (770 Hz < f < 2 kHz)
Tip noise	Interaction of tip turbulence with blade tip surface	Broadband; not fully understood
Stall, separation noise	Interaction of turbulence with blade surface	Broadband
Laminar boundary layer noise	Non-linear boundary layer instabilities interacting with the blade surface	Tonal, can be avoided
Blunt trailing edge noise	Vortex shedding at blunt trailing edge	Tonal, can be avoided
Noise from flow over holes, slits and intrusions	Unstable shear flows over holes and slits, vortex shedding from intrusions	Tonal, can be avoided

10.4.3.2 Mechanical noise

Mechanical noise originates from the relative motion of mechanical components and the dynamic response among them. The main sources of such noise include:

- Gearbox
- Generator
- Yaw drives
- Cooling fans
- Auxiliary equipment (e.g., hydraulics)

Since the emitted noise is associated with the rotation of mechanical and electrical equipment, it tends to be tonal (of a common frequency) in character, although it may have a broadband component. For example, pure tones can be emitted from the rotational frequencies of shafts and generators, and the meshing frequencies of the gears.

In addition, the hub, rotor, and tower may act as loudspeakers, transmitting the mechanical noise and radiating it. The transmission path of the noise can be air-borne (a/b) or structure-borne (s/b). Air-borne means that the noise is directly propagated from the component surface or interior into the air. Structure-borne noise is transmitted along other structural components before it is radiated into the air. For example, Figure 10.7 shows the type of transmission path and the sound power levels for the individual components determined at a downwind position (115 m) for a 2 MW wind turbine (Wagner et al., 1996). Note that the main source of mechanical noise was the gearbox that radiated noise from the nacelle surfaces and the machinery enclosure.

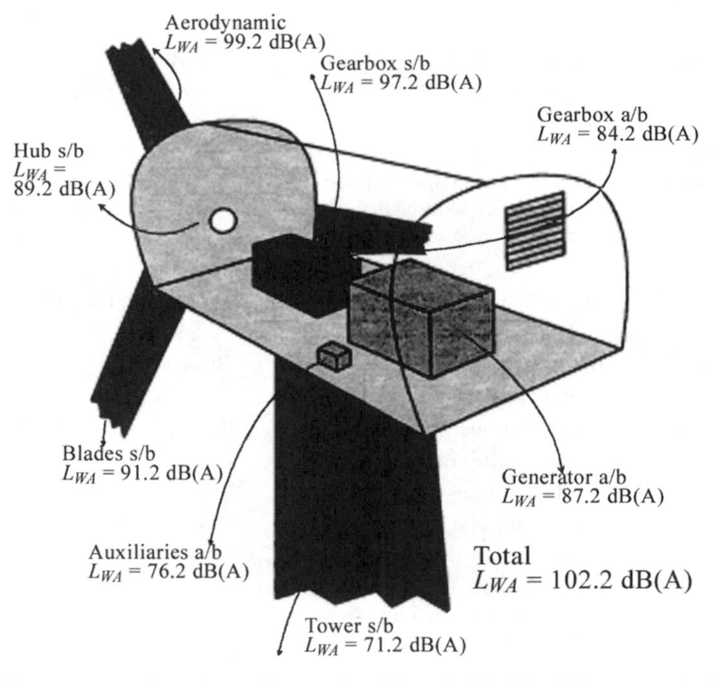

Figure 10.7 Components and total sound power level for wind turbine (Wagner et al., 1996); L_{WA}, predicted sound power level; a/b, airborne; s/b, structure borne. Reproduced by permission of Springer-Verlag GmbH and Co.

10.4.4 Noise prediction from wind turbines

10.4.4.1 Single wind turbines

The prediction of noise from a single wind turbine under expected operating conditions is an important part of an environmental noise assessment. Considering the complexity of the problem, this is not a simple task, and depending on the resources and time available can be quite involved. To complicate matters, wind turbine technology and design has steadily improved through the years, so prediction techniques based on experimental field data from operating turbines may not reflect state-of-the-art machines.

Despite these problems, researchers have developed analytical models and computational codes for the noise prediction of single wind turbines. In general, these models can be divided into the following three classes (Wagner et al., 1996):

- Class 1. This class of models gives a simple estimate of the overall sound power level as a function of basic wind turbine parameters (e.g., rotor diameter, power and wind speed). They represent rules of thumb and are simple and easy to use
- Class 2. These consider the three types of noise mechanisms previously described, and represent current turbine state-of-the-art
- Class 3. These models use refined models describing the noise generation mechanisms and relate them to a detailed description of the rotor geometry and aerodynamics

For the Class 1 models, there are empirical equations that have been used to estimate the sound power level. For example, based on turbine technology up to about 1990, Bass (1993) states that a useful rule of thumb is that wind turbines radiate 10^{-7} of their rated power as sound power. Examples of three other Class 1 models for the prediction of sound power level are summarized in Equations 10.4.4 to 10.4.6.

$$L_{WA} = 10\,(\log_{10} P_{WT}) + 50 \qquad (10.4.4)$$

$$L_{WA} = 22\,(\log_{10} D) + 72 \qquad (10.4.5)$$

$$L_{WA} = 50(\log_{10} V_{Tip}) + 10\,(\log_{10} D) - 4 \qquad (10.4.6)$$

where L_{WA} is the overall A-weighted sound power level, V_{Tip} is the tip speed at the rotor blade (m/s), D is the rotor diameter (m), and P_{WT} is the rated power of the wind turbine (W).

The first two equations represent the simplest (and least accurate today, since they were developed for older machines) methods to predict the noise level of a given turbine based on either its rated power or rotor diameter. The last equation illustrates a rule of thumb that aerodynamic noise is dependent on the fifth power of the tip speed.

The complexity of the Class 2 and 3 models is illustrated in Table 10.3, where the input details required for both models are summarized. A detailed discussion of all these models is beyond the scope of this section and is given in the text of Wagner et al. (1996).

10.4.4.2 Multiple wind turbines

Intuitively, one would expect that doubling the number of wind turbines at a given location would double the sound energy output. Since the decibel scale is logarithmic, the relation to use for the addition of two sound pressure levels (L_1 and L_2) is given as:

$$L_{total} = 10 \log_{10} \left(10^{L_1/10} + 10^{L_2/10}\right) \qquad (10.4.7)$$

Table 10.3 Typical inputs for Class 2 and 3 noise prediction models (Wagner et al., 1996). Reproduced by permission of Springer-Verlag GmbH and Co.

Group	Parameter	Class 2	Class 3
Turbine configuration	Hub height	X	X
	Type of tower (upwind or downwind)		X
Blades and rotor	Number of blades	X	X
	Chord distribution	(X)	X
	Thickness of trailing edge	(X)	X
	Radius	X	X
	Profile shape	(X)	X
	Shape of blade tip	(X)	X
	Twist distribution	(X)	X
Atmosphere	Turbulence intensity	X	X
	Ground surface roughness	X	X
	Turbulence intensity spectrum		X
	Atmospheric stability conditions		X
Turbine operation	Rotational speed	X	X
	Wind speed, alternatively: rated power, rated wind speed, cut-in wind speed	X	X
	Wind direction		X

This equation has two important implications:

- Adding sound pressure levels of equal value increases the noise level by 3 dB
- If the absolute value of $L_1 - L_2$ is greater than 15 dB, the addition of the lower level has negligible effects

This relation can be generalized for N noise sources:

$$L_{total} = 10 \log_{10} \sum_{i=1}^{N} 10^{L_i/10} \tag{10.4.8}$$

10.4.5 Noise propagation from wind turbines

In order to predict the sound pressure level at a distance from a source with a known power level, one must consider how the sound waves propagate. Details of sound propagation in general are discussed in Beranek and Ver (1992). For the case of a stand-alone wind turbine, one might calculate the sound pressure level by assuming spherical spreading, which means that the sound pressure level is reduced by 6 dB per doubling of distance. If the source is on a perfectly flat and reflecting surface, however, then hemispherical spreading has to be assumed, which leads to a 3 dB reduction per doubling of distance.

Furthermore, the effects of atmospheric absorption and the ground effect, both dependent on frequency and the distance between the source and observer, have to be considered. The ground effect is a function of the reflection coefficient of the ground and the height of the emission point.

Wind turbine noise also exhibits some special features (Wagner et al., 1996). First, the height of the source is generally higher than conventional noise sources by an order of magnitude, which leads to less importance of noise screening. In addition, the wind speed has a strong influence on the generated noise. The prevailing wind directions can also cause considerable differences in sound pressure levels between upwind and downwind positions.

The development of an accurate noise propagation model generally must include the following factors:

- Source characteristics (e.g., directivity, height, etc.)
- Distance of the source to the observer
- Air absorption
- Ground effect (i.e., reflection of sound on the ground, dependent on terrain cover, ground properties, etc.)
- Propagation in complex terrain
- Weather effects (i.e., change of wind speed or temperature with height)

A discussion of complex propogation models that include all these factors is beyond the scope of this work. A discussion of work in this area is given by Wagner et al. (1996). For estimation purposes, a simple model based on hemispherical noise propagation over a reflective surface, including air absorption, is given as:

$$L_p = L_w - 10\log_{10}(2\pi R^2) - \alpha R \qquad (10.4.9)$$

where L_p is the sound pressure level (dB) a distance R from a noise source radiating at a power level L_w (dB) and α is the frequency-dependent sound absorption coefficient.

This equation can be used with either broadband sound power levels and a broadband estimate of the sound absorption coefficient [$\alpha = 0.005$ dB(A) m^{-1}] or more preferably in octave bands using octave band power and sound absorption data.

10.4.6 Noise reduction methods for wind turbines

Turbines can be designed or retrofitted to minimize mechanical noise. This can include special finishing of gear teeth, using low-speed cooling fans, mounting components in the nacelle instead of at ground level, adding baffles and acoustic insulation to the nacelle, using vibration isolators and soft mounts for major components, and designing the turbine to prevent noises from being transmitted into the overall structure. Furthermore, if low-frequency noise is a problem in an area, the permitting agency may prohibit the installation of downwind machines (NWCC, 1998).

If a wind turbine has been designed using appropriate design procedures (as described in Chapter 6), it is likely that new noise reducing airfoils will have been used, and mechanical noise emissions will not be a problem. In general, designers trying to reduce wind turbine noise even more must concentrate on the further reduction of aerodynamic noise. It has been previously noted that the following three mechanisms of aerodynamic noise generation are important for wind turbines (assuming that tonal contributions due to slits, holes, trailing edge bluntness, control surfaces, etc., can be avoided by proper blade design):

- Trailing edge noise
- Tip noise
- Inflow turbulence noise

A review of work in these three areas is beyond the scope of this text and here the reader is referred to the text of Wagner et al. (1996). It should be noted that noise has been reduced in modern turbine designs via the use of lower tip speed ratios, lower blade angle of attack, upwind designs, and, most recently, by using specially modified blade trailing edges.

10.4.7 Noise standards or regulations

An appropriate noise assessment study should contain the following three major parts of information:

- A survey of the existing ambient background noise levels
- Prediction (or measurement) of noise levels from the turbine(s) at and near the site
- An assessment of the acceptability of the turbine(s) noise level

At the present time, there are no common international noise standards or regulations, especially ones that pertain to all of the above information. In most countries, however, noise regulations define upper bounds for the noise to which people may be exposed. These limits depend on the country and are different for daytime and nighttime. In Europe, as shown in Table 10.4, fixed noise limits are the standard (Gipe, 1995).

In the United States, although no formal federal noise regulations exist, the US Environmental Protection Agency (EPA) has established noise guidelines. Many states do have noise regulations, and many local governments have enacted noise ordinances. Examples of such ordinances for wind turbines are given in the latest edition of *Permitting of Wind Energy Facilities: A Handbook* (NWCC, 1998).

It should also be pointed out that imposing a fixed noise level standard might not prevent noise complaints. This is due to the question of relative level of broadband background turbine noise compared to changes in background noise levels (NWCC, 1998). That is, if tonal noises are present, higher levels of broadband background noise are needed to effectively mask the tone(s). Accordingly, it is common for community noise standards to incorporate a penalty for pure tones, typically 5 dB(A). Therefore, if a wind turbine meets a sound power level standard of 45 dB(A), but produces a strong whistling, 5 dB(A) are subtracted from the standard. This forces the wind turbine to meet a real standard of 40 dB(A).

Table 10.4 Noise limits of equivalent sound pressure levels, L_{eq} [dB(A)]: European countries (Gipe, 1995)

Country	Commercial	Mixed	Residential	Rural
Denmark			40	45
Germany				
day	65	60	55	50
night	50	45	40	35
Netherlands				
day		50	45	40
night		40	35	30

A discussion of noise measurement techniques that are specific to wind turbine standards or regulations is beyond the scope of this text. A review of such techniques is given in Hubbard and Shepherd (1990), Germanisher Lloyd (1994), and Wagner et al. (1996).

10.5 Electromagnetic Interference Effects

10.5.1 Overview of problem

Wind turbines can present an obstacle for incident electromagnetic waves, which may be reflected, scattered, or diffracted by the wind turbine. As shown schematically in Figure 10.8 (Wagner et al., 1996), when a wind turbine is placed between a radio, television, or microwave transmitter and receiver, it can sometimes reflect portions of the electromagnetic radiation in such a way that the reflected wave interferes with the original signal arriving at the receiver. This can cause the received signal to be significantly distorted.

Some key parameters that influence the extent of electromagnetic interference (EMI) caused by wind turbines include:

- Type of wind turbine (i.e., horizontal axis wind turbine (HAWT) or vertical axis wind turbine (VAWT))
- Wind turbine dimensions
- Turbine rotational speed
- Blade construction material
- Blade angle and geometry
- Tower geometry

In practice, the blade construction material and rotational speed are key parameters. For example, older HAWTs with rotating metal blades have caused television interference in areas near the turbine. Today EMI from wind turbines is less likely because most blades are now made from composite materials. Most modern machines, however, have lightning protection on the blade surfaces, which can increase electromagnetic interference.

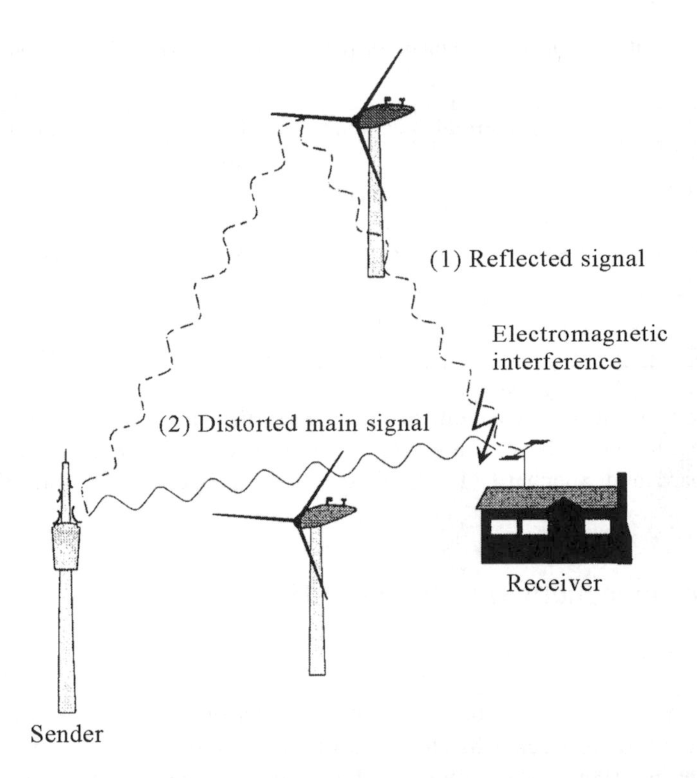

Figure 10.8 Scattering of electromagnetic signals by a wind turbine (Wagner et al., 1996). Reproduced by permission of Springer-Verlag GmbH and Co.

10.5.2 Characterization of electromagnetic interference from wind turbines

10.5.2.1 Mechanism of electromagnetic interference

Electromagnetic interference from wind turbines is generated from multiple path effects. That is, as shown in Figure 10.8, in the vicinity of a wind installation, two transmission paths occur linking the radio transmitter to the receiver. Multiple paths occur on many radio transmissions (caused by large buildings or other structures). The feature unique to wind turbines, however, is blade rotation which causes changes, over short time intervals, in the length of the secondary or scattered path. Thus, the receiver may acquire two signals simultaneously, with the secondary signal causing EMI because delay or distortion of the signal varies with time.

Effects of the signal variation change depend on the modulation scheme, that is, the manner in which the received information is coded. For example, for amplitude modulated (AM) signals, the variation in signal level can be highly undesirable. Interference may also be produced in frequency modulated (FM) signal. Furthermore, the doppler shift caused by rotation of the blades may interfere with radar, and for digital systems, the signal variation can increase the bit error rate (Chignell, 1986).

10.5.2.2 Wind turbine parameters

A wind turbine may provide a number of different electromagnetic scattering mechanisms. The turbine blades in particular can play a significant role in generating electromagnetic interference. They may scatter a signal directly as they rotate and they may also scatter signals reflected from the tower. The degree of EMI caused by wind turbines is influenced by numerous factors, including:

- The wide frequency range of radio signals
- The variety of modulation schemes
- The wide variation in wind turbine parameters

There are a multitude of possible ways whereby a wind turbine can modulate the radio signal and cause interference. Situations have either been reported or postulated where almost every design parameter of the wind turbine system may be critical to a specific radio service. The following general comments on the most important design variables can be made (Chignell, 1986):

Type of machine Different waveforms have been observed in the interference generated by horizontal and vertical axis machines. Most work in this area has concentrated on television interference (TVI) effects.

Machine dimensions The overall dimensions, particularly the diameter of the rotor, are important for establishing the radio frequency bands where interference may occur. Specifically, the larger the machine, the lower the frequency above which radio services may be affected. That is, a large machine will affect HF, VHF, UHF, and microwave bands, while a small machine may degrade only UHF and microwave transmissions.

Rotational speed The rotational speed of the wind turbine and the number of blades determine the modulation frequencies in the interfering radio or telecommunications signal. If one of these coincides with a critical parameter in the radio or telecommunications receiver, the interference is increased.

Blade Construction The blade cross-section and material can be significant here. For example, the following general observations have been made:

- The geometry of the blade should be simple; ideally a combination of simple curves which avoid sharp corners and edges
- In general, scattering from fiberglass or wooden blades is less than from a comparable metal structure. Fiberglass is partially transparent to radio waves whereas wood absorbs them, not allowing energy to be scattered
- The addition of any metallic structure such as a lightning conductor or a metal blade root to a fiberglass blade may negate the material advantages, with the composite structure scattering more effectively than an all-metal blade. To avoid this problem, the metallic components should avoid sharp edges and corners

Blade Angles and Geometry In larger machines, and with microwave signals, the angles defining the area where interference occurs become so narrow that small changes in the yaw, twist, and pitch of the blades become important along with the tilt, cone, and teeter angles.

Tower The tower may also scatter radio waves and as the wind machine rotates, the blades can 'chop' this signal. The geometry of the tower should be kept simple, but if a complicated lattice is essential, a careful choice of angles may reduce the problem. Note, however, that Sengupta and Senior (1994) have not found the tower to be a significant source of EMI.

10.5.2.3 Potential electromagnetic interference effects of wind turbines

For electromagnetic interference generated by a wind turbine to disturb a radio signal, the following conditions must be satisfied:

- A radio transmission must be present
- The wind system must modify the radio signal
- A radio receiver must be present in the volume affected by the wind system
- The radio receiver must be susceptible to the modified signal

Based on EMI experiments, field experience, and analytical modeling in the United States (Sengupta and Senior, 1994) and Europe (Chignell, 1986), a summary of the effects of wind turbine EMI on various radio transmissions is as follows:

- *Television interference* Most reports of electromagnetic interference from wind turbines concern television service. Television interference from wind turbines is characterized by video distortion that generally occurs in the form of a jittering of the picture that is synchronized with the blade passage frequency (rotor speed times number of blades). A significant amount of work on this subject has been carried out in the United States and Europe to quantify this effect
- *FM radio interference* Effects on FM broadcast reception have only been observed in laboratory simulations. They appear in the form of a background 'hiss' superimposed on the FM sound. This work concluded that the effects of wind turbine EMI to FM reception was negligible except possibly within a few tens of meters from the wind turbine
- *Interference to aircraft navigation and landing systems* The effects on VOR (VHF omnidirectional ranging) and LORAN (a long range version of VOR) systems have been studied via analytical models. Results from the VOR studies indicate that a stopped wind turbine may produce errors in the navigational information produced by the VOR. When the wind turbine is operating, however, the potential interference effects are significantly reduced. Note that existing Federal Aviation Authority (FAA) rules prevent a structure the size of many wind turbines being erected within 1 km of a VOR station. For LORAN systems, which operate at very low frequencies, no degradation in communication performance is likely to occur. This assumes that a wind turbine is not in close proximity to the transmitter or receiver
- *Interference to microwave links* Analytical work has indicated that electromagnetic interference effects tend to smear out the modulation used in typical microwave transmission systems. In Europe, experimental work has produced reports of major interference problems on microwave links

- *Interference with cellular telephones.* Since cellular radio is designed to operate in a mobile environment, it should be comparatively insensitive to EMI effects from wind turbines
- *Interference with satellite services.* Satellite services using a geostationary orbit are not likely to be affected because of the elevation angle in most latitudes and the antenna gain

10.5.3 Prediction of electromagnetic interference effects from wind turbines: analytical models

10.5.3.1 General overview

A summary of the most detailed and general models for the analysis of radio signals with electromagnetic interference from wind turbines is given by Sengupta and Senior (1994). In this report the authors summarize over 20 years of work directed toward this subject focusing on large-scale wind turbine systems. Specifically, they have developed a general analytical model for the mechanism by which a wind turbine can produce electromagnetic interference. Their model is based on the schematic system shown in Figure 10.9, which illustrates the field conditions under which a wind turbine can cause EMI. As shown, a transmitter, T, sends a direct signal to two receivers, R, and a wind turbine, WT. The rotating blades of the wind turbine both scatter and transmit a scattered signal. Therefore, the receivers may acquire two signals simultaneously, with the scattered signal causing the EMI because it is delayed in time or distorted. Signals reflected in a manner analogous to mirror reflection are defined as back-scattered (about 80% of the region around the wind turbine). Signal scattering that is analogous to shadowing is called forward-scattering and represents about 20% of the region around a turbine.

For more details of this analysis the reader is referred to Sengupta and Senior (1994), where analytical expressions are developed for the signal power interference (expressed in terms of a modulation index), and the signal scatter ratio, important EMI parameters. This work also summarizes their analytical scattering models for various wind turbine rotors. Their approach here was to develop simplified, idealized models of HAWT and VAWT rotors and to compare the model predictions of signal scattering with measured scattering. This approach is used to illustrate the basic principles of EMI from wind turbines and to provide useful equations for estimating the magnitude of potential interference in practical situations. An example of this is contained in their reports designed to assess TV interference from large and small wind turbines (Senior and Sengupta, 1983; Sengupta et al., 1983).

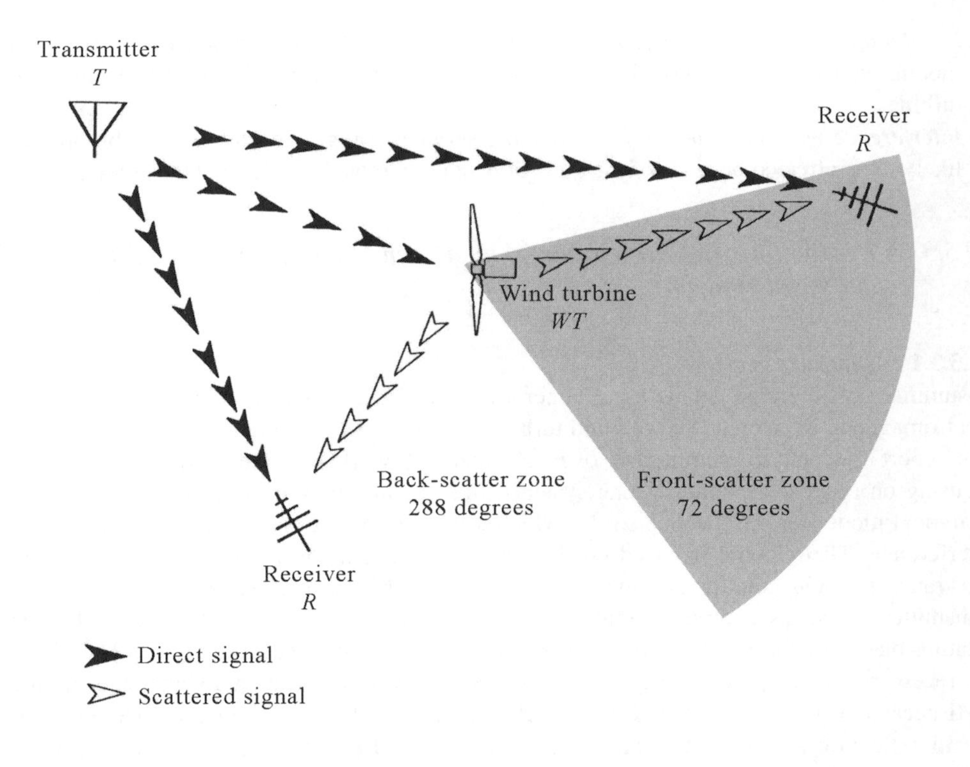

Figure 10.9 Model configuration for electromagnetic interference due to a wind turbine (Sengupta and Senior, 1994). Reproduced by permission of American Society of Mechanical Engineers

10.5.3.2 Simplified analysis

At the present time, the general, and even specialized, models developed by US researchers that have been previously described are not readily usable for wind system designers. Simplified models, however, can be used to predict potential EMI problems from a wind turbine. For example, Van Kats and Van Rees (1989) developed a simple EMI interference model for wind turbines that could be used to predict their impact on TV broadcast reception. As summarized below, results from this model produce a calculated value for the signal-to-interference ratio (*C/I*) for UHF TV broadcast reception in the area around the wind system that can used to evaluate TV picture quality.

The general geometry and definition of terms for the model is given below in Figure 10.10. This model is based on the assumption that the wind turbine behaves as an obstacle for incident electromagnetic waves, serving as a secondary source for radiation.

$L_1 + A_1$: Path loss transmitter–receiver (C-signal)
$L_2 + A_2$: Path loss transmitter–obstacle
$L_r + A_r$: Path loss obstacle–receiver (I-signal)
C/I : Signal to Interference ratio at receiver
τ : Time delay at receiver
σ_b : Bistatic radar cross-section of obstacle
ΔG : Signal discrimination of receiving aerial

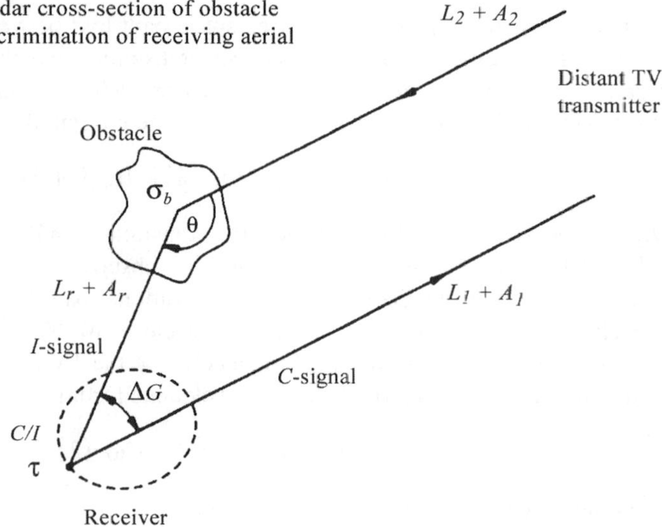

Figure 10.10 Geometry for interference of the electromagnetic path by an obstacle (Van Kats and Van Rees, 1989). Reproduced by permission of Institution of Electrical Engineers

In this model, the spreading of this secondary field over its environment is determined by the following factors:

- Size and shape of the obstacle with respect to signal wavelength, λ
- Dielectric and conductive properties of components of the wind turbine
- Position of the blades and the structure of the wind turbine with respect to the polarization of the incident wave

Because the wind turbine blades are in continuous motion, the impact of the reflected and scattered fields is complex. Thus, the calculation of these fields requires empirical correlation. To approximate the behavior of the system the model is based on radar technology, where an obstacle is characterized by its (bistatic) radar cross-section, σ_b. (This variable is defined as the imaginary surface of an isotropric radiator from which the received power corresponds to the actual power received from the obstacle.) In general, σ_b is a function of the dielectric properties and geometry of the obstacle and the signal wavelength.

When the impact of a wind turbine on radio waves is modeled using σ_b , an expression for the signal-to-interference ratio, *C/I*, can be found in the area around the wind turbine. It should be noted that for each specific radio service (e.g., TV, mobile radio, microwave links, etc.) a different *C/I* ratio will be required for reliable operation. Using the assumed geometry and nomenclature shown in Figure 10.10, the *C/I* ratio at a distance r from the turbine can be calculated by separately calculating the strength (in dBW) of the desired signal, C, and the strength of the interference signal, I. The desired signal is given by:

$$C = P_t + (L_1 + A_1) + G_r \tag{10.5.1}$$

where P_t is the transmitter power, L_1 is free space path loss between transmitter and the receiver, A_1 is the additional path loss between transmitter and the receiver, and G_r = receiver gain. In a manner similar to acoustic measurements, signal strength (in dBW) is expressed in decibels (here, dBW or dB above 1 W). The undesired signal, I, is given by:

$$I = P_t - (L_2 + A_2) + 10\log(4\pi\sigma_b / \lambda^2) - (L_r + A_r) + G_r^{'} \tag{10.5.2}$$

where L_2 is free space path loss between the transmitter and the obstacle, A_2 is the additional path loss between the transmitter and the obstacle, L_r is free space path loss between the obstacle and the receiver, A_r is the additional path loss between the obstacle and the receiver, and $G_r^{'}$ is the receiver gain (obstacle path). Next, assume that $L_1 = L$ when the distance of the wind turbine and the receiver to the TV transmitter is large. In this equation, the free space loss, L_r (in dB), can be calculated from:

$$L_r = 20\log(4\pi) + 20\log(r) - 20\log(\lambda) \tag{10.5.3}$$

If one defines the antenna discrimination factor, G, as:

$$\Delta G = G_r - G_r^{'} \tag{10.5.4}$$

Then, the *C/I* ratio (in dB) becomes:

$$C / I = 10\log(4\pi) + 20\log(r) - 10\log(\sigma_b) + A_2 - A_1 + A_r + \Delta G \tag{10.5.5}$$

From this equation, it follows that *C/I* can be improved by:

- Decreasing σ_b (reducing the effect of the wind turbine)
- Increasing ΔG (better directivity of the antenna)
- Increasing the additional loss $A_2 + A_r$ on the interference path
- Decreasing the additional loss A_t in the direct path

The control of these parameters depends strongly on the location of the receiver to the wind system. In general, two alternative situations can be distinguished:

1. A significant delay time, t, between the desired signal (C) and the undesired signal (I). The receiver is located in an area between the wind turbine and the transmitter. In this area the undesired signal is dominated by reflection and scatter from the wind turbine (back-scatter region).
2. No delay time between the desired signal (C) and the undesired signal (I). This phenomenon occurs in places where the receiver is behind the wind turbine, as seen from the transmitter. In this area the undesired signal can only be the result of scatter or refraction at the wind turbine (forward-scatter region).

In Equation 10.5.5, the distance r from the wind turbine to the receiver can be considered as a variable. When the factors A_1, A_2, and A_r are assumed constant, $\Delta G = 0$, and σ_b of the wind turbine is known, then $r = f(C/I)$ can be calculated.

When C/I is defined as that signal-to-interference ratio required for a specific radio service, a curve with radius r around a wind turbine can be drawn. This will define an area within which the required C/I will not be satisfied. Thus, using this method, criteria for the siting of a particular design of wind turbine with respect to its electromagnetic interference on TV reception can be produced, provided that:

- The C/I required for TV reception is known (experimentally determined examples of this ratio are presented in the original reference)
- The bistatic cross–section, σ_b , of the wind turbine is known (again, experimentally determined examples of this parameter for both the forward- and back-scattering regions are presented in the original reference)

10.5.4 Resources for estimation

As can be seen from the previous discussion and as noted by Chignell (1986), at present, it is impossible to provide a wind system designer with complete technical guidance on the subject of EMI. Thus, problems have to be addressed on a site-by-site basis. In Europe, the International Energy Agency (IEA) Expert Group has recommended an interim assessment procedure for identifying when electromagnetic interference may arise at a particular site. This procedure can be regarded as a means of warning when a problem may arise, but, at present, it does not make a recommendation for dealing with such situations.

For an initial step, one must establish what radio services are present in the volume occupied by the wind system - this means identifying the radio transmitters. Ideally, in each location there would be a central register of all transmitters that the wind system designer could access, but this is not usually the case. The IEA recommends that this should include approaches to radio regulatory agencies, visual observations at the site and on appropriate maps, and a site survey to monitor the radio transmissions that are actually present. Also, the survey should be long enough to include services that are only present on a part-time basis and should note the requirements of mobile users including emergency services, aircraft, shipping, and public utilities.

Once the radio services present have been identified and the transmitters have been marked on a map, the interference zones should be determined. This may be the most complex part of the process, as only the detailed analytical or experimental measurement techniques described by Sengupta and Senior (1994) are presently available. The next part of the procedure is to determine if any receivers for that service will be present in the interference zone. For television broadcasting, this essentially means establishing if there are any dwellings in the zones. If receivers exist, further advice should be sought. Note that the presence of such a receiver does not necessarily mean that EMI problem will arise - it also depends on whether the radio service is robust to major changes in its signal level.

10.6 Land-use environmental impacts

10.6.1 Overview

There are a number of land-use issues to be considered when siting wind turbines. Some of them involve government regulations and permitting (such as zoning, building permits, and approval of aviation authorities). Others may not be subject to regulation, but do have an impact on public acceptance. The following are some of the major land-use issues:

- Actual land required per energy output or capacity per unit of land area
- Amount of land potentially disturbed by a wind farm
- Non-exclusive land-use and compatibility
- Rural preservation
- Turbine density
- Access roads and erosion and/or dust emissions

It is beyond the scope of this text to go into the extensive details of these factors. Gipe (1995) presents an overview of them (and other land-use considerations). The next sections presents a summary of the most general land-use considerations for wind energy systems, and potential strategies that can be used to minimize the environmental problems associated with wind turbine land-use.

10.6.2 Land-use considerations

Compared to other power plants, wind generation systems are sometimes considered to be land intrusive rather than land intensive. The major intrusive effects, via visual impacts, were addressed in Section 10.3. On the extent of land required per unit of power capacity, wind farms require more land than most energy technologies. However, while wind energy system facilities may extend over a large geographic area, the physical 'footprint' of the actual wind turbine and supporting equipment only covers a small portion of the land.

In the United States, for example, wind farm system facilities may occupy only three to five percent of the total acreage, leaving the rest available for other uses. In Europe, it has been found that the percentage of land-use by actual facilities is less than the US wind farms in California. For example, U.K. wind farm developers have found that typically only 1% of the land covered by a wind farm is occupied by the turbines, substations, and access roads. Also, in numerous European projects, farm land is cultivated up to the base of the tower. When access is needed for heavy equipment, temporary roads are placed over tilled soil. Thus, European wind farms only occupy from one to three percent of the available land.

When one determines the actual amount of land used by wind farm systems, it is also important to note the influence of the wind turbine spacing and placement. Wind farms can occupy from 4 to 32 hectares (10 to 80 acres) per megawatt of installed capacity. The dense arrays of the California wind farms have occupied from about 6 to 7 hectares (15 to 18 acres) per megawatt of installed capacity. Typical European wind farms have the wind turbines spread out more and generally occupy 13 to 20 hectares (30 to 50 acres) per megawatt of installed capacity (Gipe, 1995).

Since wind generation is limited to areas where weather patterns provide consistent wind resources over a long season, the development of wind power in the US has occurred primarily in rural and relatively open areas. These lands are often used for agriculture, grazing, recreation, open space, scenic areas, wildlife habitat, and forest management. Wind development is generally compatible with the agricultural or grazing use of a site. Although activities in these areas may be disrupted during construction, only intensive agricultural uses may be reduced or modified during the project's operation (NWCC, 1998). In Europe, owing to higher population densities, there are many competing demands for land, and wind farms have tended to be of a smaller total size.

The development of a wind farm may affect other uses on or adjacent to a site. For example, some parks and recreational uses that emphasize wilderness values and reserves dedicated to the protection of wildlife (e.g., birds) may not be compatible with wind farm development. Other uses, such as open space preservation, growth management, or non-wilderness recreational facilities may be compatible depending on set-back requirements, the nature of on-site development, and the effect on resources of regional importance (NWCC, 1998).

In general, the variables that may determine land-use impacts include:

- Site topography
- Size, number, output, and spacing of wind turbines
- Location and design of roads
- Location of supporting facilities (consolidated or dispersed)
- Location of electric lines (overhead, or underground)

10.6.3 Mitigation of land-use problems

A wide range of actions is available to ensure that wind energy projects are consistent and compatible with most existing and planned land uses. Many of these involve the layout and design of the wind farm. For example, where wind energy development is located in or near recreational or scenic open space uses, some permitting agencies have established requirements intended to 'soften' the industrial nature of the project. As summarized by the National Wind Coordinating Committee (NWCC, 1998), these include the following (again, note that many are a result of visual impact considerations):

- Selecting equipment with minimal structural supports, such as guy wires
- Requiring electrical collection lines to be placed underground
- Requiring maintenance facilities to be off-site
- Consolidating equipment on the turbine tower or foundation pad
- Consolidating structures within a wind farm area
- Requiring the use of more efficient or larger turbines to minimize the number of turbines required to achieve a specific level of electrical output
- Selecting turbine spacing and types to reduce the density of machines and avoid the appearance of 'wind walls'
- Use of roadless construction and maintenance techniques to reduce temporary and permanent land loss

- Restricting most vehicle travel to existing access roads
- Limiting the number of new access roads, width of new roads, and avoiding or minimizing cut and fill
- Limiting placement of turbines and transmission towers in areas with steep, open topography to minimize cut and fill

From a land-use viewpoint, this complete list is really a goal for a 'perfect' wind site, and permitting agencies should consider the following points when determining which of these should be applied to a particular site:

- Cost associated with a particular strategy
- Type and level of impact
- Land-use objectives of the community
- Significance of any potential land-use inconsistency or incompatibility
- Available alternatives

Many of the previous requirements have been used in wind energy projects in Europe. Here, such projects are often located in rural or agricultural areas, and have tended to consist of single dispersed units or small clusters of wind turbines. In several European countries, wind turbines have been placed on dikes or levees along coastal beaches and jetties, or just off shore. Wind turbines sited in farming areas have also been located to minimize disturbances to planting patterns, and the permanent subsystems have been consolidated within a single right of way on field edges, in hedgerows, or along farm roads. Often, no access roads have been constructed and erection or periodic maintenance is carried out with moveable cushioned mats or grating.

In the United States, other land-use strategies associated with wind farm sites include the use of buffer zones and setbacks to separate the project from other potentially sensitive or incompatible land uses. The extent of this separation varies, depending on the area's land-use objectives and other concerns such as visual aspects, noise, and public safety. Also, on some wind farm projects, the resource managers for adjacent properties have established lease or permit conditions such that wind farm development on one site does not block the use of the wind resource on adjacent sites.

10.7 Other Environmental Considerations

This section will summarize some areas that should be considered when assessing the environmental impact of a wind energy system. These include safety, general impacts on flora and fauna , and shadow flicker.

10.7.1 Safety

10.7.1.1 The problem

Safety considerations include both public safety and occupational safety. Here, the discussion will be centered on the public safety aspects (although some occupational safety issues will be included as well). A review of occupational safety in the wind industry is contained in the work of Gipe (1995).

In the public safety area, the primary considerations associated with wind energy systems are related to the movement of the rotor and the presence of industrial equipment in areas that are potentially accessible to the public. Also, depending on the site location, wind energy system facilities may also represent an increased fire hazard. The following aspects of public safety are important (NWCC, 1998):

- *Blade throw.* One of the major safety risks from a wind turbine is that a blade or blade fragment can be thrown from a rotating machine. Wind turbines that have guy wires or other supports can also be damaged. Turbine nacelle covers and rotor nose cones can also blow off machines. In actuality, these events are rare and usually occur in extreme wind conditions, when other structures are susceptible. The distance a blade, or turbine part, may be thrown depends on many variables (e.g., turbine size, height, size of broken part, wind conditions, topography, etc.) and rarely has exceeded 500 m, with most pieces found within 100 to 200 m of the tower. A detailed example of a blade throw calculation is given by Turner (1989)
- *Falling ice or thrown ice.* Safety problems can occur when low temperatures and precipitation cause a build-up of ice on turbine blades. As the blades warm, the ice melts and either falls to the ground, or can be thrown from the rotating blade. Falling ice from nacelles or towers can also be dangerous to people directly under the wind turbine. A detailed technical review of the safety aspects of this problem is summarized by Bossanyi and Morgan (1996)
- *Tower failure.* The complete failure of wind turbine towers or guy wires usually brings the entire wind turbine to the ground if the rotor is turning, or if the problem is not detected immediately. High ice loads, poor tower or foundation design, corrosion, and high winds can increase this potential safety risk
- *Attractive nuisance.* Despite their usual rural location, many wind energy system sites are visible from public highways, and are relatively accessible to the public. Because the technology and the equipment associated with a wind turbine site are new and unusual, it can be an attraction to curious individuals. Members of the public who attempt to climb towers or open access doors or electrical panels could be subject to injuries from moving equipment during operation, electrical equipment during operation, or numerous other hazardous situations
- *Fire hazard.* In arid locations, site conditions that are preferable for development of wind sites such as high average wind speeds, low vegetation, few trees, etc., may also pose a high fire hazard potential during the dry months of the year. Particularly vulnerable sites are those located in rural areas where dry land grain farming occurs or the natural vegetation grows uncontrolled and is available for fuel. In these types of locations fires have started from sparks or flames as a result of numerous causes, including: substandard machine maintenance, poor welding practices, electrical shorts, equipment striking power lines, and lightning
- *Worker hazard.* For any industrial activity, there is the potential for injury or the loss of life to individuals. At the present time there are no statistics that can be used to compare wind facility work with work at other energy-producing facilities. There have, however,

been several fatalities associated with wind turbine installation and maintenance (see Gipe, 1995)

- *Electromagnetic fields.* Electrical and magnetic fields are caused by the flow of electric current through a conductor, such as a transmission line. The magnetic field is created in the space around the conductor, and its field intensity decreases rapidly with distance. In recent years, some members of the public have been concerned regarding the potential for health effects associated with electromagnetic fields

10.7.1.2 Mitigation of public health safety risks

The following summarizes how the public health safety risks previously mentioned can be mitigated:

- *Blade throw.* The most appropriate method for reducing the blade throw potential is the application of sound engineering design combined with a high degree of quality control. Today, it is expected that blade throw should be rare. For example, braking systems, pitch controls, and other speed controls on wind turbines should prevent design limits from being exceeded. Because of safety concerns with blade throw and structural failure, many permitting agencies have separated wind turbines from residences, public travel routes, and other land uses by a safety buffer zone or setback. Examples of typical regulations for the United States are contained in the permitting handbook of the National Wind Coordinating Committee (NWCC, 1998)
- *Falling ice or thrown ice.* To reduce the potential for injury to workers, discussion of blade throw and ice throw hazards should be included in worker training and safety programs. Also, project operators should not allow work crews near the wind turbines during very windy and icing conditions
- *Tower failure.* Complete failure of the structure should be unlikely for turbines designed according to modern safety standards (see Section 6.6.2). Additional security may be obtained by locating the turbine at least the height of the tower away from inhabited structures
- *Attractive nuisance.* Many jurisdictions have required fencing and posting at wind project boundaries to prevent unauthorized access to the site. However, other jurisdictions prefer that the land remains unfenced, particularly if located away from well-traveled public roads, so that the area appears to remain open and retains a relatively natural character. Many jurisdictions require the developer to post signs with a 24-hour toll-free emergency phone number at specified intervals around the perimeter, if the area is fenced, and throughout the wind site, if it is unfenced. Also, liability concerns dictate that access to towers and electrical equipment be locked, and that warning signs be placed on towers, electrical panels, and at the project entrance
- *Fire hazard.* The single most effective fire hazard avoidance measure is to place all wiring underground between the wind turbines and the project substation. In fire-prone areas, most agencies establish permit conditions for the project which address the potential for fire hazard, and a fire control plan may be required. In addition, most agencies require fire prevention plans and training programs to further reduce the potential for fires to escape the project and spread into surrounding areas

- *Worker hazard.* To minimize this hazard, the wind energy project should follow well-established worker protection requirements for the construction and operation of wind energy sites
- *Electromagnetic fields.* Given the rural nature of most wind farm sites, and the relatively low power levels transmitted in a typical site, this safety problem may not represent a major public safety hazard. In the United States, a few states have sought to limit field exposure levels to the levels from existing transmission lines by specifying limits on field strengths, either within or at the edge of the rights-of-way for new lines

10.7.2 Impact on flora and fauna

10.7.2.1 The problem

Since wind turbine sites are typically located in rural areas that are either undeveloped or used for grazing or farming, they have the potential to directly and indirectly affect biological resources (the effects on birds were discussed in Section 10.2). For example, since the construction of wind farms often involves the building of access roads and the use of heavy equipment it is most probable that there will be some disturbance of the flora and fauna during this phase of the project.

It should also be noted that the biological resources of concern include a broad variety of plants and animals that live, use, or pass through an area. They also include the habitat that supports the living resources, including physical features such as soil and water, and the biological components that sustain living communities. These range from bacteria and fungi through the predators on top of the food chain.

The effects on, or conflicts with, flora and fauna, if any, will depend on the plants and animals present, and the design and layout of the wind energy facility. In some cases, permitting agencies have discouraged or prevented development because of likely adverse consequences on these resources. In cases where sensitive resources are not present, or where adverse effects could be avoided or mitigated, development has been permitted to proceed.

10.7.2.2 Mitigation measures

An important consideration regarding the flora and fauna that are found at a site is the potential loss of habitat. Many species are protected state or national laws. It may therefore be useful to consult with ecological or biological specialists early in the planning stages of a project, in order to make sure that sensitive areas are not disturbed, and so that suitable mitigation measures are taken. Any survey work should be carried out at the appropriate time of the year in order to take account of the seasonal nature of some of the potential biological effects.

The wind system developer needs to meet with the local zoning or planning authority, and relevant ecological/biological consultants, in order to discuss the timing of construction and the placement of wind turbines and access roads to avoid at-risk species or habitats. In addition, there may be requirements for on-going monitoring or an overall ecological management plan during the construction period and during the lifetime of the wind system.

10.7.3 Shadow flicker and flashing

10.7.3.1 The problem

Shadow flicker occurs when the moving blades of the wind turbine rotor cast moving shadows that cause a flickering effect, which could annoy people living close to the turbine. In a similar manner, it is possible for sunlight to be reflected from gloss-surfaced turbine blades and cause a 'flashing' effect. This effect has not surfaced as a real environmental problem in the United States. It has been more of a problem in Northern Europe, because of the latitude and the low angle of the sun in the winter sky, and because of the close proximity between inhabited buildings and wind turbines.

For example, Figure 10.11 shows the results of calculations to determine the annual duration of the shadow flicker effect for a location in Denmark (European Commission, 1998). In this figure there are two houses marked as A and B that are respectively 6 and 7 hub heights from the turbine in the center of the diagram. This figure shows that house A will have a shadow from the turbine for 5 hours per year. House B will have a shadow for about 12 hours per year. Note that the results for this type of calculation vary for different geographical regions due to different allowances for cloud cover and latitude.

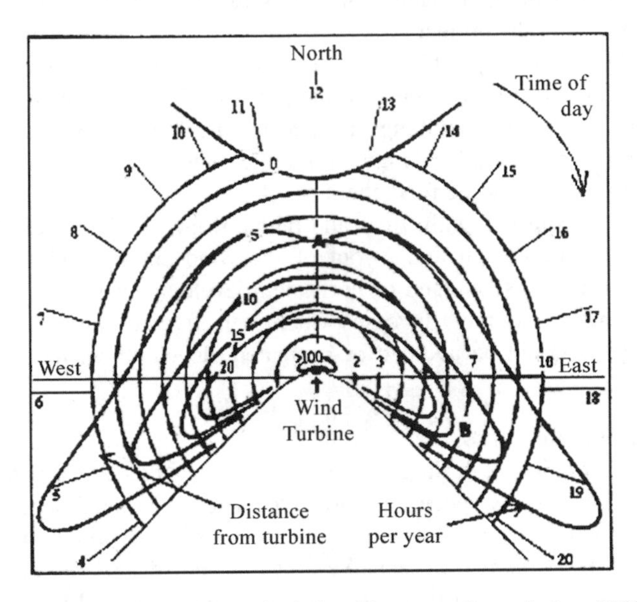

Figure 10.11 Diagram of shadow flicker calculation (European Commission, 1998); A, B houses

10.7.3.2 Mitigation measures

In the worst cases, flickering will only occur for a short duration (on the order of 30 minutes a day for 10–14 weeks of the winter season). In Europe, one proposed solution is to not run the turbines during these short time periods, while another is to site machines carefully, taking account of the shadow path on nearby residences. Flashing can be prevented by the use of non-reflective, non-gloss blades, and also by careful siting.

A common guideline used in Denmark is to have a minimum distance of 6 to 8 rotor diameters between the wind turbine and the closest neighbor. Houses located at a distance of 6 rotor diameters (about 250 meters for a 600 kW machine) from a wind turbine in any of the sectors shown in Figure 10.11 will be affected for two periods each of 5 weeks duration per year (European Commission, 1998).

References

Ackerman, T., Soder, L. (2000) Wind Energy Technology and Current Status: A Review, *Renewable and Sustainable Energy Reviews*, 4, 315–374.

Anderson, R. L., Kendall, W., Mayer, L. S., Morrison, M. L., Sinclair, K., Strickland, D., Ugoretz, S. (1997) Standard Metrics and Methods for Conducting Avian/Wind Energy Interaction Studies, *Proc. Windpower '97*, AWEA, 265–272.

Aubrey, C. (2000) Turbine Colors…Do They Have to be Grey? *Wind Directions*, March, 18–19.

Bass, J. H. (1993) Environmental Aspects – Noise, in course notes for *Principles of Wind Energy Conversion*, ed. L. L. Freris, Imperial College, London.

Beranek, L. L., Ver, I. L. (1992) *Noise and Vibration Control Engineering: Principles and Applications*, Wiley, New York.

Bossanyi, E. A., Morgan, C. A. (1996) Wind Turbine Icing – Its Implications for Public Safety, *Proc. 1996 European Union Wind Energy Conference*, 160–164.

Chignell, R. J. (1986) Electromagnetic Interference from Wind Energy Conversion Systems – Preliminary Information, *Proc. European Wind Energy Conference '86*, 583–586.

Clausager, I., Nohr, H. (1996) Impact of Wind Turbines on Birds, *Proc. 1996 European Union Wind Energy Conference*, 156–159.

Colson, E. W. (1995) Avian Interactions with Wind Energy Facilities: A Summary, *Proc. Windpower '95*, AWEA, 77–86.

Connors, S. R. (1996) Informing Decision Makers and Identifying Niche Opportunities for Windpower: Use of Multiattribute Trade off Analysis to Evaluate Non-dispatchable Resources, *Energy Policy,* 24(2), 165–176.

European Commission (1998) *Wind Energy – The Facts.*

Germanisher Lloyd (1994) *Regulation for the Certification of Wind Energy Conversion Systems, Supplement to the 1993 Edition*, Hamburg.

Gipe, P. (1995) *Wind Energy Comes of Age*, Wiley, New York.

Goodman, N. (1997) The Environmental Impact of Windpower Development in Vermont: A Policy Analysis, *Proc. Windpower '97*, AWEA, 299–308.

Hubbard, H. H., Shepherd, K. P. (1990) Wind Turbine Acoustics, *NASA Technical Paper 3057 DOE/NASA/20320-77.*

Kidner, D. B. (1996) The Visual Impact of Taff Ely Wind Farm – A Case Study Using GIS, *Wind Energy Conversion 1996*, BWEA, 205–211.

NWCC (National Wind Coordinating Committee) (1998) Permitting of Wind Energy Facilities: A Handbook, RESOLVE, Washington, DC.

NWCC (National Wind Coordinating Committee) (1999) Studying Wind Energy/Bird Interactions: A Guidance Document, RESOLVE, Washington, DC.

Orloff, S., Flannery, A. (1992) Wind Turbine Effects on Avian Activity, Habitat Use, and Mortality in Altamont Pass and Solano County Wind Resource Areas: 1989–1991, *California Energy Commission Report No. P700-92-001.*

Ratto, C. F., Solari, G. (1998) *Wind Energy and Landscape*, A. A. Balkema, Rotterdam.

Robotham, A. J. (1992) Progress in the Development of a Video Based Wind Farm Simulation Technique, *Wind Energy Conversion 1992*, BWEA, 351–353.

Sengupta, D. L., Senior, T.B.A. (1994) Electromagnetic Interference from Wind Turbines, Chapter 9 in *Wind Turbine Technology,* ed. D. A. Spera, ASME, New York.

Sengupta, D. L., Senior, T.B.A., Ferris, J. E. (1983) Study of Television Interference by Small Wind Turbines: Final Report, *Solar Energy Research Institute Report SERI/STR-215-1881*, NTIS.

Senior, T.B.A., Sengupta, D. L. (1983) Large Wind Turbine Handbook: Television Interference Assessment, *Solar Energy Research Institute Report SERI/STR-215-1879*, NTIS.

Sinclair, K. C. (1999) Status of the U.S. Department of Energy/National Renewable Energy Laboratory Avian Research Program, *Proceedings Windpower '99*, AWEA.

Sinclair, K. C., Morrison, M. L. (1997) Overview of the U.S. Department of Energy/National Renewable Energy Laboratory Avian Research Program, *Proc. Windpower '97*, AWEA, 273–279.

Stanton, C. (1994) The Visual Impact and Design of Wind Farms in the Landscape, *Wind Energy Conversion 1994*, BWEA, 249–255.

Stanton, C. (1995) Wind Farm Visual Impact and its Assessment, *Wind Directions*, BWEA, August, 8–9.

Still, D., Painter, S., Lawrence, E. S., Little, B., Thomas, M. (1996) Birds, Wind Farms, and Blyth Harbor, *Wind Energy Conversion 1996*, BWEA, 175–183.

Taylor, D. C., Durie, M. J. (1995) Wind Farm Visualisation, *Wind Energy Conversion 1995*, BWEA, 413–416.

Turner, D. M. (1989) An Analysis of Blade Throw from Wind Turbines, in *Wind Energy and the Environment*, ed. D. T. Swift-Hook, Peter Peregrinus, London.

Van Kats, P. J., Van Rees, J. (1989) Large Wind Turbines: A Source of Interference for TV Broadcast Reception, Chapter 11 in *Wind Energy and the Environment*, ed. D. T. Swift-Hook, Peter Peregrinus, London.

Wagner, S., Bareib, R., Guidati, G. (1996) *Wind Turbine Noise*, Springer, Berlin.

Wolf, B. (1995) Mitigating Avian Impacts: Applying the Wetlands Experience to Wind Farms, *Proc. Windpower '95*, AWEA, 109–116.

Nomenclature

A.1 Note on Nomenclature and Units

This text includes material from many different engineering disciplines (e.g. aerodynamics, dynamics, controls, electromagnetism, acoustics). Within each of these disciplines commonly accepted variables are often used for important concepts. Thus, for acoustics engineers, α denotes the sound absorption coefficient, in aerodynamics it denotes the angle of attack, and in one common model for wind shear it is the power law exponent. In this text, an effort has been made to ensure that concepts that are found in multiple chapters all have the same designation, while maintaining common designations for generally accepted concepts. Finally, in an effort to avoid confusion over multiple definitions of one symbol, the nomenclature used in the text is listed by chapter.

Throughout this text units are in general not stated explicitly. It is assumed, however, that a consistent set of units is used. In SI units, then, for example, length is in m, speed is in m/s, mass is in kg, density is in kg/m^3, weight and force are in N, stress is in N/m^2, torque is in Nm, and power is in W. When other forms are in common usage, such as power in kW, those units are stated. Energy is also expressed in Wh or kWh, rather than in J or kJ. Angles are assumed to be in radians except where explicitly stated otherwise. There are two possible sources of confusion: rotational speed and frequency. In this text, when referring to rotational speed n is always in revolutions per minute (rpm) and Ω is in radians/second. With regards to frequency f is always in cycles per second (Hz) while ω is always in radians/second. Finally, in other cases where the units are specific to a variable, they are indicated below.

A.2 Chapter 2

A.2.1 *English variables*

A	Area
C_p	Specific heat at constant pressure
C_f	Capacity factor
C_P	Power coefficient
CC	Capture coefficient
c	Weibull scale factor
D	Rotor diameter

E_w	Wind machine energy (Wh)
F	Coriolis parameter
F_p	Pressure force
$F(U)$	Cumulative distribution function
f	Frequency (Hz)
f_j	Number of occurrences in each bin
f_c	Coriolis force
g	Gravitational acceleration
h	Enthalpy
i	index or sample number
K_e	Energy pattern factor
k	Weibull shape factor
k	von Karman constant
L	Integral length scale
ℓ	Mixing length
m	(i) Mass
m	(ii) Direction normal to lines of constant pressure
m_i	Midpoint of bins
N	Number of long-term data points
N_s	Number of samples in short-term averaging time
N_B	Number of bins
P	Power
$\overline{P}/A$	Average wind power density
$P_w(U)$	Wind turbine power
$\overline{P}_w$	Average wind turbine power
p	Pressure
$p(\ )$	Probability density function
q	Heat transfer
R	Radius of curvature
R	Radius of wind turbine rotor
r	Lag
$S(f)$	Power spectral density function
T	Temperature
t	Time
TI	Turbulence intensity
U	Mean wind speed (mean of short term data)
$\overline{U}$	Long-term mean wind speed (mean of short-term averages)
$U*$	Friction velocity
U_c	Characteristic wind velocity
U_i	Wind speed average over period i
U_g	Geostrophic wind speed
U_{gr}	Gradient wind speed
u	Internal energy
$u(z,t)$	Instantaneous longitudinal wind speed
$\widetilde{u}$	Fluctuating wind velocity about short-term mean
u_i	Sampled wind speed

v	Specific volume
$v(z,t)$	Instantaneous lateral wind speed
$w(z,t)$	Instantaneous vertical wind speed
w_i	Width of bins
x	Dimensionless wind speed
z	Elevation
z_i	Inversion height
z_o	Surface roughness
z_r	Reference height

A.2.2 Greek variables

α	Power law exponent
Γ	Lapse rate
$\Gamma(x)$	Gamma function
Δt	Duration of short-term averaging time
δt	Sampling period
η	Drive train efficiency
λ	Tip speed ratio
ρ	Air density
σ_u	Standard deviation of short-term data (turbulence)
σ_U	Standard deviation of long-term data
τ	Shear stress
τ_0	Surface value of shear stress
τ_{xz}	Shear stress in the direction of x whose normal coincides with z
φ	Latitude
ω	Angular speed of earth
Ω	Angular velocity of rotor

A.3 Chapter 3

A.3.1 English variables

A	Projected airfoil area (chord $\times$ span), surface area, rotor swept area
a	Axial interference or induction factor
a'	Angular induction factor
B	Number of blades
C_d	Two-dimensional drag coefficient
C_D	Three-dimensional drag coefficient
$C_{d,0}$	Constant drag term
$C_{d,\alpha 1}$	Linear drag term
$C_{d,\alpha 2}$	Quadratic drag term
C_l	Two-dimensional lift coefficient

C_L	Three-dimensional lift coefficient
$C_{l,\alpha}$	Slope of lift coefficient curve
$C_{l,0}$	Lift coefficient at zero angle of attack
C_m	Pitching moment coefficient
C_P	Power coefficient
C_p	Pressure coefficient
C_T	Thrust coefficient
C_{T_r}	Local thrust coefficient
c	Airfoil chord length
d_1	Variable for determining angle of attack with simplified method
d_2	Variable for determining angle of attack with simplified method
$\mathrm{d}F_D$	Incremental drag force
$\mathrm{d}F_L$	Incremental lift force
$\mathrm{d}F_N$	Incremental force normal to plane of rotation (thrust)
$\mathrm{d}F_T$	Incremental force tangential to circle swept by blade section
$\mathrm{d}Q$	Torque on an annular control volume
$\mathrm{d}r$	Thickness
F	Tip loss correction factor
F_D	Drag force
F_N	Normal force
k	Index of blade element closest to hub.
L	Characteristic length, lift force
l	Length or span of airfoil
M	Pitching moment
m	Mass
N	Number of blade elements
P	Power
p	Pressure
Q	Torque
q_1	Variable for determining angle of attack with simplified method
q_2	Variable for determining angle of attack with simplified method
q_3	Variable for determining angle of attack with simplified method
R	Outer blade radius
Re	Reynolds number
r	Radius
r_h	Rotor radius at the hub
r_i	Radius at the midpoint of a blade section
T	Thrust
t	Time
U	Characteristic velocity, velocity of undisturbed airflow
x	Dummy variable

A.3.2 Greek variables

α	Angle of attack
ε	Surface roughness height
η_{mech}	Mechanical efficiency
η_{out}	Overall output efficiency
$\theta_{p,0}$	Blade pitch angle at the tip
θ_T	Blade twist angle,
θ_p	Section pitch angle
λ	Tip speed ratio
λ_h	Local speed ratio at the hub
λ_r	Local speed ratio
μ	Coefficient of viscosity
ν	Kinematic viscosity
ρ	Air density
σ	Rotor solidity
σ'	Local rotor solidity
φ	Angle of relative wind.
ω	The angular velocity of the wind
Ω	Angular velocity of the wind turbine rotor

A.4 Chapter 4

A.4.1 English variables

A	(i) Area
A	(ii) Constant in Euler beam equation
A	(iii) Axisymmetric flow term, $=(\Lambda/3)-(\theta_p/4)$
A_3	Axisymmetric flow term, $=(\Lambda/2)-(2\theta_P/3)$
a	i) Acceleration
a	ii) Axial induction factor
a	iii) Offset of center of mass of gyroscope
B	(i) Number of blades
B	(ii) Gravity term, $=G/2\Omega^2$
C	Constant
$C_{l\alpha}$	Slope of angle of attack line
C_l	Angle of attack
C_P	Power coefficient
C_Q	Torque coefficient
C_T	Thrust coefficient
c	(i) Maximum distance from neutral axis
c	(ii) Chord

c	(iii) Damping coefficient
c	(iv) Amplitude of sinusoid
c_c	Critical damping coefficient
D	(i) Determinant
D	(ii) Damage
$\overline{d}$	Normalized yaw moment arm $= d_{yaw}/R$
d_{yaw}	Yaw moment arm
E	(i) Modulus of elasticity
E	(ii) Energy
e	Non-dimensional hinge offset
F	External force
F_c	Centrifugal force
F_g	Gravitational force
$\tilde{F}_N$	Normal force per unit length
$\tilde{F}_T$	Tangential force per unit length
G	Gravity term, $= g\, M_B r_g / I_b$
g	Gravitational constant
$\boldsymbol{H}$	Angular momentum
H_{op}	Operating hours/yr
h	Height
I	Area moment of inertia
I_b	Area or mass moment of inertia of single blade
J	Polar area or mass moment of inertia
K	(i) Rotational spring constant
K	(ii) Flapping inertial natural frequency, $K = 1 + \varepsilon + K_\beta / I_b \Omega^2$
K_β	Spring constant in flapping direction
K_{vs}	Vertical wind shear constant
k	(i) Spring constant
k	(ii) Number of cyclic events per revolution
k_θ	Rotational stiffness
L	Length
$\tilde{L}$	Lift force per unit length
M_i	Moment applied to ith lumped mass
M	Moment
$\boldsymbol{M}$	Moment, vector
M_c	Flapping moment due to centrifugal force
M_f	Flapping moment due to flapping acceleration inertial force
M_g	Flapping moment due to gravity
M_{max}	Maximum moment
M_s	Flapping moment due to spring
M_{yaw}	Flapping yaw moment
$M_{X'}$	Yawing moments on tower
$M_{Y'}$	Backwards pitching moments on tower
$M_{Z'}$	Rolling moment on nacelle
M_β	Flapping moment
M_ζ	Lead–lag moment

m	Mass
m_B	Mass of blade
m_i	Mass of ith section of beam
N	(i) Number of teeth on a gear
N	(ii) Cycles to failure
n	(i) Gear train speed up ratio
n	(ii) Number of cycles
n_{rotor}	Rotational speed of rotor (rpm)
P	Power
Q	Torque
q	Yaw rate (rad/s)
$\overline{q}$	Normalized yaw rate, $= q / \Omega$
R	(i) Radius of rotor
R	(ii) Stress ratio
r	Radial distance from axis of rotation
r_g	Radial distance to center of mass
S	Shear force
S_β	Flapwise shear force
S_ζ	Edgewise shear force
T	Thrust
t	Time
U	Free stream wind velocity
$\overline{U}$	Normalized wind velocity, $= U/\Omega R = 1/\lambda$
U_h	Wind velocity at height h
U_P	Perpendicular component of wind velocity
U_R	Relative wind velocity
U_T	Tangential component of wind velocity
V_0	Cross-wind velocity
$\overline{V_0}$	Non-dimensional cross flow, $\overline{V}_0 = V_0/\Omega R$
$\overline{V}$	Non-dimensional total cross flow, $\overline{V} = (V_0 + qd_{yaw})/(\Omega R)$
W	Total load or weight
w	Loading per unit length
x	Distance (linear)
Y	Years of operation
y_i	Deflection of beam
Z	Non-dimensional difference between squares of rotating and non-rotating blade natural frequency

A.4.2 Greek variables

α	(i) Angular acceleration of rotor, $= \dot{\Omega}$
α	(ii) Angle of attack (radians)
α	(iii) Power law exponent

β	(i) Term used in vibrating beam solution, $\beta = \sqrt[4]{\frac{\rho\omega^2}{EI}}$
β	(ii) Flapping angle (radians)
$\dot{\beta}$	Flapping velocity (radians/s)
$\ddot{\beta}$	Second time derivative of flapping angle (radians/s^2)
β_0	Collective flapping coefficient (radians)
β_{1c}	Cosine flapping coefficient (radians)
β_{1s}	Sine flapping coefficient (radians)
β''	Azimuthal derivative of flap angle, $= \ddot{\beta}/\Omega^2$ (radians^{-1})
β'	Azimuthal second derivative of flap angle, $= \dot{\beta}/\Omega$
γ	Lock number, $= \rho c C_{L\alpha} R^4 / I_b$
$\Delta U_{P,crs}$	Perpendicular velocity perturbation due to yaw error
$\Delta U_{P,yaw}$	Perpendicular velocity perturbation due to yaw rate
$\Delta U_{P,vs}$	Perpendicular velocity perturbation due to wind shear
$\Delta U_{T,s}$	Tangential velocity perturbation due to yaw error
$\Delta U_{T,yaw}$	Tangential velocity perturbation due to yaw rate
$\Delta U_{T,vs}$	Tangential velocity perturbation due to wind shear
ε	Offset term, $= 3e/[2(1-e)]$
η	Ratio, r/R
θ	Arbitrary angle (radians)
θ_p	Pitch angle (positive towards feathering) (radians)
Θ	Steady state yaw error (radians)
λ	Tip speed ratio
λ_i	Length between sections of beam
Λ	Non-dimensional inflow, $= U(1-a)/\Omega R$
ξ	(i) Damping ratio
ξ	(ii) Aerodynamic damping ratio, $= \gamma\Omega/16\omega_\beta$
ρ	Density of air
$\tilde{\rho}$	Density per unit length
σ	Stress
σ_a	Stress amplitude
σ_e	Stress endurance limit
σ_m	Mean Stress
σ_{max}	Maximum stress
σ_{min}	Minimum stress
σ_u	Ultimate Stress
$\sigma_{\beta,max}$	Maximum flapwise stress
ϕ	(i) Angle of relative wind (aerodynamics) (radians)
ϕ	(ii) Phase angle (vibrations) (radians)
ψ	Azimuth angle, 0 = down (radians)
ω	(i) Frequency of oscillation (radians/s)
ω_i	(ii) Frequency of oscillation (radians/s) of *i*th mode
ω	(iii) Rate of precession of gyroscope (radians/s)
$\boldsymbol{\omega}$	Rate of precession (vector)
ω_d	Natural frequency for damped oscillation (radians/s)

ω_n	Natural frequency (radians/s)
ω_{NR}	Natural flapping frequency of non-rotating blade (radians/s)
ω_R	Natural flapping frequency of rotating blade (radians/s)
ω_β	Flapping frequency
Ω	Angular velocity (= $\dot{\psi}$) (radians/s)
$\boldsymbol{\Omega}$	Angular velocity vector

A.4.3 Subscripts

ahg	Axial flow, hinge spring, and gravity (blade weight)
cr	Cross wind
vs	Vertical wind shear
yr	Yaw rate
1,2	Different heights or ends of a gear train
max	Maximum
min	Minimum

A.5 Chapter 5

A.5.1 English variables

A	Area
A_c	Cross-sectional area of coil
A_m	Magnitude of arbitrary phasor
A_g	Cross-sectional area of gap
$\hat{\boldsymbol{A}}$	Arbitrary phasor
a	(i) Real component in complex number notation
a	(ii) Turns ratio of the transformer
a_n	Coefficient of cosine terms in Fourier series
$\boldsymbol{B}$	Magnetic flux density (vector)
B	Magnitude of magnetic flux density (scalar)
b	Imaginary component in complex number notation
b_n	Coefficient of sine terms in Fourier series
C	(i) Capacitance (F)
C	(ii) Constant
$\hat{\boldsymbol{C}}$	Arbitrary phasor
C_m	Magnitude of arbitrary phasor
E	(i) Energy (J)
E	(ii) Induced electromotive force EMF (V)
E	(iii) Synchronous generator field induced voltage
$\hat{E}$	EMF phasor
E_m	Stored energy in magnetic fields
E_1	Primary voltage in transformer

E_2	Secondary voltage in transformer
e_m	Stored magnetic field energy per unit volume
$\boldsymbol{F}$	Force (vector)
f	Frequency of AC electrical supply (Hz)
g	Air gap width
$\boldsymbol{H}$	Magnetic field intensity (vector) (A-t/m)
H_c	Field intensity inside the core (A-t/m)
I	Mean current
$\hat{\boldsymbol{I}}$	Current, phasor
I_a	Synchronous machine's armature current
I_f	Field current (a.k.a. excitation)
I_L	Line current in three-phase system
I_M	Magnetizing current
$I_{\max}$	Maximum value of current (AC)
I_P	Phase current in three-phase system
I_{rms}	Root mean square (rms) current
I_R	Rotor current
I_S	Induction machine stator phasor current
i	Instantaneous current
J	Inertia of generator rotor
j	$\sqrt{-1}$
k_1	Constant of proportionality in synchronous machine (Wb/A)
k_2	Constant of proportionality (V/Wb-rpm)
L	i) Inductance (H)
L	(ii) Length of coil
L	(iii) Half the period of the fundamental frequency
L	Length of face of pole peices
ℓ	Path length (vector)
ℓ_c	Length of the core at its midpoint
N	Number of turns in a coil
NI	Magnetomotive force (MMF) (A-t)
$N\Phi$	Flux linkages (Wb)
N_1	Number of coils on primary winding
N	Number of coils on secondary winding
n	(i) Actual rotational speed (rpm)
n	(ii) Harmonic number
n_s	Synchronous speed (rpm)
P	(i) Real power
P	(ii) Number of poles
P_g	Air gap power in induction machine
P_{in}	Mechanical input power input to induction generator
P_{loss}	Power lost in induction machine's stator
P_m	Mechanical power converted in induction machine
$P_{mechloss}$	Mechanical losses
P_{out}	Electrical power delivered from induction generator
P_1	Power in one phase of three-phase system,

PF	Power factor
Q	(i) Reactive power (VA)
Q	(ii) Torque
Q_e	Electrical torque
Q_m	Mechanical torque
Q_r	Applied torque to generator rotor
R	(i) Equivalent induction machine resistance
R	(ii) Reluctance of magnetic circuit (A-t/Wb)
R	(iii) Resistance
$\mathbf{R}_c$	Core reluctance (A-t/Wb)
$\mathrm{Re}\{\ \}$	Real part of complex number
R_M	Resistance in parallel with mutual inductance
R_s	Synchronous generator resistance
R_S	Stator resistance
R_R	Rotor resistance (referred to stator)
r	(i) Radial distance from center of toroid
r	(ii) Radius
r_i	Inner radius of coil
r_o	Outer radius of coil
S	Apparent power (VA)
s	Slip
t	Time
V	(i) Voltage in general
V	(ii) Electrical machine terminal voltage
$\hat{V}$	Voltage, phasor
V_{LL}	Line-to-line voltage in three-phase system
V_{LN}	Line-to-neutral voltage in three-phase system
V_{rms}	Root mean square (rms) voltage
V_{Max}	Maximum voltage
V_1	(i) AC voltage filter input voltage
V_1	(ii) Terminal voltage in primary winding
V_2	(i) AC voltage filter output voltage
V_2	(ii) Terminal voltage in secondary winding
v	Instantaneous voltage
$v(t)$	Instantaneous voltage at time t
X	Equivalent induction machine reactance
X_C	Capacitive reactance
X_L	Inductive reactance
X_{LS}	Stator leakage inductive reactance
X_{LR}	Rotor leakage inductive reactance (referred to stator)
X_M	Magnetizing reactance
X_s	Synchronous generator synchronous reactance
Y	'Wye' connected three-phase system
Y''	Admittance in resonant filter
$\hat{\mathbf{Z}}$	(i) Impedance in general
$\hat{\mathbf{Z}}$	(ii) Induction machine equivalent impedance

$\hat{Z}_C$	Capacitive impedance
$\hat{Z}_i$	Impedance with index *i*
$\hat{Z}_L$	Inductive reactance
$\hat{Z}_P$	Parallel impedance
$\hat{Z}_R$	Resistive impedance
$\hat{Z}_s$	(i) Series impedance
$\hat{Z}_s$	(ii) Synchronous impedance
$\hat{Z}_Y$	Impedance of Y connected circuit
Z_1	Series impedance in resonant filter
Z_2	Parallel impedance in resonant filter
$\hat{Z}_\Delta$	Impedance of Δ connected circuit

A.5.2 Greek variables

Δ	Delta connected three-phase system
δ	Synchronous machine power angle (rad)
η_{gen}	Overall efficiency (in the generator mode)
θ	(i) Power factor angle (rad)
θ	(ii) Rotation angle (rad)
λ	Flux linkage, $= N\Phi$
μ	Permeability, $\mu = \mu_r \mu_0$ (Wb/A-m)
μ_0	Permeability of free space, $4\pi x 10^{-7}$ (Wb/A-m)
μ_r	Relative permeability
ϕ	(i) Phase angle (rad)
ϕ	(ii) Power factor angle (rad)
ϕ_a	Phase angle of arbitrary phasor **A** (rad)
ϕ_b	Phase angle of arbitrary phasor **B** (rad)
Φ	Magnetic flux (Wb)
ω	Speed of generator rotor (rad/s)

A.5.3 Symbols

$\angle$	Angle between phasor and real axis
μF	MicroFarad (10^{-6} Farad)
Ω	Ohms

A.6 Chapter 6

A.6.1 English variables

A_c	Cross sectional area of wire rope
a	Turbulence parameter
B	Constant in composites *S–N* model

b	Width of gear tooth face
C_P	Rotor power coefficient
c	Chord length
D	Dynamic magnification factor
D_{Ring}	Diameter of ring gear
D_{Sun}	Diameter of sun gear
d	Pitch diameter
E	Modulus of elasticity
F_b	Bending load applied to gear tooth
F_c	Centrifugal force
F_d	Design values for loads
F_t	Tangential force on gear tooth
F_k	Expected values of the loads
f_d	Design values for materials
f_e	Excitation frequency, Hz
f_k	Characteristic values of the materials
f_n	Natural frequency, Hz
g	(i) Gravitational constant
g	(ii) Gearbox ratio
h	Height of gear tooth
I	Moment of inertia
I_{15}	Turbulence intensity at 15 m/s
k_g	Effective spring constant of two meshing gear teeth
L	(i) Distance to the weakest point on gear tooth
L	(ii) Height of tower
M	Moment
M_g	Moment due to gravity
m_{Tower}	Mass of tower
$m_{Turbine}$	Mass of turbine
N	(i) Number of gear teeth
N	(ii) Number of cycles
n	Rotational speed (rpm)
n_{HSS}	Rotational speed of high-speed shaft (rpm)
n_{LSS}	Rotational speed of low-speed shaft (rpm)
n_{rated}	Rotational speed of generator at rated power (rpm)
n_{rotor}	Rotor rotational speed (rpm)
n_{sync}	Synchronous rotational speed of generator (rpm)
$P_{generator}$	Generator power
P_{rated}	Rated generator power
P_{rotor}	Rotor power
p	Circular pitch of gear
P	Power
Q	Torque
R	(i) Reversing stress ratio
R	(ii) Radius
$R(f_d)$	Resistance function, normally design stress

r_{cg}	Distance to center of gravity
$S(F_d)$	Expected 'load function' for ultimate loading, normally expected stress
T	(i) Force in wire rope
T	(ii) Thrust
t	Thickness
U	Wind speed
$U(z)$	Wind speed at height z above the ground
U_{ave}	Annual average wind speed
U_{e1}	1-year extreme wind speed
U_{e50}	50-year extreme wind speed
U_{gust50}	50-year return period gust
U_{hub}	Hub-height wind speed
U_{ref}	Reference wind speed
V_{pitch}	Gear pitch circle velocity
W	Blade weight
y	Form factor (or Lewis factor)
z	Height above ground

A.6.2 Greek variables

β_i	Constants in vibrating beam
γ	Safety factor
γ_f	Partial safety factor for loads
γ_m	Partial safety factor for materials
γ_n	Consequence of failure safety factor
δ	Logarithmic damping decrement
δ_3	Delta-3 angle
η	Overall efficiency of the drive train
λ	Tip speed ratio
ξ	Damping ratio
ρ	Density of air
ρ_b	Mass density of blade
$\tilde{\rho}$	Mass density per unit length
σ	Cyclic stress amplitude
σ_a	Aerodynamically induced stress
σ_b	Allowed bending stress in gear tooth
σ_{bp}	Breaking stress
σ_c	Tensile stress due to centrifugal force
σ_g	Stress due to gravity
σ_t	Tensile stress in wire rope
σ_u	Ultimate strength.
σ_x	Standard deviation of turbulence in the direction of the mean wind
ω_n	Natural frequency (rad/s)
Ω	Rotor rotational speed (rad/s)

A.6.3 Subscripts

A	Aerodynamic
a	Air
b	Bending
bp	Breaking point
c	Cross section
cg	Center of gravity
d	Design
f	Loads
g	Gravity
k	Characteristic
m	Material
n	Consequence of failure
t	(i) Tension
t	(ii) Tangential
u	Ultimate
x	In direction of the wind

A.7 Chapter 7

A.7.1 English variables

A	Rotor swept area
B	Viscous friction coefficient
C	Capacitance
C_P	Rotor power coefficient
$C_{P,\max}$	Maximum rotor power coefficient
$e(t)$	Error
$g(t)$	Controller output
i	Index
J	Total inertia of blade and motor
J_r	Rotor inertia
K	Spring constant
K_D	Differential controller constant
K_I	Integral controller constant
K_P	Proportional controller constant
k	Slope of the torque voltage curve of a motor
m	Slope of the torque speed curve for a motor/pitch mechanism
n	Gearbox gear ratio
P_r	Rotor power
P_{el}	Electrical power
p	Number of generator poles
Q_p	Pitching moment
$Q_p(s)$	Laplace transform of pitching moment

Q_{ref} Desired rotor torque
R (i) Rotor radius
R (ii) Resistance, Ohms
s Complex frequency in Laplace transforms
s_i Sample at time i
t Time
t_i Time of the ith sample
t_0 Starting time
U_{cut-in} Cut-in wind speed
$U_{cut-out}$ Cut-out wind speed
U_{rated} Rated wind speed
$v(t)$ Voltage applied to pitch motor terminals

A.7.2 Greek variables

η Drive train efficiency
θ_p Blade position (pitch angle)
$\theta_{p,ref}$ Reference pitch angle.
$\Theta_p(s)$ Laplace transform of blade position.
λ Tip speed ratio
λ_{opt} Optimum tip speed ratio
ρ Air density
ω Frequency
Ω Rotor speed
Ω_{el} Generator speed

A.8 Chapter 8

A.8.1 English Variables

A_c Charnock constant
a (i) Non-dimensional velocity deficit (axial induction factor)
a (ii) Slope of least-squares linear fit
a (iii) Coherence decay constant
a_n Coefficient of cosine term in Fourier series
b Offset of least-squares linear fit
b_n Coefficient of sine term in Fourier series
$C_{D,10}$ Effective surface drag coefficient based on wind measured at 10 m.
C_T Turbine thrust coefficient
c Wind direction difference
c_n Magnitude of harmonic voltage of order n
cov Covariance
D Turbine diameter

D_X	Wake diameter at a distance X downstream of the rotor
f	The frequency in Hertz
g	Gravitational constant
HD_n	Harmonic distortion caused by the nth harmonic
I_F	Fault current
J	Rotating inertia
k	(i) Ratio of the standard deviations of the turbine power and the wind
k	(ii) Wake decay constant
k	(ii) Wind direction sector index
L	(i) Half the period of the fundamental frequency
L	(ii) The integral length scale of the turbulence in meters
M	Fault level
N	Number of wind turbines
n	(i) Harmonic number
n	(ii) Number of concurrent pairs of wind measurements in measure–correlate–predict (MCP) method
P	Real power
P_1	Average power of one wind turbine over a specific time interval
P_L	Power from electrical loads
P_N	Power dissipated by electrical losses
P_N	Power production from N wind turbines
P_O	Power from other generators
P_{PM}	Prime mover power
$p_i(U_c^*)$	Probability distribution of the predicted hourly wind speed where i is the index of the wind speed bin
p_k	Wind direction probability distribution
Q	Reactive power
Q_L	Torques from electrical loads
Q_O	Torques from other generators
Q_{PM}	Prime mover torque
R	Resistance
$S_1(f)$	The single point power spectrum
$S_N(f)$	Spectra of the total variance in the wind at N turbine locations
s^2	Estimate of the error sum of squares
T	Time period
t	Time
U	The mean wind speed
U_*	Friction velocity
U_0	Free stream velocity
U_{10}	Mean wind speed at an elevation of 10 m
U_c^*	Predicted wind speed at candidate site
$\overline{U}_c$	Mean hourly wind speed at the candidate site
$\overline{U}_{c,k}^*$	Predicted long-term mean wind speed in kth direction sector at the candidate site
$U_{c,i}^*$	Predicted hourly wind speed as a function of the index of the wind speed bin, i, at the reference site

$\overline{U}_r$	Long term mean wind speed at reference site
U_r	Mean hourly wind speed at reference site
U_X	Velocity in the wake at a distance X downstream of the rotor
$U_{X,1}$	Velocity in the wake at a distance X downstream of the rotor 1
$U_{X,2}$	Velocity in the wake at a distance X downstream of the rotor 2
$U(z)$	The mean wind speed at height z
V	Mean voltage
V_G	Wind turbine voltage
V_S	Grid system voltage
var	Variance
v	Instantaneous voltage
v_F	Sinusoidal voltage at the fundamental frequency
v_H	Harmonic voltage
v_n	Harmonic voltage of order I, ($n > 1$):
X	(i) Distance downstream of the rotor
X	(ii) Reactance
x	Fetch (distance from shore)
x_{ii}	The spacing between points i and j (meters)
z	Height
z_0	Roughness length
z_r	Reference height

A.8.2 Greek Variables

$\gamma_u^2(f)$	The coherence between the power spectrums at points i and j
θ_c	Mean hourly direction at the candidate site
θ_c^*	Predicted direction at candidate site
θ_r	Mean hourly direction at reference site
κ	Von Karman constant
σ	Standard deviation
σ_k	Standard deviation of wind speed in each direction bin
$\sigma_{P,1}$	Standard deviation of power from one wind turbine over a specific time interval
$\sigma_{P,N}$	Standard deviation of the wind farm electrical power output from N wind turbines
σ_u	Standard deviation of wind speed
φ_n	Phase of harmonic voltage of order n
ω	Rotational speed

A.9 Chapter 9

A.9.1 English Variables

A	Installment
AAR	Average annual return
B/C	Benefit–cost ratio
b	Learning parameter
C	Cost
$\underline{C}_c$	Capital cost of system
$\overline{C}_c$	Capital cost of system normalized by rated power
$C_{O\&M}$	Average annual operation and maintenance (O&M) Costs
$\underline{C}_{O\&M}$	Annual cost of operation and maintenance
$\overline{C}_{O\&M}$	Direct cost of operation and maintenance per unit of energy
$C(V)$	Cost of an object as a function of volume
$C(V_0)$	Cost of an object as a function of initial volume
COE	Cost of energy (cost/kWh)
COE_L	Levelized cost of energy
CRF	Capital recovery factor
E_a	Annual energy production (kWh)
F_N	Payment at end of N years
FC	Capacity factor
FCR	Fixed charge rate
FV	Future value
f_{OM}	Annual operation and maintenance (O&M) cost fraction
IRR	Internal rate of return
i	General inflation rate
j	Index for year
L	Lifetime of system
N	(i) Number of years or installments
NPV	Net present value
NPV_S	Net present value of savings
NPV_C	Net present value of costs
P_a	Annual payment
P_d	Down payment
P_e	Price obtained for electricity
PV	Present value
PV_N	Present value of future payment in year N
PWF	Present worth factor
ROI	Return on investment
r	Discount rate
S	Savings
SP	Simple payback (years)

SPW	Series present worth factor
s	Progress ratio
V	Volume of object produced
V_0	Initial cumulative volume
$Y(k,\ell)$	Function to obtain present value of a series of payments; k,ℓ : arguments

A.10 Chapter 10

A.10.1 English Variables

A_1	Additional path loss between transmitter and the receiver
A_2	Additional path loss between the transmitter and the obstacle
A_r	Additional path loss between the obstacle and the receiver
B	Number of blades
C	Signal strength
C/I	Signal to interference ratio
D	Rotor diameter
E	Electric field strength (mV m^{-1})
f	Frequency (Hz)
G_r	Receiver gain
G_r'	Receiver gain (obstacle path)
i	Noise source index
I	Interference length
L_{dn}	Day–night noise level (dB)
L_{eq}	Equivalent noise level (dB)
L_p	Sound pressure level (dB)
L_r	Free space loss (dB)
L_{total}	Total sound power level (dB)
L_w	Sound power level (dB)
L_{WA}	Predicted sound power level (dB)
L_x	A-weighted noise level exceeded x% of the time (dB)
L_1	Free space path loss (transmitter–receiver)
L_2	Free space path loss (transmitter–obstacle)
L	Free space path loss (obstacle–receiver)
P	Sound power
P_t	Transmitter power
P_T	Rated power of wind turbine
p	Sound pressure
R	Distance from noise source
r	Receiver distance from turbine
t	Time delay between desired and undesired signal
U	Wind speed
u	Velocity of sound
v_{Tip}	Blade tip speed velocity

W	Source sound power
W_0	Reference source sound power

A.10.2 Greek Variables

α	Sound absorption coefficient (dB m^{-1})
λ	Wavelength
ΔG	Antenna discrimination factor (dB)
Π	Relative position (spherical coordinates)
σ	Radar cross-section
σ_b	Bistatic radar cross-section of obstacle
τ	Time delay of signal to receiver
ω	Frequency of transmitted signal (rad s^{-1})
Ω	Turbine rotor speed (rad s^{-1})

A.11 Abbreviations

a/b	air-borne
AC	alternating current
A/D	analog-to-digital
AM	amplitude modulated
AVR	automatic voltage control
BEM	blade element momentum
CFD	computational fluid dynamic
CFM	cash flow method
CPU	computer processing unit
D/A	digital-to-analog
DC	direct current
ECD	extreme coherent gust with change in direction
ECG	extreme coherent gust
EMF	electromagnetic field
EMI	electromagnetic interference
EOG	extreme operating gust
EPA	Environmental Protection Agency, US
EPR	ethylene propylene rubber
EPRI	Electric Power Research Institute
EPRI TAG	Electric Power Research Institute, Technical Analysis Group
EWM	extreme wind speed
EWS	extreme wind shear
FAA	Federal Aviation Authority
FM	frequency modulated
GIS	geographical information system
GRP	fibreglass reinforced plastic
GTO	gate turn off thyristor

HAWT	horizontal axis wind turbine
HF	high-frequency
HVDC	high-voltage direct-current
ID	inner diameter
IEC	International Electrotechnical Commission
IGBT	insulated gate bipolar transistor
I/O	input–output
IPP	independent power producer
LCC	life cycle costing
LOLE	loss of load expectation
LOLP	loss of load probability
LORAN	long-range version of VOR
MCP	measure–correlate–predict
MMF	magnetomotive force
MOSFET	metal-oxide semiconductor field-effect transistor
NREL	National Renewable Energy Laboratory
NTM	normal turbulence model
NWCC	National Wind Coordinating Committee
NWP	normal wind profile
O&M	operation and maintenance
OD	outer diameter
PCC	point of common coupling
pdf	probability density function
pf	power factor
PI	proportional–integral
PID	proportional–integral–derivative
PNL	Pacific Northwest Laboratories
POC	point of contact
psd	power spectral density
PURPA	Public Utilities Regulatory Policies Act (US)
PV	photovoltaic
PWM	pulse width modulation (inverter)
rms	root mean square
ROI	return on investment
ROV	remotely operated vehicle
rpm	rotations per minute
rps	rotations per second
RPS	renewable portfolio standards
RSHR	rotor swept hour risk
s/b	structure-borne
SCADA	supervisory control and data acquisition
SCFF	self-contained fluid-filled
SCR	silicon-controlled rectifier
SERI	Solar Energy Research Institute
SODAR	Sonic detection and ranging acoustic Doppler sensor system
SPL	sound pressure level

SSC	system supervisory controller
TALA	tethered aerodynamic lifting anemometer
TEFC	totally enclosed, fan cooled
THD	total harmonic distortion
TVI	television interference
U&PM	unscheduled and preventive maintenance
UHF	ultrahigh-frequency
VAR	Volt–Amperes reactive
VAWT	vertical axis wind turbine
VHF	very-high-frequency
VOR	VHF omnidirectional ranging
WEST	wood–epoxy saturation technique
XLPE	cross-linked polyethylene

Problems

B.1 Problem Solving

Most of the problems in this text can be completed without reference to additional material than is in the text. In some cases, new information is introduced in the problem statement to extend the readers knowledge.

In some cases data files may be needed which are on the web site of the Renewable Energy Research Laboratory at the University of Massachusetts (http://www.ecs.umass.edu/mie/labs/rerl/index.html). This site also contains the Wind Engineering MiniCodes that have been developed at the University of Massachusetts at Amherst. A number of these codes may be useful in solving problems and, in some cases, may be needed to solve problems. The Wind Engineering MiniCodes are a complete set of short computer codes for examining wind energy related issues, especially in the context of an academic setting.

B.2 Chapter 2 Problems

B.2.1 Annual Energy Production Estimate

Based on average speed data only, estimate the annual energy production from a horizontal axis wind turbine with a 12 m diameter operating in a wind regime with an average wind speed of 8 m/s. Assume that the wind turbine is operating under standard atmospheric conditions ($\rho = 1.225$ kg/m^3). Assume a turbine efficiency of 0.4.

B.2.2 Wind Speed Variation with Height

a) Determine the wind speed at a height of 40 m over surface terrain with a few trees, if the wind speed at a height of 10 m is known to be 5m/s. For your estimate use two different wind speed estimation methods.

b) Using the same methods as part A, determine the wind speed at 40 m if the trees were all removed from the terrain.

B.2.3 Weibull Distribution Calculations

From an analysis of wind speed data (hourly interval average, taken over a one year period), the Weibull parameters are determined to be $c = 6$ m/s and $k = 1.8$.

a) What is the average velocity at this site?

b) Estimate the number of hours per year that the wind speed will be between 6.5 and 7.5 m/s during the year.

c) Estimate the number of hours per year that the wind speed is above 16 m/s.

B.2.4 Rayleigh Distribution Calculations

Analysis of time series data for a given site has yielded an average velocity of 6 m/s. It is determined that a Rayleigh wind speed distribution gives a good fit to the wind data.

a) Based on a Rayleigh wind speed distribution, estimate the number of hours that the wind speed will be between 9.5 and 10.5 m/s during the year.

b) Using a Rayleigh wind speed distribution, estimate the number of hours per year that the wind speed is equal to or above 16 m/s.

B.2.5 Annual Power Estimation Problem–Betz Type Machine

Estimate the annual production of a 12 m diameter horizontal axis wind turbine operating at standard atmospheric conditions (ρ = 1.225 kg/m^3) in a 8 m/s average wind speed regime. You are to assume that the site wind speed probability density is given by the Rayleigh density distribution.

B.2.6 Actual Data Analysis and Power Prediction

Based on the spreadsheet (MtTomData.xls) which contains one month of data (mph) from Holoyke, MA, determine:

a) The average wind speed for the month

b) The standard deviation

c) A histogram of the velocity data (via the method of bins–suggested bin width of 2 mph)

d) From the histogram data develop a velocity–duration curve

e) From above develop a power–duration curve for a given 25 kW Turbine at the Holyoke site.

For the wind turbine, assume:

$P = 0$ kW $\quad 0 < U \leq 6$ (mph)

$P = U^3/625$ kW $\quad 6 < U \leq 25$ (mph)

$P = 25$ kW $\quad 25 < U \leq 50$ (mph)

$P = 0$ kW $\quad 50 < U$ (mph)

f) From the power duration curve, determine the energy that would be produced during this month in kWh.

B.2.7 Statistical Data Analysis

Using results from Problem 2.6, carry out the following:

a) Determine Weibull and Rayleigh velocity distribution curves and normalize them appropriately. Superimpose them on the histogram of Problem 2.6.

b) Determine the Weibull and Rayleigh velocity duration and power duration curves and superimpose them on the ones obtained from the histogram.

c) Using the Weibull distribution, determine the energy that would be produced by the 25 kW machine at the Holyoke site.

d) Suppose the control system of the 25 kW machine were modified so that it operated as shown in Figure B.1 (and as detailed in Table B.1) How much less energy would be produced at Mt. Tom with the modified machine? Find a fourth order polynominal fit to the power curve. Use the Weibull distribution to calculate the productivity (in any manner you choose). Plot the power duration curves using the modified power curve and the cubic power curve from Problem 2.6 e).

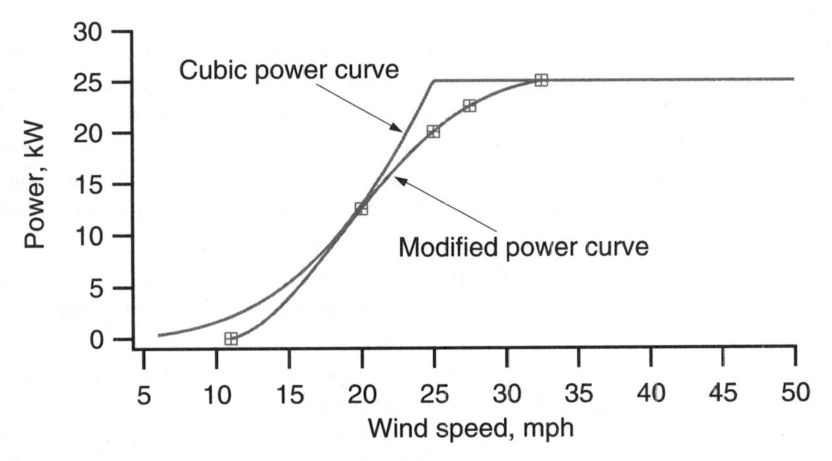

Figure B.1 Power curves for Problems 2.6 and 2.7

Table B.1 Power curve below rated power for Problem 2.7d.

Wind speed (mph)	Power (kW)
11	0
20	12.5
25	20
27.5	22.5
32.5	25

B.2.8 Power Spectral Density Estimation

Similar to Equation 2.3.17 in the Text, the following empirical expression has been used to determine the power spectral density (psd) of the wind speed at a wind turbine site with a hub height of z. The frequency is f (Hz), and n ($n = fz/U$) is a non-dimensional frequency.

$$\frac{f\,S(f)}{(2.5U^*)^2} = \frac{11.40n}{1+192.4n^{5/3}}$$

where

$$U^* = \frac{0.4U}{\ln\left(\frac{z}{z_0}\right)}$$

Determine the power spectral density of the wind at a site where the surface roughness is 0.05 m (z_0) and the hub height is 30 m, and the mean windspeed is 7.5 m/s.

B.2.9 Power Spectral Density Application

This problem uses the power spectral density (psd) to examine variance in the wind. A time series of hourly wind speeds (mph) from Mt. Tom for approximately 1 year is included in the data MtTomWindUM.txt. Routines to perform psd analysis are included with the UMass Wind Engineering MiniCodes. When the psd is graphed vs. the frequency it is hard to see features of interest. For this reason it is common to graph $fS(f)$ on the y axis vs. $\ln(f)$ on the x axis. When doing this the area under the curve between any two frequencies is proportional to the total variance associated with the corresponding range of frequencies.

a) Use the Minicodes to calculate the psd for the Mt. Tom wind data. Focus on the variations in wind over time periods of less than one month by using a segment length of 512.

b) Show from the results that the total variance as given by the integral of the psd [$S(f)$ vs. f] is approximately the same as what would be obtained in the normal way.

c) Show by equations that the area under the curve in a plot of $fS(f)$ vs. $\ln(f)$ is the same as it would be for a plot of $S(f)$ vs. f.

d) Plot $fS(f)$ vs. $\ln(f)$.

e) Find the amount of variance associated with diurnal fluctuations. Use frequencies corresponding to cycle times from 22 hours to 27 hours. How much variance is associated with higher frequency variations and how much with lower frequency variations?

B.2.10 Autocorrelation of wind speed data and data synthesis

A variety of techniques are available for creating data sets that have characteristics similar to that of real data. The Wind Engineering MiniCodes include a few of these methods. In the ARMA technique the user must input long-term mean, standard deviation, and autocorrelation at a specified lag. The code will return a time series with values that are close to the desired values. (Note: a random number generator is used in the data synthesis routines, so any given time series will not be exactly the same as any other.)

a) Find the mean, standard deviation, and autocorrelation for the Mt. Tom data, MtTom7Hzms.txt. This data is collected at a 25 m height, with a sampling frequency of 7.4 Hz. The data is in meters/second. Plot a time series of the data. Determine the autocorrelation for a lag of up to 2000 points. Determine the autocorrelation at a lag of one time step for use in synthesizing a similar data set.

b) Using the ARMA code, synthesize and plot a time series of 10,000 data points with equivalent statistics to those found in part a. Show a time series graph of the synthesized data.

c) Find the autocorrelation for a lag of up to 2000 points for both synthetic data and plot the autocorrelations of both the real and the synthesized on the same graph.

d) Comment on any similarities or differences between the two plots.

B.3 Chapter 3 Problems

B.3.1 Flying Blades

The blades of a wind turbine are ready to be installed on a turbine on top of a ridge. The horizontal blades are supported at each end by saw horses, when a storm front arrives. The turbine crew huddles in their truck as the rain starts and the wind picks up, increasing eventually to 26.82 m/s (60 mph). Realizing that the wind coming up the western slope of the ridge roughly follows the 10 degree slope, the field engineer performs a quick calculation and drives his truck upwind of the blades to disrupt the airflow around the blades, preventing them from being lifted by the wind and damaged.

The blades are 4.57 m (15 feet) long, 0.61 m (2 feet) wide, and have a mass of 45.36 kg (100 lbm). As the front arrives the temperature drops to 21.2°C (70°F). Assume that the blades are approximately symmetric airfoils (the engineer remembered that potential flow theory predicts that, prior to stall, the lift coefficient of a symmetric airfoil is approximately: $C_l = 2\pi \sin \alpha$). Assume that the center of both the lift and the drag force is concentrated over the center of mass of the blade and that the leading edge is facing into the direction of the wind. Assume the air density is 1.20 kg/m^3.

a) Was there a reason to be concerned? At what wind speed will the blades be lifted by the wind, assuming that there is no drag?

b) If they are lifted by a 26.82 m/s (60 mph) wind, how fast will they be accelerated horizontally, if the blade's lift to drag ratio, C_d / C_l , is 0.03?

B.3.2 Reynolds Numbers

The operating conditions found at two different points of a blade on a wind turbine are (Table B.2):

Table B.2

Location r/R	Relative wind velocity (m/s)	Relative wind velocity (ft/s)	Chord (m)	Chord (ft)	Angle of attack (degrees)
0.15	16.14	52.94	1.41	4.61	4.99
0.95	75.08	246.32	0.35	1.15	7.63

These conditions were determined at 0°C (32°F), for which the kinematic viscosity is 1.33×10^{-5} m^2/s. What are the Reynolds numbers found at each blade section?

B.3.3 Ideal Rotor Section Analysis

a) Find φ, θ_p, θ_T, and c for one blade section from r/R = 0.45 to r/R = 0.55 (centered on r/R = 0.50) for an ideal blade (assume $C_d = 0$, $a' = 0$). Assume λ =7, B=3, R=5 m, and C_l=1.0 and the minimum C_d / C_l occurs at $\alpha = 7$.

b) Assume that C_d / C_l actually equals 0.02 for the above blade section and that the free stream wind speed, U, equals 10 m/s. Find U_{rel}, dF_L, dF_D, dF_N, dF_T, dQ for the blade section. Don't forget to consider that the wind velocity is slowed down at the rotor. Use a = 1/3, $a' = 0$. Assume the air density is 1.24 kg/m^3 (20C).

c) For the same blade section find C_l, α and a using the general strip theory method (including angular momentum). Also find C_l, α and a if the rpm is increased such that λ = 8. Ignore drag and tip loss. Use a graphical approach. Assume that the empirical lift curve is $C_l = 0.1143\alpha + 0.2$ (α in degrees): i.e. C_l= 0.2 at α = 0 degrees, C_l= 1.0 at α = 7 degrees.

B.3.4 Ideal Rotor Design

a) Find φ, θ_p, θ_T, and c at all 10 locations (r/R = 0.10, 0.20, ..., 1.0) for the Betz optimum blade. Assume λ = 7, B = 3, R = 5 m, and C_l = 1.0 and the minimum C_d / C_l occurs at α = 7.

b) Sketch the shape (planform) of the blade, assuming that all the quarter chords lie on a straight line.

c) Illustrate the blade twist by drawing plausible airfoils with properly proportioned chord lengths, centered at the quarter chord chords for $r/R = 0.10, 0.50, 1.0$. Be sure to show where the wind is coming from and what the direction of rotation is.

B.3.5 Ideal Rotor C_P

Blades for a two-bladed wind turbine with a 24 m diameter have been designed for a tip speed ratio of 10. The 12-meter blades have the geometric and operational parameters listed in Table B.3 for operation at the design tip speed ratio. The rotor was designed assuming $C_l = 1.0$, $a' = 0$, no drag, and $a = 1/3$ using the methods outlined in the text for the design of an ideal rotor.

We want to know the rotor power coefficient for two assumed conditions: $C_d = 0$ and $C_d = 0.02$. Note that the two equations that have been derived do not serve our purpose here. Equation (3.7.11) requires a non-zero value for a' and Equation (3.7.12) has also been derived using relationships between a and a' that require non-zero values of a'.

Table B.3

Section radius r/R	Section radius (m)	Section pitch, θ_p degrees	Angle of relative wind, φ (degrees)	Section twist (degrees)	Chord, c (m)
0.05	0.60	46.13	53.13	49.32	4.02
0.15	1.80	16.96	23.96	20.15	2.04
0.25	3.00	7.93	14.93	11.12	1.30
0.35	4.20	3.78	10.78	6.97	0.94
0.45	5.40	1.43	8.43	4.61	0.74
0.55	6.60	-0.09	6.91	3.10	0.60
0.65	7.80	-1.14	5.86	2.04	0.51
0.75	9.00	-1.92	5.08	1.27	0.45
0.85	10.20	-2.52	4.48	0.67	0.39
0.95	11.40	-2.99	4.01	0.20	0.35

a) Starting with the definitions of the blade forces and the definition of C_P:

$$C_P = P / P_{wind} = \frac{\int_{r_h}^{R} \Omega \, dQ}{\frac{1}{2}\rho \pi R^2 U^3}$$

derive as simple an equation as you can for the power coefficient, C_P, of an ideal Betz limit rotor. The equation should include both lift and drag coefficients and tip speed ratio, and should assume that $a = 1/3$. Ignore tip losses

b) Using the above equation find the rotor C_P at the design tip speed ratio assuming that there is no drag (C_d = 0). How does this compare with the Betz limit?

c) For a first approximation of the effect of drag on rotor performance, find the C_p for the same rotor at the design tip speed ratio assuming the more realistic conditions that C_d/C_l actually equals 0.02. Assume that the drag has no effect on the aerodynamics and that the operating conditions assumed for the ideal rotor without drag apply. What effect does drag have on the rotor C_P, compared to the C_P assuming that $C_d = 0$?

B.3.6 Betz versus Optimum Rotor Shape

The Better Wind Turbine Company wants to start marketing wind turbines. The plans call for a 20 meter in diameter, three-bladed, wind turbine. The rotor is to have its peak power coefficient at a tip speed ratio of 6.5. The airfoil to be used has a lift coefficient of 1.0 and a minimum drag to lift ratio at an angle of attack of 7 degrees.

You, as the new blade designer, are to come up with two blade shapes as a starting point for the blade design. One shape assumes that there are no losses and that there is no wake rotation. The second design is based on the optimum rotor shape assuming that there is wake rotation (but still no losses).

Find the chord length, pitch, and twist at 10 stations of the blade, assuming that the blade extends right to the center of the rotor. How do the chord lengths and the twists compare at the tip and at the inner three blade stations?

B.3.7 Optimum Rotor C_P

The Better Wind Turbine Company wants to start marketing wind turbines. Their plans call for a turbine that produces 100 kW in a 12 m/s wind at a cold site (-22.8°C, -9°F) with an air density of 1.41 kg/m^3. They have decided on a 20 meter in diameter, three-bladed, wind turbine. The rotor is to have its peak power coefficient at a tip speed ratio of 7 in a 12 m/s wind. The airfoil to be used has a lift coefficient of 1.0 and a minimum drag to lift ratio at an angle of attack of 7 degrees.

a) You, as the new blade designer, are to come up with the blade shape as a starting point for the blade design. The design is to be based on the optimum rotor shape assuming that there is wake rotation (but no drag or tip losses). Find the chord length, pitch, and twist at 9 stations of the blade (each 1 m long), assuming that the hub occupies the inner tenth of the rotor.

b) Determine the rotor C_P assuming C_d = 0. Again determine the power coefficient assuming that the drag coefficient is 0.02, and that the aerodynamics are the same as the condition without any drag. How much power is lost due to drag? Which part of the blade produces the most power?

c) Does it look like the chosen design is adequate to provide the power that the Better Wind Turbine Company wants?

B.3.8 LS-1 Airfoil Section Analysis

A two-bladed wind turbine is designed using one of the LS-1 family of airfoils. The 13 m long blades for the turbine have the following specifications (Table B.4).

Table B.4 LS-1 Airfoil Blade Geometry

r/R	Section Radius (m)	chord (m)	twist (degrees)
0.05	0.65	1.00	13.000
0.15	1.95	1.00	11.000
0.25	3.25	1.00	9.000
0.35	4.55	1.00	7.000
0.45	5.85	0.87	5.000
0.55	7.15	0.72	3.400
0.65	8.45	0.61	2.200
0.75	9.75	0.54	1.400
0.85	11.05	0.47	0.700
0.95	12.35	0.42	0.200

Note: $\theta_{p,0}$ = -1.97 degrees (pitch at tip)

Assume that the airfoil's aerodynamics characteristics can be approximated as follows (note, α is in degrees):

For $\alpha < 21$ degrees:

$$C_l = 0.42625 + 0.11628\alpha - 0.00063973\alpha^2 - 8.712\times10^{-5}\alpha^3 - 4.2576\times10^{-6}\alpha^4$$

For $\alpha > 21$: $C_l = 0.95$

$$C_d = 0.011954 + 0.00019972\alpha + 0.00010332\alpha^2$$

For the *midpoint* of section 6 (r/R = 0.55) find the following for operation at a tip speed ratio of 8: a) angle of attack, α; b) angle of relative wind, φ; c) C_l and C_d; d) the local contributions to C_P. Ignore the effects of tip losses

B.3.9 WF-1 Rotor Analysis

This problem is based on the blades used for the UMass wind machine WF-1. Refer to Table B.5 for the blade geometry at specific locations along the blade. There is no airfoil below r/R = 0.10.

In addition, note that for the NACA 4415 airfoil: for α < 12 degrees: C_l = 0.368 + 0.0942α, $C_d = 0.00994 + 0.000259\alpha + 0.0001055\alpha^2$ (note, α is in degrees). Radius: 4.953 m (16.25 ft); No. of blades: 3; Tip speed ratio: 7; Rated wind speed: 11.62 m/s (26 mph). The pitch at the tip is: $\theta_{p,0}$ = -2 degrees.

a) Divide the blade into 10 sections (but assume that the hub occupies the innermost 1/10). For the *midpoint* of each section find the following: i) angle of attack, α; ii) angle of relative wind, φ; iii) C_l and C_d; iv) the local contributions to C_P and thrust. Include the effects of tip losses.

b) Find the overall power coefficient. How much power would the blades produce at 11.62 m/s (26 mph)? Include drag and tip losses. Assume an air density of 1.23 kg/m^3.

Table B.5 WF-1 Airfoil Geometry

r/R	Radius (ft)	Radius (m)	Chord (ft)	Chord (m)	Twist (degrees)
0.10	1.63	0.495	1.35	0.411	45.0
0.20	3.25	0.991	1.46	0.455	25.6
0.30	4.88	1.486	1.26	0.384	15.7
0.40	6.50	1.981	1.02	0.311	10.4
0.50	8.13	2.477	0.85	0.259	7.4
0.60	9.75	2.972	0.73	0.223	4.5
0.70	11.38	3.467	0.63	0.186	2.7
0.80	13.00	3.962	0.55	0.167	1.4
0.90	14.63	4.458	0.45	0.137	0.40
1.00	16.25	4.953	0.35	0.107	0.00

B.3.10 LS-1 Airfoil Section Analysis with Tip Losses

A two-bladed wind turbine is designed using one of the LS-1 family of airfoils. The 13 m long blades for the turbine have the specifications listed previously in Table B.4.

Assume that the airfoil's aerodynamics characteristics can be approximated as follows (note, α is in degrees):

For $\alpha < 21$ degrees:

$$C_l = 0.42625 + 0.11628\,\alpha - 0.00063973\,\alpha^2 - 8.712 \times 10^{-5}\,\alpha^3 - 4.2576 \times 10^{-6}\,\alpha^4$$

For $\alpha > 21$: $C_l = 0.95$

$$C_d = 0.011954 + 0.00019972\,\alpha + 0.00010332\,\alpha^2$$

For the *midpoint* of the outermost section of the blade ($r/R = 0.95$) find the following for operation at a tip speed ratio of 8: a) angle of attack, α, with and without tip losses; b) angle of relative wind, θ; c) C_l and C_d; d) the local contributions to C_P; e) the tip loss factor, F, and the axial induction factor, a, with and without tip losses. How do tip losses affect aerodynamic operation and the local contribution to C_P at this outermost section?

B.3.11 Stalled Rotor Section

A two-bladed wind turbine is operated at two different tip speed ratios. At 8.94 m/s (20 mph) (tip speed ratio = 9) one of the blade sections has an angle of attack of 7.19 degrees. At 16.09 m/s (36 mph) (tip speed ratio = 5) the same 1.22 m (4 ft) section of the blade is starting to stall, with an angle of attack of 20.96 degrees. Given the following operating conditions and geometric data, determine the relative wind velocities, the lift and drag forces, and the tangential and normal forces developed by the blade section at the two different tip speed ratios. Determine, also, the relative contribution (the fraction of the total) of the lift and drag forces to the tangential and normal forces developed by the blade section at the two different tip speed ratios.

How do the relative velocities and the lift and drag forces compare? How do the tangential and normal forces compare? How do the effects of lift and drag change between the two operating conditions?

Operating conditions are listed in Table B.6:

Table B.6 Operating conditions

λ	α	φ	a
5	20.96	22.21	0.070
9	7.19	8.44	0.390

This particular blade section is 1.22 m (4 ft) long, has a chord length of 0.811 m (2.66 ft), and has a center radius of 5.49 m (18 ft). The following lift and drag coefficients are valid for $\alpha < 21$ degrees, where α is in degrees:

$$C_l = 0.42625 + 0.11628\alpha - 0.00063973\alpha^2 - 8.712\times10^{-5}\alpha^3 - 4.2576\times10^{-6}\alpha^4$$

$$C_d = 0.011954 + 0.00019972\alpha + 0.00010332\alpha^2$$

B.4 Chapter 4 Problems

B.4.1 Rotational Energy

A wind turbine rotor turning at 60 rpm is brought to a stop by a mechanical brake. The rotor inertia is 13 558 kgm^2.

a) What is the kinetic energy in the rotor before it is stopped? How much energy does the brake absorb during the stop?

b) Suppose that all the energy is absorbed in a steel brake disc with a mass of 27 kg. Ignoring losses, how much does the temperature of the steel brake disc rise during the stop? Assume a specific heat for steel of 0.46 kJ/kg-C.

B.4.2 Shaft Deflection

A cantilevered 2 m long main shaft of a wind turbine holds a 1500 kg hub and rotor at its end. At rated power the turbine develops 275 kW and rotates at 60 rpm. The shaft is a 0.15 m in diameter cylindrical steel shaft.

a) How much does the shaft bend down at its end as a result of the load of the rotor and hub?

b) How much does the shaft twist when the turbine is operating at rated power? What is the maximum shear stress in the shaft?

B.4.3 Tower Bending

A wind turbine on a 24.38 m (80 ft) tower is subject to a thrust load of 26.69 kN (6000 lbf) during operation at 250 kW, the rated power of the turbine. In a 44.7 m/s (100 mph) hurricane the thrust load on the stopped turbine is expected to be 71.62 kN (16100 lbf).

a) If the tower is a steel tube 1.22 m (4 ft) in outer diameter (O.D.) with a 0.0254 m (1 in) thick wall, how much will the top of the tower move during rated operation and in the hurricane force winds?

b) Suppose the tower were a three-legged lattice tower with the specifications given in Figure B.2, how much will the top of the tower move during rated operation and in the hurricane force winds? Ignore any effect of cross bracing on the tower.

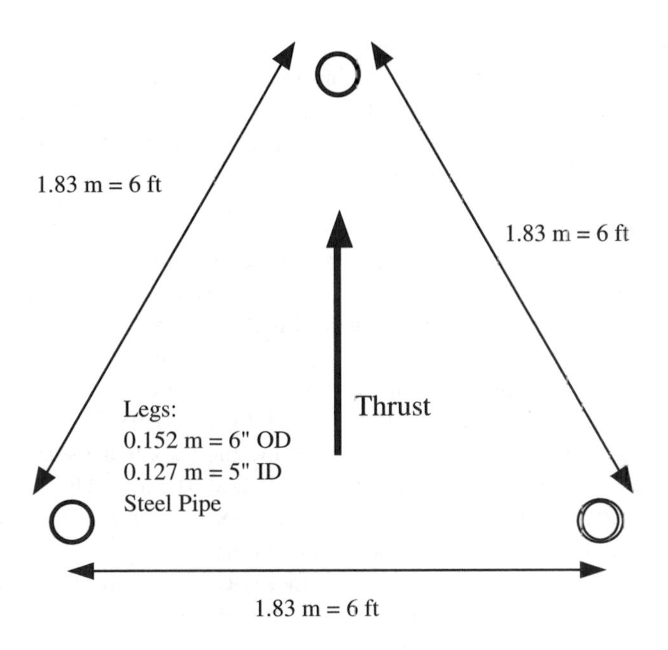

Figure B.2 Lattice tower cross-section; ID, inner diameter; OD, outer diameter

B.4.4 Gyroscopic Forces

The wind turbine rotor shown in Figure B.3 has a rotor rotation velocity, Ω, of 1 Hz (60 rpm) and is yawing at an angular velocity, ω, of 10 degrees per second. The polar moment of inertia of the rotor is 13 558 kgm^2 (10 000 slug ft^2). The rotor weighs 1459 kg (100 slugs) and is 3.05 m (10 feet) from the center of the bed plate bearing support. Centered over the bed plate bearing support are the bearings holding the main shaft. These bearings are 0.91 m (3 ft) apart. The directions of positive moments and rotation are indicated in the figure.

a) What are the bearing loads when the turbine is not yawing?
b) What are the bearing loads when the turbine is yawing?

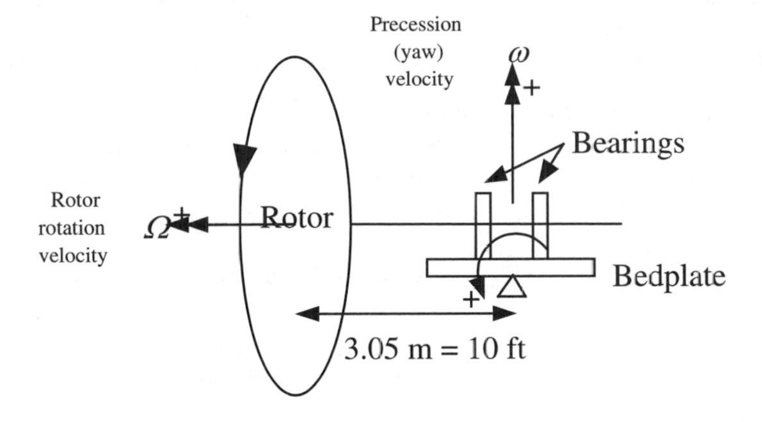

Figure B.3

B.4.5 Hinge–Spring and Offset

A 19.05 m (62.5 ft) long turbine blade has a non-rotating natural frequency of 1.67 Hz. In operation, rotating at 50 rpm, it has a natural frequency of 1.92 Hz. If the blade has a mass of 898 kg (61.53 slugs), what hinge–spring stiffness and offset would be used to model the blade dynamics with the simple dynamics model?

B.4.6 Wind Components

Blade bending moments are being measured on a research wind turbine on a day with mean hub height winds of 9.14 m/s (30 ft/s), but a wind shear results in winds of 12.19 m/s (40 ft/s) at the top of the blade tip path and 6.09 m/s (20 ft/s) at the bottom of the blade tip path. A gust of wind has started the wind turbine yawing at a steady rate of 0.1 radians per second in the $+X'$ direction (see Figure 4.17 in the text). Meanwhile a wind direction change results in a crosswind of +0.61 m/s (2.0 ft/s). The 24.38 m (80 ft) diameter turbine starts with a flap hinge angle, β, of 0.05 radians and, at the moment that measurements of are being made, the rate of change of the flap hinge angle is 0.01 rad/s. The rotor is 3.05 m/s (10 ft) from the yaw axis on this fixed speed turbine that rotates at a speed of 1 Hz. The very efficient turbine is operating with an axial induction factor of 1/3.

Ignoring tower shadow and transient effects, if these operating conditions were to persist for one revolution of the rotor, what would the perpendicular and tangential wind as a function of azimuth angle be half way out on the blades? What would the magnitude of the contributions to the perpendicular wind be from yaw rate, shear, crosswind, and blade flapping? How would the angle of attack vary at this part of the blade as the azimuth changes? Assume that the blade pitch angle, θ_p, is 0.05 radians (2.86 degrees).

B.4.7 Steady State Flap Angle

When only blade rotation and the hinge–spring and offset are included in the calculation of the flap angle, the only non-zero term is the steady state flap angle, β_0.

a) What is the derivative of the steady state flap angle with respect to wind velocity? Assume that the axial induction factor is also a function of wind velocity.

b) Assuming that the Lock number is positive, what would the effect of a negative value of da/dU be on the steady state flap angle?

c) Suppose a = 1/3 and $da/dU = 0$. In this case, what is the expression for the derivative of the steady state flap angle with respect to wind velocity?

d) What is the derivative of the steady state flap angle with respect to blade pitch?

B.4.8 Steady State Lead–Lag Motion

Just as the flap angle can be represented by a the sum of a constant term, a sine term, and a cosine term, the lead-lag angle in the simplified dynamics model can be represented by the sum of a constant term, a sine term, and a cosine term:

$$\zeta \approx \zeta_0 + \zeta_{1c} \cos(\psi) + \zeta_{1s} \sin(\psi)$$

The following matrix equation for the solution of the lead–lag motion can be derived by substituting this solution into the lead–lag equations of motion (note that flap coupling terms have been omitted for simplicity):

$$\begin{bmatrix} 2B & (K_2-1) & 0 \\ 0 & 0 & (K_2-1) \\ K_2 & B & 0 \end{bmatrix} \begin{bmatrix} \zeta_0 \\ \zeta_{1c} \\ \zeta_{1s} \end{bmatrix} = \begin{bmatrix} -\frac{\gamma}{2}\left[K_{vs}\overline{U}A_4 - \frac{\theta_P \Lambda}{2}\left(\overline{V}_0 + \overline{q}\overline{d}\right)\right] \\ -2B - \frac{\gamma}{2}\overline{q}A_4 \\ \frac{\gamma}{2}\Lambda A_2 \end{bmatrix}$$

where:

K_2 = inertial natural frequency (includes offset, hinge–spring) $= \varepsilon_2 + \dfrac{K_\zeta}{I_b \Omega^2} = \left(\dfrac{\omega_\zeta}{\Omega}\right)^2$

A_2 = axisymmetric flow term (includes tip speed ratio and pitch angle) $= \dfrac{\Lambda}{2} + \dfrac{\theta_p}{3}$

A_4 = axisymmetric flow term (includes tip speed ratio and pitch angle) $= \dfrac{2}{3}\Lambda - \dfrac{\theta_p}{4}$

B = gravity term $= \dfrac{G}{2\Omega^2}$

γ = Lock number (ratio of aerodynamic force to moment of inertia) $= \dfrac{\rho C_{L\alpha} c R^4}{I_b}$

$\overline{V}_0$ = normalized cross flow term ($= V_0 / \Omega R$)

$\overline{q}$ = normalized yaw rate term ($= q / \Omega$)

$\overline{d}$ = normalized yaw moment arm $= \dfrac{d_{yaw}}{R}$

$\overline{U}$ = normalized wind velocity $= \dfrac{U}{\Omega R}$

Λ = non-dimensional inflow $= \dfrac{U - u_i}{\Omega R}$

K_{vs} = linear wind shear constant.

θ_p = pitch angle

a) Write the matrix equation for lead–lag motion, including the terms for the steady mean wind and the hinge spring model, but assuming that gravity is zero and that other aerodynamic forcing functions are zero.

b) Solve the equations and find the expression for the steady state lead–lag angle as a function of the Lock number, the non-dimensional wind speed, the lead–lag natural frequency, and the blade pitch.

c) Suppose the pitch were 5.7 degrees, the rotor speed were 50 rpm and the diameter were 9.14 m (30 ft). By what factor would the steady state lead–lag angle increase if the wind speed increased from 7.62 m/s (25 ft/s) to 15.24 m/s (50 ft/s)? Assume that the axial induction factor decreases from 0.30 to 0.25 as the speed increases to 15.24 m/s (50 ft/s).

B.4.9 Lead–Lag Motion with Wind Shear

Using the lead–lag equations described in Problem 4.8:

a) Write the matrix equation for lead–lag motion if all of the terms are ignored except those that have to do with steady winds and vertical wind shear.

b) Solve the equations and find the expression for the lead–lag angle with steady winds and vertical shear. What is the effect of vertical wind shear on each term of the lead–lag angle in the absence of yaw rate, gravity, and crosswind?

B.5 Chapter 5 Problems

B.5.1 Simple DC circuit

A 48 V DC wind turbine is hooked up to charge a battery bank consisting of four 12 V batteries in series. The wind is not blowing, but the batteries must supply 2 loads. One is a light bulb, rated at 175 W at 48 V; the other is a heater, rated at 1000 W at 48 V. The loads are in parallel. The batteries may be considered to be constant 12 V voltage sources, with an internal series resistance of 0.05 Ohm each. How much power is actually supplied to the loads? How does that power compare to that which is expected? (Ignore the effect of temperature on resistance of filament and heating elements.)

B.5.2 Electromagnet

An electromagnet is used to hold a wind turbine's aerodynamics brakes in place during operation. The magnet supplies 30 lbs (133.4 N) of force to do this. The magnet is supplied by a 60 V DC source. The coil of the electromagnet draws 0.1 A. The core of the electromagnet is assumed to have a relative permeability of 10^4. The diameter of the core is 3 in (7.62 cm).

Find the number of turns in the coil and the wire size. Assume that the relation between force, magnetic flux, and area of the core is given by $F = 397\ 840\ B^2 A_c$:
where: F= force (N), B = magnetic flux (Wb/m^2), A_c = area of core (m^2). Also, assume that the length of each turn is equal to the circumference of the core. The resistivity of copper is $\rho = 1.72 \times 10^{-6}\ \Omega\,\text{cm}$.

B.5.3 Phasors

Consider the following phasors: $\mathbf{X} = 10 + j14$, $\mathbf{Y} = -4 + j5$. Find the following, and express in both rectangular and polar form: $\mathbf{X+Y}$, $\mathbf{X-Y}$, $\mathbf{XY}$, $\mathbf{X/Y}$.

B.5.4 Three-Phase Power

Show that the magnitude of the line-to-neutral voltage in a balanced, Y connected three phase system (V_{LN}) is equal to the line-to-line voltage (V_{LL}) divided by the square root of 3, i.e.

$$V_{LN} = V_{LL} / \sqrt{3}.$$

The line feeding a Y-connected three-phase generator has a line-to-line voltage of 480 V. What is the line-to-neutral voltage?

B.5.5 Simple AC circuit

An AC circuit has a resistor, capacitor and inductor in series with a 120 V, 60 Hz voltage source. The resistance of the resistor is 2 Ohms, the inductance of the inductor is 0.01 Henry; the capacitance of the capacitor is 0.0005 Farads. Find the following: reactance of the capacitor and inductor, current, apparent power, real power, reactive power, power factor angle and power factor.

B.5.6 Transformer

A transformer rated at 120 V/480 V, 10 kVA has an equivalent circuit as shown in Figure B.4. The low voltage side is connected to a heater, rated at 5 kW, 120 V. The high voltage side is connected to a 480 V, 60 Hz, single-phase power line. Find the actual power transferred, the magnitude of the measured voltage across the heater, and the efficiency (power out/power in) of the transformer.

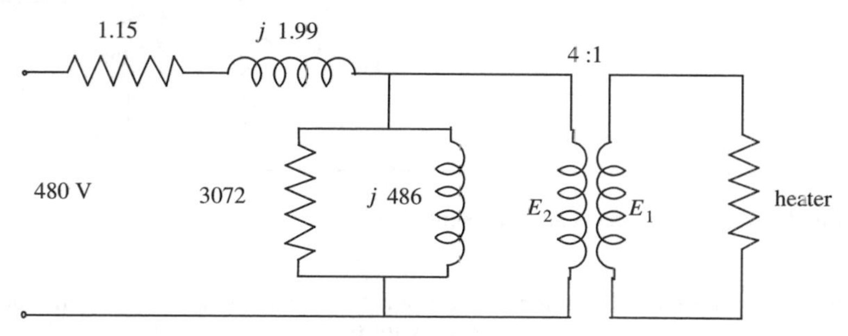

Figure B.4 Transformer equivalent circuit

B.5.7 Rectified AC voltage

A small wind turbine generator (single phase) produces a 60 Hz voltage at 120 V rms. The output of the generator is connected to a diode bridge full-wave rectifier, which produces a fluctuating DC voltage. What is the average DC voltage? A silicon-controlled rectifier (SCR) is substituted for the diode rectifier. Under one condition the SCRs are turned on at 60 deg after the beginning of each half cycle. What is the average DC voltage in that case?

B.5.8 Induction Machine Parameters

The parameters of an induction machine can be estimated from test data taken under a few specified conditions. The two key tests are the no-load test and blocked-rotor test. Under the no-load test, the rated voltage is applied to the machine and it is allowed to run at no-load (i.e. with nothing connected to the shaft). In the blocked-rotor test the rotor is prevented from turning and a reduced voltage is applied to the terminals of the machine. In both cases the voltage, current and power are measured. . Under the no-load test, the slip is essentially equal to 0, and the loop with the rotor parameters may be ignored. The magnetizing reactance accounts for most of the impedance and can be found from the test data. Under blocked-rotor conditions, slip is equal to 1.0 so the magnetizing reactance can be ignored, and the impedance of the leakage parameters can be found. A third test can be used to estimate the windage and friction losses. In this test the machine being tested is driven by a second machine but it is not connected to the power system. The power of the second machine is measured, and from that value is subtracted the latter's no-load power. The difference is approximately equal to the test machine's windage and friction.

In a simplified version of the analysis, all of the leakage terms may be assumed to be on the same side of the mutual inductance, as shown in Figure B.5. In addition, the resistance in parallel with the magnetizing reactance is assumed to be infinite, the stator and rotor resistances, R_S and R'_R, are assumed equal to each other and the leakage inductances are also assumed to be equal to each other.

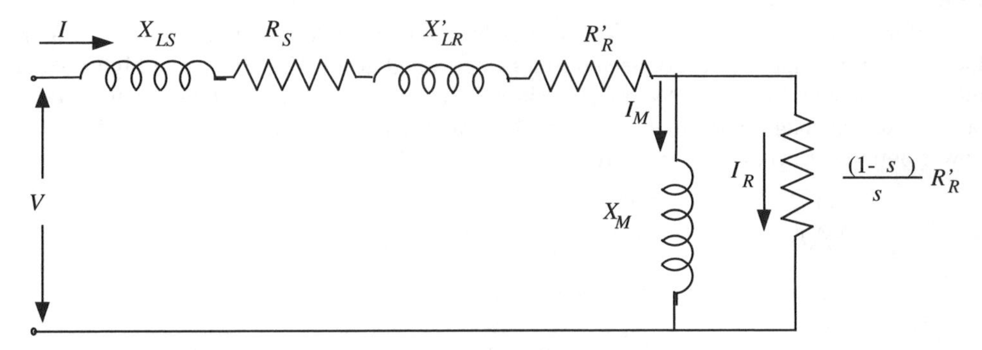

Figure B.5 Simplified induction machine equivalent circuit

The following test data are available for a wye-connected, three-phase induction machine:

No-load test: $V_0 = 480$ V, $I_0 = 46$ A, $P_0 = 3500$ W

Blocked rotor test: $V_B = 70$ V, $I_B = 109$ A, $P_B = 5600$ W

Windage and friction losses equal 3000 W.

Find the parameters for the induction generator model.

B.5.9 Induction Generator Performance

A four pole induction generator is rated at 300 kVA and 480 V. It has the following parameters: $X_{LS} = X_{LR} = 0.15\ \Omega$, $R_{LS} = 0.014\ \Omega$, $R_{LR} = 0.0136\ \Omega$, $X_M = 5\ \Omega$. How much power does it produce at a slip of -0.025? How fast is it turning at that time? Also, find the torque, power factor and efficiency. (Ignore mechanical losses.)

Suppose the generator is used in a wind turbine, and the torque due to the wind is increased to a value of 2100 Nm. What happens?

B.5.10 Synchronous Generator

A resistive load of 500 kW at 480 V is supplied by an eight-pole synchronous generator, connected to a diesel engine, and a wind turbine with the same induction generator as that in Problem 5.9. The synchronous machine is rated to produce 1000 kVA. Its synchronous reactance is 0.4 Ohms. A voltage regulator maintains a constant voltage of 480 V across the load and a governor on the diesel engine maintains a fixed speed. The speed is such that the grid frequency is 60 Hz. The power system is three-phase, and the electrical machines are both Y-connected. The power from the wind is such that the induction generator is operating at a slip of -0.01. At this slip, the power 153.5 kW and the power factor is 0.888.

a) Find the synchronous generator's speed, power factor, and power angle.

b) Confirm that the power and power factor are as stated.

B.5.11 Harmonics from Inverter

A six-pulse inverter has a staircase voltage, two cycles of which are shown in Figure B.6. The staircase rises from 88.85 V to 177.7 V, etc. The frequency is 60 Hz.

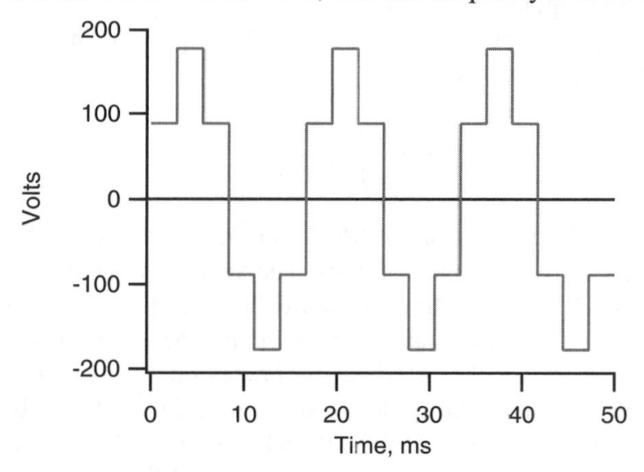

Figure B.6 Inverter staircase voltage

a) Show that the voltage can be expressed as the following Fourier series:

$$V_{inv}(t) = A\left[\sin(\omega t) + \frac{1}{5}\sin(5\omega t) + \frac{1}{7}\sin(7\omega t) + \frac{1}{11}\sin(11\ \omega t) + \ldots\right]$$

where $A = (3/\pi)(177.7)$ and $\omega = 2\pi f$

b) A series parallel resonance filter is connected across the terminals. The reactance of each component is the same, so that

$$f(n) = \left| n^2 \Big/ \left\{ \left(n^2 - 1\right)^2 - n^2 \right\} \right| .$$

Show how the filter decreases the harmonics. Find the reduction to the filter of the fundamental and at least the first three non-zero harmonics. Illustrate the filtered and unfiltered voltage waveform.

Hint: in general, a Fourier series for a pulse train (a square wave) can be expressed as:

$$V(t) = \frac{4}{\pi}H\left[\sin(\omega t) + \frac{1}{3}\sin(3\ \omega t) + \frac{1}{5}\sin(5\omega\ t) + \frac{1}{7}\sin(7\omega t) + \frac{1}{9}\sin(9\omega t) + \ldots\right]$$

where $\omega = 2\pi f$ and H = the height of the square wave above zero.

B.6 Chapter 6 Problems

B.6.1 Brake Return Spring

For a helical coil spring, the spring constant, k, can be expressed in terms of the spring's dimensions and its material properties:

$$k = \frac{Gd^4}{64R^3 N_c}$$

where: G = modulus of elasticity in shear, N/m^2 (lb/in^2); d = diameter of wire, m (in); R = radius of coil, m (in), and N_c = number of coils. The average value of G for spring steel is 11.5x10^6 lb/in^2 (7.93x10^{10} N/m^2). The coil radius is based on the distance from the longitudinal axis of the spring to the center of the wire making up the coil.

A spring is needed to return the aerodynamic brakes on a wind turbine blade to their closed position after deployment. The spring should exert a force of 147 lb. (653.9 N) when extended 6 inches (0.1524 m) (brake deployed) and a force of 25 lb. (111.2 N) when the brake is closed. Space limitations dictate that the diameter of the spring be equal 1.5 inches (0.0381 m). A spring of wire gage 7 (0.177 inches or 4.5 mm diameter) is available. How many coils would be required for this spring to have the desired force? How much would the spring coil weigh? Assume a density for steel of 498 lb/ft^3 (76 815 N/m^3).

B.6.2 Gearbox ball bearings

The basic rating load for ball bearings, L_R, is that load for which a bearing should perform adequately for at least one million revolutions. A general equation relates the basic rating load for ball bearings [with balls up to 1 inch (0.083 m diameter)] to a few bearing parameters:

$$L_R = f_c\left[n_R \cos(\alpha)\right]^{0.7} Z^{2/3} D^{1.8}$$

where f_c = constant, between 3,500 and 4,500 (260,200 to 334,600 for ball diameter in m), depending on $D\cos(\alpha)/d_m$; D = ball diameter, inches (m); α = angle of contact of the balls (often 0 deg); d_m = pitch diameter of ball races [= (bore + outside diameter)/2]; n_R = number of rows of balls; Z = number of balls per row.

If two bearings are to operate for a different number of hours, the rating loads must be adjusted according to the relation:

$$L_2 / L_1 = \left(N_1 / N_2\right)^{1/3}$$

A bearing used on the output shaft of a wind turbine's gearbox has the following characteristics: OD = 4.9213 inches (0.125 m), ID (bore) =2.7559 inches (0.07 m). The bearing has 13 balls, each 11/16 inches (0.01746 m) in diameter. Find the basic rating load of the bearing. Suppose that the gearbox is intended to run for 20 years, at 4000 hrs/yr. How does the bearing load rating change? Assume that f_c = 4500. The angle of contact of the balls is equal to zero. The shaft is turning at 1800 rpm.

B.6.3 Tower bolts

A wind turbine is to be mounted on a three-legged truss tower 30 m high. The turbine has a mass of 5000 kg. The rotor diameter is 25 m. The tower is uniformly tapered and has a mass of 4000 kg. The tower legs are bolted in an equilateral pattern, 4 m apart, to a concrete foundation. Each leg is held in place by six ¾ inch bolts with coarse threads. The wind is blowing in line with one of the tower legs and perpendicular to a line through the other two legs (which are downwind of the single leg). The wind speed is 15 m/s. Find the torque in the bolts such that the upwind tower leg is just beginning to lift off the foundation. Assume that the rotor is operating at the maximum theoretical power coefficient. Hint: the torque in a bolt of this type can be approximately related to the load in the bolt by the following equation:

$$Q_B = 0.195 dW$$

where Q_B = bolt torque, Nm; d = nominal bolt diameter, m; W = load in bolt, N.

B.6.4 Design Operating Gust

A turbine is being designed to operate in winds up to 15 m/s (at a hub height of 40 m). The turbine is to have a rotor diameter of D = 45 m. Consider an extreme operating gust which it might be expected to experience at least once every 50 years. What is wind speed at the peak of the gust? For how long would the wind speed exceed that of the nominal maximum operating wind speed? Graphically illustrate the hub height wind speed during the gust. Assume a higher turbulence site.

The IEC 50-year gust for turbines with hub heights greater than 30 m is defined by:

$$U(t) = \begin{cases} U - 0.37 \ U_{gust50} \sin(3\pi t/T)\left[1-\cos(2\pi t/T)\right], & \text{for} \quad 0 \le t \le T \\ U, & \text{for} \quad t < 0 \ \text{ and } \ t > T \end{cases}$$

where T = 14 s and

$$U_{gust50} = 6.4\left(\frac{\sigma_x}{1+0.1(D/21)}\right).$$

B.6.5 Scaling

Show that maximum stresses due to flapping in a wind turbine blade are independent of size for turbines of similar design. For simplicity, assume a rectangular blade and an ideal rotor.

B.6.6 Blade Vibrations

An inventor has proposed a multi-blade wind turbine. The downwind rotor is to have a diameter of 40 ft (12.2 m). Each blade is 17.6 ft (5.36 m) long, with a thickness of 3 inches (7.62 cm) and width of 8 inches (20.3 cm). Assume that the blades have a rectangular shape. The blades are to be made of wood, which can be assumed to have a modulus of elasticity of 2.0 $\times 10^6$ lb/in^2 (1.38×10^{10} Pa) and a weight density of 40 lb/ft^3 (6280 N/m^3). The rotor turns at 120 rpm. Ignoring other potential problems, are any vibration problems to be expected? (Ignore rotational stiffening.) Explain.

B.6.7 Gears

A gearbox is to be selected for providing speedup to a generator on a new turbine. The turbine is to have a rotor diameter of 40 ft (12.2 m) and to be rated at 75 kW. The generator is to be connected to the electrical grid, and has a synchronous speed of 1800 rpm. A parallel shaft gearbox is being considered. The gearbox has three shafts (an input, intermediate, and output shaft) and four gears. The gears are to have a circular pitch of 0.7854 inches (1.995 cm). The characteristics of the gears are summarized in Table B.7. Find the diameter, rotational speed and peripheral velocity of each gear. Find the tangential forces between gears 1 and 2 and between gears 3 and 4.

Table B.7 Gear teeth

Gear	Shaft	Teeth
1	1	96
2	2	24
3	2	84
4	3	28

B.6.8 Clutch brake

Plate clutches are often used as brakes on wind turbines. According to "uniform wear theory", the load carrying capability of each surface of a plate clutch can be described by the following equation for torque. The torque, Q, is simply the normal force times the coefficient of friction, μ, times the average radius, r_{av}, of the friction surface.

$$Q = \mu \frac{r_o + r_i}{2} F_n = \mu r_{av} F_n$$

where: r_o = outer diameter; r_i = inner diameter; μ = coefficient of friction; F_n = normal force.

A clutch type brake is intended to stop a wind turbine at a wind speed 50 mph (22.4 m/s). The brake is to be installed on the generator shaft of the turbine. The turbine has the following characteristics: rotor speed = 60 rpm, generator power at 50 mph (22.4 m/s) = 350 kW, generator synchronous speed = 1800 rpm. The brake will consists of four surfaces, each with an outer diameter of 20 inches (0.508 m) and an inner diameter of 18 inches (0.457 m). The coefficient of friction is 0.3. The brake is spring applied. It is released when air at 10 lb/in^2 (68.95 kPa) is supplied to a piston which counteracts the springs. The effective diameter of the piston is 18 inches (0.457 m). What is the maximum torque that the brake can apply? Should that be enough to stop the turbine if the generator is disconnected from the grid? Ignore inefficiencies.

B.6.9 Tower Natural Frequency

A cantilevered steel pipe tower, 80 ft (24.38 m) tall is being considered for a new wind turbine. The weight of the turbine is 12•000 lb (53.4 kN). The pipe under consideration has a constant outer diameter of 3.5 ft (1.067 m) and a wall thickness of 3/4" (1.905 cm). The turbine is to have three blades. The nominal rotor is 45 rpm. Would the tower be considered stiff, soft or soft–soft? Suppose that a tapered tower were being considered instead. How would the natural frequency be analyzed, using the methods discussed in this book? Assume that the density of steel is 489 lb/ft^3 (76.8 kN/m^3) and its elasticity is 30 x 10^6 lb/in^2 (2.069 x 10^{11} Pa).

B.6.10 Power Curve

A wind turbine is being designed to supply power to a load which is not connected to a conventional electrical grid. This could be a water pumping turbine, for example. This problem concerns the estimate of the power curve for this wind turbine–load combination.
The power coefficient vs. tip speed ratio of many wind turbine rotors can be described by the following simple third-order polynomial:

$$C_p = \left(\frac{3C_{p,\max}}{\lambda_{\max}^2}\right)\lambda^2 - \left(\frac{2C_{p,\max}}{\lambda_{\max}^3}\right)\lambda^3$$

where: $C_{p,\max}$ = maximum power coefficient and $\lambda_{\max}$ = the tip speed ratio corresponding to the maximum power coefficient.

Using the assumptions that a) the power coefficient = zero at a tip speed ratio of zero [i.e. $C_p(\lambda = 0) = 0$)], b) the slope of the power coefficient curve is zero at $\lambda = 0$ and c) the slope of the power coefficient curve is also zero at tip speed ratio $\lambda_{\max}$, derive the above relation.

The load the wind turbine is supplying is assumed to vary as the square of its rotational speed, N_L, which is related to the rotor speed by the gear ratio, g, so that $N_L = g\,N_R$. Using the rotor speed as reference, the load power is:

$$P_L = g^2\,k\,N_R^2$$

where k is a constant.

A closed-form expression can be derived for the power from the turbine to the load, as a function of rotor size, air density, wind speed, etc. Find that expression. Ignore the effect of inefficiencies in the turbine or the load.

For the following turbine and load, find the power curve between 5 mph (2.24 m/s) and 30 mph (13.4 m/s): $C_{p,\max} = 0.4$, $\lambda_{\max} = 7$, R = 5 ft (1.524 m), gear ratio = 2:1 speed up, rated load power = 3 kW, rated load speed = 1800 rpm.

B.7 Chapter 7 Problems

B.7.1 Yaw Control System

Wind power has been used for centuries for productive purposes. One early yaw control system for windmills, invented by Meikle about 1750, used a fan tail and gear system to turn the rotor into the wind. The windmill turret supported the rotor that faced into the wind. The fan tail rotor, oriented at right angles to the power producing rotor, was used to turn the whole turret and rotor (see Figure B.7). The gear ratio between the fan tail rotor shaft rotation and the turret rotation was about 3000:1. Explain the operation of the yaw orientation system, including the feedback path that oriented the rotor into the wind.

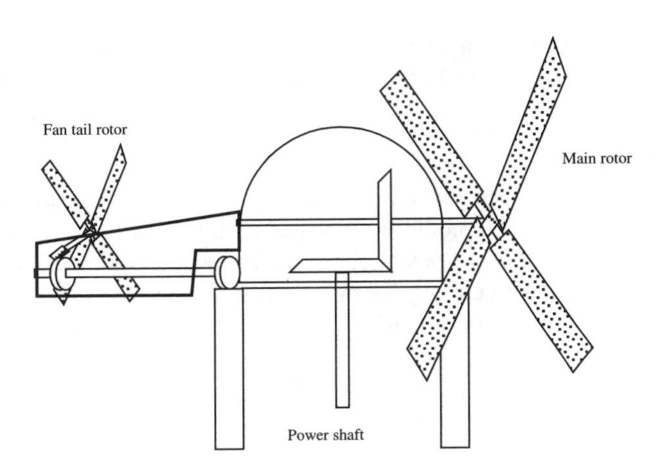

Figure B.7 Meikle yaw control

B.7.2 Supervisory Control System Needs

A supervisory control system, as part of its tasks, is to monitor gearbox operation and the need for gearbox maintenance or repairs. What information should be collected by the supervisory controller and what information should be reported to the system operators?

B.7.3 Variable Speed Wind Turbine Control

A variable speed wind turbine control system uses blade pitch and speed variations to provide constant power to the grid. What are the tradeoffs between fluctuations in rotor speed and the time response of the pitch control system?

B.7.4 Temperature Control System Components

An accelerometer on an experimental wind turbine is used to measure tower vibration. Measurements indicate that the sensor is very sensitive to temperature variations. To complete a series of tests in cold weather the test engineers rig up a quick electrical heating element and controller to keep the sensor's temperature constant. The system includes 1) a small electronic chip that provides a millivolt output that varies with temperature, 2) an electronic circuit, 3) a transistor, 4) a resistance heater, and 5) a housing to encase the chip, the heater, and the accelerometer. The electronic circuit provides a voltage output related to the difference between the voltages at the two inputs. A small current (milliamps) flows into the transistor that is a function of the circuit output voltage. The transistor uses that current to provide up to two amps of current to the heater. The heater maintains the temperature in the enclosure at 70•F.

a) What is the process that is being influenced by the control system?

b) What elements play the roles of the sensor(s), the controller, the power amplifier, and the actuator?

c) What if any are the disturbances in the system?

B.7.5 Temperature Control System Response

An accelerometer on an experimental wind turbine is used to measure tower vibration. Measurements indicate that the sensor is very sensitive to temperature variations. To complete a series of tests in cold weather the test engineers rig up a quick electrical heating element and controller to keep the sensor's temperature constant. The closed-loop transfer function of the system is:

$$\frac{T}{T_{ref}} = \frac{0.1}{s^2 + 0.5s + 0.1}$$

Plot the step response of the system to a step increase of the reference temperature of one degree F.

Hint: The step response is formed by multiplying the transfer function by T_{ref} = 1/s. The solution is determined by taking the inverse Laplace transform of the resulting equation:

$$T = \frac{0.1}{s^2 + 0.5s + 0.1}\left(\frac{1}{s}\right)$$

First it must be converted into a sum of terms, using the method of partial fraction expansion:

$$T = \frac{0.1}{s^2 + 0.5s + 0.1}\left(\frac{1}{s}\right) = \left(\frac{A}{s}\right) + \frac{Bs + C}{s^2 + 0.5s + 0.1}$$

Then the inverse Laplace transform of each of the terms can be determined. The fraction with the second order denominator may need to be broken into two terms to find the inverse Laplace transform.

B.7.6 Pitch System Step Responses

The transfer function for a pitch control system is:

$$\frac{\Theta_m(s)}{\Theta_{m,ref}(s)} = \left(\frac{K}{s^3 + s^2 + s + K}\right)$$

The closed-loop system response to a step command to pitch the blades 1 degree is:

$$\Theta_m(s) = \left(\frac{K}{s^3 + s^2 + s + K}\right)\left(\frac{1}{s}\right) = \frac{K}{(s + a)\left(s^2 + bs + c\right)\ s}$$

a) Calculate and plot the closed-loop time domain system step response for K = 0.5, the initial choice of the designer. Comment on the closed-loop system response, including damping, overshoot, and response time.

b) For no good reason, the control system designer decides to increase the gain of the system, *K*. Calculate the closed-loop time domain system step response for K = 3, the new choice of the designer.

c) What differences are evident between the time responses, each with a different gain?

Hint: perform a partial fraction expansion of the general form of the closed-loop response:

$$\frac{d}{(s+a)\left(s^2+bs+c\right)s}$$

and find the inverse Laplace transform of each term, using the a, b, c, d variables. This symbolic form will be handy, as it will be used twice in the solution.

For parts a) and b), find the real root, *a*, of the denominator by graphing s^3+s^2+s+d, using the appropriate *d*. The values of *s* at the zero crossing is $-a$. The second order root is found using long division:

$$\frac{s^3+s^2+s+d}{(s+a)}=s^2+(1-a)s+\left(1-a+a^2\right)$$

Thus, the roots of the denominator are *s*, $(s+a)$ and

$$s^2+bs+c=s^2+(1-a)s+\left(1-a+a^2\right)$$

Insert the solutions for a, b, c, d into the general solution and plot the result.

B.7.7 Yaw Drive Design Issues

A wind turbine manufacturer wants to design a yaw drive control system. To minimize wear on the drive gears the yaw is to be locked with a yaw brake until the 10 minute time-averaged yaw error is more than some specified amount (the "yaw error limit"). At that point the yaw drive would move the turbine to face the previously determined 10 minute time-averaged wind direction.

a) What consequences does the choice of yaw error limit and averaging time have on machine operation?

b) What approach would you take to determining the quantitative tradeoffs between yaw error limit and other factors?

B.7.8 Pitch Control Power

A pitch control system is being designed for a wind turbine. The response of the pitch control system to a unit step command has been determined to be:

$$\theta_p=\theta_{p,ref}\left[1-1.67e^{-1.6t}\cos(1.2t-0.93)\right]$$

where θ_p is the pitch angle change and $\theta_{p,ref}$ is the magnitude of the commanded pitch angle change. In order to shut down the turbine in high winds the pitch must change by 16 degrees. This motion is opposed by three torques, those due to friction, the pitching moment and inertia. The total friction torque, $Q_{friction}$, is assumed to be a constant 30 Nm, and the total pitching moment, $Q_{pitching}$, in high winds is assumed to be a constant 1000 Nm. Finally, the inertial torque is a function of the inertial moments: $Q_{inertia}=J\ddot{\theta}_p$. The total moment of inertia, *J*, of the blades is 100 kg m^2. The total power needed to pitch the blades is:

$$P_{total} = \dot{\theta}_p \left(J\ddot{\theta}_p + Q_{friction} + Q_{pitching} \right)$$

What is the peak power required to pitch the blades 16 degrees (.279 radians)?

B.7.9 Grid-Connected Turbine Model

The transfer function for the shaft torque, $Q_S(s)$, of a 250 kW grid-connected, fixed-pitch wind turbine with an induction generator as a function of the aerodynamic torque, $\alpha W(s)$, is:

$$\frac{Q_S(s)}{\alpha W(s)} = \frac{49(s+96.92)}{(s+97.154)(s^2+1.109s+48.88)}$$

a) This transfer function can be characterized as being the product of a first-order system, a second-order system and a term of the form $K(s+a)$. The transfer function for second-order systems can be expressed as:

$$G(s) = \frac{\omega_n^{\ 2}}{s^2 + 2\zeta\omega_n s + \omega_n^{\ 2}}$$

where ω_n is the system natural frequency and ζ is the damping ratio. What is the natural frequency and damping ratio of the second-order system component?

b) The transfer function for first-order systems can be expressed as:

$$G(s) = \frac{1/\tau}{s + 1/\tau}$$

where τ is the time constant of the first-order system. What is the time constant of the first-order system component?

B.7.10 Analog Control System

A proportional–integral–derivative (PID) controller can be designed using the electrical circuit pictured in Figure B.8. The differential equation for the controller is:

$$g(t) = -\left[K_P e(t) + K_D \dot{e}(t) + K_I \int e(t)\mathrm{d}t\right] = -\left[\left(\frac{R_2}{R_1} + \frac{C_1}{C_2}\right)e(t) + (R_2C_1)\dot{e}(t) + \frac{1}{R_1C_2}\int e(t)\mathrm{d}t\right]$$

where $g(t)$ is the controller output, and R and C are the resistances and capacitances of the respective circuit elements, e(t) is the error signal that is input to the controller and the controller constants are K_P, K_I, and K_D.

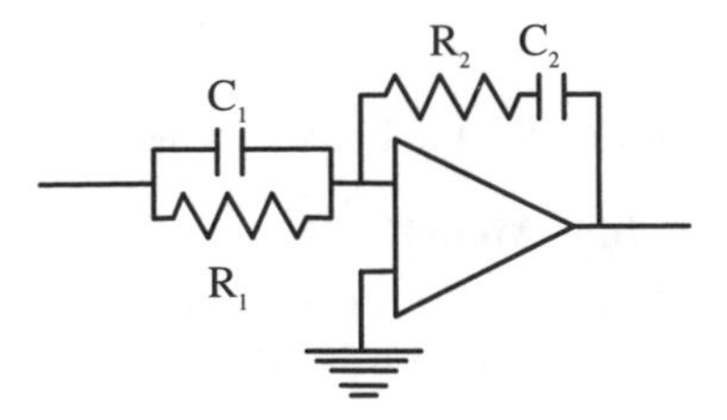

Figure B.8

If R_1 is 10000 Ohms, what values of R_2, C_1 and C_2 would be required for K_P, = 10, K_I = 100 and K_D = 100. Note resistances are in Ohms and capacitances are in Farads.

B.7.11 Digital System Time Response

If it is required that a digital control system determine the system behavior at frequencies of at least 10 Hz, a) what is the minimum sampling frequency that can be used and b) what cut-off frequency should be used to filter the input data?

B.8 Chapter 8 Problems

B.8.1 Wind Farm Siting Issues

A wind farm is being considered for a ridge top site. Name 10 or more issues that might be considered in evaluating this site.

B.8.2 Wake Modeling

Four identical wind turbines that are lined up in a row 12 rotor diameters apart are experiencing wind parallel to the row of wind turbines. Use Katić's wake model to determine the speed of the wind approaching each of the wind turbines. Assume that k = 0.10 and that the thrust coefficient is 0.7.

B.8.3 Wind Farm Design

A wind farm developer has identified a site with unique winds. They blow all of the time from one direction and at one speed, 15 m/s. The site has enough room for two rows of turbines perpendicular to the prevailing winds. She is wrestling with which size of turbines and how many to put in each row. All turbines being considered have the same hub height. She has developed two options.

Option 1 – This is an array of 24 1.5 MW turbines in two rows. The turbines just fit onto the site:

- Front row – 12 turbines, each with a rotor diameter of 60 m and a rated power output of 1.5 MW in 15 m/s wind speeds.
- Second row 500 m behind the first one (only 8.33 rotor diameters) – 12 turbines, each with a rotor diameter of 60 m and a rated power output of 1.5 MW in 15 m/s wind speeds.

Option 2 – This option includes slightly smaller turbines in the first row. This allows a few more turbines in the first row and reduces the velocity deficit at the second row because of the smaller rotor diameters in the first row. In addition, because of the reduced concerns of abutters, these smaller turbines can be moved 100 m closer to the front edge of the site, further reducing the wake effect. The second row is the same as in Option 1:

- Front row – 15 turbines, each with a rotor diameter of 50 m and a power output of 1.0 MW in 15 m/s wind speeds.
- Second row 600 m behind the first one –12 turbines, each with a rotor diameter of 60 m and a power output of 1.5 MW in 15 m/s wind speeds.

To evaluate her options, the developer uses Katić's wake model to determine the wind speed behind the first row of turbines. She assumes that $k = 0.10$ and that the thrust coefficient is 8/9. She also assumes that in winds lower than the rated wind speed of 15 m/s, the power output of the turbines can be approximated by:

$$P = P_{rated}\left(\frac{U_X}{15}\right)^3$$

a) For both options, what are the power production for each row and the total power production?
b) If the turbines cost $800/installed kW, how much will each option cost to install (assume that no loans are needed)?
c) If the developer's annual operation and maintenenace (O&M) costs are 7% of the installed cost plus $10/MWh for maintenance, what are the total annual O&M costs?
d) If the developer's income from the sale of energy is $30/MWh, what is her net income for a year (Sales income - annual O&M costs)?
e) If net income is the deciding factor, which option should she pursue?

B.8.4 Power Smoothing

A wind farm consists of two turbines, located 100 m from each other. They are operating in winds with a mean speed of 10 m/s and an integral time scale of 10 s.

a) What does the graph of the Von Karman spectrum look like (using log–log coordinates)?

b) What is the function for the wind farm filter for these two turbines?

c) Suppose that the spectrum of the local wind can be expressed by:

$$S_1(f) = \sigma_U^2 \, 40e^{-40f}$$

Assume also that each wind turbine sees the same mean wind speed, U, and that the total average wind turbine power, P_N, can be expressed as $P_N = NkU$ where k is equal to 10 and N is the number of wind turbines. What is the standard deviation of the wind power from the two wind turbines? How does this compare to the standard deviation of the power from two wind turbines that experience exactly the same wind?

B.8.5 Grid Fault Level and Voltage Fluctuations

Consider a wind turbine generator connected to a grid system (see Figure B.9) with a line-neutral system voltage, V_S. The voltage at the wind turbine, V_G, is not necessarily the same as V_S. The distribution system resistance is R, and the distribution system reactance is X.

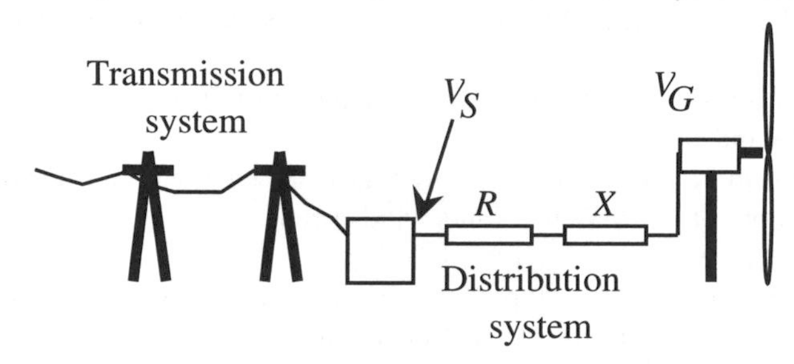

Figure B.9

The real generated power is P and reactive power consumed by the wind turbine generator is Q. Suppose $R = 8.8\ \Omega$, $X = 4.66\ \Omega$, $Vs = 11$ kV, $P = 313$ kW, $Q = 162$ kVAr and the power factor at the generator is pf = 0.89. Determine the voltage difference between the grid and the generator, the voltage drop as a percent of the grid voltage, the fault level at the generator and the installed generator capacity as a percent of the fault level.

B.8.6 Harmonic Distortion

A wind turbine is connected to an 11 kV distribution line. The magnitude of the harmonic voltages, c_n, at the point of common connection of the wind turbine and other grid users are listed in TableB.8.

Table B.8

n	1	2	3	4	5	6	7	8	9	10	11	12	13	14	15
c_n, Volts	11000	120	0	85	15	0	99	151	0	12	216	0	236	80	0

What is the harmonic distortion of each harmonic and the total harmonic distortion? Are these within the allowable IEEE 519 limits?

B.8.7 Offshore Winds

The 10 minute average wind speed at an offshore buoy is measured to be 8.5 m/s at an elevation of 10 m. The buoy is 10 km from land in the direction of the oncoming wind.

a) If the surface roughness length is assumed to be 0.0002 m, what is the mean wind speed at 80 m?

b) If the Charnock constant, A_C, is assumed to be to be 0.018 and $C_{D,10}$ is 0.0015, what is the mean wind speed at 80 m, using Charnock's model for offshore wind shear?

c) What is the mean wind speed at 80 m if $C_{D,10}$ is 0.0015 and Johnson's model is assumed to be correct?

B.8.8 Hybrid System Design Rules

An isolated power system serving the community of Cantgettherefromhere uses a 100 kW diesel generator. The community plans to add 60 kW of wind power. The hourly average load and power from the wind turbine over a 24 hour period is detailed in Table B.9. The diesel generator has a no-load fuel usage of 3 liter per hour and an additional incremental fuel use of 1/4 liter per kilowatt-hour.

For the day in question, using the hybrid system design rules, determine:

a) The maximum renewable energy that can possibly be used in an ideal system.

b) The maximum renewables contribution without storage, and the maximum with storage.

c) The maximum fuel savings that can be achieved (with intelligent use of controls, storage and renewables).

d) The minimum diesel fuel use that can be achieved with intelligent use of storage and controls (and without renewables).

Table B.9

Hour	Load (kW)	Wind (kW)	Hour	Load (kW)	Wind (kW)
0	25	30	12	85	45
1	20	30	13	95	45
2	15	40	14	95	50
3	14	30	15	90	55
4	16	20	16	80	60
5	20	10	17	72	60
6	30	5	18	60	48
7	40	5	19	74	50
8	50	15	20	76	55
9	70	20	21	60	60
10	80	25	22	46	60
11	90	40	23	35	55

B.8.9 Wind/Diesel/Battery System Operation

Based on the analysis in Problem 8.8 and other input, the community of Cantgettherefromhere has upgraded its power system. The hybrid power system includes a 100 kW diesel generator, 100kW of installed load, 60 kW of wind power, a 100 kW dump load, and energy storage with 100 kWh capacity. The hourly average load and power from the wind over a 24 hour period have remained the same and are listed in Table B.9.

Assume that the mean hourly data accurately describe the load and power from the wind and that fluctuations about the mean load are handled by the energy storage. Determine the hourly energy flows in the system for the following system control and operating approaches.

a) Diesel-only system – In this baseline system, the diesel provides all of the power to the load. The diesel provides power down to 0 kW with no provision to limit the ensure a minimum diesel load to avoid diesel engine wear.

b) Minimum diesel – In this system minimizing diesel power takes priority, except that the diesel may not be shut off and must run at a minimum of 30 kW to ensure long diesel engine life. Thus, the system operates by the rules:

1. The minimum diesel power is 30 kW
2. Only the minimum diesel power that is needed above 30 kW is used (no power above 30 kW is used to fill up the storage)
3. Storage can only be used between 20 and 95 kWh of capacity, to maximize storage efficiency and battery life
4. Storage capacity starts at 50 kWh
5. The sum of the energy into the power sources in the system (diesel, wind, battery) must equal the sum of the energy into the power sinks in the system (the load, dump load, and battery).
6. If there is excess energy in the system it is first stored, if possible, and dumped only if necessary.

c) Diesel shut off – In this system a rotary inverter between the batteries and the grid provides reactive power and the diesel can be shut off. It shuts off whenever possible, but when running, needs to be at a minimum load of 30 kW. When running it is also used to fill up the storage to use the fuel most efficiently. The battery level is maintained between 20 and 90 kWh. This insures that there is adequate capacity to handle fluctuating loads when there is no diesel in the system. Thus, the system operates by the rules:

1. The diesel can be shut off
2. The minimum diesel power is 30 kW when the diesel is running
3. When the diesel is running, the storage is also filled up if possible to improve diesel fuel efficiency
4. Storage can only be used between 20 and 90 kWh of capacity, to maximize storage efficiency and battery life
5. Storage capacity starts at 50 kWh
6. The sum of the energy into the power sources in the system (diesel, wind, battery) must equal the sum of the energy into the power sinks in the system (the load, dump load, and battery)
7. If there is excess energy in the system it is first stored, if possible, and dumped only if necessary

Specifically, for each operating approach, determine how much energy is supplied by the diesel over the 24 hour period and the reduction in diesel power compared to the diesel only case.

B.9 Chapter 9 Problems

B.9.1 Learning Curve Problem

The first unit of a new wind turbine costs $100,000. The system is estimated to follow an $s = 0.83$ learning curve. What is the cost of the 100th unit?

B.9.2 Progress Ratio Estimate for Utility-scale Wind Turbines

Estimate the progress ratio (s) of modern utility-scale wind turbines. Use the time period of about 1980 to the present.

B.9.3 Simple Cost of Energy Problem

A small 50 kW wind turbine with an initial cost of $50,000 is installed in Nebraska. The fixed charge rate is 15% and the annual operation and maintenance ($C_{O\&M}$) is 2% of the initial cost. This system produces 65,000 kWh/yr (E_a). Determine the cost of energy (COE) for this system.

B.9.4 Small Wind Turbine Economics

A small wind machine, with a diameter of 6 m, is rated at 4 kW in an 11 m/s wind speed. The installed cost is $10,000. Assuming Weibull parameters at the site to be $c = 8$ m/s and $k = 2.2$, it is claimed that the capacity factor is 0.38.

The interest rate, b, is 11% and the period of the loan, N, is 15 years. Find the cost per unit area, per kW, and the levelized cost per kWh. The lifetime of the turbine, L, is assumed to be 20 years and the discount rate, r, is 10%. You do not have to include factors such as inflation, tax credits, or operation and maintenance costs.

B.9.5 Spreadsheet Problem

The estimated cost and savings from the purchase and operation of a wind machine over 20 years are given in Table B.10. a) Compare the total costs with the total savings. b) Find the present values of the cost and savings. Use an annual discount rate of 10%. Would this be a good investment? c) Determine the pay back period for the wind machine. d) What is the breakeven cost of the machine? e) Estimate the cost of energy produced by the wind machine if it is expected to produce 9,000 kWh annually.

Table B.10

Year	Costs	Savings	Year	Costs	Savings
0	$9000		11	227	669
1	127	2324	12	241	719
2	134	1673	13	255	773
3	142	375	14	271	831
4	151	403	15	287	894
5	160	433	16	304	961
6	168	466	17	322	1033
7	180	501	18	342	1110
8	191	539	19	362	1194
9	202	579	20	251	1283
10	214	622			

B.9.6 Wind–Diesel System Economics

This problem is intended to assess the relative economic merits of installing a wind–diesel system rather than a diesel-only system at a hypothetical location. Assume that the existing diesel is due for replacement and that the annual system demand will remain fixed during the lifetime of the project.

Your problem is to calculate the levelized cost of energy for the diesel only and for the wind–diesel system. The following system and economic parameters hold:

System	
Average system load	96 kW (841,000 kWh/yr)
Diesel rated power	200 kW
Mean diesel usage- no wind turbine generator	42.4 l/hr (371,000 l/yr)
Wind turbine rated power	100 kW
Average annual wind speed	7.54 m/s
Annual wind speed variability	0.47
Average wind turbine power	30.9 kW
Useful wind turbine power (based on 25% lower limit on diesel loading)	21.1 kW
Mean diesel usage with WTG	34.8 l/hr (305,000 l/yr)
Economics	
Wind turbine cost	$1500/kW installed
Diesel cost	$350/kW installed
System life	20 yr
Initial payment	$30,000
Loan term	10 yr
Interest	10%
General inflation	5%
Fuel inflation	6%
Discount rate	9%
Diesel fuel cost	$0.50/l
Wind turbine O&M cost	2% of capital cost/yr
Diesel O&M cost	5% of capital cost/yr

B.10 Chapter 10 Problems

B.10.1 Visual Impact Assessment

Assume that you are planning the development of a small wind farm (say 10 turbines of 500 kW) in a rural area. Describe how you would reduce the visual impact of this project. Also prepare a list of the tools that could be used for assessing the impact of this project on the surrounding environment.

B.10.2 Turbine Noise Source Problem

Estimate the sound power level for a three-bladed, upwind wind turbine with the following specifications: rated power = 500 kW, rotor diameter = 40 m, rotor rotational speed = 40 rpm, wind speed = 12 m/s.

B.10.3 Turbine Noise Distance Problem

Suppose the sound pressure level at a distance of 100 m from a single wind turbine is 60 dB. What are the sound pressure levels at a distance of 300, 500, and 1000 m?

B.10.4 Multiple Turbine Noise Problem

Assuming the same turbine as in the previous problem, what are the sound pressure levels at a distance of 300, 500, and 1000 m for 2, 5, and 10 turbines?

B.10.5 Electromagnetic Interference for TV Reception

For electromagnetic interference problems once the (bistatic) radar cross-section (σ_b) is known, the polar coordinate r can be calculated for a required interference ratio (C/I). For a TV transmitter operating in an urban environment, σ_b was experimentally determined to be 24 and 46.5 dBm^2 in the backscatter and forward scatter regions. Determine the value of r in both the backscatter and forward scatter regions for required values of C/I equal to 39, 33, 27, and 20 dB.

Index